Crime Prevention and Security Management

Series Editor

Martin Gill, Perpetuity Research, Tunbridge Wells, UK

It is widely recognized that we live in an increasingly unsafe society, but the study of security and crime prevention has lagged behind in its importance on the political agenda and has not matched the level of public concern. This exciting new series aims to address these issues looking at topics such as crime control, policing, security, theft, workplace violence and crime, fear of crime, civil disorder, white collar crime and anti-social behaviour. International in perspective, providing critically and theoretically-informed work, and edited by a leading scholar in the field, this series will advance new understandings of crime prevention and security management.

Mark Stokes

Designing Out Crime

A Handbook for Practitioners, Managers and the Community

palgrave
macmillan

Mark Stokes
University of the West of England
Bristol, UK

ISSN 2946-3513 ISSN 2946-3521 (electronic)
Crime Prevention and Security Management
ISBN 978-3-031-99839-3 ISBN 978-3-031-99840-9 (eBook)
https://doi.org/10.1007/978-3-031-99840-9

Cover credit: Mark Stokes

This Palgrave Macmillan imprint is published by the registered company Springer Nature Switzerland AG
The registered company address is: Gewerbestrasse 11, 6330 Cham, Switzerland

Series Editor's Introduction

There is now a wealth of independent evidence that security measures work, and this is especially true of the concept of *Designing Out Crime*,[1] an approach to crime reduction that focusses on the design and management of buildings and places that are as diverse as pedestrian subways, toilets, parks, organisations and homes. It is championed and managed via 'Police CPI' a not-for-profit organisation that works on behalf of the UK police service. The 'Secured by Design' award is widely recognised as a quality symbol by a variety of stakeholders including some local authorities and architects.

In practice it involves a multifaceted approach, and a Handbook to guide practitioners and others including academics through these has much to commend it. The author, Mark Stokes, is to be congratulated in taking on a mammoth task and condensing the contents into digestible sections that both inform and guide in equal measure.

The book itself discusses theory and includes a focus on Crime Science, and incorporates a focus on many topics that inform the application of CPTED principles from 'naming and shaming' to 'naming and praising' (Stokes favours the latter); the value of commensurate risk; the environmental visual audit report; defensive planting and an extensive discussion of the merits of different types of

---

[1] For example, see review by Armitage, R. and Tompson, L. (2022) The Role of Crime Prevention Through Environmental Design (CPTED) in Improving Household Security. In M. Gill (ed) *The Handbook of Security* (third edition), Palgrave: Basingstoke; for a more general discussion on the effectiveness of security see: Tilley, N. and Farrell, G. (2022) Security and International Crime Drops. In M. Gill (ed) *The Handbook of Security* (third edition), Palgrave: Basingstoke.

lighting, intruder alarms and CCTV (to name just some). I mention these merely to whet the appetite of the reader conscious that the content is so vast they don't do justice to the enormity of insights that follow in the pages ahead.

What perhaps can best be said is that this book provides detailed insights into anyone interested in understanding all aspects of the Secured by Design award process and for that matter related processes such as *Secured Environments* and *Safer Stations*. As the author acknowledges it is more a text that will be digested in parts rather than the type that will be read from start to finish. It is difficult to imagine how anyone indeed everyone involved in the practice of the Secured by Design award—or for that matter the thinking behind it—would not view this book as a 'must have reference' and at least occasional aide memoir, and let me suggest, an essential one.

April 2025 Martin Gill

Academic Overview by Professor Ken Pease OBE

There is now a high level of scientific consensus about the fact and likely disastrous consequences of anthropogenic climate change. And yet, to date, there has been a near total unwillingness to acknowledge the contribution that crime itself, together with state and individual responses to crime, makes to humanity's carbon footprint. I have previously referred to this as the 'Carbon-Crime Blind Spot'.

In England and Wales alone recorded crime accounts for a minimum of 6 million tonnes of CO_2, a figure that can be multiplied by a factor of six when such issues as unreported crime and decisions to move home to avoid crime are taken into account. As a consequence, the aspiration to prevent crime has a new-found validity and urgency beyond the distress of victims, in terms of the reduction in greenhouse gas emissions. Environmental manipulation does reduce crime. The argument that crime is simply displaced is demonstrably wrong and hence inexcusable in its implied argument for inaction. The spectacular inequalities in the rates of victimisation and the inherent failures in distributive justice, provide moral underpinning for the enterprise of *Designing Out Crime.*

The idea that the burden of crime should be allocated fairly, is rarely debated. Indeed, the annually produced crime statistics primarily count offences and offenders, rather than victims and victimised communities.

Once focus is placed on victims, the grotesque inequality and the prevalence of chronic victimisation of the same victims, becomes evident. Roughly 40 per cent of all crime takes place in the most vulnerable 10 per cent of areas. Victims of crime often suffer subsequent offences. Such levels of chronic victimisation caused Giddens (1998) and many others to call for a renewed emphasis on crime reduction.

There is already in place a mechanism that guarantees the pre-empting of so much crime: the *Designing Out Crime* process delivering the Secured by Design award. Independent research into the effectiveness of Secured by Design has repeatedly demonstrated its capacity to reduce crime over the long term. Indeed, Brown, Pascoe (1999) and Armitage (2000) evidenced sustained reductions in property crime in excess of 40 per cent. A decade later, Armitage and Monchuk (2010) demonstrated how very long-term reductions of close to 60 per cent had been realised on social housing estates in West Yorkshire, where the greatest concentrations of repeat victimisation take place. Perversely, the apparent relative indifference of many of those in the private house building industry to the free of charge service of Secured by Design, has allowed, indeed required, the police to concentrate on such victims.

The author of this handbook has invested monumental time and effort into describing how the key tenets of physical security can be used to deliver long-term crime reduction. This provides his key motivation in these cash-strapped times where *High-Crime* rates no longer attract additional direct funding. The message is clear: investment in approaches already shown to be effective will yield sustainable crime reduction, allow a better service in response to those crime events which remain and make for progress towards the ordered and sustainable society which policing aspires to achieve.

Ken Pease OBE

Thanks and Appreciation

I would like to thank Secured by Design (for their support and awarding me the financial means, courtesy of the Secured by Design Innovation Awards) and all the academics, architects, CPTED officers/advisers, crime scientists, DOCOs, planners, security specialists and International CPTED Association (ICA) members who have assisted me (however unwittingly!) in writing this handbook.

In addition, special thanks are given to Dr. Saif Al Kaabi, Jim Aldridge, Dr. Rob Annable, Professor Rachel Armitage, Dr. Bill Baker, Phil Ball, Raj Bansal, Mike Barton CBE, Steve Bates, Calvin Beckford, Bill Bellman, Phil Bennett, Dr. Tim Brain QPM, Councillor Marje Bridle, Peter Broad, Steve Broatch, Michael Brooke, Gerry Brophy, Jon Brown, Dr. Stephen Brown, Rick Burgess, Phil Butler, Michael Butters, Peter Cain, Ben Castell, Janet Caton, Linda Chandler, Nigel Charlesworth, Richard Childs QPM, John Christophers, Michael Clare, Professor Ron Clarke, Colin Cockfield, Pat Cogan, Jon Cole, Guy Collyer OBE, Professor Ian Colquhoun, Mollie Cooke, Dave Cotton, Michael Crofts, Dave Cross, Dr. Caroline Davey, Barry Davidson, Ray Dawson, Kate Downs, Mark Dowse, Jack Dromey MP, Paul Dyton, Professor Paul Ekblom, Professor Graham Farrell, Professor Marcus Felson, Peter Fry, Barry Fulford, Professor Lorraine Gamman, Michael Gardener, Chris Gibson, Chris Gilbert, Professor Martin Gill, Dr. William Gisby, John Gisbourne, Hazel Goss MBE, Andy Gregory, Bill Guest, Heather Gurden, Kim Guthrie, Bill Hannis, Paul Hardy, Ian Heward, Brian Hewett, Mike Hodge, Alfie Hosker, Jason Hunter, Art Hushen, David Jacobs, Keith James, Howard Jones, Ian Jones, Professor Phil Jones, Armando Jongejan, Tinus Kruger, Peter Kane, Dave King, Professor Stuart Kirby, Bob Knights

MBE, Kamilla Kristiansen, Tony Lake, David Lancaster, Greg Lawrence, Jenny Lawrence MBE, Kimberley Leary, Freda Mason, Kenny McHugh, Alan McInnes, Ken Meanwell, Bob Meath, Roy Meddings, John Meek, Graham Mitchell, Dr. Leanne Monchuk, Grant Moss, Professor Sabah Mushatat, Sgt Naheed Mushtaq, Charles Nayler, Dr. Andrew Newton, Stu Nicholls, Michael O'Connor, Julie Onwukegu, Artur Pankowski, Dannie Parkes, Glynn Parkes, Dr. Tim Pascoe, Professor Ken Pease, Gareth Pemberton, Roy Porter, Dr. Barry Poyner, Michael Quinney, Sue Ramsay, Tim Ramskill, Mick Reynolds, Darren Robbins, Tudor Roberts, Paul Robinson, Dr. Michelle Rogerson, Nick Rubery, Sandie Ruffett, Phil Russon, Kassie Rushton, Greg Saville, Henry Shaftoe, Insp Ritu Sharma, David Sheffield, Mark Silvester, Mick Simpson, Chris Sims OBE QPM, Doug Skins, John Smith, Mark Smith MBE, Insp. Phillipa Smith, Sean Smith, Gillian Smyth, Paul van Soomeren, Dave Squires, Mark Stephenson, Cheryl Stevenson, Baroness Stewart of Edgbaston, Ron Stokes, Elaine Stone, Dave Stubbs, Anne Summers, Dave Thorneywork, Joe Tildesley, Professor Nick Tilley, Joanne Tolley, Alan Townsend MBE, Alan Walker, Terry Webb, Jackie White, John White, Derek Williams, Neil Williamson, Dr. Andrew Wootton and Dr. Martin Wright. Can I also apologise to those people who were similarly instrumental in my writing this handbook but are not acknowledged by name.

Finally, sincere appreciation is given to my family and friends, for encouraging me to write this handbook and suffering my necessarily imposed absence in pursuit of that goal. The three decades I devoted to the discipline of *Designing Out Crime* proved to be the most rewarding of my police career and beyond. This was largely due to witnessing the long term, sustainable effects, rather than processing (as a custody sergeant for many years) the same criminals (and eventually the sons of criminals!) time and time again. Writing this handbook has in many ways been a labour of love. Moreover, it provides a body of good practice for the corporate memory of policing and in so doing, helps to satisfy Dr. Ivan Hill and Professor Ken Pease's (2000) desire for self-sustaining Crime Prevention initiatives. I hope you believe it was worth the effort.

Mark Stokes

Contents

About the Author

Mark Stokes left school at 16 and worked as a Laboratory Technician, before beginning his career with West Midlands Police as a Police Cadet. A year later he was 'sworn-in' as Police Constable 6131, patrolling the streets of Solihull—home of Landrover Cars. Mark was subsequently promoted to Sergeant in the City of Coventry.

He was appointed Force Architectural Liaison Officer (DOCO) Sergeant in 1992. Returning to front-line police duties Mark was promoted to Inspector and thereafter, Force Designing Out Crime Inspector followed by West Midlands Police Crime Reduction Manager. By stealth (and sheer longevity) he was one of the longest serving officers in the disciplines of Crime Reduction and Designing Out Crime, until his retirement in 2011. In that summer and during a House of Commons debate on policing, Jack Dromey, the then Member of Parliament for Birmingham Erdington described Mark as "a specialist in Crime Prevention and an expert at Designing Out Crime with a deserved international reputation."

Between 2007–2009 Mark was awarded an unprecedented trio of Secured by Design Innovation Awards for his submissions in respect of The Four Towers at Duddeston Manor (see Sect. 9.1); Brindleyplace: City Centre Commercial, Retail and Leisure Complex (Sect. 9.2) and Cottesbrooke Infant School (Sect. 9.3). The financial sums awarded were used to produce this handbook.

Since retiring as a police officer, Mark has been a Consultant with Secured by Design; University Lecturer in Designing Out Crime and Partnership Working; War Crimes Investigator; and Crime Reduction Supervisor with British Transport Police—where he co-authored the Goldstein award-winning 'Reducing Suicide: a Problem-solving and Partnership Approach between British Transport Police and the Rail Industry', which he presented in Providence, Rhode Island. A trainer and assessor at the Police Crime Prevention Academy, he then became the Rural and Business Crime Officer for Herefordshire, writing the Safer Streets 3 Environmental Visual Audit, which secured £500,000 to combat violence against women and girls. Since 2022 Mark has been a Senior Lecturer on the Police Constable Degree Apprenticeship programmse at the University of the West of England, Bristol.

He has a Bachelor of Science Honours Degree from the Open University, Master of Arts Degree in Public Order from the University of Leicester and in 2022 was awarded his doctorate by Huddersfield University for a PhD research project investigating the long-term sustainability of Designing Out Crime and the Secured by Design award process. Mark has delivered presentations on the theme of Designing Out Crime across the UK and in Abu Dhabi, Ajman, Amsterdam, Brisbane and Calgary.

A keen gardener (see Chapter 28 Defensive Planting), Mark was a pioneer of long-distance cycling for charity—beginning in 1979 with Land's End to John o' Groats (twice); and thereafter San Francisco to New York; Lima to Machu Picchu; Solihull to Istanbul; and to the works of his architect heroes, Charles Rennie Mackintosh and Frank Lloyd Wright. A founder member of the West Midlands Police and British Police Ski and Snowboard Sections, he continues to ski competitively (only just!) for the 'Old Bill' and is an occasional mountaineer—having climbed extensively in the world's highest mountain ranges.

Background 1

1.1 Using this Handbook

This handbook has not been written in the belief or with the intention that it should be read from cover to cover and inwardly digested. It would be a major conceit on the part of the author to imagine that anyone has the time or inclination to do so, least of all the Designing Out Crime Officer also known as Crime Prevention Design Adviser, Architectural Liaison Officer or Design for Security Consultant in the UK and CPTED Officer/Adviser in much of the world beyond. However, throughout this handbook, the term Designing Out Crime Officer (DOCO) will be used for all these different names. Furthermore, each section and chapter are structured to be read in relative isolation albeit with suggested links to other chapters, evidence from Crime Science and references—together with a full glossary, bibliography and index. As a result the layout is designed to facilitate the reader being able to 'dip in and out', rather than having to wade through pages of often complex and at times near unintelligible data in order to discover a key point, a much-repeated task for the author—and naturally the proper course of thorough research!

The handbook is deliberately first and foremost for the use of practitioners—people already convinced of the spectacular success of Designing out Crime (DOC). For such 'super controllers' (Sampson & Eck, 2008; Kirby, 2018), novice and experienced practitioners alike, it can act as an aide memoir, providing suitable case examples and the evidence (where available) from Crime Science. However, it should also prove to be invaluable to Community Safety managers and anyone in the wider world (especially those operating within the professions of the built environment) who has an interest in how design can influence crime,

M. Stokes, *Designing Out Crime*, Crime Prevention and Security Management,
https://doi.org/10.1007/978-3-031-99840-9_1

fear of crime and anti-social behaviour to positive effect. And there is good reason to be interested: it cannot be over-emphasised that Designing Out Crime is extremely effective, with little evidence of 'Displacement of Crime' and more likely a 'Diffusion of Benefits' (see Sect. 2.1).

As the author is British, there is an automatic bias within this manual to practice and experience inside the UK. However, and thereby illustrating that this is an international discipline, practical examples and research findings from around the world are also included. Hopefully, this international and more evidence-based approach (drawing down from Crime Science and Environmental Criminology) will make it of greater relevance to a far wider audience.

Most importantly, the key message of this handbook, is the real and genuine capacity to Design Out Crime on a long term, sustainable basis—and not just in terms of opportunist, volume property crime. With a growing wealth of academic research and practical experience to support this statement, at its most blunt this form of highly effective 'Situational Crime Prevention' (SCP—Carstensen & Frederiksen, 1997) saves money and allows the police service and other social agencies more time and resources to concentrate on those forms of crime (and offenders) whose behaviour it has proved more difficult to prevent. In the UK, police-sponsored SCP initiatives owe their ancestry to the Willink Royal Commission on the Police (1962), Police Act (Home Office, 1964) and the Cornish Committee on the Prevention and Detection of Crime (Home Office, 1965). More recently, SCP has embraced carbon reduction and other forms of sustainability to alleviate the otherwise characteristic vulnerability of the built environment. As such, the practical, financial, moral and green considerations to invest in DOC are both self-evident and proven.

1.2 Introduction

Designing Out Crime (DOC), a discipline within Situational Crime Prevention (SCP), has a positive evidence base spanning close to half a century. Within the UK, the police-owned and managed Secured by Design scheme is the main delivery mechanism for DOC, with police officers and police staff in all forces responsible for its application. Indeed, it is a universally accepted wisdom that prevention is better than cure.

This is not a revolutionary new idea. In 1829, one of the first joint Commissioners of the Metropolitan Police in London, Sir Richard Mayne, penned the 'Nine Principles of Policing', foremost of which was:

> To prevent crime and disorder, as an alternative to their repression by military force and severity of legal punishment.

Mayne was undoubtedly influenced by the founder of modern policing and the man who appointed him, Home Secretary Sir Robert Peel. But the important point is that prevention was its foremost priority and this has subsequently evolved to embrace Designing out Crime.

The primary rationale for this handbook is DOC, a more specialised mechanism of what is usually referred to as Situational Crime Prevention—'situational' because in the first instance it is the location and/or target that is considered vulnerable and therefore suitable for influence in order to prevent crime at a practical level.

This handbook draws heavily on the lessons learnt from current best practice and wherever possible, Crime Science and Environmental Criminology, to support recommendations made in the pursuit of DOC. The days of police practitioners (and many others) relying on anecdotal and at times highly subjective (if not speculative) value judgements to influence the practical application of SCP, should by now have reached their endgame and be discarded to history. Opinion alone is simply insufficient. Moreover, this dovetails perfectly with 'Evidence-Based Crime Prevention' (Sherman et al., 2002) and policing.

The work of DOC practitioners should be based on academically proven evidence (discussed in later chapters), with the profession challenging the aforementioned anecdotal and speculative opinion made by their colleagues and others. Practitioners provide evidence of successful development projects that support theories of what works in practice. These successes increase the evidence base—a priority consideration of this handbook. Indeed, Randy Atlas (1999) describes how the arena of Crime Prevention Through Environmental Design (CPTED, Jeffery, 1971—see Chapters 25–37) is woefully under-researched. Consequently, this handbook details specific areas where practitioners can assist in improving the evidence base.

Within the British policing model, DOC has been in existence since the mid-1980s. It was built on the firm foundation of traditional Physical Security 'Target Hardening' Measures (Newman, 1972, 1973: stronger doors, additional locks, bars on windows etc. see Sects. 4.2.1–4.2.4) and Technological Innovations (see Sects. 4.3.1–4.3.3). The emergence of DOC marked the incorporation of those same Physical Security 'Target Hardening' Measures and Technological Innovations within C. Ray Jeffery's (1971) more holistic concept of CPTED and to a great extent utilised Oscar Newman's (1972, 1973 and 1996) concepts of

Defensible Space, Natural Surveillance and Territoriality (see Sects. 4.4.1–4.4.6, the Environmental Elements of CPTED).

By the late 1980s, all police forces in the UK had fully trained and designated staff working on DOC projects. The descriptions Architectural Liaison Officer (ALO), Crime Prevention Design Adviser (CPDA) and Design for Security Consultant (DfSC) have previously been used. However, the term Designing Out Crime Officer is now recognised nationally to describe the role and is utilised throughout this handbook. It is the responsibility of these DOCOs to deliver the fusion of Physical Security 'Target Hardening' Measures, Technological Innovations and the Environmental Elements of CPTED—together with evidence from design ways of thinking and formal impact evaluations of what works and does not work in a practical setting. In both theory and evaluation DOC has become heavily influenced by Crime Science.

Since 2005 Designing Out Crime has been the title of the training programme delivered for the UK police service by the National Police Improvement Agency (NPIA), the College of Policing and more recently the Police Crime Prevention Academy.

Returning to its genesis in the mid-1980s, DOC has been a broad church. Within the British policing model, this has provided a natural career path for traditional Crime Prevention/Reduction specialists, whilst simultaneously recruiting into the service professionals from the built environment and beyond, e.g. architects, planners and surveyors. As a further example, Greater Manchester Police has always engaged police staff (described as 'non-sworn' officers in North America and other parts of the world). Collectively, this has provided a broad range of skills and robust spectrum of opinion and debate at the cutting edge of practical application.

Most importantly and to repeat, it is fundamental to appreciate that DOC is a crime function—and not one of the planning processes. Local government planning departments involved in the design of the built environment are statutorily obliged to consider a raft of competing issues. One of the key responsibilities of the DOCO is to advocate to these same professionals the necessity to prioritise issues that can engineer DOC and in so doing, provide sustainable benefits that inherently protect their other considerations and responsibilities.

Key Issue

Designing Out Crime is concerned with improving the built environment in order to improve the quality of life of our citizens. DOCOs realise that if they were to create a 'Fortress Mentality' (a charge sometimes laid against them) decreased quality of life—specifically an increased Fear of Crime, is likely to result. The debate concerning appropriate/excessive Permeability (often linked to the Fortress Mentality charge) is discussed in Sect. 4.4.6. See also Farrell and Tilley (2022) 'Elegant Security'.

During the early years of DOC this was often a difficult task, with planning chiefs repeating the highly logical observation, 'crime is but one of the issues we have to consider'. However, the publication in 1994 of Department of the Environment Circular 5/94 Planning Out Crime; Safer Places: the Planning System and Crime Prevention (2004); Planning Policy Statement 1 (PPS1) Delivering Sustainable Development (2005) and the National Planning Policy Framework (2012), (2018), (2019), (2021), (2023), (2024) and (2025) have raised the profile and relevance of DOC. Furthermore, after more than three decades, we now have a wealth of independent academic research that repeatedly points to the efficacy of DOC and, in particular, the hugely successful delivery mechanism known as Secured by Design (SBD—see Chapter 5).

Launched in 1989 and rolled out across the UK over the following three years, SBD was originally devised as a marketing scheme for the police DOC service. Initially targeted at the private, for sale housing industry, it rewards developers who have incorporated specified standards of Physical Security 'Target Hardening' Measures, Technological Innovations and the Environmental Elements of CPTED. Subsequently expanded to include all forms of residential and non-residential development, close to four decades later, SBD has grown to become an internationally respected brand with an undisputed claim to being the most effective and Sustainable Crime Prevention initiative in the history of policing—worldwide.

Through sheer longevity, DOC can claim to being the strongest and most sustainable mechanism within the Crime Reduction/Prevention toolbox. Consequently, failing to solicit the advice of an experienced and dedicated DOCO will inevitably miss that once in a 30+ year window of opportunity to DOC on a long term and truly sustainable basis. There already exists considerable evidence that testifies to this having taken place time and time again, with the effects on local crime, FOC and ASB being felt within months rather than years. This results in both community disquiet and the potential for unwanted attention by more senior police colleagues and elected representatives. Moreover, as Professor Ted Kitchen predicted (2007), in the future, this will undoubtedly result in the victims of crime taking legal redress through the civil courts.

Finally, Designing Out Crime often requires subjective thought and application of new and revolutionary ideas (especially in relation to Threat and Risk Assessment (TARA—see Sect. 1.5.15). We should relish such challenges, rather than shy away from them. And by extension, this handbook does not shy away from the current 'hot issues' in the world of DOC.

1.3 A Brief History of Designing out Crime

Self-evidently, Designing out Crime (DOC) is a relatively new discipline that has only truly existed in name (in the UK originally as Architectural Liaison or Crime Prevention Design) since the mid-1980s. However, its origins and motivation can be traced back to the dawn of humanity—and identified within the animal kingdom (e.g. the armadillo's toughened skin, the horn of the rhinoceros and venomous snakes) and within the world of nature in general (e.g. stinging insects, cacti thorns and poisonous plants). Moreover, the Iron Age hill forts of the Ancient Britons, pyramids of Egypt, castles and forts of Arabia, Japan and Mediaeval Europe, plus the Great Wall of China, point to humanity's desire to protect her and himself by means of physical barriers and what are by modern definition, security measures (see Toy, 1955).

In the UK what are most usually referred to as Physical Security 'Target Hardening' Measures (a term first used by the architect, author and pioneer Oscar Newman in his seminal works of 1972, 1973)—meaning stronger doors, bars on windows, specific types of glazing, better lock security etc., (see Sects. 4.2.1–4.2.4) has existed since at least Victorian times (i.e. the latter half of the nineteenth century) within the British policing model and can be similarly identified across the English-speaking world in the police services of Australia, Canada, Ireland, New Zealand, the USA and beyond.

Traditional histories within police Crime Prevention cite Ernest Burgess (1925) and the Chicago School of Sociology in the 1920s, leading to Jane Jacobs' most famous work, The Death and Life of Great American Cities (1961). Jacobs' concept of 'Eyes Upon the Street' can be identified as a direct influence on the subsequent concept of Natural Surveillance (see Sect. 4.4.2). However, readers should be aware that contrary to her own request, some of Jacobs' work has been selectively appropriated by those who desire a highly permeable built environment, where the evidenced criminogenic capacity of such Permeability (see Sect. 4.4.6) is ignored. This very much champions 'Access and Movement' within design layouts and from a (DOC) perspective such Permeability almost always designs in 'Crime Opportunity'—as evidenced throughout this handbook.

DOC began to become established following the incorporation of Physical Security 'Target Hardening' Measures, Technological Innovations and Environmental Elements within C. Ray Jeffery's (1971) concept of Crime Prevention Through Environmental Design (CPTED), together with the simultaneous development by the aforementioned Oscar Newman in his early works, Defensible Space: Crime Prevention Through Urban Design (1972) and Defensible Space: People and Design in the Violent City (1973). Newman stressed the importance of Defensible Space, Natural Surveillance and Territoriality (plus other concepts—see Section E) following his analysis of public housing in New York. Initially, Newman received positive reviews but soon attracted criticism from within mainstream criminology and beyond of 'Architectural Determinism'.

Nevertheless, in the UK and much of the world the work of Jeffery and Newman had a seminal effect, leading directly to the incorporation of CPTED within the Crime Prevention disciplines of the police forces of the UK and those other countries prepared to look outside the box of solely criminal justice solutions. An indication of this motivation can be seen in Barry Poyner's book, Design against Crime: beyond defensible space (1983). In his foreword to Poyner's book, the then Commissioner of the Metropolitan Police (Greater London), Sir Kenneth Newman, observed:

> ...the Police, faced with virtually static resources and ever-increasing crime rates, are turning to crime prevention through environmental design as an approach that promises to be more effective than any traditional Police methods.

During the mid to late 1980s this adoption manifested itself in the UK with the creation of the posts of Crime Prevention Design Adviser or Architectural Liaison Officer (CPDA/ALO), but referred to throughout this handbook as Designing out Crime Officers (DOCOs): see especially Chapter 4, What's in a Name: Designing Out Crime v. Architectural Liaison/Crime Prevention Design/Design for Security/CPTED. And incidentally, this also happened to provide a natural career path from Crime Prevention/Reduction Officer (occupied primarily with Physical Security 'Target Hardening' Measures and Technological Innovations) to that of CPDA/ALO—now DOCO.

After the police services of the UK had recruited their first CPDA/ALOs, 1989 witnessed what can now be recognised as the truly historic advent of Secured by Design (SBD—see Chapter 5). Initially launched in the South-East region of England, by 1992 SBD was available throughout the UK and now has an unrivalled claim to being the most effective, sustainable and successful Crime Prevention/Reduction initiative in the history of policing. In 2000 management of the award and sponsorship arrangements were passed to the police operated ACPO CPI Limited, now known as Police Crime Prevention Initiatives (Police CPI).

In addition to the work of Jeffery and Newman, Designing Out Crime has found support in the fields of Environmental Criminology, see especially the work of Vancouver based Paul and Patricia Brantingham (1975, 1995 and 2008) in relation to Crime Pattern Theory (see Chapter 10) and more recently what is now known as the discipline of Crime Science. Furthermore, a series of independent academic assessments—Brown (1999) Pascoe (1999); Armitage (1999); Teedon et al. (2009); Armitage and Monchuk (2009); Jones et al. (2016) and most recently Stokes (2022) have testified to the long-term effectiveness of SBD in reducing crime and by default, DOC.

This brings us to an inconvenient truth: namely, that DOCOs (CPTED officers/ advisers outside the UK) are painfully few in number. As of October 2025 the International CPTED Association (ICA) now has members in 37 countries across the globe. However, in terms of service delivery, DOC is very much a function of the English-speaking world—most especially in Australia, Canada, Ireland, New Zealand, the UK and the USA. Moreover, there have been some worrying developments during the early twenty-first century.

Following the 9/11 (11 September 2001) terrorist atrocities in the USA, many US police CPTED officers/advisers transferred into the newly created Department for Homeland Security. And in the UK, from in excess of 500 police practitioners at the turn of the millennium, by 2009 this figure had reduced to under 200 in post, three quarters of whom had other police responsibilities (University of Salford and Huddersfield University, 2009). Professor Rachel Armitage (2016) then deduced that this figure had fallen to 137. Cuts in police budgets; the closure of the Home Office Crime Reduction College at Easingwold, North Yorkshire and with it, delivery of training courses free of financial cost to UK police forces; the growth of short-term performance culture within policing (now largely and thankfully consigned to history) and a failure to appreciate the long-term sustainable value of DOC (and Crime Prevention/Reduction in general), led directly to this shortfall. Indeed, there have been parts of the UK where however temporarily, there was no DOCO in post. This is a perplexing situation considering: the all but legal requirement for such a role; the opportunities to DOC that are being lost and the training/experience required by any subsequent practitioner parachuted in or tasked as a novice to deal with the aftermath. However, research conducted by this author during 2020 established this total had improved to 199 DOCOs in post across the UK (Stokes, 2022).

The irony is that demand for this service has never been greater, central government having repeatedly underscored the value of DOC—for example, in the Modern Crime Prevention Strategy (Home Office, 2016). Indeed, this last document states how better home security has reduced the number of residential burglaries (p.4). The value of DOC from the built environment—with specific reference to SBD (see Chapter 38) is detailed on page 17. Furthermore, in 2017 the UK government expressed a desire to build 300,000 homes each year (restated by the successor government in 2024).

The Home Office and Campaign for Architecture in the Built Environment (CABE) funded the Residential Design and Crime project (Applied Criminology Centre, University of Huddersfield, 2011). They also sold the efficacy of DOC to sister ministries: Department for Communities and Local Government (DCLG)—Building Regulations that now incorporate minimum-security standards, Closure of Public Rights of Way on Grounds of Crime and Public Space Protection Orders (see Sect. 6.4); Department of Health—hospital security and the Department for Education—security at schools and higher education establishments.

The prospect of a National Police Crime Prevention Service (NPCPS) taking up responsibility for the DOC function is now a fading memory—although it may resurface in the future. And having faced unprecedented cuts in funding over recent years, the police service is struggling to maintain delivery of a range of different functions. Combine this with a desire by central government to place no additional financial demands on business and it is easy to appreciate how in some forces, the DOC role has itself been under threat. Nor is this a tragedy without consequence, because where there has been a failure to provide the DOC service, the after effects of inherent Crime Opportunity (see Sect. 2.3) will last for decades thereafter. See also Sect. 10.1, The Future of Designing Out Crime.

1.4 What's in a Name: Designing Out Crime—Architectural Liaison/Crime Prevention Design/Design for Security/CPTED

It is vitally important to understand that whether the terms Architectural Liaison, Crime Prevention Design, Design for Security, CPTED or indeed Designing Out Crime (DOC) are used, this is essentially a criminological function. As a direct consequence, it possesses the responsibility to draw down evidence from Crime Science and Environmental Criminology in support of the practical application of these disciplines within the built environment.

DOC overlaps into areas where other professions are perceived to have established territorial claim, e.g. Architecture, Planning, Development and indeed Design. But at its core lie crime, fear of crime (FOC) and anti-social behaviour (ASB—disorder), the removal or at least substantial reduction of which will by default strengthen those aforementioned other disciplines. In other words, DOC is a mutually symbiotic process that delivers advantages for all the parties involved—unless you are intent on crime or ASB. And as a criminological function, a powerful argument can be made that it is first and foremost the responsibility of the police or a devolved body like the once proposed National Police Crime Prevention Service (NPCPS), to articulate and deliver this function/service.

However, this does not explain how practitioners within the police services of the UK were encumbered with a title like Architectural Liaison Officer (ALO) or the more accurate if wordy Crime Prevention Design Adviser (CPDA). For this you need to go back to the mid-1980s and the creation of the ALO/CPDA role in the wake of seminal works by C. Ray Jeffery (1971); Oscar Newman (1972, 1973) and Ron Clarke and Patricia Mayhew (1980). Indeed, the title of the latter's published work in 1980 was Designing Out Crime.

The original publicity for the Secured by Design award (see Chapter 5) described how it had been drawn up by the 'Home Office and Senior Crime Prevention Managers'. It was the latter, courtesy of the then ACPO Crime Prevention Technical Committee—which has since evolved to become the UK Crime Prevention Design Group (CPDG), who used first and foremost the ground breaking work of Oscar Newman (1972 and 1973) primarily Defensible Space, Natural Surveillance, Territoriality; and C. Ray Jeffery's (1971) Crime Prevention Through Environmental Design (CPTED) and chose in the main to adopt the term Architectural Liaison Officer to describe those tasked with delivering this 'radical new discipline' (Secured by Design, 1989).

Nevertheless, from the outset the Metropolitan Police in London and some other police forces in south-east England did not choose this title, preferring instead the description Crime Prevention Design Adviser (CPDA). To this day, a minority of UK police forces continue to use the terms ALO or CPDA. This demonstrates a difference of opinion, the result of which has persisted for almost four decades. And in recent years, former ALOs in the Greater Manchester Police area now use the title 'Design for Security Consultants'.

In order to standardise the role name across the UK, the description Designing Out Crime Officer (DOCO) first appeared in SBD—New Homes 2014. Moreover, in 2016 the UK wide CPDG formally adopted the term DOCO but acknowledged that the terms CPDA and ALO would remain in a number of SBD Development Guides, UK governments' literature and planning guidance for the foreseeable future. More recently, in 2021 the National Police Chiefs' Council (NPCC) designated the DOCO as a specialist role within policing. Consequently, the term Designing Out Crime Officer/DOCO is used throughout this handbook.

Not that Architectural Liaison Officer was the original chosen title. According to little more than legend within the discipline, the term originally intended to describe this function was 'Architect Liaison'. However, the Royal Institute of British Architects (RIBA), fearing the title architect might be devalued in the same way that engineer has been appropriated by many in the UK, objected. The description 'Architectural' was thus the compromise result.

Unfortunately, this use of the term Architectural Liaison Officer (quite apart from engendering occasional hoots of laughter from police colleagues), also allowed too many observers to openly state that the discipline had nothing to do with policing. After all, the all-important word 'crime' is missing from the title and as a result, there is nothing to suggest such a role would assist in delivering reductions in crime or improve neighbourhood policing, community cohesion or performance targets (be they for volume property crime or more serious forms of criminal behaviour—often the key priority for policing in recent decades).

Furthermore, adoption of the word 'Architectural' also suggests a fundamental misunderstanding as to what this specialism entails. As illustrated later in this handbook (see especially Sect. 2.8, Environmental Cues), it is naive to believe that design of the building alone (the architecture), together with the inclusion of Physical Security 'Target Hardening' Measures and Technological Innovations (see Sects. 4.2.1–4.3.3) will deter criminal attack. The essence of the Environmental Elements of Crime Prevention Through Environmental Design of CPTED (see Sects. 4.4.1–4.4.6), is that it is the environment beyond the physical building that is crucial in a criminal's decision-making process (and incidentally, this is as true of professional/travelling/career criminals as it is of opportunists—see Sect. 2.3). By extension, use of the word 'Architectural' is both inadequate and inappropriate in describing the DOC function that is ultimately delivered.

If the word Architectural is therefore considered inappropriate, it should also be appreciated that the words 'Design/Designing' and 'Planning' have a perceived right of use that other professions already lay claim to. Consequently, this is pertinently the right moment to assert that DOC is no threat to the academic, architect, designer, planning officer or any other profession. It is complementary to them and should be both perceived and treated as such.

In the UK context, further misunderstanding exists in relation to the training blocks necessary to become a qualified Designing Out Crime Officer (DOCO). The normal career path is for individuals to first receive training in Crime Prevention/Crime Reduction (both descriptions are in effect all but interchangeable in this context) and for there then to follow a break of at least nine months, in order to flatten (however marginally) the learning curve. As a result, there is a natural hierarchy within the discipline and whilst historically almost all DOCOs would have been trained Crime Prevention/Reduction Officers, the reverse is not the case. Furthermore, where individuals are in post as DOCOs they should be described as such and not simply as Crime Prevention or Crime Reduction Officers. To fail to acknowledge their advanced training, qualifications and experience does them a great disservice and could cause potential problems at any subsequent planning inquiry when they are first extensively cross-examined regarding their professional background (see Sect. 7.6, Appearing Before Planning Hearings, Public Inquiries, Inquests etc.)

Please also notice how the descriptions Expert Witness and Statutory Consultee are being deliberately ignored in this context—at least for now. Suffice to suggest that over the past quarter century there has been a gradual movement in the UK context to a situation where the DOCO has either reached or is close to being recognised as existing within the bounds of such definitions.

Finally, there is of course a universal meaning to Designing out Crime, which can be defined as:

Key Issue

Designing out Crime (DOC) combines Physical Security 'Target Hardening Measures, Technological Innovations and the Environmental Elements of CPTED—and in addition to incorporating 'design ways of thinking', attempts to draw on evidence from formal impact evaluations of what works and does not work in a practical setting. In both theory and evaluation, DOC has become heavily influenced by Crime Science.

As such, the terms Architectural Liaison, Crime Prevention Design and Designing Out Crime should be considered interchangeable, especially when used in this handbook. Furthermore, a North American and beyond perspective might contend that Crime Prevention Through Environmental Design (CPTED) is the paramount description—having itself evolved to incorporate Second and Third Generation versions of CPTED (see Sect. 4.1 the Introduction to Chapter 4).

In addition, a number of membership groups for DOC professionals exist around the world. In the UK there was the Designing Out Crime Association (DOCA); in Europe, Dutch-DOCA, e-DOCA; and the worldwide (and authoritative) International CPTED Association—the ICA, which holds biannual conferences at diverse locations around the world.

1.5 Role of the Designing Out Crime Officer (DOCO): Core Attributes

At its most straightforward, the role of the DOCO is to assist our colleagues in partner agencies and other professions within the built environment, to Design Out Crime (DOC), fear of crime (FOC) and opportunities for anti-social behaviour (ASB). Since 1989 the most effective and sustainable mechanism for delivering that service in the UK has been Secured by Design (see Chapter 5). Furthermore, the following skills will help to facilitate and deliver this overall desire.

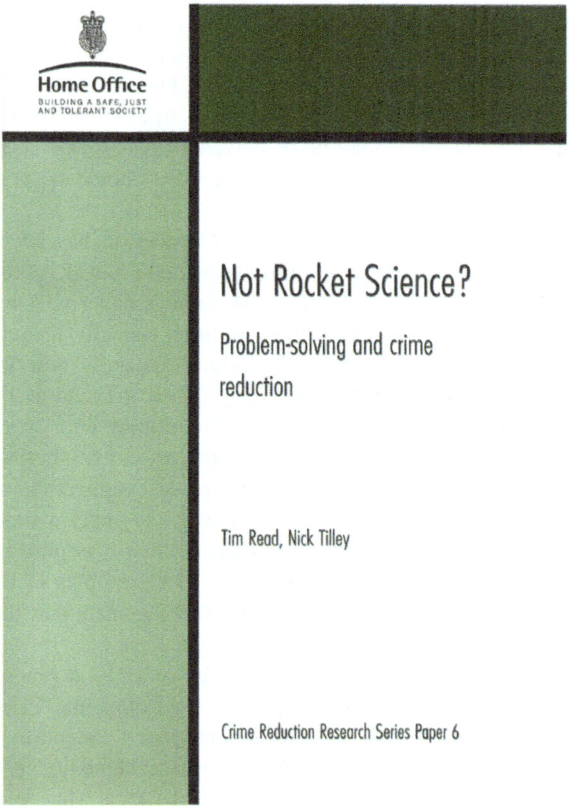

Not Rocket Science? Problem-solving and crime reduction

1.5.1 Not Rocket Science?—Well Almost

Not Rocket Science? Problem-solving and crime reduction (2000, see Sect. 2.14) is the title of a Home Office paper written by Professors Tim Read and Nick Tilley. They powerfully contend that Problem-Solving does not call for the skills of astrophysics. However, the difficulty in putting into practice relatively minor Crime Prevention/Reduction initiatives might suggest otherwise. Indeed, Professor Paul Ekblom (2011) convincingly asserts that whilst crime and crime prevention are undoubtedly complex interventions, they *'...can be tamed providing we develop suitable frameworks and languages to handle it within research and practice'* (2011, p.279).

It is of fundamental importance the DOCO never underplays the capacity and relevance of their role in the design of the built environment, together with its proven capacity to reduce if not eliminate Crime, fear of crime (FOC) and anti-social behaviour (ASB). What appears simple and straightforward to both novice and experienced DOCO alike, may be perceived as near rocket science by those outside the discipline. Indeed, this is a message that should be practiced by all those involved in the delivery of DOC.

As previously described, Crime Prevention/Reduction has been a specialist function within the police service for little more than half a century, but sadly it has never been an attractive discipline and perhaps never will be—even once the evidential connection with Crime Science is universally recognised. Nevertheless, that is no reason to denigrate its value, especially now that it has (at least in part), evolved into DOC by embracing Crime Prevention Through Environmental Design (CPTED), Environmental Criminology and Crime Science in general. Furthermore, the drawing down of data, tested hypotheses, conclusions and independent academic support effectively places a corresponding responsibility on the part of the individual DOCO. Namely, to apply a similar scientific mind and reject much of the anecdotal reference when it is portrayed as established fact within some professions. One of the best examples of this misplaced over-emphasis concerns the concept of Crime Displacement—at the expense of its positive cousin, Diffusion of Benefits (see Sect. 2.1).

And whilst considering how the fundamentals of DOC do indeed owe their credibility to research within the disciplines of Environmental Criminology and Crime Science, this is probably an appropriate time to mention that DOCOs have their critics. The charge of exhibiting a 'Fortress Mentality' is the one most usually levelled at the DOCO, although rarely to the face. That said, this author can personally testify to one senior civil servant at the former Office of the Deputy Prime Minister who railed that, 'the police aren't going to get fortress Britain!'

(whatever that is?). Indeed, the 'fortress' tag can be easily rebuffed by asserting that if after the DOCO's involvement a particular project is a fortress (visible and/or real), then the mission will have failed—in that visual fortresses will raise the fear of crime (FOC). One of the few exceptions to this rule concerns that of a military or police armoury. But surely everyone would expect a building in which firearms and ammunition are stored to be in both appearance and reality, something akin to a fortress?

A more worrying form of criticism, is to suggest that the DOCO is an 'Architectural Determinist'. This translates as implying that within the DOC fraternity, we believe that the built environment ALONE influences criminal behaviour. Once again, this strange assertion dissolves under the slightest scrutiny. No DOCO would suggest that the environment in which an individual resides or socialises, is solely responsible for their committing crime or ASB, although we do contend it can be a contributory factor especially when it provides easy Crime Opportunities (see Sect. 2.3). Furthermore, DOC powerfully argues that often minor changes to that environment (see Sect. 2.17, Routine Activity Theory) can deliver spectacular results in terms of reduced crime, ASB and thereafter, reductions in the FOC. The irony is these very same critics often cite a poor-quality built environment as encouraging the original offending behaviour, and then conveniently forget this particular nugget of truth.

The above examples provide some indication of the prejudices that lie in wait for the DOCO. They are rarely articulated in public, but scratch beneath the surface and they certainly exist. And of course, historically the DOCO was not expected to cite research findings, let alone quote from Crime Science. That's the price you pay for working with the 'practical application of criminological theory' (said to this author in person by Baroness Gisela Stuart, former Member of Parliament for Birmingham, Edgbaston whilst providing a criminogenic description and analysis of her own constituency's housing layout). But it is also the reward for being at the leading edge of putting theory into practice and we should all learn to relish the cut and thrust of the academic debate—including the insults and brickbats aimed at this all but 'rocket science'.

1.5.2 Crime Prevention/Reduction Officer

In the UK, many newly appointed DOCOs will already possess the skills of a Crime Prevention Officer/Crime Reduction Officer (CPO/CRO)—having a background of experience in those areas of business and attended training courses that delivered/developed those attributes. Many of those skills will relate to Physical Security 'Target Hardening' Measures (see Sects. 4.2.1–4.2.4), a major part of the overall discipline of DOC. However, even those previously trained and experienced in the delivery of these skills should not underestimate the near vertical learning curve that will confront them when they and become a DOCO—including Police Crime Prevention Academy initial training and RPEL continuing professional development.

For those entering the discipline with no previous Crime Prevention/Reduction experience, a real baptism of fire may lie in wait. That said, a willingness to embrace the parameters of DOC and its cutting edge application (especially when drawing down wisdom from Environmental Criminology and Crime Science), will almost certainly guarantee a rich tapestry of rewards and immense job satisfaction.

1.5.3 Street Level Criminologist/Crime Scientist

A crucial point to repeat and reinforce is that DOC is first and foremost a crime function: in essence we are seeking to influence criminal behaviour, ASB—and reduce the fear of crime amongst the overwhelming majority of ordinary citizens. It is not a function of the built environment—although that is not to say professionals from such disciplines cannot similarly become what Ron Glensor (1996), former Deputy Chief of Police in Reno, Nevada, describes as 'Street Level Criminologists'.

In the UK context we might prefer the description 'Street Level Crime Scientists', but the message remains clear: it is a necessary skill to both understand the rationale for much of Crime Science (including Environmental Criminology) and to be able to apply it to the design of the built environment.

1.5.4 Enthusiasm

A powerful argument can be made that the single most important attribute in being an effective *DOCO*, is enthusiasm for the discipline itself. If you do not believe in the capacity to *DOC* or are less than wholly committed to being an advocate for this relatively 'new skill in the built environment', then in all honesty please find another avenue for your talents.

Because, with so few practitioners overall, *DOC* needs true and ardent ambassadors who champion this wonderfully effective and profound intervention.

Having escaped the straight-jacket of purely tick the box assessment, *DOC* necessitates independent thinking—especially in respect of the *Environmental Elements* within *CPTED*, concepts including *Defensible Space*, *Natural Surveillance*, *Territoriality* and *Permeability* (see Sects. 4.4.1–4.4.6). Consequently, the specialism requires disciples who do not shy away from utilising all their knowledge and experience (tempered by the necessary diplomatic skills—see Sect. 1.5.9 below) to comment in these subject areas.

1.5.5 Tenacity and Resolve

This can be best described as the ability not to let go of a particular issue. 'Can-do' professions like those of the emergency services and military have a reputation, however ill-deserved, of rapidly becoming bored of a problem that cannot be resolved in a matter of days if not hours. However, many other bodies see the resolution process as taking weeks, months and possibly years.

For the DOCO, this means developing the skill whereby they can differentiate between the 'normal' time it takes to deliver on particular projects and initiatives, as opposed to being delayed by inertia—if not worse. Indeed, there is an increasing amount of anecdotal evidence illustrating how a minority of key decision-makers outside the discipline, will deploy the tactics of delay and obfuscation in the face of an emergency service like the police whose initiative would deliver immediate results and benefits. In such circumstances, the capacity to influence and deploy the enthusiasm of truly effective partners within sister organisations may prove to be highly beneficial (see 2.14, Problem-solving, Problem-Oriented Policing and Partnerships).

Similarly, briefing and marshalling the efforts of senior management within your own organisation, can be sufficient to persuade reluctant individuals. Members of the Community Safety Partnerships (CSPs)—formerly the Crime and Disorder Reduction Partnerships (CDRPs) and elected representatives (most usually Councillors and Members of Parliament in the UK context) may also be useful allies. However, a serious word of caution here. Your own senior management may take a less than favourable view of any approaches you may direct to elected members or the CSPs. You are strongly advised to comply with local management practice over such matters and as always, past experience and the development of a network of contacts will prove to be invaluable.

In the final analysis, if you are the only person (or one of very few) who supports the initiative in question, it may be time to walk away—or at least store it for future use. There is certainly no merit in flogging the proverbial dead horse.

However, repeated past experience suggests the appropriate time will eventually arrive for your original idea, in which case having it immediately ready to update and submit may accrue all manner of plaudits.

1.5.6 Accuracy

Nothing destroys the credibility of a DOCO more than the failure to be accurate. This is especially the case when quoting the relevant British Standard or similar quality assurance reference. Ensure upper case letters are used when required and the correct spacing in the body of the report being written. Section 3.4, Writing the Report/EVA: The 'Crown Jewels' of its Written Presentation, covers these crucial elements in the DOCO's toolbox—including sentence construction and punctuation, repeated key failings.

Similarly, in an age when plans, diagrams, comments and recommendations are increasingly being received and forwarded by electronic means, there is no excuse to resort to mobile phone text culture in the body of an e-mail. Remember, your observations and criticisms might ultimately be presented at a full planning hearing or appeal and counsel for the plaintiff will be highly critical and cause embarrassment if the words in your e-mail are reminiscent of a school student's joke text. See also see 7.6, Appearing Before Planning Hearings, Public Inquiries, Inquests etc.

1.5.7 'Regardez!'

On her deathbed the final word said by the celebrated French author, Sidonie-Gabrielle Colette, was reported to have been47 'Regardez!' (Lottman, 1991). This roughly translates as an instruction to 'look!' Colette had just observed a butterfly enter the room and for a woman who had spent a lifetime observing and writing about the French people and nation, it was perhaps an appropriate response with which to leave this world.

It is also an appropriate skill for the DOCO and Street Level Crime Scientist, because very often there will be evidence at the scene of a potential development that will influence the recommendations they make (see Sect. 3.3, Site Surveys and Environmental Visual Audits—EVAs). This evidence will most usually take the form of Environmental Cues (see Sect. 2.8) which are usually visual (graffiti, communal doors propped open by a brick, car tyre tracks across a lawned area, a pool of broken car window Toughened Safety Glass etc.), although not always. For example, loud music emanating from a dwelling and causing distress to neighbours would indicate at the very least, ASB. And the smell of urine in a communal hallway/lobby area will undoubtedly provide evidence of casual intrusion and worse, plus the lack of true Defensible Space (see Sect. 4.4.1).

The important point to note is that during every site visit, including the one made on completion of the project, the DOCO should be looking for such Environmental Cues and observing whether they change (are in effect 'Designed Out') as the development proceeds. One note of caution though: this is not an instruction to develop some Sherlock Holmes style investigatory powers and read too much into the 'evidence' before you. Conclusions need to be drawn from established evidence within Crime Science/Environmental Criminology, or at the very least repeated past experience/personal witness.

1.5.8 Photograph Now!

Developing the theme of the French author Colette's instruction to observe (see the previous sub-chapter), one of the first realisations which often dawns during training for the role of DOCO, is the value of photographs. Once you know where and how to look, the necessity to photograph good examples of DOC (and unfortunately, more often bad examples of where crime, FOC and ASB appear to have been designed-in) can become irresistible.

The built environment 'speaks to us all' and as previously indicated, this whole subject is discussed in Sect. 2.8, under the heading of Environmental Cues. Most often, these cues are visual. Developments in camera and mobile phone technology, especially image quality and digital recording, mean that even the most unskilled of amateurs (like this author) can quickly build a portfolio and eventually a library of visual reference and evidence.

In the first instance, these pictures will probably be for continuing professional development and as an aide memoire. However, they will also serve additional functions: as evidence in planning reviews; Secured by Design (see Chapter 5) applications and final site inspections; and when delivering presentations (most usually by way of PowerPoint) for the training of new or potential converts, practitioners and as an ambassador for the DOC function. In such circumstances, a photograph can literally replace a thousand words—and is far more likely to maintain the attention of the audience, rather than text (especially if you win the 'graveyard slot' after lunch or at the end of the day!)

Because the built environment is constantly changing (it is in effect part of our own living history), the advice is to photo whilst you have the opportunity. It is all too easy to snap away at a completed project, but our discipline is littered with observations of the type, *'I wish I'd photographed that before they pulled it down/when it was being built'*.

One advantage for the newly appointed *DOCO* is that they should not have to suffer the task of transferring old style 35 mm prints or slides to digital recording media, plus former problems with storage/quality of the images. Consequently, new recruits to the discipline should find it a relatively easy task to create their own library from purely digital images.

Some further words of advice in relation to digital photographs:

- Store your images on at least two WORM (write once read many) mediums. That way, if one is lost you will always have a back-up
- Keep an index of the images, which should include the date the photograph was taken and what you believe the pictures illustrate. It is far too easy to forget the underlying message contained in a particular photo
- Exercise care when photographing drunken or other forms of ASB. In other words, conduct your own Dynamic Risk Assessment at the time (see Sect. 3. 2, Health and Safety of the DOCO when Surveying Sites) and take due regard of the provisions of the Data Protection Act, 2018 and UK GDPR
- When publishing or otherwise displaying images of crime or ASB, be similarly aware of the requirements of the Data Protection Act, 2018, UK GDPR and Freedom of Information Act, 2000. In addition, be mindful of the necessity, where appropriate, to pixelate faces or other identifying features such as vehicle registration numbers.

1.5.9 Diplomatic Skills

There is some anecdotal evidence (no doubt inflated by urban myth) relating to ALOs/CPDAs who in the early years of the discipline were prone to enter the offices of an architect, developer or planning officer and declare that the plans before them represented some hideous monstrosity. Needless to say, they were deservedly shown the door and never invited back!

This is rarely an issue where police staff (non-sworn DOCOs), recruited from the professions of the built environment to perform the DOC function, or with police officers who have some experience in the role. However, all new recruits to the role of a DOCO are strongly advised to remember the following: architects, planning officers, structural engineers and indeed other professionals from within the built environment are most usually university graduates who have undertaken considerable additional training, often lasting many years (at least seven years for architects) and almost always will have specialised in that discipline all their working life.

As a result, they do not take kindly to members of the police service (who they perceive as having received minimal training and experience) descending on their offices and voicing their amateur, tabloid-influenced opinions! In other words, you may consider such a professional's work to be ugly or inappropriate. But the crucial advice is to keep that view to yourself—especially when harbouring a desire to have some influence over the development in question, or any future project where DOC (or specific CPTED measures) will be in contention.

1.5.10 The Art of Compromise and Negotiation

Hand in hand with diplomatic skills, the DOCO also requires the skill of what might be described as the art of compromise and negotiation. When in negotiation with an architect, developer, planning officer or fellow professional from the built environment, you will rarely (if ever) be able to convince them of the necessity for every amendment you desire. If you do not possess the ability to compromise (and this has been a repeated failing by a minority of DOCOs in the past), it may mean you lose the whole package of recommendations, a majority of which the other party is willing to incorporate.

Consequently, if by the very essence of your nature you do not have the capacity for compromise, then in practical terms the world of DOC is not for you. Because, this discipline is not about a simple 'tick the box' culture. Indeed, it necessitates the DOCO thinking outside those boxes and having the confidence to express both verbally and in writing their considered and authoritative (if necessarily subjective) opinion—influenced by evidence from the worlds of Environmental Criminology and Crime Science in general. Historically, a few DOCOs have found it difficult to understand that the role is essentially about shades of grey—especially in respect of the Environmental Elements of CPTED (see Sects. 4.4.1–4.4.6) and both Threat and Risk Assessment in general, rather than black and white certainties.

That said, nor is DOC a world for 'shrinking violets'. A resolute stance is often a necessary attribute. The true skill is being able to develop a background of theory and practical knowledge that allows the individual DOCO to realise (often instinctively, as a result of being nurtured over time) on what points she or he can be prepared to compromise and negotiate.

Moreover (and this is the true beauty of DOC), because much of the work deliberated by the DOCO is at the cutting edge of the practical application of Crime Science and Environmental Criminology, there will often be diverse opinions as to what is most effective (and ineffective). We should not be frightened of such evolution, but revel (enjoy) this debate and application. Furthermore, some of these conclusions have a habit of changing over time and as new evidence is produced. Some describe that as true wisdom.

1.5.11 Truly Effective Partnership Working

The efficacy of cross-agency Partnership (collaborative) Working has been a constant rejoinder in recent years, if not decades. Indeed, who could argue against such an ideal that dovetails accurately into the philosophy of Problem-Solving, Problem-Oriented Policing and Partnerships (POP) see Sect. 2.14 and especially Read and Tilley (2000); Bullock, Erol and Tilley (2006). However, whilst there is a surfeit of evidence concerning such highly effective Partnership Working, the following warnings need to be given.

Be especially wary of ‘partners who talk the talk but don’t walk the walk’. Claims of Partnership Working (known as ‘Collaborative Working’ within the health professions) are legion, but it is first necessary to sort the effective initiatives from those that exist in name alone. Truly Effective Partnership Working is the only really valuable type. Those that pay lip service to the DOC approach should be treated with the utmost diplomatic care. Unless you have some statutory power with which to compel them (and subject to management approval), it is better to walk away from the development rather than compromise to such an extent that the recommendations accepted become meaningless in the overall context of the project.

Once again, it is a skill to be able to differentiate between partnerships in name alone and more importantly, those individuals who are receptive to such overtures from DOC. Similarly, a distinct minority (thankfully) of supposed partners are well versed in the arts of delay and obfuscation. They can play these tactics with calculated skill, especially when confronted with members of the ‘wider police family’ who they believe are easily prone to becoming bored and moving on to another issue. Tenacity and resolve (see Sect. 1.5.5) are the best responses to such individuals, ever mindful of the warning contained in the previous paragraph. See also Sect. 9.1, The Four Towers at Duddeston Manor, for a detailed example of highly effective Partnership Working in practice.

1.5.12 Administration System

For far too many years the absence of a well-managed administration system has proved to be the downfall of a few DOCOs. Moreover, on those occasions when appeals have been made regarding decisions to refuse the SBD award, the appeal has been successful simply due to the absence of an audit trail of copies of reports and/or letters sent to the client. Such deficiencies are unforgivable and can be easily avoided.

It is the responsibility of the individual DOCO to maintain a full and efficient administration system, covering all the current projects with which they are engaged. This will often include the reports and letters relating to the projects initiated by their predecessor. And when the current occupier of the role moves on to pastures new, his or her documentation will need to be passed to their successor. Any attempt at a 'scorched earth' (industrial sabotage) policy when staff change, should be regarded as grossly inappropriate by any professional DOCO and one that may warrant civil or indeed criminal prosecution.

A paper or preferably computer-based indexing and referencing system, detailing the progress of individual development projects, with sufficient storage capacity for hard copies of plans and supporting documentation, is a minimum requirement. Furthermore, since 2009 all current *Secured by Design* projects should have been recorded on the police *SBD* managed and computerised database.

1.5.13 Naming and Praising

'Naming and Praising' is the adult version of that age-old call to arms 'Naming and Shaming'. On occasion, the latter may be suggested as an appropriate response to repeated failings by a third party. The advice of this handbook is to strongly counsel AGAINST such a strategy, which can easily backfire. Indeed, whilst extremely tempting, the DOCO can fall into a trap and become the unwitting victim, by being persuaded to fire the metaphorical bullets moulded by others.

The rationale here is quite straightforward and can be best explained by the biblical quotation, 'let him without sin cast the first stone'. Your argument must be entirely without fault and even then, however right and proper your case, do not expect the party you are Naming and Shaming to necessarily play by the rules of honesty and integrity. Indeed, Naming and Shaming is a dangerous political minefield littered with a history of unintended consequences that have a habit of exploding in the face of the innocent party.

As an example of the academic research, consummate skill and tenacity required before embarking on an exercise of Naming and Shaming, read Professor Gloria Laycock's (2004) paper, The UK Car Theft Index: an example of government leverage. This highly successful exercise (culminating with the major car manufacturers being 'invited' to the Home Office by the then Home Secretary) was a catalyst to 'incentivising' the motor vehicle industry to incorporate enhanced security measures (such as immobilisers and alarms) in new cars. However, it also demonstrates why such an initiative should not be embarked upon without massive care and in-depth preparation.

It is far easier to adopt a strategy of Naming and Praising, a device that can be almost as effective as Naming and Shaming, but without the potentially dangerous political fall-out. Naming and Praising is designed to identify good practice and advertise it to a wider audience, often through positive use of the media. Bad practice is in effect ignored by such a strategy and thereby isolated. And whilst this might at first glance appear a less macho form of attack, the benefits it can deliver are profound. For special interests, be they private or public sector, have no desire to be ignored and thereby marginalised, or advertised as deficient by a lack of inclusion. Such a subterfuge has a proven track record of success in neutralising formerly hostile elements, if not converting them to your perspective.

The most successful (not to mention long-term sustainable) Naming and Praising process is of course the Secured by Design award system (see Chapter 5).

1.5.14 Learning Log

From a twenty-first century viewpoint, it is strongly advised that the DOCO maintains what might be best described as a Learning Log. This could be an electronic document, although many continue to prefer a written record, for example, a simple A4-sized 'blue book' that (for this author) lasts around six months.

The Learning Log is a multi-purpose document that records (on a daily basis), the Designing Out Crime work undertaken—and most pertinently, any new lessons learnt. Whilst this book can also perform the function of a daily diary, its primary role is to detail work in progress recommendations made and conclusions drawn—especially in the worlds of CPTED and SBD where subjective interpretations are likely to evolve and change over time. And, at least in theory, they might be required at a subsequent Planning Appeal which in the context of English law is akin to the legal judicial process. Also include CPDA

RPEL and other continuing professional development. Such a document should be kept for at least 7 years once it is completed.

One key point: the Learning Log is not an alternative to the administration system detailed under Sect. 1.5.12 above. The latter should work in tandem with the Learning Log, detailing letters and e-mails sent to clients, making phone calls etc. And include an index—advice that this author has ignored to his immense cost in terms of time spent searching for the all-important missing data!

1.5.15 Commensurate Risk

A major skill for the *DOCO* (and one that tends to be developed over time and from experience) is that of making recommendations based on *Commensurate Risk*. In real terms, this can be best described as follows:

Key Issue
Commensurate Risk concerns the ability to deliver security recommendations that are believed to be appropriate to the risk perceived to exist in a specific location at a given time and as they apply to a given task(s), function(s) and/or personnel/goods that need to be protected in that particular environment.

For example, Secured by Design (SBD—see Chapter 5) is a minimum standard. But local circumstances (such as existing high levels of crime or types of crime) may necessitate additional CPTED measures being incorporated into a particular development. These might take the form of enhanced perimeter protection, intruder alarms and/or CCTV—considered appropriate for the perceived site specifics of the project in question.

Of course, it would be extremely easy for the DOCO to simply recommend that enhanced security measures should always be incorporated, regardless of the perceived risk. However, ignoring the additional and unnecessary economic and carbon costs of such a practice, it might also fall into the trap of demonstrating a 'Fortress Mentality'.

Most importantly, Commensurate Risk demands that the DOCO uses their skill and judgement (considering all the evidence before them, accumulated knowledge and any additional factors) to make recommendations considered to be appropriate and cost-effective to DOC, FOC and ASB at that location.

This is sometimes achieved by use of a Risk Assessment/Threat Vulnerability Matrix, where the severity, likelihood and overall vulnerability to the risk are considered—often by way of a numbered scoring system. However, the use of such a matrix or the production of a Threat and Risk Assessment (TARA) is unnecessarily complicated for the majority of simple and straightforward DOC projects.

See especially Sect. 5.4 SBD—Residential Guide 2025; Sect. 5.19 'Park Mark' Safer Parking Scheme and Sect. 6.1 Domestic Violence Sanctuary Rooms as examples of where assessment of Commensurate Risk is of instrumental concern.

1.5.16 Short-Term Hits

As previously indicated, the very nature of DOC (including the time it takes from making recommendations to them being realised in the built environment) means that the discipline most frequently delivers results in the medium to long term. This helps to explain why in the past, DOC has on occasion been marginalised, especially where short-term performance culture has been of primary concern—and exacerbated by massively shrinking budgets.

However, there are measures that with the necessary level of Truly Effective Partnership Working (see Sect. 1.5.11 above), can deliver immediate results in the short-term.

For example: the removal of an element of public seating that is acting as a congregation point for ASB (and worse); increasing the level of Street Lighting (see Sect. 4.3.1) in a particular environment to deter crime, ASB and/or reduce FOC; pruning back vegetation that is masking Natural Surveillance (see Sect. 4.4.2) or Formal Surveillance opportunities around an ATM (Automatic Teller or Cashpoint Machine—see 6.11); and 'tweaking' the ability of a CCTV camera in order that it can see a particular part of the built environment to where drug-dealers have been 'micro-displaced' (see Chapter Sect. 4.3.3 CCTV—Closed Circuit Television).

Such short-term hits will satisfy the statistic hungry manager, elected members, elements amongst the media and the public at large. Most importantly, this will advertise the true value of the role of DOCO and her or his multiple skills and potential applications.

1.5.17 Avoiding Gimmicks

For the Crime Prevention/Crime Reduction practitioner, the resort to a scatter-gun approach using interventions of questionable pedigree and effectiveness, might be justified on the basis of management direction or to satisfy political/media curiosity that a problem is being addressed. Indeed, the capacity to fire repeated magic bullets, regardless as to whether there is any proof of past success, is often appreciated as a key skill for such practitioners. Furthermore, there is considerable evidence that keeping criminals 'on their toes' can have highly beneficial effects in terms of preventing/reducing crime—see especially Smith, Clarke and Pease (2002) and their work in relation to 'Anticipatory Benefits'.

However, there is no space for gimmicks in DOC. Never recommend dummy CCTV cameras or intruder alarm boxes. Quite apart from the legal perils in relation to dummy CCTV cameras and the absence of evidence to suggest such devices have any deterrent value, there is also a moral imperative. Namely, as Street Level Crime Scientists we will not support interventions that we know to be false, useless or worse—counter-productive. Moreover, there is also the legal necessity to use all our professional skills to protect the reputations of both the organisation for which we work and ourselves.

This may require considerable moral courage by the DOCO, in challenging a direction to recommend a given intervention. The advice is to consult with colleagues and stick to your principles, ever mindful of the earlier Sects. 1.5.9 Diplomatic Skills; and 1.5.10 The Art of Compromise and Negotiation.

1.5.18 Use of Post-Numerals

This is a highly subjective issue and one the newly-appointed DOCO may need to decide. Namely, do they use their academic post-numerals in correspondence and on their business card—where available. This seems such an innocent question yet is loaded with presumptions and potential dangers for the DOCO.

In the world of academia (and certainly within the UK context), the use of post-numerals is often frowned upon—at least culturally. Moreover, this embargo often extends into certain other professions. Set against this is the desire to register the credibility and qualifications of the individual DOCO. Architects, planners, surveyors etc. performing this role may find it unnecessary to display their post-numerals. Historically, police officers and police staff, having encountered sheer disbelief that they might possess a degree, may think differently. Certainly, this author's opinion is that if you have genuinely earned these qualifications, display them.

One note of warning in this regard. If you do decide to use your post-numerals, restrict yourself to the ones that really matter. To list the entire alphabet after your name (it has been witnessed!) however well deserved, may be interpreted as insecurity, pomposity or worse.

In the final analysis the decision should rest with the individual DOCO, subject to management approval and practice in the organisation for which they work.

1.5.19 Marketing Yourself

Following on from the previous sub-chapter, it might be regarded as singularly unnecessary for the DOCO to even consider having to market her or himself. After all, their increasing tally of highly prestigious SBD awards and all manner of DOC projects should be testament to their effectiveness in delivering long term, sustainable results. However, conscious that not everyone is an instant convert to this message, there is no harm in utilising the age-old maxim that it pays to advertise. Moreover, if you don't advertise your abilities, neither will anyone else.

This is a key skill area in itself, if only because of the danger in over-exposure. Meaning, DOCOs should not court media coverage on too regular a basis. Therefore, the ultimate skill lies in deciding when to bring your efforts to the attention of management and the media.

As a minimum, the SBD award ceremony merits a report to local management. They should be informed that you have seized the once in a 30 years opportunity to DOC, FOC and ASB from a particular development. And whilst guided by the views of the client and local policy, it may be appropriate for the DOCO to issue a press release regarding the award. Indeed, in one UK police force it is their Chief Constable signs each SBD award certificate.

Some DOCOs are uncomfortable with the concept of having to sell themselves in this manner. The advice from this author is that they should banish such misgivings, especially in the current economic climate where managers are tempted to eliminate what they perceive to be superfluous functions. As previously described, DOC is no place for shrinking violets!

Finally, in any media interview, report or press release it is wise to have an arsenal of statements rehearsed in advance. For example, the long-term, sustainable effectiveness of SBD and the DOC process (see Professor Rachel Armitage's research of 2000 and 2010—the latter with Dr. Leanne Monchuk); how it reduces the carbon effects of crime—see Sect. 2.11 and Professors Ken Pease and Graham Farrell's The Carbon Cost of Crime and Its Implications (2009); the absence of

Crime Displacement and more likely a Diffusion of Benefits (Sect. 2.1); Professors Sean Johnson and Kate Bowers' research paper Permeability and Burglary Risk: are Culs-de-Sac Safer? (2009) see Sects. 4.4.6 and 6.2); and any other factors drawn down from Crime Science and/or statistical analysis that support the efficacy of your case e.g. your advocacy of Elegant Security (see Farrell and Tilley, 2022).

1.5.20 Timeliness and Prioritisation

When planning applications are published, in the UK interested parties (like the DOCO) will usually have a maximum of 21 days to make observations. Moreover, should the DOCO fail to pass comment on such plans, it is not uncommon for the planning authority in question to endorse the application with: *'The police had no comment to make'*—all but suggesting police approval.

This last example underscores the necessity for the DOCO to prioritise their workload in order that such planning applications are dealt with expeditiously. It also demonstrates how the DOCO must ensure that when planning applications are referred to the police for comment, they are addressed directly to her or himself. There are countless examples from across the UK of planning applications lying dormant on the desk of the wrong police officer/staff member and the 21 days (or other designated time period) response date being missed as a result.

Of course, such scenarios also point to the necessity for creating an efficient networking system with the relevant Local Authority Planning Department, thereby helping to eliminate administrative blockages and ensuring 'response by' dates are met. Furthermore, this liaison will often disclose the response date has been extended.

Most importantly—NEVER discard a planning application simply because the response date has expired. First contact the relevant Planning Department and enquire whether the date has been/can be extended and/or evidence submitted.

Key Issue

In recent years and as the number of DOCOs within the UK reduced, it became apparent that an increasing number of planning applications were being endorsed by the Local Authority with the statement, 'the police have no comment to make'. Elected members have used the media to criticise the police for this failure to comment. Consequently, it may be pertinent for DOCOs to draw this issue to the attention of police managers when attempts are being made to reduce service delivery in the arena of DOC, because failure to pass comment will eventually attract civil litigation.

References

Armitage, R., & Monchuk, L. (2009). *Re-evaluating secured by design (SBD) housing in West Yorkshire*. University of Huddersfield.

Armitage, R. (1999). *An evaluation of secured by design housing schemes throughout the West Yorkshire area. The applied criminology group*. The University of Huddersfield.

Armitage, R. (2016). Crime prevention through environmental design. In R. Wortley & M. Townsley (Eds.), *Environmental criminology and crime analysis*. Routledge.

Carstensen, N., & Frederiksen, K. (1997). Situational Crime Prevention. In L. Ravn (Ed.), *Kriminalistisk Arbog*. Criminal Procedure Institute.

Department for Communities and Local Government (DCLG). (2012a). *National Planning Policy Framework (NPPF)*. DCLG.

Department for Communities and Local Government (DCLG). (2012b). *National Planning Practice Guidance (NPPG)*. DCLG.

Farrell, G., & Tilley, N. (2022). Elegant Security: Concept, Evidence and Implications. *European Journal of Criminology, 19*(5), 932–953.

Jacobs, J. (1961). *The keath and life of great American cities*. Random House.

Jones, A., Valero-Silva, N., & Lucas, D. (2016). *The effects of 'secure warm modern' homes in Nottingham: Decent homes impact study*. Nottingham City Homes.

Kirby, S. (2018). Chair's opening address given to UK wide Vehicle Crime Problem Solving Event hosted by West Midlands Police in Birmingham on 24 July 2018.

Newman, O. (1972). *Defensible space: Crime prevention through Urban Design*. Macmillan.

Newman, O. (1973). *Defensible space: People and Design in the Violent City*. Architectural Press.

Pascoe, T. (1999). *Evaluation of secured by Design in Public Sector Housing–Final Report*. Building Research Establishment.

Poyner, B. (1983). *Design against crime: beyond defensible space*. Butterworth.

Sampson, R. & Eck, J.E. (2008). '*Super Controllers: Can I be your Superman? Presentation given at the 19th POP (Problem-Oriented Policing)* Conference held in Bellevue, Washington, September 2008.

Stokes, M. (2022) *An Investigation of the Sustainability of Crime Prevention in the Built Environment: Impact and Implementation Factors*. Published Doctoral Thesis. University of Huddersfield.

Teedon, P., Reid, T., Griffiths, P., Lindsay, K., Glen, S., McFayden, A., & Cruz, P. (2009). *Secured by design impact evaluation final report*. Glasgow Caledonian University.

Tilley, N., Pease, K., Hough, M. and Brown, R. (1999) *Burglary prevention: Early lessons from the Crime Reduction Programme*, Policing and Reducing Crime Unit, Crime Reduction Research Series Paper 1, London: Home Office.

Toy, S. (1955). *A history of fortification: 3000 B.C. to a.D. 1700.* Heinemann.

Evidence from Crime Science

2

2.1 Crime Displacement and Diffusion of Benefits

Key Issue

Understanding the relative unimportance of Crime Displacement (where crime moves in time, location, target, tactics, type or perpetrator) and the importance of its reverse and positive cousin, Diffusion of Benefits, is probably the most pertinent mental skill for the DOCO to first comprehend. Indeed, it can be powerfully argued that such an understanding is the key that opens the door to Crime Science in practice, thereby providing the academic evidence with which to support recommendations subsequently made.

Background

To begin, Crime Displacement certainly exists, but by no means on the scale believed by many people—including those involved in law enforcement. Hill and Pease (2000, p. 131) describe Crime Displacement as the 'wicked issue', wicked because it acts simultaneously as a counsel of despair (in other words, what is the point in Situational Crime Prevention(SCP) using CPTED and DOC interventions, if they only serve to displace crime) and as an excuse for inaction. It also ignores its polar opposite, Diffusion of Benefits. Moreover, if Crime Displacement did indeed operate on the scale imagined by so many, those of us involved in DOC would be reduced to little more than deck-chair re-arrangers on the sinking Titanic! More succinctly and as Johnson et al. (2014) convincingly assert, economic expenditure on this approach to Crime Prevention would be pointless if we were simply displacing crime.

M. Stokes, *Designing Out Crime*, Crime Prevention and Security Management,
https://doi.org/10.1007/978-3-031-99840-9_2

To comprehend why Crime Displacement is so often misdiagnosed, first requires an understanding of the concept itself. To this end, there are currently believed to be six forms of Crime Displacement (that can often be perceived to overlap each other). The first five of these were described by Felson and Clarke (1998) as:

- Geographical Displacement (also known as Spatial Displacement)—where crime is moved from one location to another (and incidentally, the form of displacement most commonly misdiagnosed, asserted and researched)
- Temporal Displacement—when crime takes place at a different time of the day or night. For example, a police initiative aimed at arresting thieves breaking into cars during the hours of darkness, might in practice cause them to change their criminal activities to daylight hours. Guerette and Bowers (2009) discovered that only 5% of the studies they examined had investigated the potential for Temporal Displacement
- Target Displacement (also known as Target Switch)—here the criminal decides to choose a different target. Rather than spraying graffiti onto bus shelters, he might choose shop front roller shutters as an alternative. Guerette and Bowers (2009) established that just 14% of the studies in their review had included this form of displacement
- Tactical Displacement—involving a different method to commit the crime. For example, instead of using cloned credit cards to extract banknotes from ATMs, a fork-lift truck might be used to remove the whole ATM (Automatic Teller/Cashpoint Machine) from the wall. Guerette and Bowers (2009) ascertained that whilst 49 studies had considered this type of Crime Displacement, the evaluations suffered from being inadequately designed
- Type Displacement (also known as Crime Switch)—where one form of crime is substituted for another. This can be an especially insidious form of displacement, if it counsels against SCP, CPTED and DOC interventions on the basis that (for example) preventing residential burglary might lead instead to the more serious crime of robbery. Barr and Pease (1990) described this as Malign Displacement and the reverse as Benign Displacement. Little of the research to date has investigated this form of *Crime Displacement*, although personal witness recalls that the malign variant had considerable support from barrack room theorists during the 1990s, despite an absence of scientific evidence regarding its causation.

Barr and Pease (1990), further developed by Hill and Pease (2000), went on to describe a sixth form of Crime Displacement, not of the crime but the perpetrator:

- Perpetrator Displacement—where a Crime Opportunity is so appealing it will attract an endless queue of potential perpetrators. Hill and Pease (2000) provide the powerful example of drug importation and how third-world poverty creates an unlimited pool of volunteers to be mules. And other examples can be identified: a dwelling repeatedly targeted by burglars —presumably because it exudes the same Environmental Cues (see Sect. 2.8) that influence the criminal's decision-making process. Indeed, once you start looking for such 'honey-pots' of Perpetrator Displacement (or at least magnets to criminal activity and often Repeat Victimisation) they are easy to locate: the elderly person's bungalow routinely subject to distraction burglars; the public bench vandalised time and again because it is where disaffected young people congregate; the hospital car park where vehicles are broken into on a regular basis. However, as previously alluded to, such features may be more Crime Generators or Crime Attractors (Brantingham & Brantingham, 2008) than displacers.

Major Criticisms

Barr and Pease (1990, p. 280) observed how scepticism regarding the value of Crime Prevention initiatives '*...is often and frustratingly based on the presumption of total displacement...*' and perpetrated by '*extreme case pessimists*'. Criticism also originates within mainstream criminology, on the basis that Situational Crime Prevention interventions do not challenge the individual's desire or motivation to commit crime. Hill and Pease (2000, p. 131) convincingly assert that such a debate as '*ill-informed and misleading*'. This is, of course, a variation on the counsel of despair, do nothing reaction and it helps to explain why certain elements of DOC, CPTED and SCP are on occasion unfairly labelled as Architectural Determinism by ill-informed critics.

The Suicide Analogy

One of the essential beauties of Crime Displacement research (and many other elements of SCP) is how a cross-discipline approach has provided evidence that directly challenges the extent to which it takes place. Surprisingly, this comes from research into suicide, which has not been a crime under English law since 1961 and never in Scotland (unlike assisted suicide across the UK). Furthermore, the suicide analogy is predicated on the extent to which the removal of a primary causal means (coal gas/carbon monoxide in this instance) reduced the overall numbers of those who took their own lives.

Until the 1960s, coal gas was a major energy source in the UK and accounted for almost half of all suicides committed in England and Wales. This form of suicide, by 'leaving the gas on' and/or 'putting your head in the oven' (as it was described in everyday speech) had been available since Victorian times. However, during the 1960s onwards the conversion to first oil-based and then 'Natural' or 'North Sea Gas' (methane), effectively deprived potential suicides of an easily available and pain-free death. Carbon monoxide is actually poisonous (attaches to the haemoglobin in the blood in preference to oxygen). Whereas methane (like CO_2) excludes oxygen from the air inhaled, causing the victim to retch.

The net result: between 1958 and 1977 suicides by means of domestic gas fell from 2637 to 8—roughly 50% of all suicides to 0.2%. Moreover, this had simultaneously led to a reduction in all suicides of some 25% during the same time frame—from 5298 to 3944 (Felson & Clarke, 1998). Clarke and Mayhew (1988) describe how this reduction took place during a time of economic decline when suicide might be expected to increase—as it did in many European countries.

Consequently, it can be concluded that presuming all other factors remained the same (and of course they didn't, as the decriminalisation of the offence in 1961 demonstrates—albeit without the increase many predicted), a quarter of all potential suicides (1300 people) were not displaced into choosing another means of taking their lives.

Diffusion of Benefits

In the same way that people who quote the economic law of 'Economies of Scale' forget its sister maxim of 'Diminishing Returns', those who cite Crime Displacement are prone to ignore its positive opposite, Diffusion of Benefits. In essence, rather than fearing crime might be displaced in one of the six ways detailed above, what if instead an SCP intervention caused positive benefits to spread to (for example) neighbouring geographical areas? This is not fanciful stuff, because there now exists considerable evidence to demonstrate that this regularly takes place. It is just that it does not gel with the doom-laden, hand-wringing prognosis of its critics.

Displacement and Diffusion Research

Clarke and Weisburd (1994) were the first to articulate this revolutionary concept—that interventions could reduce Crime Risk to targets not directly subject to such interventions, (e.g. those located nearby). This was originally described as a 'diffusion of crime control benefit', although it is also referred to as the 'Free Rider' or 'Halo' effect and now as Diffusion of Benefits.

The authors described two mechanisms through which such benefits could occur: discouragement and deterrence. Deterrence takes place when an offender is indecisive regarding the increased risk of offending (overestimates the 'reach' of an intervention). Whilst discouragement occurs when changing targets would involve extra effort or reduced reward.

Bowers and Johnson (2003) observed how changes in a nearby catchment area and those in the associated treatment area (and a suitable control area) suggested that when Crime Displacement did occur, it was only partial and a Diffusion of Benefit was a slightly more likely outcome. More recently, a systematic review of Crime Displacement associated with SCP interventions was conducted by Johnson et al. (2012). From this study, the authors conclude that a Diffusion of Benefit was as likely to be observed as Spatial Displacement. Whilst Johnson et al. (2014) assert, investigation of displacement and diffusion should become integral to all future evaluations of Crime Prevention outcomes.

Previously, Ekblom's Safer Cities evaluation (1996) reported that whilst burglary reduction did indeed lead to some Crime Displacement, moderate or higher intensity anti-burglary action (estimated at over £4 per household) led to a Diffusion of Benefit. However, despite the sophisticated methods employed it was not possible to estimate how the savings in the target area and some adjacent areas, balanced against the losses from Crime Displacement in others.

Bowers et al. (2011) acknowledged the earlier research of Barr and Pease (1990), Eck (1993), Hesseling (1994), Weisburd et al. (2006), Cornish and Clarke (2011) and many others investigating the propensity for Crime Displacement and its positive cousin, Diffusion of Benefits. More recently, Johnson et al. (2014) revisited all the research to establish a more systematic and statistically rigorous review of the evidence. This included the present methodological limitations and quantifying effects, spatial patterns, Crime Switch, Temporal Displacement, Target Switch and Tactical Switch—concepts considered in the overwhelming majority of previous studies, albeit with varying degrees of adequate design. They concluded that *crime* rarely relocates and that a *Diffusion of Benefits* is far more likely. Furthermore, this has now been repeatedly observed in successive studies: Weisburd et al. (2006) and Guerette and Bowers (2009). Braga et al. (2012), Telep et al. (2014) and the National Police Improvement Agency (NPIA) funded systematic review by Bowers et al. (2011) which analysed the extent to which geographically focused police initiatives displaced *crime*.

Moreover, because studies typically compare crime rates before and after intervention, they have in general not investigated whether there is a temporal pattern of occurrence. For example, the benefits or displacement activity might be identified at the outset, or materialise or fade over time. Similarly, the two types of side-effects might occur simultaneously, or more complex scenarios may be disclosed. Most pertinently, the authors describe how '…generally, the sustainability of the effect of crime prevention is a neglected topic in the evaluation literature' (Johnson et al., 2014, p. 15).

Certain crimes are more prone to displacement. Rengert (1990), Caulkins (1992) and Eck (1993) found that Drug Dealing fell into this category. And in the world of CCTV (Closed Circuit Television—see Sect. 4.3.3) 'micro-displacement' (this author's description—and something of a 'Higgs Boson' awaiting evidence) of drug dealing to just outside the camera's field of vision, is repeatedly reported by public area CCTV managers—albeit anecdotally. However, Cromwell et al. (1991) discovered that even drug-addicted burglars were not automatically displaced when situational measures prevented them committing crime.

In recent decades, many studies have similarly demonstrated how Crime Displacement is a relatively minor feature. For example, Sherman et al. (1995) reported no displacement of gun crime in Kansas City after a crack-down in Hot-Spot areas that reduced such offences by 49% and gun seizures increase by 65%. Hesseling and Aron (1995) saw no displacement of car crime to surrounding areas following a Crime Prevention programme. And similarly, Mativat and Tremblay (1997) found no evidence of Crime Displacement in a Canadian initiative designed to combat credit card fraud. On a related theme, Felson and Clarke (1998) concluded that new identification procedures were highly successful in reducing cheque frauds in Sweden and with no displacement to a range of conceivable alternative crimes.

It is also theoretically possible that Crime Prevention interventions might produce both Crime Displacement and Diffusion of Benefits. Indeed, Eck (1993) discovered only 9% of the 33 studies examined indicated a substantial degree of potential Crime Displacement. Whilst Hesseling (1994) reported there was no potential Crime Displacement in 40% and a potential Diffusion of Benefits in some.

Despite the inadequacies of displacement research and design, Johnson et al. (2014) contend that there is much that can be used for research-informed decision-making. However, they also stress the relative absence of research relating to Tactical, Target and Temporal Displacement and the necessity to improve methods to detect and measure Crime Displacement—including concern about the lack of standardised test statistics and information. Most importantly, as Crime Opportunity is concentrated across time and place and offenders choose between differential opportunity structures, this helps to explain why crime does not simply 'move around the corner' following focused prevention efforts.

A further consideration rests with the sheer illogicality that all Crime Prevention activity would automatically lead to Crime Displacement—the 'hydraulic' model of human motivation. For example, Bennett and Wright (1984) reported from interviewing convicted burglars who when prevented from committing their crimes, less than half (43%) said they would search for alternative targets. Indeed, some burglars attributed this to being 'unlucky' and said they would not attempt the crime again on that particular day.

Routine Activity Theory (Cohen & Felson, 1979—see Sect. 2.2) and offender mobility provide further explanations for the absence of Crime Displacement. Brantingham & Brantingham (1981—see Sect. 2.5, Environmental Criminology and Crime Pattern Theory) contend that criminals exploit the 'mental maps' and rewards to be found within their environment, built up during routine activities. And as not all criminals will have the ability to utilise all of the Crime Opportunities (see Sect. 2.3) they discover, at least some will be forced to seek such opportunities further afield—and that will require additional effort.

Eck (1993) developed the 'Familiarity Decay Hypothesis' to describe this situation—and powerfully contends that the likelihood an offender will target an opportunity will be inversely related to the distance it is located from their routine activity space(s). If he cannot find suitable targets near to known/preferred locations, Crime Displacement is less likely to occur.

More recently, reviews utilising systematic search criteria, meta-analytic and other techniques have concentrated on Situational Crime Prevention and geographically focused policing initiatives (Guerette & Bowers, 2009). Indeed, Braga et al. (2012) reviewed police-led efforts to control Crime Hot-Spots. Whilst Bowers et al. (2004) and Johnson et al. (2014) convincingly advocate that a better understanding of intervention effects could inform policy and tactical decision-making.

Secured by Design and Crime Displacement

In the UK Designing Out Crime has historically been more associated with the prevention of Residential Burglary than any other crime. Indeed, substantial and sustainable reductions of this crime have been achieved (see Armitage & Monchuk (2009) and Sect. 9.1, The Four Towers at Duddeston Manor). Furthermore, Hesseling's (1994) review of Crime Displacement research cites Spickenheuer (1983), Forrester et al. (1988), Schneider (1988), Lindsey and McGillis (1988), Pease (1991) and Chenery et al. (1997) where no evidence of the displacement of Residential Burglary was discovered.

The key Crime Prevention priority of Designing Out Crime in the UK and by extension, Secured by Design (see Chapter 5), continues to be Residential Burglary. Three major evaluations of SBD appeared at the turn of the century: by Dr Tim Pascoe (1999) at the Building Research Establishment; Jon Brown (1999) of Rachel Armitage (1999) at Huddersfield University. These all concluded that SBD had led to substantial reductions in Residential Burglary (at least 40% was the recognised norm), no Crime Displacement was identified by Pascoe and Brown. Whilst Armitage discovered very minor Crime Displacement in six areas and a Diffusion of Benefits in nine.

Teedon et al. (2009) investigated the highly positive effects of SBD on Residential Burglary in Glasgow. Whilst Rachel Armitage revisited her earlier research and with Leanne Monchuk, published (2009) an updated evaluation of SBD on housing estates in the West Yorkshire Police area. In the wake of enhanced target hardening measures (in particular PAS 24 doorsets) they discovered not only increased levels of long-term, Sustainable Crime Prevention, but also no evidence of Crime Displacement. Stokes (2022) also confirms the long-term sustainable effectiveness of SBD.

Conclusion

All told, the obvious question to ask is why so much crime is NOT displaced—especially as this goes against the supposed 'common sense' presumption of many. The answer may possibly lie in two explanations from Crime Science: Crime Opportunity (see Sect. 2.3) and Rational Choice Theory (Cornish & Clarke, 1986—see Sect. 2.4). The latter contends that offenders evaluate (albeit briefly) the risks, effort and rewards they perceive as associated with committing a crime. Therefore, by manipulating these perceptions crime occurrence can be reduced—the rationale behind much of Situational Crime Prevention (Clarke, 1983).

A fundamental understanding of how in practice Crime Displacement is nowhere near the all-pervasive outcome imagined by those armchair experts (and that if anything, a Diffusion of Benefits is more likely to result from the introduction of Situational Crime Prevention/CPTED measures) leads to an even greater conclusion: namely, that if the police and similar social agencies search for solutions outside the box of the criminal justice system, *'large vistas are opened for exploration'*, ones that provide a *'far-reaching and imaginative search for alternative ways to commonly recurring problems, uncurtailed by prior thinking'* (Herman Goldstein, 1990, p. 44—see Sect. 2.14).

The relevance of this for the DOCO is self-evident, in that it provides the very rationale for their existence—producing interventions that deliver Sustainable Crime Prevention/Opportunities, whereas the criminal justice approach is hampered by The Carbon Cost of Crime (see Sect. 2.11) and the Attrition Rate of Crime (see Sect. 2.15). Furthermore, with a track record of success evidenced over more than four decades, the DOCO is now in the enviable position of being able to champion the manifest financial, carbon-reducing and moral cost-effectiveness of her or his involvement in a particular project to their managers, architects, developers, planners, elected members, communities and the public at large. In this context, Designing Out Crime is a powerful initiative with an evidenced track record of Crime Prevention and little if any chance of making things worse or simply moving such crime problems elsewhere. See also Guerette and Bowers (2017) for a review of Situational Crime Prevention evaluations and the extent of Crime Displacement.

2.2 Routine Activity Theory

Routine Activity Theory (or RAT as it is often pronounced and abbreviated) is perceived by many—although by no means every observer—as the most all-embracing theory (a meta-theory) within the discipline of Crime Science. Developed by criminologists Lawrence Cohen and Marcus Felson and first published in 1979, it has since been further developed and extended by Professor Felson (2002) and others. Routine Activity Theory had a seismic impact on criminological thinking (and by default, Crime Prevention), because it contradicts the dominant assumption that links social problems with crime. In essence, if there is such a link then what could explain how in the UK (for example) between 1921 and 1991 there was an upward curve in the number of crimes committed, yet simultaneous to this time period poverty decreased? The answer is that progress (be it socio-economic, technological and/or political), assists in causing crime and creates new Crime Opportunities (see the following Sect. 2.3).

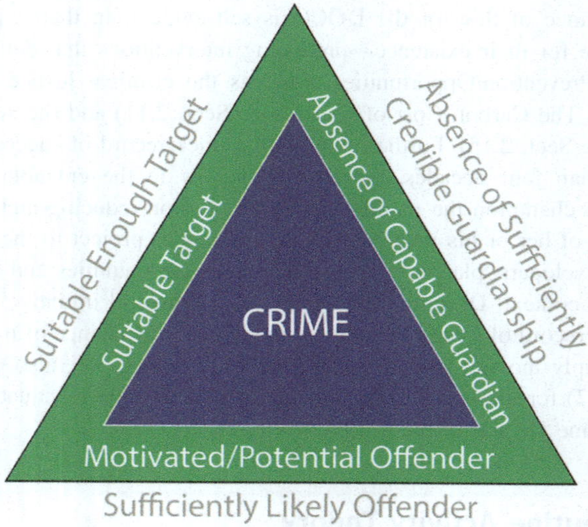

Routine Activity Theory (RAT) triangle

Furthermore, Cohen and Felson contend that for a crime to take place three elements must coincide:

1. **Likely Offender**—sometimes the terms **Motivated Offender** and **Potential Offender** are used—in that given the opportunity any of us might commit or be tempted to commit a crime, whilst even professional/career/travelling criminals do not commit Crime all the time. However, **Sufficiently Likely Offender** better describes this element as it incorporates offender capability/ resources as stated in the original model;
2. **Suitable Target**—meaning a person, product, building, vehicle etc. Suitability depends on the context, e.g. not all criminals are robbers, and thus what is suitable for one criminal is not to another. Now regularly described as **Suitable Enough Target**;
3. **Absence of a Capable Guardian**—originally, someone who was able to protect the Suitable Target (the absence of Guardianship). The overall title of **Absence of Sufficiently Credible Guardianship** is now often used (Tilley, 2009).

The absence or presence of an Intimate Handler (Felson, 1986) as a variation on the Capable Guardian element, has since been added to the equation. For example, whilst a bus conductor could be regarded as a Capable Guardian, a disapproving mother, father or other guardian might fulfil the role of Intimate Handler in respect of the potential actions by their child.

Routine Activity Theory has been further developed by Professors Ron Clarke and John Eck (2003) to produce the feature with which most Designing Out Crime Officers will already understand, the Problem Analysis Triangle (PAT—see Sect. 2.14). In reality, this has also evolved to become two triangles, the first inside the second. The three sides of the first triangle translate as **Target/Victim, Offender and Place.** Those of the second as **Guardian, Handler and Manager.** This is not a simple convergence with Rational Choice Theory (see Sect. 2.4), whereas the PAT is closely related to Routine Activity Theory.

From the DOC perspective of, RAT is important because it implies that for a crime to take place, these three elements need to converge. Moreover, the DOCO is in a uniquely powerful position in that she or he has the capacity to influence any of the elements on one of the three sides of the triangle. For example:

- Motivated/Potential Offender—through the Physical Security 'Target Hardening' Measures, Technological Innovations and Environmental Elements of CPTED, Crime Opportunities are removed and the Motivated/Potential Offender's decision-making process suitably affected. However, having decided (for example) not to burgle one particular dwelling, he still remains motivated unless enough vulnerable/attractive targets are secured causing the potential reward to be reduced. This may cause the offender(s) to abandon their more tactical choice to become involved in crime. Indeed, Clarke (2000) refers to two levels of decision-making: criminal involvement (shall I start/keep on committing crime?) and tactical level (shall I break into this house?)
- Suitable Target—the strongest intervention in the DOCO's toolkit, as the CPTED measures make the target or Target Enclosure unsuitable
- Guardianship—whilst the capacity to secure new forms of human Guardianship is all but impossible for the DOCO, her or his recommendations will often provide surrogates for Guardianship: e.g. Intruder Alarm or CCTV systems (see Sects. 4.3.2 and 4.3.3 respectively and Reynald, 2009).

At the practical level, RAT tells us that we only need to influence any one of these three sides of the triangle to prevent the crime from taking place. Nevertheless, in a practical setting it is desirable to influence more than one element on the basis of redundancy/belt and braces approach, especially as we are in reality engineering a reduction in risk, rather than providing an absolute block. And as always, there is every reason to make reference to this theory and its practical application when producing reports and statements in respect of Designing Out Crime issues.

However, as previously described, not all Crime Scientists believe in the simplicity of RAT. For Armitage (2013, p. 18), Routine Activity Theory is but one of the 'Opportunity Theories' that includes Rational Choice Theory (see Sect. 2.4) and Crime Pattern Theory (see Sect. 2.5). Whilst Ekblom (2011a) observes that there has been little attempt to integrate these different theories. Moreover, he asserts that elements like Surveillance and Territoriality '*...have surface simplicity, but in fact confusingly overlap*' (2011a, p. 25). This theme is further developed in Sect. 2.6, Conjunction of Criminal Opportunity (CCO) with its multiplicity of potential interventions, in the form of 11 proximal causal pathways (immediate causes of criminal events) and 11 counterpart principles of intervention/prevention. For an investigation into the application of RAT to Cybercrime, see Leukfeldt and Yar (2016).

2.3 Crime Opportunity

The importance of opportunity in shaping behaviour (human and across the world of nature) is self-evident. Therefore, it is hardly surprising that it features as a primary cause in the commission of crime and anti-social behaviour (ASB). Ignoring the fact that until 60 years ago Suicide was a crime in England, Wales and Northern Ireland, it was demonstrated in Sect. 2.1, Crime Displacement and Diffusion of Benefits, how the changeover from coal gas (carbon monoxide) to natural gas (methane) also delivered an unexpected bonus in terms of the drastic reduction in the total number of Suicides in England and Wales. Put simply, coal gas provided a relatively easy, painless and soporific means of Suicide, denied by methane. And as Clarke and Mayhew (1988) contend, opportunity will often be a seminal factor in the commission of crime.

Opportunity Makes the Thief: Practical theory for crime prevention

Professor Nick Tilley (2009) describes how there are distinct similarities between the Opportunity and Routine Activity Theories (see the previous Sect. 2.2). Situational Crime Prevention powerfully contends that opportunity is a major determinant in causing crime—especially Property Crime. Remove the opportunity and the crime will be more difficult or impossible to commit.

As human beings we are genetically programmed and nurtured to adopt behaviour that avoids crime and threats to our wellbeing. Perhaps the most universal of these (and one that can be traced back to the dawn of humanity), is our fear of the dark. The advent of Lighting (see Sect. 4.3.1) and especially Street Lighting) helped to reduce this fear. Furthermore, when the former Deputy Chief of Police for Reno, Nevada, Ron Glensor (1996) writes about police officers becoming *'Street Level Criminologists'* (see Sect. 1.5.3), we must also accept that the majority of citizens already possess the acumen to avoid danger from crime. For example, if a pile of broken car window glass is seen in a car park, we may infer (often accurately) that a vehicle was broken into at that location and as a result, decide to park elsewhere. Similarly, passengers on double-decker buses choose to sit on the lower deck, especially at night; we increasingly drive with the door locks engaged (in many cars this is now automatically and electronically applied); and children are more often taken to school by car than they were 30 plus years ago. Of course, much of our avoidance behaviour owes more to the fear of crime (FOC) than its reality. But imagine how Crime Opportunities might soar without such caution?

Professor Ron Clarke, '…the main architect of the situational approach to crime prevention' (Tilley, 2009, pp. 105–106), writes:

'Situational Crime Prevention comprises opportunity-reducing measures that:

1. Are directed at highly specific forms of crime;
2. Involve the management, design or manipulation of the immediate environment in as systematic and permanent a way as possible;
3. Make crime more difficult and risky, or less rewarding and excusable as judged by a wide range of offenders (Clarke, 1995)'.

Clarke's reference to the 'immediate environment' is vitally important, because it mirrors the attention to detail required by the Designing Out Crime Officer (DOCO). Yes, Secured by Design (see Chapter 5) reflects the minimum standard of Physical Security 'Target Hardening' Measures, Technological Innovations and Environmental Elements of Crime Prevention Through Environmental Design (CPTED—see Sects. 4.2.1–4.4.6) required for a given project. But a Crime Risk Assessment (CRA) may suggest additional, site-specific security measures are necessary to produce long term, Sustainable Crime Prevention.

Mayhew et al. (1976) stress the power of opportunity in determining behaviour and describe how both offenders' and victims' Crime Opportunities vary according to age, gender and lifestyle—factors which we may find obvious, yet they had previously been rarely if ever articulated with reference to crime.

One of the most revealing examples they considered involved the results of steering column locks becoming a mandatory fitting on all new cars in the UK from 1971. Crime data from the Metropolitan Police area (Greater London) for 1969 and 1973 disclosed no overall change in the number of offences, yet there appeared to have been some Crime Displacement from newer cars (with steering locks) to older cars. Similarly (and with a direct relevance to Guardianship), Mayhew et al. (1976) quote a Manchester study where the upper decks on double-decker buses suffered 20 times as many incidents when the bus was single-crewed—no conductor or 'clippie'—Capable Guardian (Cohen and Felson, 1979).

Felson and Clarke's 'Opportunity Makes the Thief: Practical Theory for Crime Prevention' (1988), powerfully contends that opportunity is a root cause of crime. Moreover, opportunity can occur in many different forms: the car owner who leaves keys in the ignition; the householder failing to close windows when they leave the house; the availability of firearms and most relevant to the DOCO, the layout of a development that facilitates easy Attack and Escape Routes and anonymity for the criminal.

Professors Felson and Clarke (1998) set out their *Ten Principles of Opportunity and Crime*:

1. Opportunity plays a role in causing all crime
2. Crime Opportunities are highly specific
3. Crime Opportunities are concentrated in time and space
4. Crime Opportunities depend on everyday movements in activity
5. One crime produces opportunities for another
6. Some products offer more tempting Crime Opportunities
7. Social and technological changes produce new Crime Opportunities
8. Crime can be prevented by reducing opportunities
9. Reducing opportunities does not usually displace crime
10. Focused opportunity reduction can produce wider declines in crime.

Consequently, it does not necessarily follow that opportunity will lead to the evolution of criminal activity that many have imagined. In this last regard, the research of Professors Paul Wiles and Andrew Costello (2000) *The 'road to nowhere': the Evidence for Travelling Criminals*, is especially pertinent. There is a commonly held belief in the UK and beyond that the opportunity for increased mobility has led criminals to travel further to commit crime (Crime Travel Footprint/Opportunist Radius)—especially into affluent suburbs and rural areas.

However, the Wiles and Costello research disclosed that the majority of offenders tended to commit crime within the areas in which they lived or socialised. Indeed, their main conclusions were:

- The vast majority of offender movements are relatively short
- Most travel associated with crime is largely not motivated by planning to offend and is much more dependent on opportunities becoming available during normal routines, rather than long-range search patterns
- When offenders travel to commit crime, it is overwhelmingly local
- Long-range crime is mainly to places with strong connections to the offender's home location
- There was little evidence that offenders' travelling to commit crime was significantly increasing over time, or of new travel opportunities being exploited.

See also Sect. 6.8, Rural Crime.

Much of the previous research relating to offender travel comes from North America. Wiles and Costello (2000) describe how in general, these distances do not appear to vary by time of day or year. However, such distances did increase with age (and criminal experience?).

For Residential Burglary, the following average travel distances were reported: White (1932) 1.76 miles in Indianapolis; Repetto (1974) 0.5 miles in Boston and a nearby small city; Pyle (1974) 2.48 miles (2.34 miles for Non-Residential Burglary) in Akron, Ohio; Philips (1980) 1.05 miles in Lexington-Fayette, Kentucky; Rhodes and Conly (1981) 1.62 miles in Washington D.C.; Gabor and Gottheil (1984) 0.35 miles in Ottawa. Similarly, for Auto-Theft: White (1932) 3.43 miles in Indianapolis; Philips (1980) 1.15 miles in Lexington-Fayette, Kentucky; Gabor and Gottheil (1984) 1.24 miles in Ottawa.

Wiles and Costello (2000) utilised police recorded crime data: UK PNC (Police National Computer) DNA data and interviews with convicted burglars and 'TWOCers' (twockers—Taking Without Owner's Consent and a specific type of Vehicle Crime where it cannot be proved there was an intention to keep the vehicle—AKA 'Joyriding'). Their research was centred around the cities of Sheffield in South Yorkshire, York and the rural area of Hambleton, both the latter in North Yorkshire.

Key findings for Sheffield disclosed a 1.93 mile average travel distance for all offences together with an average offender age of 22.5 years; 1.88 miles for Residential Burglary, average age 21.3 years; 1.83 miles Non-Residential Burglary, average age 21.8 yearsand 2.36 miles for TWOC, average age 18.8 years—a 'young man's offence' (p. 14), but then most offending is very local and committed by young males. Shoplifting had the highest travel distance of 2.46 miles and was the only offence with a high proportion of female offenders.

Of the offenders who travelled into Sheffield, 46% did so for shoplifting, presumably attracted by its large city centre and retail parks—especially Meadowhall which accounts for 41% of travel into city crime, 80% being shoplifting. Moreover, between 1966 and 1995 there was virtually no change in the percentage of offenders living within and outside the city: 85.9% and 14.1% in 1966; 85.6% and 14.4% in 1995.

Similar travel distances and offender ages were reported for those offenders residing in York and rural Hambleton. And perhaps unsurprisingly, over half those interviewed were heroin addicts: 40% of TWOCers and 69% of the burglars. Indeed, the main motivation for burglary was the need for money, primarily to purchase drugs. The burglars were also of the opinion that their necessity for drugs made them reckless in the actions they took.

It should also be noted that Al-Kahtani (1996) reported similar patterns of behaviour involved in the commission of Residential Burglary in Saudi Arabia. This suggests that such relatively short travel distances may be consistent beyond England and North America.

For the DOCO, the target (or at least the Target Enclosure (see Ekblom, 2000) selection by burglars is revealing: 63% chance; 31% passing and looked easy (poor security); 26% passing and looked easy (unoccupied); 26% passing and looked easy (isolated/quiet); 20% had been noticed previously; 17% tipped off; 14% passing and looked wealthy; 6% revenge; 6% other; 3% burgled before.

Similarly, having established that a high proportion of offenders travel relatively short distances to commit crime, it could be argued that enhanced Permeability (see Sect. 4.4.6) assists their target selection—in that they will pass more properties suitable for burglary within their Crime Travel Footprint/Opportunist Radius. This will be especially true where they are travelling to the target on foot (see also Sect. 6.2, 'Leaking' Culs-de-Sac and Sect. 6.4, Closure of Rights of Way on Grounds of Crime and Public Space Protection Orders).

As a result, is it surprising that the overwhelming majority of offenders do not travel long distances to commit crime? After all, long-range travel requires knowledge, confidence, skills and resources—which many offenders do not possess. Moreover, and as in nature, human endeavour seeks to minimise effort (conserve energy). This might also explain why the least desirable areas in which offenders tend to live possess high offence and victim rates.

The 'New Opportunity Theories' (Tilley & Clarke, 2013, pp. 4–8) point to known offenders committing High Volume Crime, dominated by opportunistic offending during routine and limited travel patterns. Indeed, Wiles and Costello (2000, p. 45) '*...feel confident in asserting that generally, high volume crime is a highly localised phenomenon, especially for offences like residential burglary and criminal damage*'. Moreover, Houghton (1992) describes how offenders often live in poorer social housing containing a surplus of suitable targets ('impacted crime areas'). This facet of criminal behaviour, where victims are often poorer than offenders (localised victimisation), is especially dispiriting.

In *Environmental Criminology and Crime Pattern Theory*, Brantingham and Brantingham (1995) describe the proximity of the crime location to where the offender resides, works or socialises in terms of a 'node' or a series of 'nodes' (often referred to as Activity Nodes). They also refer to the 'awareness spaces' (1995, p. 18) of large numbers of people (see Sect. 2.5).

Opportunity reduction is at the heart of Professors Cornish and Clarke's (2003) Twenty-Five Techniques of Situational Prevention (see the following Sect. 2.4). And in the Conjunction of Criminal Opportunity (see Sect. 2.5), Professor Ekblom brings together all the Opportunity Theories under a single theoretical framework. See also Clarke and Felson (2017); Wilcox and Cullen (2018) 'Situational Opportunity Theories of Crime'. And Alabi, Adeleke and Olajide (2021) regarding the influences of opportunity in residential neighbourhood crime.

2.4 Rational Choice Theory

All Designing Out Crime Officers (DOCOs) will be already aware of Rational Choice Theory (RCT), if only because of the Twenty-Five Techniques of Situational Prevention encountered during their initial training for the role (Cornish & Clarke, 2003). It is unfortunate that because the acronyms for Routine Activity Theory (RAT) and Rational Choice Theory (RCT) are very similar, the two are often confused. Nevertheless, we need to persist with our comprehension of each as they are fundamental in understanding how the built environment can be manipulated to Design Out Crime, fear of crime (FOC) and anti-social behaviour (ASB).

Rational Choice Theory is concerned with the immediate considerations (underlying mechanisms) that cause a criminal to decide to commit or not commit a crime. Professors Derek Cornish and Ron Clarke (1986, 2008) were the authors of Rational Choice Theory, which has in turn born fruit in a variety of different forms of opportunity reduction—some alluded to in the previous chapter.

Professor Nick Tilley (2009, p. 110) describes Rational Choice Theory as '*weak*' (2009, p. 110)—meaning the rationality displayed by the criminal is very limited. The offender does not systematically consider all the possible options for his actions, but he does take at least a passing interest in the potential risk, effort and reward of his crime. Such *'...behaviour is not mechanically driven by external factors'* and consequently there is the capacity to alter the conditions for choices, to *'...produce preventive benefits at the margin'* (Tilley, 2009, p. 110). From the DOCO's perspective this can be achieved through Physical Security 'Target Hardening' Measures, Technological Innovations and the Environmental Elements of Crime Prevention Through Environmental Design (CPTED—see Sects. 4.2.1–4.4.6).

Key Issue

In the same way that understanding how Crime Displacement v. Diffusion of Benefits (see Sect. 2.1) was described as the key that opens the door to understanding the relevance of and evidence from Crime Science to the discipline of Designing Out Crime, Rational Choice Theory complements that evidence and provides a rationale for where the DOCO can intervene and disrupt the 'Crime Script' (Cornish, 1994)—see below.

Cornish and Clarke (1986, 2008) contend that the commission of a crime will usually involve a series of actions involving decisions made at each point along the timeline. Cornish (1994) coined the term Crime Script to describe the individual elements of this series of actions taken by the offender along the timeline.

For example, in the crime of Residential Burglary the Crime Script might be as follows. Preparation: acquire jemmying implements and select co-offenders; Entry: enter the neighbourhood; Pre-condition: search for suitable unoccupied properties; Instrumental Pre-condition: identify and select a dwelling; Instrumental Initiation: disable any alarm system; Instrumental Actualisation: break into the dwelling; Doing: stealing jewellery; Post-Condition: note the address for a repeat attack and Exit: walk away from the scene.

Furthermore, he convincingly contends that interventions at any of these points (disrupting the Crime Script) have the capacity to prevent the Crime from taking place, either per se or by making the Crime more difficult, risky or by reducing the reward. Clarke (1999) utilises the 'Pinch Point' concept to identify the most practical and cost-effective stages to intervene in the Crime Scriptand an important means of analysis for the DOCO/CPTED Officer/Adviser.

Clarke subsequently created his Twelve Techniques of Situational Prevention (1995). Originally, there were 4 techniques under three headings: Increasing the Effort; Increasing the Risk and Reducing the Reward. These evolved to become Increasing Perceived Effort, Increasing Perceived Risks and Reducing Anticipated Rewards. And as a response to Wortley's (1996) critique, were joined by Reduce Provocations and Remove Excuses. Similarly, the number of techniques under each heading increased to five, thereby producing the Twenty-Five Techniques of Situational Prevention (Cornish & Clarke, 2003).

Cornish and Clarke (2003) contend that Rational Choice Theory is a constituent part of how Situational Crime Prevention (SCP) measures can influence offending behaviour. Whilst, Ekblom and Tilley (2000) maintain that 'Removing the Resources for Crime' emphasises the value of 'Controlling Facilitators' within the Twenty-Five Techniques. Thus, it is not the offender's choices they want to affect, as his sheer ability to commit the crime.

In a similar vein, Professor Gloria Laycock (1985, 1997) investigated Property Marking campaigns in three Welsh villages and the necessity for ongoing publicity. She discovered that reductions in Residential Burglary were due more to the publicity campaign, rather than the actual Property Marking. In reality, the vulnerability of offenders had not increased—but their perceptions made them believe that following the Property Marking campaign they were at far greater risk of being apprehended. This is reflected in Professor Clarke's incorporation of perception into the techniques of SCP. Moreover, there is a cross-reference with the Anticipatory Benefits identified by Smith et al. (2002) where perception of risk by the offender is of seminal importance.

On occasion, Situational Crime Prevention measures can lead to certain types of crime being all but abolished. For example: 'safe-cracking' (due to much stronger safes equipped with electronic, time delay locking systems); lock-picking (many locking systems incorporate measures to prevent this Modus Operandi, such as those present in BS 3621 five lever mortise deadlocks); and obscene phone calls (the advent of the '1471' caller identification system).

Finally, Professor Clarke has given permission for his Twenty-Five Techniques of Situational Prevention to be utilised and adapted for specific crime types. Eight of these are reproduced as Appendices A–H to this handbook for the following present day crime types:

1. Garage Forecourt Bilkings (Making Off Without Payment or Gasoline Drive-Offs)
2. Cannabis Farms (Commercial Cannabis Cultivation)
3. Car-Key Burglary

4. C/ViT (Cash/Valuables in Transit) Robbery
5. Metal Theft
6. Night-Time Economy (Pubs, Clubs and Alcohol Licensing)
7. Serious Acquisitive Crime
8. Violent Crime

Assiduous readers will notice that not all Twenty-Five Techniques can be applied to every crime type. Number 18, 'Reduce Emotional Arousal', is especially problematic in this regard. However, Professor Clarke explains (personal witness) that this is inevitable in view of the massive spectrum of crime types that exist.

Similarly, this author's application of the matrix in respect of the eight forms of crime listed above, is neither de rigueur nor final. Readers and especially those who wish to use any of these matrices, are recommended to pick and choose, develop and replace those techniques which they believe will be the most effective in tackling a local crime issue. They can then substitute their own technique where they think it is theoretically and practically plausible (and ideally based on experience) than the one quoted in the relevant matrix. Furthermore, additional matrices in respect of Cyber Crime and Safeguarding and Vulnerability (Threat, Risk and Harm) are at present being developed. See also Akers (2017); Clarke and Felson (2017); and Sect. 2.6, Conjunction of Criminal Opportunity (CCO).

2.5 Environmental Criminology and Crime Pattern Theory

The ancestry of Environmental Criminology can be traced back 200 years (Burgess, 1925) and more recently to the same Ernest Burgess and the Chicago School of Sociology (and subsequently Criminology) in the 1920s and 30s. Environmental explanations for crime, fear of crime (FOC) and anti-social behaviour (ASB) lie at the heart of C. Ray Jeffery's (1971) concept of CPTED (see Chapter 4). Similarly, Oscar Newman's Defensible Space (1972, 1973 and 1996) is fundamentally located within the environmental arena.

Jeffery used the behavioural psychology ideas of B.F. Skinner (1953) who diagnosed that the results of actions lead to their repetition/non-repetition. As Tilley (2009, p. 123) describes, 'Jeffery stressed the reinforcement of behaviour as the key determinant of crime'. Jeffery's contention was that a burglar will repeatedly break into houses unless he is apprehended or otherwise prevented (thus the value of CPTED measures) and providing he finds items worth stealing.

Crime Pattern Theory (CPT—or the Geometric Theory of Crime) emerged from Environmental Criminology and owes its origins to the work of the Vancouver-based wife and husband team of Patricia and Paul Brantingham & Brantingham and Brantingham (1981, 2008). By examining the routine activities and 'awareness spaces' of offenders, CPT aims to explain its geographical distribution. Crime is generally understood to cluster at specific geographical locations according to the availability of suitable targets (Sherman, Gartin and Buerger, 1989). In addition, Brantingham and Brantingham (1995) contend that certain locations will act as:

- Crime Generators—places with a large number of users and targets, e.g. shopping centres, transport hubs, sports arenas, music venues etc.
- Crime Attractors—places providing multiple Crime Opportunities that lure an increasing number of those with criminal motivation, e.g. drug and prostitution areas, certain festivals, specific sports venues etc.
- Crime Enablers—develop where there is an absence of behaviour regulation including Guardianship, e.g. upstairs on double-decker buses since conductors were withdrawn.

Lynch (1960) suggests that we all appreciate and use our knowledge to navigate through the main zones of activity—where we reside, socialise, shop, are educated or work. We do this courtesy of our 'mental maps' of five spatial elements: paths (including roads, transport lines); edges (walls, fencing, hedges); districts (neighbourhoods); nodes (focal points such as bus and train stations, shopping centres and other hubs); and landmarks (physical structures). In Environmental Criminology these zones are described as Activity Nodes and the routes between them as Travel Paths.

Frank et al. (2011) describe how the interaction with our own environments is predictable. This Activity Space is determined by age, gender, residential location (rural, suburban, urban), personal mobility and socio-economic status. It possesses cross-discipline application and is used in criminology, geography, public health and transportation.

Furthermore, we each possess some knowledge of what lies beyond our usual Activity Space—known as Awareness Space (Brown et al., 1977) and subject to a concept known as Distance Decay. This translates as meaning the further away you travel from your Activity Space, there is a consequent reduction (decay) in your knowledge of the more distant areas that constitute your Awareness Space.

CPT maintains higher rates of crime will be experienced at the 'edges' of such Awareness Space, where the offender is not recognised (anonymity). Think of housing that backs onto interconnecting ('Leaking') public footpaths, canal tow-paths, clusters of shops etc.). Whilst crime will more readily occur at certain times and locations within offenders' Awareness Spaces.

Attractiveness, accessibility and Crime Opportunities (see Sect. 2.3) of an area are believed to influence target selection by offenders (Bernasco & Luykx, 2003) and apply to different offender age groups (Groff, 2005). The actual location where offenders reside (see Wiles & Costello, 2000) combined with knowledge of their Activity Space, Awareness Spaces and the location of potential targets, is highly influential in determining offending behaviour.

As previously described, Activity Nodes and Travel Paths are intrinsic elements of CPT and can be applied to our daily activities: when we leave the home, travel to educational establishments or work, shop, visit friends or recreation venues. During a normal day our Travel Paths will take us from and to several Activity Nodes, before most often returning home. Some Activity Nodes attract large numbers of people (e.g. recreation venues) whilst others (like the home) attract relatively few. The 'pull' of such venues is further amplified by what are described as 'spatial temporal transitions'. For example, a horse racecourse is likely to generate very little crime on days when no meeting is taking place. And as Bromley and Nelson (2002) report, alcohol-related crime in the city of Worcester displayed distinct 'spatial temporal transitions'. Hot-Spots of such crime peaked around midnight, but once these venues had closed alcohol-related crime was concentrated in the suburbs and close to residential areas.

Roads, footpaths, canal tow-paths etc. constitute the Travel Paths between our Activity Nodes. Moreover, over time these individual routine travel patterns reinforce our Activity Space and Awareness Spaces—equally true for offenders. The target selection patterns of serial rapists were investigated by Alston (1994) who discovered that initial contact with victims was often made close to routine Travel Paths. Rengert (2004) reported that an offender's 'Journey to Crime' will consist of three elements: a starting location (usually their home); direction of travel and the distance to the Crime location. As described in Sect. 2.3 and by Wiles and Costello (2000), most crime takes place close to the offender's home and is subject to a Distance Decay Pattern. Frank et al. (2011) concentrated on the directionality of the Journey to Crime.

Expanding on this theme, Brantingham and Brantingham (1995) contend that crime consists of two general categories: Property Crime—most often location specific. And those committed Against the Person (assaults, woundings etc.)—and usually not tied to a specific location. Directionality is location specific and for the most part determined by offenders' Activity Nodes and Travel Paths. If the offender's criminal activities take place along a single trajectory, directionality is strong. As an example, Rengert and Wasilchick (1985) reported that in the instance of burglary, they discovered a strong directional preference towards the offender's place of work—or along the Travel Paths between the home and work Activity Nodes.

Meanwhile, Madensen and Eck (2015) maintain that Crime Place consists of Pooled Places, Proximal Places and Proprietary Places—and the latter of Crime Sites, Convergent Settings (Felson, 2002); Comfort spaces (Hammer, 2011) and Corrupting Spots.

Such research is important because it can identify where the greatest Crime Prevention impact can be made and thereby influence future policy-making and practice delivery. To date, little Crime Prevention activity has been based on scientific evidence of where the need is greatest and what interventions will deliver the biggest benefit. In a Crime Prevention discipline with scant resources and infinite demands, maximising the impact of the DOCO's recommendations ('getting the biggest bang for our buck', Barton, 2017) should dominate service delivery decision-making in the future. See also Eck and Weisburd (1995; Brantingham and Brantingham (2016); van Sleeuwen, Ruiter and Steenbeek (2021).

2.6 Conjunction of Criminal Opportunity

The Problem Analysis Triangle (Clarke & Eck, 2003—see Sect. 2.14) is frequently used as a framework for causes and interventions within Crime Prevention. However, it is a relatively simple model that does not illustrate the vast range of potential interventions, any one of which may prevent a crime from taking place. The Conjunction of Criminal Opportunity (CCO—Ekblom, 2011a) provides a more complex model, one with additional scope, integration and detail. CCO consists of 11 proximal causal pathways (those that are close to the crime situation in time and space) that collectively cause criminal events to happen and 11 counterpart principles of intervention designed to subdue those causes.

Professor Paul Ekblom (2011a), describes how CCO emerged in the early 1990s from the Safer Cities Programme and evaluation of a challenging range of projects. This was compounded by the *'fragmentation'* of Crime Prevention into situational and offender-oriented explanations—ones that remain poorly integrated and cause ongoing issues for performance of prevention.

Originally entitled 'Proximal Circumstances' (Ekblom, 1994, 1996), CCO aims to convey the specific *'causes acting via the offender in the crime situation'* (Ekblom, 2011a, p. 134). Early childhood experiences, inequality or market forces (though important) are excluded for being too varied, complex, hard to measure and define—a common theme across Crime Science. These can only act via their influence on the proximal causes.

Evolving from the academic to the practitioner friendly, CCO draws on Routine Activity Theory, Crime Pattern Theory and Rational Choice Theory. It was also at the heart of the Home Office Crime Prevention Toolkits (2001).

CCO provides a *'mechanism map'* of the 11 *'proximal causal preconditions'* required for a crime to take place. Its preventive role is in blocking, weakening or diverting one or more of these generic, mechanism-based intervention principles. The 11 *'causal components'* consist of living 'agents' and non-living *'entities'*. Agents will include offenders, Crime Preventers (e.g. guardians and place managers—a cross-reference with Routine Activity Theory), Crime Promoters (those who increase the Crime Risk—e.g. antiques dealers buying stolen goods)—but not victims (a role that only emerges once the criminal event is under way or has finished).

Entities include crime targets (inherently criminogenic persons or objects—vulnerable, valuable or provocative). Target Enclosures (safes, buildings and gated compounds situated in a wider environment—shopping mall, park, housing estate etc.) This is an important distinction, because in, for example, the crime of Residential Burglary, Ekblom is differentiating the dwelling that is being broken into (the Target Enclosure) from the Target stolen, e.g. a cache of jewellery (Ekblom, 2011a).

Similarly, the distinction between Crime Generators and Crime Attractors (Brantingham & Brantingham, 2008—see Sect. 2.5) will include High-Crime locations—due to heavy routine use or deliberate criminal attack.

CCO provides psychological depth to offenders. It starts with a predisposition to offend then adds: resources to aid or to avoid offending; readiness to offend; perception and anticipation of offence and the presence of an offender in the crime situation. Moreover, CCO involves two perspectives of the offender: as agents with goals decisions and actions (e.g. planned, Serious Organised Crime and Terrorism); or as caused agents (e.g. criminals who are socialised, motivated, prompted, provoked etc.)

Each of the 11 causal components of CCO has an equivalent preventive intervention. From the perspective of Designing Out Crime, Ekblom (2011a, p. 142) provides the highly pertinent example of pruning the bushes in a hospital car park. This modifies the crime environment situation, removing the capacity for ambush by street robbers, but may also increase car park use, producing additional revenue, more visits to patients, shorter hospital stays and most importantly, fewer robberies.

Whilst CCO is analytic, the Twenty-Five Techniques of Situational Prevention (Cornish & Clarke, 2003) amount to practical examples of such methods, organised around columns representing risk, effort, reward, provocation and excuses. However, not only are the 25 Techniques limited to Situational Crime Prevention interventions: there are other restrictions. For example, they concentrate on the offender's decision-making and capacity to be provoked, rather than including the preventer's perspective. And only a single causal mechanism is attributed to each intervention, which underestimates the inherent preventive effect of such mechanisms. This may also inadvertently suggest that there is only one solution to a given problem and indeed, deny the capacity to comprehend the extent to which each of the risk-effort reward mechanisms is working in a certain situation.

Similarly, the Problem Analysis Triangle (Clarke & Eck, 2003) that emerged from Routine Activity Theory (Cohen & Felson, 1979), whilst less prescriptive than the 25 Techniques, does not provide the six aspects of offender included in CCO (indeed it added *'guardian'* but lost *'place'*), or distinguish between the 'target' as contents or the Target Enclosure of those same contents. CCO also includes *'remote interventions'*, such as youth clubs and school holiday sports facilities, designed to divert young people away from crime or ASB.

There is also an understated, dynamic slant to CCO—for example, the shifting relationship between offenders, preventers and possibly promoters. Similarly, perception and anticipation are linked to offender decision-making and might include simple 'weak' rational choice through to the complex mix of self-control and moral choice described by Wikstrom (2006). Crime Scripts are intrinsic to CCO, e.g. the proximal circumstances necessary for stealing the getaway car, obtaining a false passport, weapons and the actual robbery. Whilst the decision-making of offenders/preventers/promoters and the design of situational interventions denote *'script clashes'*. Serious and Organised Crime has also provided a worthy field for the CCO approach (Ekblom, 2003b).

Collectively, this is of importance for DOCOs because it demonstrates how their interventions are not limited to any one side of the RAT/Crime or Problem Analysis Triangles, or indeed the Twenty-Five Techniques of Situational Prevention. Instead, there is a multiplicity of interventions covering the 11 intervention principles, any one of which is a plausible candidate for preventing the crime from taking place. And at the coalface of practical application, absorbing and utilising new concepts like *'Arms Races'* between offenders and preventers (Ekblom, 1997) and theories like CCO, demonstrate that DOCOs are willing to embrace advances and evolutions in Crime Science.

Ekblom's famous shield and oar diagram illustrates the 11 *'causal components of crime'*, radiating from the *'Crime or Disorder Event'* splash. By adding Crime

Prevention shields to the 11 oars it becomes ‘Crime Prevention and the Conjunction of Criminal Opportunity’ with a new splash of *‘Absence of Crime or Disorder Event’*.

> A criminal event happens when a predisposed, motivated and equipped offender encounters, seeks or engineers a situation conducive to crime, perceives opportunity and decides, or is provoked to act. The situation comprises a human, material or informational target that is vulnerable, attractive or provocative, perhaps within an insecure Enclosure, and in any case located in a wider environment that motivates and/or gives tactical advantage to the offender, who may also be aided deliberately, recklessly or innocently by people acting as crime promoters, and insufficiently hindered by people acting as crime preventers. (Ekblom, 2011a, p. 136)

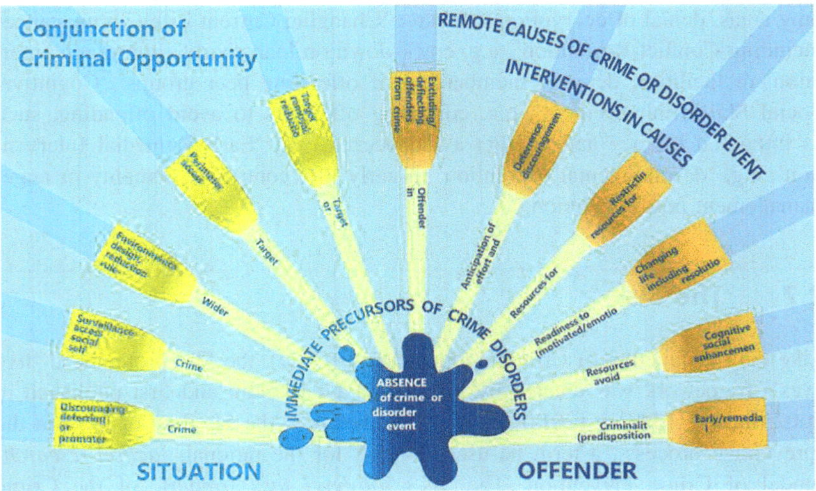

Conjunction of Criminal Opportunity

Those crime prevention shields that relate to the ‘situation’ are obviously more relevant to the DOCO those on the ‘Offender’ side. ‘Discouraging/Deterring Careless or Deliberate Promoters’, e.g. encouraging residents to secure windows, or antiques dealers not to buy stolen goods; ‘Surveillance, Access Control, Social Control, Self-Protection’, e.g. CCTV, Neighbourhood Watch, PAS 24 doors and windows, locks and electronic door entry systems, security guards, Personal Safety measures; ‘Environmental Design, Conflict Reduction, Rule-Setting’, e.g.

enhancing Surveillance opportunities, denying Attack and Escape Routes, alcohol restriction zones; 'Perimeter Access Security', e.g. securing the Target Enclosure by installing fencing, anti-ram-raid posts, door and window security, firewalls to prevent computer hackers; 'Target Hardening/Removal or Value Reduction', e.g. building/vehicle security; making banknotes more difficult to forge; encrypting data; equipping security guards with body armour.

Whilst on the 'Offender' side '… Excluding/Deflecting Offenders from Crime Situation', e.g. stopping more than two children from entering a shop together, imprisonment, electronic tagging/curfew, diversionary activities like motor vehicle clubs; 'Deterrence and Discouragement': increasing perceived risk of arrest and punishment, marking property to render unsalable if stolen, confiscating profits from convicted criminals; 'Restricting Resources for Crime', e.g. control of firearms (see Zimring, 1972), removing bricks/bottles from Night-Time Economy areas, denial of decryption software; 'Changing Current Life Circumstances including Conflict Resolution', e.g. crack-down on loan sharks, improving entertainment facilities, denying membership of offending peer groups; 'Cognitive/ Social Skills Enhancement', e.g. educating offenders to avoid offending, such as improved literacy and conflict avoidance classes; 'Early/Remedial Intervention', e.g. 'developmental prevention' in early life, control of sexuality or anger management post-conviction.

2.7 The 5Is

Like the Conjunction of Criminal Opportunity (CCO), the 5Is (pronounced 'five eyes') framework was created by Professor Paul Ekblom and first published in 2002. Indeed, Ekblom (2011a, p. 83) describes how the 5Is developed from the 'preventive process', a term he used in 1988 for the rational, *'action research'* model of Crime Prevention. The 5Is (*'top-level task streams'* of the Crime Prevention process) interventions are:

- **Intelligence**—gathering and analysing the nature, causes and negative consequences of a particular crime, in order to influence the Crime Prevention and Community Safety aims and priorities of the practitioners/organisations addressing it
- **Intervention**—design and planning practical methods to block, divert or weaken the causes of future and ongoing crime, or mitigating harm already done. It provides focus by combining local evidence from Intelligence and generic evidence/knowledge of what works in practice

- **Implementation**—the practical and management tasks necessary to deliver the Intervention methods (e.g. recruitment, training and management)
- **Involvement**—intertwined with Implementation, the Involvement of other people and agencies to appreciate, accept, undertake, share or support the tasks, roles and responsibilities involved in the implementation of preventive interventions, or by otherwise providing a receptive climate
- **Impact**—harvesting evidence of the effectiveness of the preventive action. This can then be used to improve performance; guide continuation, expansion and replication; accountability; and transfer into the 'collective evidence base'.

Ekblom observes how the 5Is resemble and indeed are compatible with the SARA (Scanning, Analysis, Response and Assessment) process synonymous with Problem-solving and Problem-Oriented Policing and Partnerships (POP—see Sect. 2.14). Where they differ is that in the 5Is, Intelligence covers both the Scanning and Analysis elements of SARA, whilst the 'amorphous' Response phase of SARA is further sub-divided into Intervention, Implementation and Involvement. Consequently, it can be powerfully argued that the 5Is has improved upon SARA.

This is important because as with the CCO, both SARA and the Problem Analysis Triangle (PAT) are relatively simple frameworks that necessarily do not reflect the *'rich complexity of preventive action'* (Ekblom, 2011a, p. 86). The 5Is develops SARA to provide systematic and detailed analysis, together with design of Intervention, Implementation and Involvement options. As such, the 5Is and CCO more accurately reflect the DOCO's massive skill set in terms of practical application and in measuring the effectiveness of Designing Out Crime. And please be aware, Professor Ekblom has immense experience in dissecting and reviewing the effectiveness of a vast range of Crime Prevention projects.

5Is is an action-oriented knowledge management and application framework, focusing on the work of ground level practitioners, like DOCOs. However, it can also assist delivery managers, policy makers and has great potential as a research tool—as used by this author for his investigation at The Four Towers (see Sect. 9.1). For practitioners like the DOCO, the 5Is tasks aim to combine evidence and experience of the crime problem and consequently should assist to:

- Identify and clarify the Crime and Community Safety problems, causes or risk factors they are attempting to prevent (e.g. Residential Burglary)
- Search and select appropriate good practice from the body of existing knowledge and experience (e.g. Secured by Design—see Chapter 5)

- Replicate the preventive process, adapting this (where necessary) to the specific problem, causes or risk factors and context (e.g. recommending enhanced security measures where the Crime Impact Statement/Crime Pattern Analysis suggests these are necessary)
- Innovate intelligently, especially where there is no well-documented and well-evaluated volume of good practice examples (e.g. in the absence of a specific SBD Development Guide). Such innovation should be based on tested theoretical principles (e.g. via CCO interventions) and plausible practice knowledge.

At this functional delivery level, the 5Is framework is largely concerned with capturing, assessing, consolidating and sharing good practice with fellow practitioners. However, for policymakers it should assist in assembling and organising the knowledge that connects policy to practice through delivery, together with the selection and design of policies so that they can be delivered at acceptable cost, timescale and risk.

5Is is essentially a process model of Crime Prevention and Community Safety. A structured series of tasks (and purpose) is involved in this process and every task requires management/planning and performance dimensions. Unfortunately, as all DOCOs will be only too aware, there is an inherent 'messiness' about action (putting decisions into practice). For example: Convoluted implementation chains as evidenced by a residential development sanctioned by the local authority housing department, but where its sister planning, highways and other departments may require additional measures be incorporated

- Parallel actions under the different 'Is', e.g. identifying and responding to a specific crime issue would fall under the Implementation heading, whereas researching the causes of the crime would be classified as Intelligence
- The 5Is procedures can act on each other, thus in developing a partnership, Involvement may have a purpose of pooling Intelligence; whilst Intelligence actions may be undertaken to help identify those partners to involve
- Combining reproducible and interchangeable action elements with progressive detail to support customisation and innovation
- Feedback—in designing and trialling methods of Intervention and Involvement
- Initiation of action (tasks) may take place at different points in the cycle. For example, POP approaches begin with Intelligence. Design-based solutions commence even earlier, e.g. an application for a Secured by Design award ideally should be made at the concept stage of a project.

5Is is designed to facilitate the systematic capture and transfer of knowledge. Familiarisation with the 5Is is a case of investing additional complexity in how we structure our knowledge and know-how to enable us to better think, communicate and act in the face of the immense complexity of real-life prevention.

For those possessing considerable experience of Designing Out Crime, this last issue possesses special resonance. Indeed, Ekblom quotes Brown and Scott's (2007, p. 45) powerful observation *'...organizational memory about particular interventions can be short, and there can be danger that mistakes made in implementation are repeated time and again because the response knowledge is not disseminated'*. This is especially significant at a time when experienced practitioners (DOCOs) have left and continue to leave the service, taking their knowledge with them. And sadly, attempts to maintain the corporate memory regarding that departing knowledge are most often absent.

2.8 Environmental Cues

Professor Amos Rapoport (1982) contends that people act and behave differently depending upon the setting, implying that the built environment provides cues for our individual and collective behaviour. The environment can thus be perceived as a form of non-verbal communication and indeed, Pascoe (1996) applied such Environmental Cues to the decision-making processes and Crime Scripts (Cornish, 1994) of offenders involved in Residential Burglary. For humans, Environmental Cues are largely visual in context, but not exclusively so and may include auditory, haptic (touch) and olfactory (smell) elements. Indeed, in the past the smell of urine was often pervasive in public telephone kiosks and UK ground floor communal areas of high-rise tower blocks. And of course, Visual Environmental Cues lie at the heart of Professors Kelling and Wilson's Broken Windows theory (1982). The rapid removal of graffiti from the New York subway system being the example most often quoted for the need to challenge small elements of anti-social behaviour (ASB). See also Andresen (2010).

Environmental Cues can be identified in Professor Richard Wortley's (1996) emphasis on the role of 'situational inducements' to decision-making by the perpetrator. For him, such behavioural precipitators consist of four sub-types: prompts, pressures, permissibility and provocations. Furthermore, Wortley's critique of Situational Crime Prevention describes a two-stage model, where opportunity (see Sect. 2.3) works in tandem with an activated motivational state.

Professors Derek Cornish and Ron Clarke (2003) provided a response to Wortley, incorporating a number of the latter's ideas under the headings of 'Reduce

Provocations' and 'Remove Excuses' to create the *Twenty-Five Techniques of Situational Prevention* (see Sect. 2.2 and Appendices A–H).

Not that Environmental Cues are solely of relevance to criminal and anti-social elements. We all respond to such cues and they are integral to human decision-making and across the animal kingdom, in the search for food, reproduction and survival. As humans, we are predisposed and nurtured to avoid dangerous situations—thus the general antipathy towards Pedestrian Subways (see Sect. 6.14), narrow passageways and dark environments—Suicide being the tragic exception, see Sect. 10.1. And as previously indicated, the overwhelming majority of these Environmental Cues will be visual—although noise and smell will also feature, if not taste and rarely touch.

There is also a cross-reference with Rational Choice Theory (see Sect. 2.4). Environmental Cues can signal risk (hence they can deter), effort and reward (discourage or encourage). The capacity of such Environmental Cues to influence an offender's decision-making process will be influenced by alcohol—thus the failure to act rationally. This may explain why there is little evidence that CCTV (see Sect. 4.3.3) has any positive effect in reducing alcohol or drug-fuelled crime.

Perhaps the true value of such cues is best described with the instance of a pile of broken car window glass seen in a car park. Many of us will instinctively decide to park elsewhere when we see such glass, often accurately perceiving that a car had been broken into at that same location. This might demonstrate how we are behaving as Street Level Crime Scientists, to paraphrase Deputy Chief of Police Ron Glensor (see Sect. 1.5.3). Fear of crime (FOC) avoidance behaviour can also be identified in: crossing the road to avoid a group of young people (regardless of any real threat they might pose); refusing to use an ATM (cashpoint machine) situated in a location perceived to be dangerous; and failing to enter at night the supposed cafe culture environs of a pedestrianised town or city centre, because we do not believe it to be a safe environment. The list of such situations could be all but endless. But notice how the Environmental Cues that influence our FOC (however wrongly perceived) are a constituent part of decision-making.

As a further example of the relevance and application of Environmental Cues, De Andrade et al. (2016) investigated alcohol consumption on perceptions of violence and crime at Licensed Premises in Queensland, Australia. Bouncer friendliness, patron gender composition and (bar) room temperature were used as experimental variables. The authors reported that in general, perceived fear of victimisation, likelihood and frequency of crime and severity of injuries was highest when the bouncer was unfriendly, bar room temperature hot and the majority of patrons male.

Most importantly, Environmental Cues are of crucial importance in the decision-making processes of those criminals intent on the most serious forms of crime, e.g. C/ViT (Cash/Valuables in Transit) Robbery. This helps explain the necessity to 'Target Harden' all elements of the collection and delivery process along the potential Crime Script—from the cash handling centre to and from the recipient. Indeed, the advent of the cash-dock/airlock delivery/collection systems and reduction of the distance between where C/ViT vehicles can park and the intended recipient, demonstrate how the Security Industry and Crime Prevention professionals like DOCOs are constantly required to introduce new preventive measures to keep ahead of the criminal. For further evidence in this regard see especially Clarke (1990); Morrison and O'Donnell (1996); Gill (2000, 2001); Willis (2006); Ekblom (2012).

In this highly permeable environment, the owner parks his motorcycle outside the bedroom window—telling us the area is unsafe?

Parked outside the owner's bedroom window, note the yellow-coloured curly motorcycle alarm cable—a Visual Environmental Cue (see Chapter 2.8)

In reverse, the picture above of a motorcycle parked outside the owner's bedroom window and with the alarm set (the curly cable which operates as a Visual Environmental Cue for any potential vehicle thief) demonstrates his fear of theft—reinforced by there being no secure location in which it could be parked. Indeed, the owner confirmed this rationale, a situation made worse in this example by the New Urbanist/Urban Villagers deliberate unfettered Permeability (see Sect. 4.4.6).

2.9 Hot-Spots and Tipping Theory

The originators of Crime Pattern Theory, Patricia and Paul Brantingham (1995) contend that Hot-Spots develop in specific locations, ones that can be labelled as Crime Generators (high volume routine encounters of routine activity elements). Entertainment venues, sports stadia and shopping centres/malls are obvious examples. There are a number of reasons why such facilities might generate crime e.g. the release from prison of a prolific offender and nearby resident. Or an event, such as the annual fair, carnival or football match.

Historically, Hot-Spot environments have always attracted police resources and attention, due to their detrimental effect on the local crime figures. Thus, in addition to the default response of High Visibility Policing, often highly innovative methods have been utilised to target offenders. For example, the deployment of non-CID plain clothes officers to target pickpockets at horse race meetings since the Victorian era; and more recently, redeployable CCTV cameras, or 'sting' operations where CRAVED—Concealable, Removable, Accessible, Valuable, Enjoyable, Disposable (Cornish & Clarke, 2003) items such as laptop computers ('Hot Products') are left on prominent display. The overall logic is that the removal of one or more prolific offenders will (both in theory and practice) produce multiple dividends in terms of Crime Prevention and arrests made. See also Sherman (2011).

However, this chapter is more concerned with the type and extent of crime in a given location that can cause a Crime Generator to pass a Tipping Point (exponential rise) and become a Crime Attractor. Gladwell (2000) relates how the expression Tipping Point emerged from within the medical profession to describe a contagious disease that might 'tip over' and become an epidemic. It was subsequently applied in the theatre of war and defence. However, the application of Tipping Point to crime is more recent and open to debate. In essence, a location subject to an excess of low level (usually Volume Property Crime) and already a Crime Generator reaches a Tipping Point and becomes a Crime Attractor where both higher levels and more serious types of crime now exist. Most importantly, the term Hot-Spot includes both Crime Generators and Crime Attractors.

Professor Bill Hillier (2004) provides the logical contention that you might expect to see high levels of crime at larger venues, thereby challenging Professor Alice Colman's DICE (Design Improvement Controlled Experiment) conclusions in Utopia on Trial (1985). But neither explains the dynamics of what is causing the increase and evolution of the crime taking place.

This author's contribution to a report completed by Dr Tim Pascoe of the Building Research Establishment (2002) for Government Office of the West Midlands, involved four deliberately large (to enable sufficient statistical resilience) Hot-Spot facilities. These were located across the region and each exhibited the characteristics of both a Crime Generator and Crime Attractor: an extensive hospital complex; exhibition and conference centre; large retail environment; and a hypermarket. Recorded Crime figures for all four venues were obtained and the results analysed. From this data it was the type and seriousness of the crimes recorded that suggested the Volume Property Crime had passed a Tipping Point and evolved to include far more serious types of crime.

The 'best' example of this Tipping phenomenon was found at the extensive hospital complex. Analysis of recorded crime over a three-year period disclosed: very high levels of Volume Property Crime (criminal damage, theft from motor vehicles, other thefts); high levels of Residential Burglary (at the nurses' home); assaults and woundings committed against patients and staff and the attempted murder of a police officer!

At the hypermarket, inordinately high levels of shoplifting had caused the neighbourhood police inspector to liaise with his local Member of Parliament about how on any given afternoon, all his officers were engaged in arresting shoplifters from that store. Moreover, crime committed on the hypermarket car park appeared to have migrated inside the store with assaults on staff. There was even a kidnapping reported to the police as taking place on the car park.

The exhibition and conference centre were repeatedly targeted by professional/career/travelling car (and motorcycle) thieves, selecting the same make of vehicle from those parked there (either to steal or take the contents—laptop computers from the boot/trunk being a repeated favourite). However, the most serious incident was the abduction of a woman as she returned to her car, before being driven to an ATM (cashpoint machine) in the city centre and forced to withdraw cash.

Meanwhile, the large retail complex suffered an ostrich-like mentality by its management—thankfully long since replaced. They refused to display Crime Prevention signage (see Sect. 2.13) arguing that these would raise the fear of crime (FOC) amongst their customers! And this despite inordinately high Vehicle Crime and Retail Crime taking place within its confines.

Mining down into the crime data (including the arrest details), all four venues disclosed primarily opportunist but also professional/career/travelling criminals being attracted to these venues. Either by word of mouth, previous experience or the Environmental Cues (detailed in Sect. 2.8), such targets were attractive to all types of criminals—possibly influenced to at least some extent by a general belief they were unlikely to be apprehended. The Building Research Establishment (BRE, 2002) investigation also considered the 'trigger factors' that might lead to such a Tipping Point being reached. Within the influence of Designing Out Crime these included High-Crime, FOC, anti-social behaviour (ASB), bad design/layout, excess concentration of bars and pubs, graffiti, litter and abandoned cars. It concluded that to avert the triggers, action should be taken before that point is reached. Similarly, attention was given to the role of trigger factors as a predictive tool. However, such trigger factors may also be causal influences increasing the Crime Risk and if they accumulate, 'Tipping' the location; or a particular subset of influences with a special power to flip the site.

Perhaps the most cogent argument in persuading security and other managers of the necessity for intervention, revolves around these Hot-Spots and Tipping Point concepts. Where an increasing level of Volume Property Crime can be identified, such Crime Prevention measures should be introduced before (ideally) the Tipping Point is passed and more serious forms of crime take place and become embedded.

For Designing Out Crime Officers and police managers in general, Hot-Spots and Tipping Points provide a rationale for the necessity of specific levels of security (like those contained within the SBD Development Guides—see Chapter 5). In this context they can be used as a warning to architects, developers, planning officers and all professionals involved with the built environment at the design stage of what might take place if the DOCO's recommendations are ignored.

In conclusion, whilst the existence of Hot-Spots and both Crime Generators and Crime Attractors is overwhelmingly acknowledged within Crime Science, the evidence base for Tipping Point is at present far weaker and open to debate. Indeed, it can be argued that a Crime Generator becomes a Crime Attractor without any Tipping Point being breached—especially true when a large increase in simple Volume Property Crime is witnessed.

2.10 Space Syntax

Is it possible to mathematically calculate the propensity of a given location to generate crime? Professor Bill Hillier and Julienne Hanson (1984) together with many others believe it can and describe the mechanism to deliver this as Space Syntax.

First developed in the 1980s at The Bartlett, UCL (University College London), its original role was to assist architects in simulating how their designs might produce intended and unintended social effects. Indeed, Hillier (1999) contends that many places that suffer high levels of crime are very often isolated and clinically zoned. Furthermore, he believes it is possible to predict criminal behaviour on the basis of a pre-determined geographical matrix.

Johnson and Bowers (2010, p. 7) describe Space Syntax as '*a mathematical approach which takes explicit account of the street network and how each street segment connects to every other at the local and wider area level*'. In this context, space is an intrinsic aspect of human behaviour. Space Syntax consists of three conceptions of space:

- Isovist—the field of view from a given point (often spiky)

- Axial line—longest possible straight line that penetrates at least one convex space
- Convex space—occupiable void, where you are in the system.

Human space involves the interrelations between the many spaces that constitute the spatial layout of a building, village, estate, town, city etc. Space Syntax analysis most often utilises one or more of the analysis methods of Integration, Choice or Depth Distance:

- Integration—the number of turns within a street segment using the shortest paths. On a map of the street layout, colour coding is then used to identify the Hot-Spots, with red usually indicating those street segments that require the least number of turns to reach all other streets—the 'most integrated'
  - Choice—imagines street movement in terms of flow. At each intersection that flow is divided and further divided at each successive intersection. This is repeated from each street section and mathematical calculations used to produce a graph of highest choice values
  - Depth Distance—the linear distance from the centre point of each street segment to all other segments. A graph of accumulated total scores denotes those with the lowest Depth Distance.

Space Syntax is used to assess the navigability of space in the design of airports, hospitals, universities and other premises. It is also utilised in architecture, archaeology, human geography, planning and transport. However, it is in the field of crime research that Space Syntax has come under considerable scrutiny.

Hillier and Shu (2000) investigated burglary and Space Syntax in three socially and spatially different UK locations. They concluded most crime took place in visually broken up and relatively segregated spaces. Shu and Huang (2003) reported similar findings in Taiwan and how segregated areas with fewer passers-by (less Permeability) suffered higher levels of Residential Burglary.

The suggestion that Culs-de-Sac were more susceptible to crime was seized on by certain elements including the media. Yet it is both counter-intuitive and contrary to the experience of Designing Out Crime Officers—many of them serving or former police officers who before migrating into this discipline repeatedly attended far more burglaries committed on 'busy, connected routes', than in Culs-de-Sac. Furthermore, it became apparent this early research did not distinguish between a 'Leaking' Culs-de-Sac (see Sect. 6.2) and 'True' Culs-de-Sac. Hillier later recognised that the media had presented this as a simplistic case of Culs-de-Sac versus through roads whilst ignoring the 'difficult bits' (2004, p. 9). See also the work of Professors Shane Johnson and Kate Bowers (2010) in this last regard (Sects. 4.4.6 and 6.2).

Hillier and Sahbaz (2005) analysis of Street Robbery and Residential Burglary in a London street network, reported how urban integration and increased Movement (Permeability) had a double effect: increased Natural Surveillance and safety in numbers so as to reduce crime; whilst simultaneously enabling potential criminals to utilise those same integrated and accessible streets.

However, Schneider and Kitchen (2002) had already declared that Space Syntax is unable to conduct fine grain analysis which accounts for 'types of surveillance, street patterns or local context' (p. 40). And indeed, Space Syntax research by Nubani and Wineman (2005) in the city of Ypsilanti, Michigan discovered that local integration and connectivity led to increased incidence of crime. Neighbourhoods with highly accessible routes for residents also provided criminals with more Attack and Escape Routes. This supports the DOC approach to Permeability. More recently, Hillier and Sahbaz (2005) found that flats had the lowest risk of burglary (purpose-built flats were safer than converted properties), whilst detached houses possessed the highest risk.

As indicated, the issue of Space Syntax is inextricably linked to that of Permeability (see Sect. 4.4.6)—and thus the debate between the efficacy of 'Leaking' Culs-de-Sac, true Culs-de-Sac and Sustainable Transport Links. It is easy to fall into the trap of seeing this as a battle—with the New Urbanists/Urban Villagers and planning professionals of that persuasion on one side, whilst on the other we have the whole Designing Out Crime community including practitioners, academics (with a growing wealth of evidence from Crime Science that supports our discipline) and other professionals who actively work and participate in the built environment.

Nevertheless, it can be accurately presumed that everyone involved in this leading-edge debate has as their aim the best interests of society and its members. Consequently, we should (as always) relish the cut and thrust of such discussion yet simultaneously refrain from non-objective criticism. Indeed, Summers and Johnson (2017) used Space Syntax to test Crime Pattern Analysis in the context of street network configuration and the location of serious outdoor violence. Similarly, Mara and Cutini (2023) utilised Space Syntax and crime theories to model the crime dynamics of risky places.

2.11 The Carbon Cost of Crime

Many UK DOCOs can bear witness to how the correlation between crime and sustainability has been difficult to champion in the early years of the twenty-first century. Overtures is this direction were repeatedly rebuffed, causing many to conclude that whilst sustainability and the need to reduce the carbon footprint (a sub-section of industrial ecology, Weideman et al., 2008) of humanity had risen up the political and social agendas, proponents of these concerns ignored such costs (the synergy) as might be attributable to crime, fear of crime (FOC) and anti-social behaviour (ASB). This included the capacity to reduce The Carbon Cost of Crime by means of Designing Out Crime (DOC).

The Carbon Cost of Crime and your Home

That was until January 2009 when at the UK National CPDA/ALO (DOCO) Training Event held in Nottingham, Professor Ken Pease articulated his concerns in a presentation entitled The Carbon-Crime Blind Spot. Six months later, The Carbon Cost of Crime and its Implications (Farrell and Pease, 2009) was published followed by the synopsis, The Carbon Cost of Crime and Your Home (SBD, 2009).

The economic and social costs of crime have always been accepted (see Brand & Price, 2000). However, Professors Pease and Farrell (2011) detail the failure to recognise the connection between crime and highly profligate lifestyles. Such recognition would require a major shift in policy favouring Primary Crime Prevention through the design, implementation and maintenance of products less prone to crime.

Moreover, they conservatively estimate the Carbon Cost of Crime in England and Wales alone amounts to an annual minimum of 6 million tonnes of CO_2 per year, although this figure could be multiplied by a factor as much as 6 if unrecorded crime and costs of moving home were factored in—some 6% of the UK's total output of CO_2. See also Skudder et al. (2016) who subsequently recalculated the total figure at 4 million tonnes, albeit they suggest elements such as Cybercrime are likely to be underestimated.

Pease and Farrell (2011) illustrate how the fiscal and Carbon Costs of Crime can be reduced, witness the growing number of developments (especially residential) that have successfully Designed Out Crime. Furthermore, at current levels of Residential Burglary, the marginal carbon cost of building a home to SBD standards would be recovered within four years. Developments in predictive patrolling, enhanced longevity of police vehicles and reduced court and criminal justice costs, could reduce carbon even further. And, whilst ecological economics aims to maximise use of raw materials, CPTED (Jeffery, 1971) has already (at least in part) proved its potential.

Daly's (2002) definition of Sustainable Development has been incorporated into the discipline of eco-accounting. This considers the Earth's biological capacity compared to current consumption patterns. The argument being, the current lifestyle of First World countries requires several Earths in order to be indefinitely sustainable. The emergent scientific consensus holds that the consequences (especially the climate implications of current energy profligacy) are apocalyptic (Inter-Governmental Panel on Climate Change, IPCC, 2008; and the Stern Report, 2007).

The authors contend the climate change issue is now so important that it should influence all policy areas. Yet, the connection between crime and carbon profligacy remains largely unacknowledged. They cite numerous climate change texts and websites where crime is either ignored or barely mentioned. Nevertheless, the green agenda emphasises minimising extraction of natural resources and their maximisation once extracted. Pease and Farrell powerfully conclude crime is but one route causing the longevity of natural resources to be reduced. Former senior economist at the World Bank, Herman Daly, supports application of his principles to Designing Out Crime, which in itself evidences the secondary if increasingly relevant green and carbon saving remit of the DOCO.

Pease and Farrell (2011) convincingly argue that there is a Carbon-Crime Blind Spot. Even the aforementioned Stern Report (2007) '…the most encyclopaedic economic analysis of climate change available' (2009, p. 11), makes no mention of Crime and Disorder (ASB). They conclude crime and Crime Prevention have to date rarely featured in terms of carbon cost and suggest a number of reasons for this state of affairs: mental fixation on criminal justice solutions; degree of contingency and consequent mental accessibility of carbon reduction techniques; costing of remedies but not problems and the grossly mistaken belief that effective Crime Prevention would inevitably lead to a police state. This is an extension of the Fortress Mentality and Architectural Determinism slurs articulated towards the DOCO (see Sect. 1.3).

A financial cost breakdown of crime in 2003/4 was provided for the Home Office by Dubourg et al. (2005). Pease and Farrell use this to estimate The Carbon Cost of Crime, using a formula provided by the International Energy Agency (2007) which has calculated that (translated from US dollars into £ sterling), £1034.60 worth of goods and services yields 1 tonne of emitted CO_2. The authors contend that crime-related goods and services have the same carbon footprints as ordinary goods and services.

Collectively, these figures demonstrate crime is most definitely not carbon neutral. Moreover, Pease and Farrell (2011) maintain these statistics represent a massive underestimation of the real Carbon Cost of Crime. For example, it fails to include:

- The 58% of crime not reported to the police (Kershaw et al., 2008) amounting to almost 0.58 million tonnes
- The carbon cost of arson (criminal fire starting) is some 1.3 million tonnes—not including the direct emission of carbon during the burning process
- Defensive expenditure and insurance administration costs
- The carbon-profligate costs emanating from the premature demolition of crime-challenged housing developments (e.g. 1950s, 60 s and 70 s sink estates)
- Moving home due to crime victimisation. Over five years 5.3 million people claimed crime was a primary motivation in moving home. Estimated £5,800

cost on a £200,000 house, this equates to 5.6 tonnes of CO_2 per house move, thereby multiplying by a factor of more than 5 the original estimated Carbon Cost of Crime (see Dugan, 1999)

- The Carbon Cost of Crime in respect of stolen motor vehicles. Whilst a Volkswagen Golf may emit 27 tonnes of CO_2 and 77 kg of Nitrogen Dioxide during its lifetime, if stolen these costs could multiply depending on a range of different factors
- And as an unrelated example, the trade in fake pharmaceuticals is uncosted. Yet, as between one third and half a billion people now contract malaria every year, the increasing fiscal and carbon costs of counterfeiting new drugs could amount to £ billions.

There is now a wealth of evidence demonstrating that design of the built environment can be crime reductive: Eck and Weisburd (1995); Eck (1997); Felson (2002) and Poyner (2006). However, it is evidence relating to the effectiveness of the Secured by Design initiative which they find particularly attractive, as reported by: Brown (1999); Pascoe (1999); Armitage (1999); Teedon et al. (2009); Armitage and Monchuk (2009); Jones et al. (2016) and most recently Stokes (2022).

Similarly, in their analysis of developments built since the original research programme, Armitage and Monchuk (2009) are able to demonstrate '...that SBD as an evolving standard confers ever greater Crime Reduction advantage'. Furthermore, on the basis of their figures, the fiscal and carbon costs of including Secured by Design measures in a development, would be recovered by Crime Reduction alone within four years—not to mention other savings such as healthcare from more contented residents. Pease and Farrell (2009, p.21) also refer to the *'...vexed question of Permeability and Crime Risk'* (see Sect. 4.4.6).

Unlike all previous research in this area, Johnson and Bowers (2010) were able to absorb the multi-level structure of the data under analysis—most especially road segment type. They powerfully demonstrate how increased Permeability is connected with increased Crime Risk. And that burglary is lower in culs-de-sac, especially those that have no footpath links to other roads (do not 'leak') something every DOCO has always known instinctively through personal experience. Most pertinently, Pease and Farrell state:

> The Johnson and Bowers work, in the writers view, closes the book on the Permeability debate in favour of the SBD characterisation... (Pease and Farrell, 2009, p. 22)

Finally, the authors call for Secured by Design to be a mandatory requirement for all new developments and that SBD should continue to evolve, conscious of the Johnson and Bowers (2010) conclusions.

The SBD synopsis (2009) capitalises on the Pease and Farrell document. It observes how this independent and academically scrutinised research demonstrates SBD developments are half as likely to suffer from Residential Burglary and have two thirds less Vehicle Crime and 25% less Criminal Damage. Consequently, the capacity to Design Out Crime also delivers immediate benefits in terms of The Carbon Cost of Crime. Moreover, as the financial cost of SBD was assessed at between £70 and £240 depending upon the size of the dwelling (Davis Langdon for Secured by Design, 2010a, 2010b), there is also a financial (as well as a Carbon Cost of Crime) imperative for it being a mandatory requirement for all new and refurbished homes. SBD status and the use of SBD accredited products, also assist in increasing the lifespan of buildings. And finally, by improving the security of premises and reducing unnecessary maintenance work (often caused by criminal attack) SBD allows residents to feel safer.

In their calculations, Skudder et al. (2016) include indirect or embedded emissions of products stolen or damaged, together with knock-on service costs: insurance, health, legal, police and prisons. They use environmentally extended input–output analysis (EE-IOA—Weidmann et al., 2009) and the 'Economic and Social Costs of Crime' (Brand & Price, 2000; Dubourg and Hamed, 2005; Home Office, 2011). Such costs are sub-divided into: anticipation (defensive expenditure and insurance); consequence (damaged and stolen property, emotional and physical impacts, health and victim services) and response (police, legal, defence, police, jury and prison services).

Detailed carbon footprints are then calculated, amongst which homicide and woundings are the most costly, due to their likelihood of being reported/recorded, subsequent police investigations, trials and prison sentences. Burglary and other property offences possess a larger anticipation of crime carbon footprint, due to the emissions caused by insurance costs and defensive (Crime Prevention) expenditure. Whereas robbery has a large emission base due to additional carbon costs relating to victims' subsequent use of health services.

Skudder et al. (2016) have further developed the Pease and Farrell research to include the 'Rebound Effect' and how reducing crime may not necessarily lead to less carbon. They describe how a 5% reduction in Residential Burglary may reduce emissions by 3%, or raise the carbon footprint by 23%—the result of money saved being spent in more carbon-intensive ways, e.g. holidays, a new car or kitchen.

At almost 1.5 million tonnes of CO_2 (37%) the value of property stolen constitutes the greatest source of emissions. Health services 16%; damaged/destroyed property 15%; police activity 9% (1000 tonnes per day); prisons 7%. Nevertheless, even if The Carbon Cost of Crime is revised down to 4 million tonnes, Skudder et al. (2016) calculate that equates to the energy used by 900,000 UK homes.

The replacement of goods stolen totals 51% of the carbon footprint, helping to explain why burglary possesses the highest footprint total of any crime at 30%—albeit this does not include the reuse of such items after theft. But compare this to homicide at only 1% due to the small number of offences. In this context, emissions due to the consequences of crime constitute 67% of the overall footprint. This points to the value of Crime Prevention (the DOCO's role)—especially in relation to burglary. The UK was the first country in the world to investigate the Carbon-Crime Blind Spot. And the criminal justice services constitute an important aspect of offences against the person, e.g. homicide, woundings and assaults.

Secured by Design: Carbon Reduction and Cost Efficiency Saving Model
Pease and Farrell (2011) also reference the former DOCO with Northamptonshire Police and subsequent SBD Consultant, David Lancaster's then forthcoming Carbon Reduction and Cost Efficiency Saving Model (2010) and production of a Carbon Cost of Crime Calculator (2011).

In a succinct four pages it explains how each police force can calculate The Carbon Cost of Crime as a result of the three offences of Residential Burglary, Theft of Motor Vehicle and Theft from Motor Vehicle. When the world is ready to accept the full impact of The Carbon Cost of Crime, such calculations should have a profound effect in shining light on the Carbon-Crime Blind Spot.

2.12 Symbolic (Psychological) Barriers

The capacity of Symbolic Barriers (sometimes referred to as Psychological Barriers) to act as a deterrent of crime, fear of crime (FOC) and anti-social behaviour (ASB), has been a crucial component within Designing Out Crime (DOC) since its infancy. Indeed, the original Secured by Design 'Estate' guidance document (1989) states:

> By using real or symbolic barriers such as a change of road surface in either texture or colour—brick pillars, narrowed entrances, areas are 'defined' in such a way as to give the impression that an area within is private to a particular household or a small community. These visual impediments imply a boundary past which many intruders will be reluctant to step.

Glasgow, Scotland, train station platform gate with anti-trespass symbolic barriers. See also Sect. 2.12. Symbolic (Psychological) Barriers

However, the evidence from Crime Science to support the efficacy of such measures is somewhat sparse. Indeed, as Shaftoe and James (2004, p. 447) describe, this appears *'...to be based more on an act of faith rather than hard empirical testing'*.

Those of us who undertook our Designing Out Crime Officer (Crime Prevention Design Adviser/Architectural Liaison Officer—CPDA/ALO) training in the early 1990s, may well remember a piece of film called 'Circle on the Beach'. Originally broadcast in 1973 as part of a BBC TV programme presented by Professor Laurie Taylor, this recording at Bridlington Sands featured a woman lying on a sun-drenched beach and within a circle drawn in the sand. Over the next few hours, the camera recorded in time-lapse mode, how people went out of their way to avoid entering this same circle. One exception was a young toddler who crawled into the circle, only to be immediately removed by a parent.

Not only did this film purport to demonstrate the value of Symbolic Barriers and by default, certain elements of Crime Prevention Through Environmental Design (CPTED—see Chapter 4), it also appears to have been instrumental in convincing early proponents of DOC and those who created Secured by Design (SBD), of its relevance and efficacy.

In their paper, Do Symbolic Barriers Prevent Crime and Offer an Increased Sense of Safety? (2004), Professor Henry Shaftoe and Dr Sarah James describe the increasing tendency in the UK and beyond to incorporate Symbolic Barriers such as gateways, archways, changes in surface texture or colour at the approach to residential premises. Furthermore, the results of their 'modest pilot study' indicated such Symbolic Barriers affected the behaviour of only some individuals—whilst simultaneously acting as a provocation to others.

The authors suggest Newman's (1972, 1973, 1996) concepts of Defensible Space and Territoriality (see Sects. 4.4.1 and 4.4.3) may have provided the motivation for such Symbolic Barriers. Indeed, they cite Newman's contention that the success of Symbolic Barriers was dependent upon four factors:

1. Ability of the intruder to read the symbols for their intended meaning
2. Evident ability of the residents to control and reinforce 'symbolic space definition' through surveillance
3. Ability of internally defined space to necessitate the intruder make his intentions obvious (space must have a 'low tolerance for ambiguous use')
4. Ability of the residents to challenge the intruder and take appropriate action.

Such a theory has inherent cost advantages: real physical barriers (e.g. a high wall, fence, hedge or lockable gates) are unnecessary. Instead, people are expected to alter their behaviour due to the psychological reception of 'signals' emanating from symbolic changes to the built environment (Rapoport, 1982). The theory suggests the same degree of 'environmental sensitivity' (Taylor, 1973) by both the criminal and respectable middle classes who approach them. However, Shaftoe and James describe that unfortunately, no one had tested the theory of how Symbolic Barriers deter criminals or make people feel safer.

The authors used a modest 'field experiment in auto-shaping' (Zimbardo, 1973) to test the actual effect of such Environmental Cues (see Sect. 2.8). A piece of green-coloured 'Astroturf' was used to change the surface texture on a footpath at the University of the West of England, Bristol before observing the behaviour of those who approached. The Astroturf was left in place for 20 minutes, followed by a further 20 minutes once a large red diagonal cross had been added. Both video-recording and photography were used to observe what took place during these time frames. And one of the authors holding a clipboard also stood nearby for part of the time.

'An unpredictable range of behaviours' were witnessed:

- The vast majority of pedestrians actively sought to avoid walking on the Astroturf
- Many who did walk on it indicated a level of discomfort
- Some people consciously and deliberately walked on it, often with a level of defiance: young males in particular stamped on it—often to the admiration of their peers
- Both avoidance and defiant behaviour was especially pronounced once the large red cross was added
- The reverse psychological effect was particularly noticeable after the large red cross had been added
- An official observer (one of the authors bearing a clipboard) did not appear to have any effect.

From this evidence Shaftoe and James conclude:

1. Variations in floor surface texture and colour can influence people's behaviour in the desired direction, but also cause the opposite reaction in others;
2. The design, size, colour and location of symbolic floor surfaces may determine people's reactions.

The latter is largely ignored in the previous literature and by those who encourage such measures (Schneider & Kitchen, 2002). Moreover, as only 75% of the path width was covered by Astroturf (and real grass lay on either side) those walking in pairs had to choose which Symbolic Barrier to ignore—Astroturf or real grass, unless walking in single file. Similarly, the cumulative effect; time delay in obeying the Symbolic Barrier and effects of specific colours (red denoting danger or 'stop', pastel shades to produce a calming effect) need to be considered.

Crouch et al. (1999) assert that psychological signals will only work if the defined neighbourhood is predominantly law-abiding, otherwise it may work in reverse. Moreover, a hierarchy of Symbolic Barriers may exist: with changes in floorscape surface texture the weakest; ceremonial gateposts, archways and shallow moats the strongest.

Compare the above research to that of Tony Holt and Jon Spencer (2005) regarding 'Little Yellow Box' (Privacy Zones or Personal Defensible Space) in front of ATMs (see Sect. 6.11). In the pilot study area these '...had a significant effect on the number of Robberies committed at or in close proximity to the ATM' (2005, p. 15).

They also found that street robbers saw ATMs as good locations to target cash-rich victims, distracted (whilst withdrawing cash) and unaware of their immediate environment. The lack of spatial engagement and body language increases vulnerability and capacity for victimisation. Indeed, where personal space is intruded it creates shock, inability to react and compliance.

A yellow box, hatched area Privacy Zone was used to create Personal Defensible Space and raise awareness/protect the user, by preventing unwanted attention and making the area more hostile to the offender.

Holt (2000) discovered that ATMs account for as much as 25% of all Street Robbery and snatch-type theft. The Manchester study used 22 ATMs in the pilot project area that was both a Crime Generator and Crime Attractor (Brantingham & Brantingham, 2008—see Sect. 2.5). Robbery and Theft from the Person within 150 metres of each ATM Activity Node was counted before and after a yellow box had been painted in front of the ATM.

The results were especially encouraging, there being an average reduction in Street Robbery and Theft from the Person of 34% during the pilot period across ALL ATM locations (which also suggests a potential Diffusion of Benefits—see Sect. 2.1). The exception was one unmarked ATM that was already suffering the highest level of crime amongst the 22 machines.

Surprisingly, users respected personal space within the Privacy Zone even at night when there were large groups of people, some of whom (when viewed from CCTV recordings) appeared to be intoxicated. People rarely breached the Privacy Zone and as a consequence, it appears that the defining of such Personal Defensible Space may act as a formidable psychological barrier.

Holt and Spencer (2005) contend ATM Privacy Zones raise users' awareness of their immediate vicinity, thereby increasing spatial engagement and lessening the level of distraction. It may also deliver a heightened sensitivity about their potential for victimisation, resulting in individuals becoming more alert to their surroundings.

However, this author's PhD research project involving The Four Towers at Duddeston Manor (see Sect. 9.1) demonstrates how over three decades, Symbolic Barriers like brick pillar gateposts, gates (that are never closed), fencing (easily climbed) and changes in road surface texture and colour acted (at best) as only a partial deterrent to crime and ASB. This was witnessed in the undamaged former semi-circular painted steel seat and rose arbours, festooned with flowers each June. Nevertheless, Vehicle Crime has increased over time. See also Landman and Liebermann (2005).

Recent applications of Symbolic Barriers can be seen in Pawnbrokers' shops, where roped barriers are being used to slow the escape of potential robbers—or more appropriately influence their decision-making processes. And with British Transport Police in preventing Suicide on the railways of Great Britain and whose advocation of Symbolic Barriers in platform design are detailed in Sect. 10.1, The Future of Designing Out Crime.

2.13 The Value of Signage and Notices (It Pays to Advertise)

Does Signage and Notices (be they in pictographic or written form or a combination of the two) have any effect in the arenas of Crime Prevention, fear of crime and ASB? Evidence of its effectiveness is somewhat sparse. And indeed, signage is often displayed due to simple legal requirement, e.g. as a result of Health and Safety legislation, the Information Commissioner's Code of Practice (2008) in respect of public area CCTV cameras (see Sect. 4.3.3) or in an attempt to escape civil litigation. Under English law the notice 'Trespassers Will Be Prosecuted' is often understood to be a 'wooden lie'—trespass alone most usually being insufficient to secure conviction. Similarly, signs displaying the message 'No Ball Games' have a history of ineffectiveness.

Nevertheless, amongst Designing Out Crime Officers (DOCOs) and in the wider world, the use of Signage and Notices to denote rule setting has a strong currency. Indeed, £millions are spent on advertising every year and for a purpose all but identical to that provided by Signage and Notices: selling a message/ product. For example, at Cottesbrooke Infant School (see Sect. 9.3) highly relevant Signage and Notices are displayed at the school entrance, primarily to advise parents and other visitors. The display of these is a positive indication that the school environment is excellently managed—a highly visible example of the Good Housekeeping Measures advised in Sect.8.3 and helping to adhere to the Broken Windows philosophy of Kelling and Wilson (1982—see Sect. 2.8). Similarly, rule setting and the posting of instructions are two of the Twenty-Five Techniques of Situational Prevention (Cornish & Clarke, 2003—see Appendices A–H in this handbook).

See Nettle, Nott and Bateson (2012) regarding the use of signage to prevent bicycle theft. And also Professor Rob Mawby (2016) 'Policing policies in cities dependent on tourism: developing an evidence-based approach'. In addition, Armitage, Joyce and Monchuk (2018) provide a fascinating examination of offender perspectives about risk and protective factors (including signage) in the the design and layout of retail environments.

Crime Prevention poster warning Japanese tourists in Stockholm, Sweden, of the danger in having their luggage stolen from hotel reception areas. This is a highly specific Crime Prevention message and the Swedish Police confirmed Japanese tourists had been repeatedly targeted by thieves.

Meanwhile, Mihinjac reports (2015) that in the Queen Street Mall, Brisbane, Australia 'proper wayfinding and maintenance was achieved' during the 1999 refurbishment. This involved decluttering the Mall, improved legibility, clear sightlines and open design to enhance both safety and feelings of safety—all elements that fall within the remit of the Environmental Elements of CPTED (see Chapter 4).

Outside the arena of road traffic law and Environmental Crime (see Sect. 8. 2), such rule setting through Signage and Notices has gradually disappeared from many environments in the UK during the past half century. For example, it was once common to see notices on trains, buses and in parks prohibiting a range of activities, e.g. spitting and the use of transistor radios were banned, together with the fines for contravention. And whilst there are instances of signage stating, 'No Feet on Seats', these are relatively rare—and repeatedly ignored.

No doubt a liberal perspective would contend the absence of such Signage and Notices denotes an increasingly tolerant society. However, for law enforcers the response, 'I didn't know that was an offence', is a constant rejoinder and excuse. Moreover, if rules are established but not displayed or enforced, their very existence is compromised if not nullified.

It is interesting to compare this liberalism with attitudes in Australia and the USA. In those countries there appears to be no such bar on rule setting by Signage and Notices. Certainly, on public transport the full spectrum of prohibited activities is usually displayed, together with the financial penalty for failing to obey them. Meanwhile, the traditionally tolerant Benelux countries have no qualms regarding the display of Signage and Notices about the actions deemed to be illegal. This author's favourite is the combined 'No Toking' (smoking of cannabis) and 'No Drinking of Alcohol' in public signs in the t'Bartjes and other suburbs of Amsterdam!

No toking of cannabis or drinking alcohol street signs in t'Bartjes district of Amsterdam, The Netherlands

Beware Hotel Thieves poster—displayed inside hotel receptions in Stockholm, Sweden and aimed at Japanese tourists

For the DOCO asked to voice an opinion on this issue, the following points are advised:

- If required by law (e.g. public area CCTV cameras), Signage and Notices must always be displayed

- Displaying Signage and Notices is not at variance with liberal ideals. Indeed, informing the public that certain actions are prohibited and/or anti-social, serves a positive function
- The message contained on Signage and Notices should be enforceable if the need arises
- The wording should not sound to officious or provocative (Wortley, 2001)
- Signage and Notices should be fixed high enough not to be easily removed or defaced
- Signage and Notices should be appropriately and sparingly used. They should not emulate the 'confetti' like preponderance of road traffic signage
- Signage and Notices should be kept in pristine condition. Old/decaying Signage and Notices suggest no one cares about the issue in question and falls into the Broken Windows trap described by Wilson and Kelling (1982).

Whilst this chapter has been overwhelmingly concerned with visible Signage and Notices, other human senses can be informed in a similar fashion, e.g. it is commonly known that householders who wish to sell their home are advised that the smell of newly baked bread or freshly brewed coffee is a powerful tool in convincing potential buyers. However, no doubt more pertinent to the DOCO, 'audible' Signage and Notices can be similarly useful. Loudspeakers attached to CCTV cameras or Intruder Alarm systems have proved to be especially effective in warning potential offenders that they are being watched and to desist when the police or other authority has been summoned. However, it is also vital that all such messaging is regularly updated to prevent them becoming visual or aural 'wallpaper', that no one notices or heeds.

2.14 Problem-Oriented Policing and Partnerships

It seems logical that cross-discipline Partnership Working (known as 'Collaborative Working' in the health professions) will deliver improved and sustainable results across a whole field of human endeavour. However, with the growth of new professions, institutions and further specialisms, it was perhaps inevitable that failures of communication would result. This has been compounded by high profile examples of ineffective Partnership Working amongst the 'caring' services. Thankfully, there exists a well-evidenced solution: Problem-Oriented Policing delivered through Problem-Solving (Berry et al., 2011; Hinkle et al. (2020).

Herman Goldstein (1979) provided the description Problem-Oriented Policing (POP). He believed the police service concentrated on organisational efficiency to the detriment of its core outcomes. Whereas, Goldstein contends the following key areas are paramount:

1. Community issues the public expects the police to deal with
2. Problem reduction, elimination, better handling or lessening
3. Pro-activity based on careful enquiry
4. Disaggregation of incidents into separately identifiable problems
5. Commitment to systematic enquiry
6. A broad, uninhibited and imaginative quest for tailor-made solutions
7. Higher levels of accountability
8. Acceptance that efforts to address problems involve risk-taking
9. Routine, skilful, methodologically informed and ideally, independent evaluation
10. Encouraging rigorous methods in making enquiries
11. Full use of police skills and expertise
12. Acknowledgement of the limitations of the criminal justice system
13. Identification and attendance to multiple interests in a problem.

In, *Not Rocket Science? Problem-solving and Crime Reduction* (2000), Professors Tim Read and Nick Tilley describe how from 266 responses, 42% of the POP initiatives were deemed to have been failures by the police service concerned. Moreover, in 40% of UK schemes, the police were the sole partner (2000, p. 5) suggesting many respondents did not understand the rationale that enables POP to work. Another concern was the 95% who reported the police were the major partner.

Actually, such a high percentage might be expected when the police are the initiating agency—for which purpose Designing Out Crime, CPTED and Secured by Design (see Chapter 5) are effective delivery strategies. Furthermore, whilst on occasion it might be wise for the police to adopt a less dominant role, examples like the former West Midlands Arson Task Force (a combined initiative by West Midlands Fire Service and West Midlands Police) demonstrate the evidenced success in a partnership when two or more 'can do' organisations proactively work together with shared aims, goals and a Problem-solving approach. See especially their management and good practice documents, Keep Your Business in Business (2005, 2009) and Keep Your School in Business (2006).

The desire to reduce the perception that the police would automatically dominate the partnership process, can be identified in the title, Problem-oriented Policing and Partnerships (Bullock et al., 2006). This is not about devaluing the role of the police, but like Section 17 of the Crime and Disorder Act 1998, an encouragement for partner agencies to accept their role as part of the corporate approach to Crime Prevention and Community Safety (Morgan Report, 1991; Moss & Pease, 1999).

Bullock et al., (2006) provide an overview of best practice and evolution in Problem-Oriented Partnerships. They describe the key conditions through which Truly Effective Partnership Working (see Sect. 1.5.11) can deliver demonstrable results. Too many partnerships exist in name alone, providing a veneer of respectability to an otherwise poor situation. But where such obstacles as career protection, indifference and inappropriate professional behaviour are overcome, the rewards can be immense—especially for the communities we all serve. Indeed, Goldstein (1997) subsequently produced an impressive 'Hierarchy of Levers' (see Bullock, Erol and Tilley, 2006, p.131) designed to influence partners about their role in the POP process. These range from the production of advice leaflets through to taking civil action in the courts.

The key element in the delivery of POP is the Problem-solving approach. A number of different models of Problem-solving exist, including the 5Is (see Sect. 2.7) and SARA (Scanning, Analysis, Response and Assessment—Eck & Spelman, 1987). These are seen as integral to Problem-solving, as is reference to Routine Activity Theory (RAT—see Sect. 2.2) and the Problem Analysis Triangle (PAT). It also places great emphasis on 'Evidence-Based Policy and Practice'.

Unfortunately, scrutiny of many POP projects indicates the absence of such evidence and the rarity of full implementation of POP principles—even amongst the best projects! Indeed, assessment/impact is often an infrequent commodity. The immediacy of crime, fear of crime (FOC) and anti-social behaviour (ASB) might provide a fig-leaf of excuse for such lapses. However, their serial nature and lack of improvement over time (Bullock et al., 2006) suggests something far more serious. Namely, a culture unwilling to accept that POP demands holistic engagement, rather than a pick and choose mentality.

One very good POP example and a Tilley Award winner (named after Crime Prevention guru, Professor Nick Tilley) was the Kirkholt Burglary Reduction Project. Established in the wake of exceptionally high levels of Residential Burglary (24.6% in 1985—or an all but one in four chance of being burgled) on the Kirkholt estate in the Rochdale area of Greater Manchester, the Kirkholt project (Pease, 1991) involved a number of coordinated interventions including: property post-coding (zip codes); removing coin-payment gas and electricity metres; upgrading Physical Security 'Target Hardening' Measures; Home Watch; Cocoon Neighbourhood Watch; and computerised monitoring and evaluation. Genuinely impressive reductions in Residential Burglary (first 40% and ultimately 76%)

were achieved, especially in respect of Repeat Victimisation (see also Tilley et al., 1999).

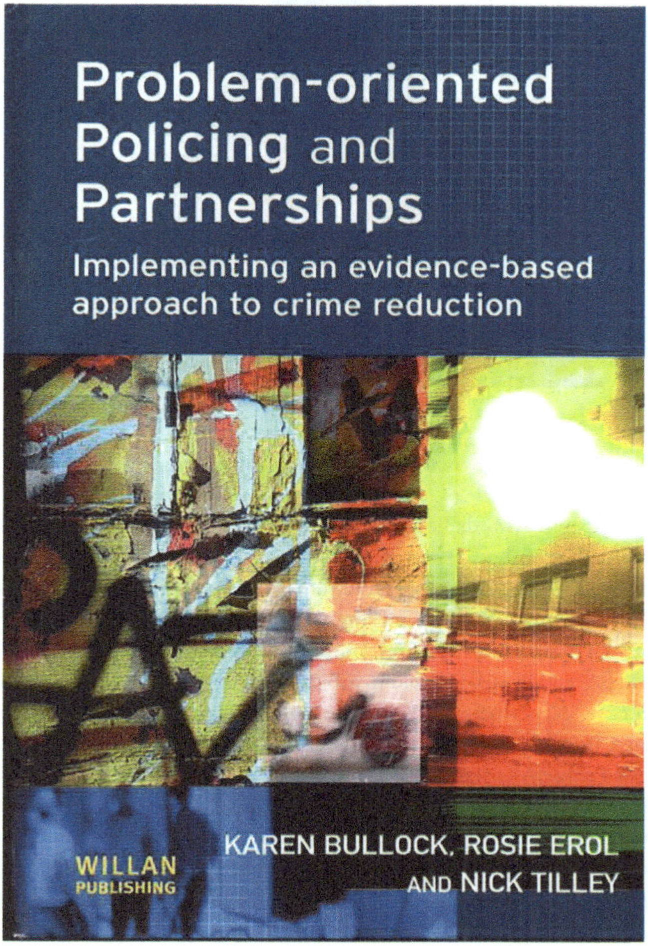

Problem-oriented Policing and Partnerships

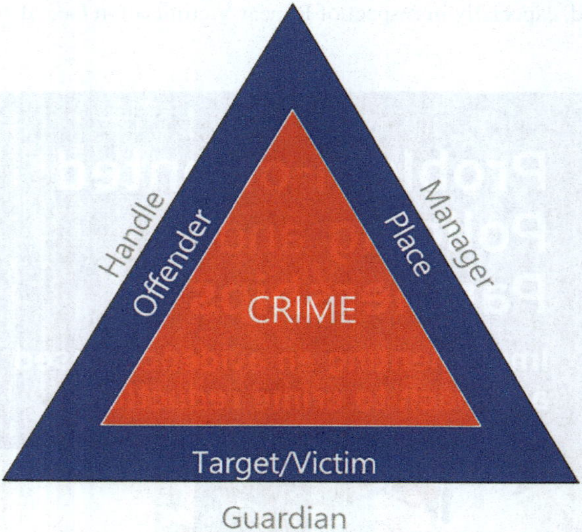

Problem analysis triangle, Revised by Clarke and Eck (2009)

At the Goldstein Awards held in 2012 at Providence, Rhode Island, USA, Professors Michael Scott and Stuart Kirby advocated 'Becoming a Problem-Solving Police Service'. In the UK, the pre-eminent example of such root and branch adoption of POP and Problem-solving, has been Durham Constabulary. There, the then Chief Constable Michael Barton (a champion of POP for more than three decades), inculcated Problem-solving throughout his police force via the SARA process. Nevertheless, he counsels that no one should underestimate the challenge of implementing POP. See Implementing POP: Leading, Structuring, Managing a Problem-Oriented Police Agency (2012).

For the Designing Out Crime Officer (DOCO), their role is self-evidently one that automatically involves Problem-solving and POP working. Indeed, in the UK Secured by Design can be viewed as the obvious mechanism whereby Problem-solving has been mainstreamed into police Crime Prevention delivery. However, this does not contradict Bullock et al.'s (2006, p. 17) observation that as a rule POP '*...is probably best described as pockets of isolated good practice which tend to be associated with highly motivated individuals*' and first postulated by Read and Tilley (2000); Kirby and Reed (2004).

Problem-Solving and POP are also essential mechanisms for the DOCO to apply to new and unusual challenges—especially where partners who have no experience of the DOC process are encountered or perceived hostility lurks behind the scenes. Often, the skill is in first identifying that the Partnership Approach is required (the easy bit). For this purpose, Professor Scott (2018) contends that '*limited police resources should be concentrated on places, people and things*'. And second, being able to assemble the core group of enthusiastic individuals (the difficult bit—not all partners being willing to participate and contribute positively: see Sect. 1.5.11). Ultimately, there is little merit in attempting to flog the metaphorical dead horse if partners are unforthcoming or do so under duress. But even if an SBD award or a similar accolade is not achieved, *Partnership Working* is likely to have engineered a network of relationships and structures that can be highly valuable at a later date.

For those individuals who persist with the partnership process, their success as a *POP* initiative is recognised by the 'Tilley Awards'. First launched in 1999 by the then Home Office Policing and Reducing Crime Unit, these aim to encourage police forces to identify and submit those projects demonstrating excellent practice in *Problem-Oriented Policing and Partnerships*. Emulating the success of the Goldstein Awards in the USA, Tilley Award winners are invited to attend and be judged on an international basis at the Goldstein Awards annual conferences, hosted by police forces across the USA.

2.15 Intervention Sustainability Spectrum

It is generally appreciated that different forms of Crime Prevention intervention will have varying degrees of success and sustainability. A primary reason for advocating Designing Out Crime is that a sizeable portfolio of evidence (see especially Brown (1999); Pascoe (1999); Armitage (1999); Teedon et al. (2009); Armitage and Monchuk (2009); Jones et al. (2016); Stokes (2022) and best practice examples (see Chapter 9) now exist. These point to the long-term effectiveness and sustainability of SBD interventions—and by default their inherent cost-effectiveness, both financially and in terms of green issues (see Sect. 2.11, The Carbon Cost of Crime).

Taking into account all the currently available evidence in respect of macro level interventions to prevent or reduce crime, in ascending order of sustainability these will include:

- The arrest, prosecution and incarceration of offenders
- The post imprisonment/management of offenders
- Electronic tagging of offenders
- High visibility foot patrols by police officers, community support officers, street wardens etc.
- Bolt-on, retro-fitted, post event Physical Security 'Target Hardening' Measures and Technological Innovations (see Sects. 4.2.1–4.3.3)
- The substitution of products for ones less attractive to criminals and the vulnerable (e.g. methane instead of coal gas; polymer or bitumen roofing products rather than copper and lead)
- The design of products (e.g. Design Against Crime and Designing Out Crime initiatives) to make them less attractive to criminals. For example: steering wheel locks, alarms and immobilisers on motor vehicles; theft-resistant bicycles, shoulder bags, robbery-resistant chairs etc.
- Designed-in Physical Security 'Target Hardening' Measures and Technological Innovations of CPTED
- Designing Out Crime/Counter Terrorism Design utilising Physical Security 'Target Hardening' Measures, Technological Innovations and the Environmental Elements of CPTED at the post, pre-planning and (ideally) concept stage of a development.

Such interventions will have varying degrees of effectiveness in terms of sustainability. For example, the Attrition Rate of Crime (Clarke & Eck, 2007) and other research (e.g. Burrows et al., 2005; MOPAC, 2023) suggests that for every 100 offences committed, only 3% will result in a criminal conviction (Home Office Standing Committee on Crime Prevention, 1991). Less than a decade later that figure had fallen to 2.2% with 0.3% leading to a custodial sentence (Barclay & Tavares, 1999; Garside, 2004).

Consequently, whilst imprisonment might feature at the highest end of effectiveness in terms of preventive value in the short to medium term, in the long term, it cannot be described as truly sustainable—unless prisoners are never released! Moreover, incarceration of whatever form is an extremely expensive option—both financially and in terms of other resources. Indeed, it is a constituent part of The Carbon Cost of Crime (Pease & Farrell, 2009). And in addition, there is the powerful argument which contends that prisons and similar institutions act as 'Universities of Crime' for at least a proportion of the inmates—and across a wide range of different offence categories.

The Intervention Sustainability Spectrum (very much a title for this author's work in progress) is relevant, primarily because it acts as a visual means of helping to explain to decision-makers outside the world of Designing Out Crime just why it is such a valuable commodity. With Designing Out Crime at one extreme of the spectrum (indicating long-term effectiveness and sustainability), imprisonment and incarceration in general are located at the opposite end of the scale. High Visibility Policing (delivered by uniformed police officers, community support officers and the 'wider police family') is similarly expensive, both financially and in terms of the resources necessarily redeployed.

The location of other interventions at various points along the spectrum, is dependent upon their perceived relevance in terms of sustainability. All too obviously, this is a less than truly scientific assessment, but the spectrum was never intended to be and were it so described would necessitate considerable research and ultimately, highly subjective interpretation.

For example, the imprisonment of convicted offenders is considered by some to be the only effective long-term solution to crime. And indeed, as any police officer can confirm the arrest and detention of a known and highly active/productive offender (in terms of the crime he commits—see Sect. 2.3, Crime Opportunity) can have an immediate and positive effect on the local crime figures. However, this situation is rarely, if ever sustainable, either because he is eventually released and as previously indicated in respect of Property Crime, will probably reoffend; or other criminals will take his place. Whereas, we now possess evidence that demonstrates how Designing Out Crime interventions can have long term (meaning potentially decades) of sustainable effectiveness.

To further demonstrate that arrest and detention are not alone in having a potentially very short sustainable shelf life, the repeated mantra of many in public office around the world advocating the validity of High Visibility Policing (meaning street patrols by uniformed officers) can be even more transient in effectiveness. Evidence demonstrates that whilst it may have an immediate and indeed Anticipatory Benefit when announced (see Smith et al., 2002), after it is inevitably withdrawn crime and ASB have a nasty habit of rapidly returning to their former levels. Moreover, High Visibility Policing is in itself unsustainable, both police and community support officers having to be transferred from other areas, with the ultimate potential of leading to increased crime in those locations deprived of staff.

Between Designing Out Crime, Counter Terrorism Design and the incarceration of offenders, come the post imprisonment management of offenders; electronic tagging of offenders and bolt-on, retro-fitted (post construction) Physical Security 'Target Hardening' Measures. These are all financially expensive and their capacity for sustainability is often highly flawed. For example, adhesive anti-shatter film that requires replacing on windows every 10 years. Or, additional locking systems installed into a wooden, non-PAS 24 door that will warp and shrink over time.

What we can assert is how those most sustainable and effective (in terms of reduced crime and both economic and carbon cost) elements involve the removal, or at least the reduction, of opportunity (see Sect. 2.2, Routine Activity Theory, and 2.3, Crime Opportunity). Consequently, Designing Out Crime (from developments and/or products) and Design Against Crime (the substitution and design of products to make them less attractive to criminals) represent the best forms of Sustainable Crime Prevention. See also the Design Council's (2003) Think Thief: A Designer's Guide to Designing Out Crime.

Conclusion
The Sustainability Intervention Spectrum remains in development. However, its main purpose is to act as an explanatory tool in describing to colleagues and potential converts within the professions of the built environment and partner agencies, why Designing Out Crime, CPTED, Situational Crime Prevention and SBD are effective and sustainable mechanisms—often possessing an inherent durability that will last for several decades.

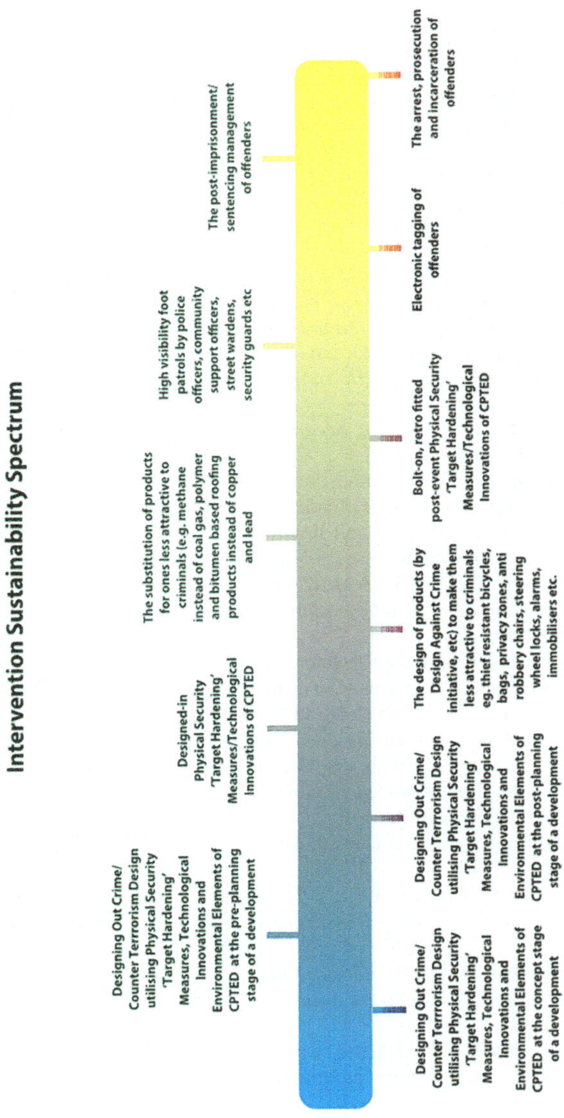

Intervention Sustainability Spectrum

References

Akers, R.L. (2017) Rational Choice, Deterrence, and Social Learning Theory. In *Criminology: The Path Not Taken*. Routledge.

Alabi, O.T., Adeleke, M.A. & Olajide, S.E. (2021). Measuring the influences of opportunity in residential neighbourhood crime. *International Journal of Multidisciplinary Research and Development*. *8*(6), 17–26.

Al-Kahtani, S. (1996). *Travel to Crime Distances of Burglars in Saudi Arabia*. University of Manchester.

Alston, J. D. (1994). *The Serial Rapist's Spatial Pattern of Target Selection* (Unpublished thesis). Burnaby, B.C: Simon Fraser University.

Andresen, M.A. (2010). The Place of Environmental Criminology within Criminological Thought, in *Classics in Environmental Criminology*. 1st Edition. Routledge.

Armitage, R. (1999). *An Evaluation of Secured by Design Housing Schemes Throughout the West Yorkshire Area*. The Applied Criminology Group, The University of Huddersfield.

Armitage, R. (2013). *Crime Prevention Through Housing Design: Policy and Practice*. Palgrave Macmillan.

Armitage, R., & Monchuk, L. (2009). *Re-evaluating Secured by Design (SBD) Housing in West Yorkshire*. University of Huddersfield.

Armitage, R., Joyce, C., Monchuk, L. (2018). Crime Prevention Through Environmental Design (CPTED) and Retail Crime: Exploring Offender Perspectives on Risk and Protective Factors in the Design and Layout of Retail Environments. In V. Ceccato & R. Armitage (Eds)., *Retail Crime. Crime Prevention and Security Management*. Palgrave Macmillan, Cham. https://doi.org/10.1007/978-3-319-73065-3_6

Barclay, G. C., & Tavares, C. (1999). *Digest 4: Information on the Criminal Justice System in England and Wales*. Home Office.

Barton, M. (2017). 'Delivering Problem Oriented Policing (POP)' Presentation given by Chief Constable Michael Barton at the Designing Out Crime Officers (DOCOs) Annual Training Event, Whittlebury Hall, Northamptonshire, 15 March 2017.

Barr, R., & Pease, K. (1990). Crime Placement, Displacement and Deflection. In M. Tonry & N. Morris (Eds.), *Crime and Justice* (Vol. 12, pp. 28–37). Chicago University Press.

Bennett, T., & Wright, R. (1984). *Burglars on Burglary*. Gower.

Bernasco, W., & Luykx, F. (2003). Effects of Attractiveness, Opportunity and Accessibility to Burglars on Residential Burglary Rates of Urban Neighborhoods. *Criminology, 41*, 981–1001.

Berry, G., Briggs, P., Erol, R., & Van Staden, L. (2011). *The Effectiveness of Partnership Working in a Crime and Disorder Context: A Rapid Evidence Assessment* (Research Report 52). Home Office.

Bowers, K., Johnson, S., & Hirschfield, A. (2004). Closing Off Opportunities for Crime: An Evaluation of Alley-Gating. *European Journal on Criminal Policy and Research, 10*, 285–308.

Bowers, K.J., Johnson, S.D., Guerette, R.T. et al. (2011) Spatial displacement and diffusion of benefits among geographically focused policing initiatives: a meta-analytical review. *J Exp Criminol* 7, 347–374. https://doi.org/10.1007/s11292-011-9134-8

Braga, A. A., Hureau, D. M., & Papachristos, A. V. (2012). An Ex Post Facto Evaluation Framework for Place-Based Police Interventions. *Evaluation Review, 35*(6), 592–626.

Brand, S., & Price, R. (2000). *The Economic and Social Costs of Crime* (Home Office Research Study No. 217). Home Office.

Brantingham, P. J., & Brantingham, P. L. (1981). Notes on the Geometry of Crime. In P. J. Brantingham & P. L. Brantingham (Eds.), *Environmental Criminology*. Sage.

Brantingham, P., & Brantingham, P. (1995). Criminality of Place: Crime Generators and Crime Attractors. *European Journal on Criminal Policy and Research 3,* 3, special issue on Crime Environments and Situational Prevention, 5–21.

Brantingham, P., & Brantingham, P. (2008). Offender Mobility and Crime Pattern Formation from First Principles. In L. Liu & J. Eck (Eds.),*Artificial Crime Analysis Systems: Using Computer Simulations and Geographical Information Systems* (pp. 193–208). Idea Press.

Brantingham, P.J. and Brantingham, P.L. (2016). 'The geometry of crime and crime pattern theory.' *Environmental Criminology and Crime Analysis.* 2nd edition. Routledge

Bromley, R. D. F., & Nelson, A. L. (2002). Alcohol-Related Crime and Disorder Across Urban Space and Time: Evidence from a British City. *Geoforum, 33,* 239–254.

Brown, J. (1999). *An Evaluation of the Secured by Design Initiative in Gwent, South Wales* (Unpublished MSc dissertation). Scarman Centre for the Study of Public Order. Leicester: University of Leicester.

Brown, L. A., Malecki, E. J., & Philliber, S. J. (1977). Awareness Space Characteristics in a Migration Context. *Environment and Behavior, 9*(3), 335–348.

Brown, R., & Scott, M. (2007). *Implementing Responses to Problems.* Problem-Oriented Guides for Police (Problem-Solving Tools Series No. 7). Centre for Problem-Oriented Policing.

Bullock, K., Erol, R., & Tilley, N. (2006). *Problem-Oriented Policing and Partnerships: Implementing an Evidence-Based Approach to Crime Reduction.* Willan.

Burgess, E. W. (1925). The Growth of the City. In R. E. Park, E. W. Burgess, & R. D MacKenzie (Eds.), *The City.* University of Chicago Press.

Burrows, J., Hopkins, M., Hubbard, R., Robinson, A. Speed, M. & Tilley, N. (2005). *Understanding the attrition process in volume crime investigations.* Home Office Research Study 295. London: Home Office Research, Development and Statistics Directorate

Caulkins, J. (1992). *Thinking About Displacement in Drug Markets: Why Observing Change of Venue Isn't Enough.* Journal of Drug Issues.

Chenery, S., Holt, J., & Pease, K. (1997). *Biting Back II: Preventing Repeat Victimisation in Huddersfield* (Crime Detection and Prevention Paper 82). Home Office.

Clarke, R. V. (1983). Situation Crime Prevention: Its Theoretical Basis and Practical Scope. *Crime and Justice, 4,* 225–256.

Clarke, R. V. (1990). *Armed Robbery Interim Report 2: Target Hardening of Banks and Displacement of Robberies.* National Police Research Unit.

Clarke, R. V. (1995). Situational Crime Prevention. In Crime and Justice Vol. 19, *Building a Safer Society: Strategic Approaches to Crime Prevention* (pp. 91–150). University of Chicago Press.

Clarke, R. V. (1999). *Hot Products: Understanding, Anticipating and Reducing Demand for Stolen Goods* (Police Research Series 112). Home Office.

Clarke, R. V., & Eck, J. (2003). *Become a Problem-Solving Crime Analyst in 55 Small Steps.* Jill Dando Institute, University College London. Available online at: www.jdi.ucl.ac.uk/publications/manual/crime_manual_content.php

Clarke, R. V., & Eck, J. (2007). *Understanding Risky Facilities.* Center for Problem-Oriented Policing. Available online at: http://www.popcenter.org/tools/PDFs/risky_facilitiespdf.

Clarke, R. V., & Eck, J. E. (2009). *Crime Analysis for Problem Solvers in 60 Small Steps.* Center for Problem-Oriented Policing and U.S. Department of Justice, Office of Community Oriented Policing Services.

Clarke, R.V. & Felson, M. (2017). *Routine Activity and Rational Choice.* Volume 5. https://doi.org/10.4324/9781315128788.

Clarke, R., & Mayhew, P. (1988). The British Gas Suicide Story and Its Implications for Prevention. In M. Tonry & N. Morris (Eds.), *Crime and Justice: A Review of Research* (Vol. 10). Chicago University Press.

Clarke, R. V., & Weisburd, D. (1994). Diffusion of Crime Control Benefits: Observations on the Reverse of Displacement. In R. Clarke (Ed.), *Crime Prevention Studies* (Vol. 2). Criminal Justice Press.

Cohen, L. E., & Felson, M. (1979). Social Change and Crime Rate Trends: A Routine Activity Approach. *American Sociological Review, 44*, 488–608.

Cornish, D. (1994). The Procedural Analysis of Offending and its Relevance for Situational Prevention. In R. V. Clarke (Eds.), *Crime Prevention Studies* (Vol. 3). Lynne Rienner.

Cornish, D., & Clarke, R. (1986). *The Reasoning Criminal.* Springer-Verlag.

Cornish, D., & Clarke, R. (2003). Opportunities, Precipitators and Criminal Decisions: A Reply to Wortley's Critique of Situational Crime Prevention. In *Crime Prevention Studies* (Vol. 16). Criminal Justice Press.

Cornish, D., & Clarke, R. (2008). The Rational Choice Perspective. In R. Wortley & L. Mazarolle (Eds.), *Environmental Criminology and Crime Analysis.* Willan Publishing.

Cornish, D. B., & Clarke, R. V. (2011). *Understand Crime Displacement: An Application of Rational Choice Theory.* Routledge.

Cromwell, P., Olson, J., & Avary, D. (1991). *Breaking and Entering: An Ethnographic Analysis of Burglary.* Sage.

Crouch, S., Shaftoe, H., & Fleming, R. (1999). *Design for Secure Residential Environments.* Longman.

Daly, H. (2002, April 30). *Sustainable Development: Definitions, Principles, Policies.* Invited Address World Bank.

de Andrade, D., Homel, R., & Mazerolle, L. (2016). Boozy Nights and Violent Fights: Perceptions of Environmental Cues to Violence and Crime in Licensed Venues. *Journal of Interpersonal Violence*, 1–22. https://doi.org/10.1177/0886260516657910.

Design Council. (2003). *Think Thief: A Designer's Guide to Designing Out Crime.* Design Council.

Dubourg, R., & Hamed, J. (2005). *The Economic and Social Costs of Crime against Individuals and Households 2003/4* (Home Office RDS Online Report 30/05).

Dugan, L. 1999. The effect of criminal victimization on a household's moving decision. *Criminology 37* (4): 903–930.

Eck, J. E. (1993). The Threat of Crime Displacement. *Criminal Justice Abstracts, 25*, 527–546.

Eck, J. W. (1997). Preventing Crime in Places. In L. W. Sherman, et al. (Eds.), *Preventing Crime: What works, What Doesn't, What's Promising?* National Institute of Justice.

Eck, J. E., & Spelman, W. (1987). *Solving Problems: Problem-Oriented Policing in Newport News*. Police Executive Research Forum.

Eck, J.E. and Weisburd, D.L. (1995). *Crime Places in Crime Theory*. Criminal Justice Press.

Eck, J., & Weisburd, D. L. (2015). Crime Places in Crime Theory. *Crime and Place: Crime Prevention Studies* (Vol. 4, pp. 1–33). Criminal Justice Press.

Ekblom, P. (1994). Proximal Circumstances: A Mechanism-Based Classification of Crime Prevention. *Crime Prevention Studies* (Vol. 2). Criminal Justice Press.

Ekblom, P. (1996). Towards a Discipline of Crime Prevention: A Systematic Approach to its Nature, Range and Concepts. In T. Bennett (Ed.), *Preventing Crime and Disorder: Targeting Strategies and Responsibilities*. Cropwood Series. Institute of Criminology.

Ekblom, P. (1997). Gearing up Against Crime: A Dynamic Framework to Help Designers Keep up with the Adaptive Criminal in a Changing World. *International Journal of Risk, Security and Crime Prevention, 214*, 249–265.

Ekblom, P. (2000). The Conjunction of Criminal Opportunity—A Tool for Clear, 'Joined-Up' Thinking and Community Safety and Crime Reduction. In S. Ballintyne, K. Pease, & V. McLaren (Eds.), *Secure Foundations: Key Issues in Crime Prevention, Crime Reduction and Community Safety*. Institute of Public Policy Research.

Ekblom, P. (2003b). Organised Crime and the Conjunction of Criminal Opportunity Framework. In A. Edwards & P. Gill (Eds.), *Transnational Organised Crime: Perspectives on Global Security*. Routledge.

Ekblom, P. (2011a). *Crime Prevention, Security and Community Safety Using the 5Is Framework*. Palgrave Macmillan.

Ekblom, P. (2011b). Deconstructing CPTED… and Reconstructing It for Practice. Knowledge management and research. *European Journal on Criminal Policy and Research, 17*(1), 7–28.

Ekblom, P. (2012). *Design Against Crime: Crime Proofing Everyday Products*. Lynne-Reiner.

Ekblom, P., & Tilley, N. (2000). Going Equipped: Criminology, Situational Crime Prevention and the Resourceful Offender. *British Journal of Criminology, 40*(3), 375–398.

Felson, M. (1986). Routine Activities, Social Controls, Rational Decisions, and Criminal Outcomes. In D. Cornish & R. V. G. Clarke (Eds.), *The Reasoning Criminal*. Springer-Verlag.

Felson, M. (2002). *Crime and Everyday Life* (3rd ed.). Sage.

Felson, M., & Clarke, R. (1998). *Opportunity Makes the Thief: Practical Theory for Crime Prevention, Police Research Series Paper 98*. Home Office.

Forrester, D., Chatterton, M., & Pease, K. (1988). *The Kirkholt Burglary Prevention Project*, Rochdale (Crime Prevention Unit: Paper 13). Home Office.

Frank, R., Dabbaghian, V., Reid, A. A., Singh, S. K., Cinnamon, J., & Brantingham, P. (2011). Power of Criminal Attractors: Modelling the Pull of Activity Nodes. *Journal of Artificial Societies and Social Simulation, 14*(1), 6.

Gabor, T., & Gottheil, E. (1984). Offender Characteristics and Spatial Mobility: An Empirical Study and Some Policy Implications. *Canadian Journal of Criminology, 26*, 267–281.

Garside, R. (2004). *Crime, Persistent Offenders and the Justice Gap*. Crime and Society Foundation.

Gill, M. (2000). *Commercial Robbery*. Blackstone.

Gill, M. (2001). The Craft of Robbers of Cash-in-Transit Vans: Crime Facilitators and the Entrepreneurial Approach. *Internation/al Journal of the Sociology of Law, 29*(3), 277–291.

Gladwell, M. (2000). *The Tipping Point, How Little Things Can Make a Big Difference*. Little Brown.

Glensor, R. (1996). Community Problem Solving in the USA: A New Synergy. In *Focus on Police Research and Development, No. 7* (pp. 14–15). Home Office Police Research Group.

Goldstein. (1979, April). Improving Policing: A Problem-Oriented Approach. *Crime and Delinquency* (pp. 235–58).

Goldstein, H. (1990). *Problem-Oriented Policing*. McGraw-Hill.

Goldstein, H. (1997). The Pattern of Emerging Tactics for Shifting Ownership of Prevention Strategies in the Wave of Change in Policing: Their Implications for both Environmental Criminology and the Police. Paper presented at the sixth *Annual Seminar on Environmental Criminology and Crime Analysis*. National Police Academy, Oslo.

Groff, E. R. (2005). *The Geography of Juvenile Crime Place Trajectories* (Masters Thesis). University of Maryland.

Guerette, R. T., & Bowers, K. (2009). Assessing the Extent of Crime Displacement and Diffusion of Benefits: A Review of Situational Crime Prevention Evaluations. *Criminology, 47*(4).

Hesseling, R. (1994) Displacement: A Review of the Empirical Literature. In *Crime Prevention Studies* (Vol. 3). Criminal Justice Press.

Hesseling, R. and Aron, U. (1995). *Car Burglaries: Reduction or Displacement? The Results of a Rotterdam Campaign to Curb Thefts from Cars*. Arnhem: Gouda Quint Bv.

Hill, I., & Pease, K. (2000). The 'Wicked' Issues: Displacement and Sustainability. In S. Ballantyne, et al. (Eds.), *Secure Foundations: Key Issues in Crime Prevention, Crime Reduction and Community Safety*. Institute for Public Policy Research.

Hillier, B. (1999). *Space is the Machine: A Configurational Theory of Architecture*. Cambridge University Press.

Hillier, B. (2004). Can Streets be Made Safer? *Urban Design International, 9*, 31–45.

Hillier, B., & Hanson, J. (1984). *The Social Logic of Space*. Cambridge University Press.

Hillier, B., & Sahbaz, O. (2005). High Resolution Analysis of Crime Patterns in Urban Street Networks: An Initial Statistical Sketch from an Ongoing Study of a London Borough. In *Proceedings of the Fifth International Space Syntax Symposium*, held in Delft, The Netherlands.

Hillier, B., & Shu, S. (2000). Crime and Urban Layout: The Need for Evidence. In S. Ballintyne, K. Pease, & V. McLaren (Eds.), *Secure Foundations: Key Issues in Crime Prevention, Crime Reduction and Community Safety*. Institute of Public Policy Research.

Hinkle, J.C., Weisburd, D. Cody W., Telep, C.W. & Kevin Petersen, K. (2020). *Problem-oriented policing for reducing crime and disorder: An updated systematic review and meta-analysis*. First published: 15 June 2020. https://doi.org/10.1002/cl2.1089

Holt, T. (2000). *Baits His Hook and Takes Your Cash*. Greater Manchester Police: Unpublished Report.

Holt, T., & Spencer, J. (2005). *A Little Yellow Box: The Targeting of Automatic Teller Machines as a Strategy in Reducing Street Robbery*. Perpetuity Press.

Home Office Standing Conference on Crime Prevention. (1991). *Safer Communities: The Local Delivery of Crime Prevention Through the Partnership Approach.* Home Office.

Houghton, G. (1992). *Car Crime in England and Wales: The Home Office Car Theft Index; Crime Prevention Unit Paper 33.* Home Office.

Information Commissioner's Office. (ICO–2008). *CCTV Code of Practice.* Office of the Information Commissioner.

Intergovernmental Panel on Climate Change. (2008). *Climate Change: Synthesis Report. Summary for Policymakers.* IPCC.

International Energy Agency. (2007). *CO2 Emissions from Fuel Combustion 1971–2005.* IEA.

Jeffery, C. R. (1971). *Crime Prevention Through Environmental Design.* Sage.

Johnson, S. D., & Bowers, K. J. (2010). Permeability and Burglary Risk: Are Culs-de-Sac Safer? *Journal of Quantitative Criminology, 26*(1), 89–111.

Johnson, S. D., Guerette, R. T., & Bowers, K. (2012). Crime Displacement and Diffusion of Benefits. In B. C. Welsh & D. P. Farrington (Eds.), *The Oxford Handbook of Crime Prevention* (pp. 337–353). Oxford University Press.

Johnson, S. D., Guerette, R. T., & Bowers, K. (2014). Crime Displacement: What We Know, What We Don't Know, and What It Means for Crime Reduction. *Journal of Experimental Criminology, 10*(4), 549–571.

Jones, A., Valero-Silva, N., & Lucas, D. (2016). *The Effects of 'Secure Warm Modern' Homes in Nottingham: Decent Homes Impact Study.* Nottingham City Homes.

Kershaw, C., Nicholas, S., & Walker, A. (2008). Crime in England and Wales 2008/8. *Home Office Statistical Bulletin 07/08.*

Kirby, K. & Reed, D. (2004). *Developing a Problem-oriented Organisation.* London: Association of Chief Police Officers.

Kelling, G., & Wilson, J. (1982). Broken windows: The police and neighbourhood safety. *The Atlantic Monthly* (pp. 29–38). Washington DC.

Lancaster, D. (2010). *Carbon Reduction and Cost Efficiency Saving Model.* Secured by Design.

Landman, K. & Liebermann, S. (2005). Planning against crime: Preventing crime with people not barriers. *SA Crime Quarterly* No. *11*. Pretoria. Institute for Security Studies.

Laycock, G. (1985). *Property Marking: A Deterrent to Domestic Burglary?* Crime Prevention Unit Paper 3. Home Office.

Laycock, G. (1997). Operation Identification, or the Power of Publicity? In R. Clarke (Ed.), *Situational Crime Prevention: Successful Case Studies* (2nd ed.). Harrow and Heston.

Leukfeldt, E.R. & Yar, M. (2016). Applying Routine Activity Theory to Cybercrime: A Theoretical and Empirical Analysis, *Deviant Behavior, 37*(3), 263–280. https://doi.org/10.1080/01639625.2015.1012409

Lindsey, B., & McGillis, D. (1988). Citywide Community Crime Prevention: An Assessment of the Seattle Program. In D. P. Rosenbaum (Ed.), *Community Crime Prevention: Does It Work?* Sage.

Lynch, K. (1960). *The Image of the City.* MIT Press.

Madensen, T. and Eck, J. (2008). Violence in Bars: Exploring the Impact of Place Manager Decision-Making. *Crime Prev Community Saf 10*, pp. 111–125. https://doi.org/10.1057/cpcs.2008.2

Mara, F. and Cutini, V. (2023). Modelling Crime Dynamics and Risky Places: A Basic Combination of Crime Theories and Space Syntax to Evaluate Spatial Crime Impedance. In Gervasi, O., et al. *Computational Science and Its Applications*—ICCSA 2023 Workshops. ICCSA 2023. Lecture Notes in Computer Science, vol 14110. Springer, Cham. https://doi.org/10.1007/978-3-031-37123-3_18

Mativat, F., & Tremblay, P. (1997). Counterfeiting Credit Cards: Displacement Effects, Suitable Offenders and Crime Wave Patterns. *British Journal of Criminology, 37*, 165–183.

Mawby, R. (2016). Policing Policies in Cities Dependent on Tourism: Developing an Evidence-Based Approach. *International Journal of Police Science and Management, 18*(3), 195–205.

Mayhew, P., Clarke, R., Sturman, A., & Hough, J. (1976). *Crime as Opportunity* (Home Office Research Study 34). HMSO.

MOPAC—Mayor's Office for Policing and Crime. (2023). *The London Victim Attrition Review*. London City Hall. Accessed online 12 May 2025.

Morgan, J. (1991). *Safer Communities: The Local Delivery of Crime Prevention Through the Partnership Approach*. Home Office.

Morrison, S. A., & O'Donnell, I. (1996). *Armed Robbery: A Study in London* (Occasional paper no. 15). Oxford: Centre for Criminological Research.

Moss, K., & Pease, K. (1999). Crime and Disorder Act 1998, Section 17: A Wolf in Sheep's Clothing. *Crime Prevention and Community Safety, 1*(4), 15–19.

Nettle, D., Nott, K. and Bateson, M. (2012). 'Cycle Thieves, We Are Watching You': Impact of a Simple Signage Intervention against Bicycle Theft. PloS one, 2012—Journals.plos.org. https://doi.org/10.1371/journal.pone.0051738

Newman, O. (1972). *Defensible Space: Crime Prevention Through Urban Design*. Macmillan.

Newman, O. (1973). *Defensible Space: People and Design in the Violent City*. Architectural Press.

Newman, O. (1996). *Creating Defensible Space*. U.S. Department of Housing and Urban Development.

Nubani, L., & Wineman, J. (2005). The Role of Space Syntax in Identifying the Relationship Between Space and Crime. In A. van Nes (Ed.), *Proceedings of the Fifth International Space Syntax Symposium* (pp. 413–422). The Netherlands.

Pascoe, T. (1996). *Predicting the targets of household burglars—developing risk-based models from a case study of recent experience in Redditch, United Kingdom* (PhD Thesis). Cranfield University School of Management. Accessible at: www.vellig-ontwerp-beheer.nl/publicaties/predicting-the-targets-of-household-burglars

Pascoe, T. (1999). *Evaluation of Secured by Design in Public Sector Housing—Final Report*. Building Research Establishment.

Pease, K. (1991). The Kirkholt Project: Preventing Burglary on a British Housing Estate. *The Security Journal, 2*, 73–77.

Pease, K. and Farrell, G. (2011) 'Climate Change and Crime.' *European Journal on Criminal Policy and Research 17* (2): pp. 149-62.

Pease, K., & Farrell, G. (2009). *The Carbon Cost of Crime and Its Implications*. Secured by Design.

Philips, P. D. (1980). Characteristics and Typology of the Journey to Crime. In D. E. George-Abeyie & K. D. Harris (Eds.), *Crime: A Spatial Perspective*. Columbia University Press.

Pyle, G. (1974). *'The Spatial Dynamics of Crime', Department of Geography Research Paper 159.* University of Chicago Press.
Rapoport, A. (1982). *The Meaning of the Built Environment. A Nonverbal Communication Approach.* Sage.
Read, T., & Tilley, N. (2000). *Not Rocket Science? Problem-solving and Crime Reduction* (Crime Reduction Research Series 6). Home Office.
Rengert, G. (1990). *Drug Purchasing as a Routine Activity of Drug Dependent Property Criminals and the Spatial Concentration of Crime,* paper presented at the American Society of Criminology 1990 Annual Conference.
Rengert, G. F. (2004). The Journey to Crime. In G. Bruinsma, H. Elffers, & J. D. Keijser (Eds.), *Punishment, Places and Perpetrators: Developments in Criminology and Criminal Justice Research.* Willan.
Rengert, G. F., & Wasilchick, J. (1985). *Suburban Burglary: A Time and a Place for Everything.* Charles C. Thomas.
Reynald, D. M. (2009). Guardianship in Action: Developing a New Tool for Measurement. *Crime Prevention and Community Safety: An International Journal, 11*(1), 1–20.
Rhodes, W. M., & Conly, C. C. (1981). Crime and Mobility. In P. J. Brantingham & P. L. Brantingham (Eds.), *Environmental Criminology.* Sage.
Schneider, A. (1988). Neighborhood-Based Antiburglary Strategies: An Analysis of Public and Private Benefits from the Portland Program. In D. P. Rosenbaum (Ed.), *Community Crime Prevention: Does It Work?* Sage.
Schneider, R., & Kitchen, T. (2002). *Planning for Crime Prevention: A Transatlantic Perspective.* Routledge.
Scott, M. S. (2018). *How Can We Collaborate More Effectively?* Presentation given at Problem Solving Conference hosted by Durham Constabulary in the City of Durham on 26 February 2018.
Secured by Design. (1989). *Secured by Design—The Concept,* South East Region Senior Crime Prevention Officers Conference. Rea Publicity.
Secured by Design. (2009). *The Carbon Cost of Crime and your Home.* Secured by Design.
Secured by Design. (2010a). *The Capital Costs of Secured by Design accreditation, by David Langdon Associates.* Secured by Design.
Secured by Design. (2010b). *Design and Access Statements: How to Use Them to Prevent Crime.* Secured by Design.
Shaftoe, H., & James, S. (2004). *Do Symbolic Barriers Prevent Crime and Offer an Increased Sense of Safety?* Routledge.
Sherman, L. (2011). The Pennsylvania School of Criminology: Building Tipping Points for a Discipline. in *The Origins of American Criminology.* Cullen, F.T., Lero Jonson, C., Myer, A.J. and Adler, F. (eds.) Routledge.
Sherman, L., Shaw, J., & Rogan, D. (1995). *The Kansas City Gun Experiment, Research in Brief Series.* National Institute of Justice.
Shu, S., & Huang, J. (2003). Spatial Configuration and Vulnerability of Residential Burglary: A Case Study of a City in Taiwan. In *Proceedings of the 4th International Space Syntax Symposium.* London.
Skinner, B. F. (1953). *Science and Human Behavior.* Macmillan.

Skudder, H., Druckman, A., Cole, J., McInnes, A., Brunton- Smith, I., & Ansaloni, G. P. (2016). Addressing the Carbon-Crime Blind Spot: A Carbon Footprint Approach. *Journal of Industrial Ecology*. https://doi.org/10.1111/jiec.12457

Smith, M.P. and Clarke, R.V. (2000). 'Crime and Public Transport.' *Crime and Justice* Vol. 27: University of Chicago Press.

Smith, M., Clarke, R. V., & Pease, K. (2002). Anticipatory Benefits in Crime Prevention. *Crime Prevention Studies, 13*, 71–88. Criminal Justice Press.

Spickenheuer, J. (1983). untranslated, Voetsurveillance en Preventie- voorlichting in Amsterdam Osdorp: een Onderzoek naar de Effecten op de Criminaliteitsontwikkeling, Gravenhage, Netherlands: Staatsuitgeverij.

Stern, N. (2007). *The Economics of Climate Change*. Cambridge University Press.

Stokes, M. (2022). *An Investigation of the Sustainability of Crime Prevention in the Built Environment: Impact and Implementation Factors* (Published Doctoral Thesis). University of Huddersfield.

Summers, L., & Johnson, S. D. (2017). Does the Configuration of the Street Network Influence Where Outdoor Serious Violence Takes Place? Using Space Syntax to Test Crime Pattern Theory. *J Quant Criminology* 33, 397–420. https://doi.org/10.1007/s10940-016-9306-9.

Taylor, L. (1973). The Meaning of the Environment. In C. Ward (Ed.), *Vandalism*. Architectural Press.

Teedon, P., Reid, T., Griffiths, P., Lindsay, K., Glen, S., McFayden, A., & Cruz, P. (2009). *Secured by Design Impact Evaluation Final Report*. Glasgow Caledonian University.

Telep, C. W., Mitchell, R. J., & Weisburd, D. (2014). How Much Time Should the Police Spend at Crime Hot Spots? Answers from a Police Agency Directed Randomized Field Trial in Sacramento, California. *Justice Quarterly, 31*, 905–933.

Tilley, N. (2009). *Crime Prevention*. Willan.

Tilley, N. and Farrell, G. (2013). The Reasoning Criminologist: Essays in Honour of Ronald V. Clarke. Nick Tilley and Graham Farrel editors. Routledge.

Tilley, N., Pease, K., Hough, M., & Brown, R. (1999). *Burglary Prevention: Early Lessons from the Crime Reduction Programme, Policing and Reducing Crime Unit, Crime Reduction Research Series Paper 1*. Home Office.

Weideman, B. P., Thrane, M., Christensen, P., Schmidt, J., & Lokke, S. (2008). Carbon Footprint: A Catalyst for Life Cycle Assessment. *Journal of Industrial Ecology, 12*(1), 3–6.

Weisburd, D., Wyckoff, L., Ready, J., Eck, J. E., Hinkle, J. C., & Gajewski, F. (2006). Does Crime Just Move Around the Corner? A Controlled Study of Spatial Displacement and Diffusion of Crime Control Benefits. *Criminology, 44*, 549–591.

White, R. C. (1932). The Relations of Felonies to Environmental Factors in Indianapolis. *Social Forces, 10*, 498–509.

Wikstrom, P.-O. (2006). Individuals, Settings, and Acts of Crime: Situational Mechanisms and the Explanation of Crime. In P.-O. Wikstrom & R. Sampson (Eds.), *The Explanation of Crime: Context, Mechanisms and Development*. University Press.

Wilcox, P. and Cullen, F.T. (2018). Situational Opportunity Theories of Crime. *Annual Review of Criminology* 1, 123–148. (Volume publication date January 2018)

Wiles, P., & Costello, A. (2000). *The 'Road to Nowhere': The Evidence for Travelling Criminals* (Home Office Research Study 207). Home Office Research, Development and Statistics Directorate.

Willis, K. (2006). Armed Robbery: Who Commits It and Why? *Trends and Issues in Crime and Criminal Justice* No. 328. Australian Institute of Criminology.

Wortley, R. (1996). 'Guilt, Shame and Situational Crime Prevention', in R. Homel (ed.) *The Politics and Practice of Situational Crime Prevention.* Crime Prevention Studies, 5. Monsey: Criminal Justice Press.

Wortley, R. (2001). 'A Classification of Techniques for Controlling Situational Precipitators of Crime', *Security Journal, 14*: 63–82.

Zimbardo, P. (1973). A Field Experiment in Auto Shaping. In C. Ward, (Ed.) *Vandalism.* Architectural Press.

Zimring, F. (1972). The Medium is the Message: Firearm Caliber as a Determinant of Death from Assault. *Journal of Legal Studies.*

3 Delivery of the Designing Out Crime Function

3.1 Reading Plans

The receipt of plans for a proposed development can cause even the most experienced DOCO's/CPTED Officer's/Adviser's heart to sink. The pack of plans can seem inordinately thick and heavy. Where it concerns a large-scale development, this package might even be several inches thick. And just because plans are slowly being transferred into electronic format and submission, don't expect any less of a shock when it comes to unzipping the attachment and realising the extent of the designs contained therein—or how just one plan consumes a massive amount of memory space. However, even enormous sized developments and the accompanying plans can be quickly and efficiently cut down to a manageable size.

3.1.1 First Actions on Receipt

On first receipt of the plans the primary task should be to log their arrival (see Sect. 1.5.12. Administration System). At the very least a single page entry (either as a hard copy or in a computer log) should be created and the development registered in the index. UK police-employed DOCOs should register the development on the SBD registration website.

Thereafter, it is time to reduce the plans to a realistic number. The key ones relate to the site layout and 'footprint' of the proposed development. Very often the remainder are supernumerary to the DOC process. However, front, side and rear elevations (views) assist in conceptualising the finished building(s) in the DOCO's mind. Lower ground, first floor, second floor etc., may be important

M. Stokes, *Designing Out Crime*, Crime Prevention and Security Management,
https://doi.org/10.1007/978-3-031-99840-9_3

where specific, risk-attached functions are designed to take place at that location. For example: a pharmacy (storage and dispensing of drugs) in a hospital, healthcare centre or doctors' surgery. Or an ICT (information communication technology) classroom and/or storeroom for computers and associated hardware on school, college or university premises. Or where such floor levels provide additional external access and exit (including emergency exits).

This is not to say that those other than the crucial plans detailed above should be discarded. Keep the whole package, but separate and store in an 'active' file, the site plan, types previously listed and those associated elements that assist in interpreting the development (e.g. a landscaping, Lighting or CCTV plan).

3.1.2 Reading the Plans

Once again, there is no great mystery in reading plans although to the uninitiated it may cause a degree of apprehension. For the most part, plans continue to be a two-dimensional guide to the eventual layout and design of the completed development. However, computer aided design (CAD) has brought both colour and three-dimensional perspectives to the way in which a development will ultimately appear once constructed, courtesy of 'flythroughs' and multiple viewing angles.

Where there is either a single combined site and development plan, or separate versions of each, they will almost without exception be shown from above—in effect a 'bird's eye view'. The obvious parallel to be drawn here is with conventional maps and indeed, if you can read a map, it is in effect a simple transfer of skills to read a plan and understand what the Architect(s) is intending to create. It is sometimes (though by no means always) the practice for the plan drawer to supply a 'key' to the plans that explains the meaning of specific lines, symbols and forms of black or grey shading—although and as previously mentioned, different colours are increasingly becoming the norm when plans are initially produced.

Most importantly, this is where the Layered Security ('Onion Peeling') process begins. Because, the easiest way to interpret and understand plans (and deliver recommendations), is to start from outside the development and move inwards. And by 'outside' we mean the area surrounding the whole site, its Geographical Juxtaposition (Newman, 1972, 1973; Cozens et al., 2019; and Sect. 4.4.5). For example, on a housing development it is the land and/or existing properties beyond the proposed development that are first scrutinised, then the development perimeter, followed by: conduits for public and private access; boundaries of each plot; space between that boundary and the building and, finally, the building/target itself.

Ideally, the process of 'reading' plans should be undertaken before any meeting with the architect(s), developer, project manager etc. That way the DOCO has time to formulate her or his questions, proposed recommendations and those issues that appear in advance to be mandatory requirements from the perspective of DOC and SBD. Furthermore, her or his own local knowledge of the area in question may be invaluable. One word of caution at this stage. The increasing submission of plans online is generally appreciated as a positive development. However, where the DOCO only has access to a small computer monitor or cannot print off a hard copy in anything larger than A3 size, they are advised to request a hard copy of the relevant plans (especially the overall site plan) in order to examine the granular detail that might otherwise be missed.

3.1.3 Danger Signals

In the same way that a map can disclose danger signals (such as mine-shafts, cliffs and power lines), a plan can also alert the DOCO to potential Crime Attractors/Crime Generators/Crime Enablers (Brantingham and Brantingham, 1995) FOC and ASB. Be especially aware of: uncontrolled public seating (where no one has ownership of the bench or similar); expanses of vegetation (blocking Natural Surveillance and Technological Surveillance opportunities—e.g. by CCTV cameras); footpath links between Culs-de-Sac ('Leaking' Culs-de-Sac—see Sect. 6.2); rear access footpaths ('Radburn' footpaths), rear parking areas and pram stores, residents' gardens and other Activity Nodes (Eck et al., 2005) that encourage anti-social elements to gather and cause distress to residents, shoppers etc.). Often such design features are added to plans in order to demonstrate community cohesion, yet in the short-term they can act as magnets/generators of ASB and worse.

'Water features' should attract special scrutiny. Ponds in a residents' garden possess inherent danger for young children and easily become a dumping ground for shopping trolleys, wheelie bins and all manner of rubbish. However, the world-wide hike in the value of metal had the unintended consequence of clearing the environment of scrap metal items. Nor was this solely a feature in the UK, as witnesses from around the globe have testified.

Meanwhile, in the UK canals (man-made inland waterways) have correctly morphed from being perceived as conduits for criminal behaviour, into desirable leisure and residential frontages. Historically, housing and the majority of commercial buildings in urban areas would only ever back onto canals. Whereas, most often they now face such waterways. Unfortunately, some criminals have managed to combine the two and continue to regard canals as useful Attack and Escape Routes for offences like burglary (see Sect. 4.4.6, Permeability).

Of even greater importance, where family housing faces onto canals, rivers and ponds, due consideration must be given to the aforementioned safety of young children living in close proximity to such water features. Fencing between the water and a public footpath will be a justifiably necessary recommendation on the part of the DOCO—where it does not already appear on the plans.

Another potential danger signal or Environmental Cue (see Sect. 2.8), concerns unnecessary footpaths—especially where there is an indication that such footpaths will be lit. The purpose of such a footpath should be explained fully (by the architect or client) to the DOCO and where it is isolated (to the extent that people will only use it during the hours of darkness if it is lit) counsel against it being illuminated strongly given. The rationale being, there is little point in encouraging people to put themselves in danger in an isolated environment through the use of street lighting (see Sect. 4.3.1).

Furthermore, considerable anecdotal evidence suggests that such footpaths generate FOC and once again, act as Activity Nodes for ASB. See also Beckford et al. (1996); Sect. 6.3; Alleygating; and Sect. 6.4, Closure of Rights of Way on Grounds of Crime and Public Space Protection Orders.

Which brings us to the issue of 'Leaking' Culs-de-Sac (see Sect. 6.2). There can be few if any DOCOs who do not believe that 'Leaking' Culs-de-Sac facilitate crime. Indeed, they are a menace that has provided Attack and Escape Routes for burglars, car thieves and other criminals for many decades past. Yet, most often they are rarely used by the residents they were designed to assist. They simply appeared a pleasant design feature to be included in the original master-plan.

Unfortunately, the issue of'Leaking' Culs-de-Sac is tied with that of Permeability and in recent years, the desire to produce Sustainable Transport Links. Few would doubt the validity of facilitating ease of access for residents to shops, schools, bus stops and other social facilities. But where such routes can be confidently predicted (for example, by means of local case examples, existing levels of crime and evidence from Crime Science) to produce crime, FOC and ASB, they should be opposed—especially where there is little evidence to suggest they will actually be/or are being used. Furthermore, in law it is extremely difficult to close off an existing Public Right of Way on grounds of crime, once it has become established (see Sect. 6.4).

In conclusion, plans are an important and advanced indication of how the completed development will manifest itself in physical form, once it has been built. They can also alert the DOCO as to problem areas in terms of crime, FOC and ASB. And in pursuit of the SBD award (see Chapter 5), they also provide the necessary basis on which recommendations can be made to design out or at least drastically minimise such negative consequences.

3.2 Health and Safety of the DOCO/CPTED Officer When Visiting Sites

Before conducting the actual site survey/inspection (the subject of the next chapter), it is timely to consider Health and Safety issues in this regard. Many will no doubt groan at the very mention of this increasingly important and litigious issue. But common sense dictates that construction sites are inherently dangerous places (witness in the UK increasing scrutiny by the Health and Safety Executive). As a result, the Health and Safety of the individual DOCO, must be paramount both when considering and during such a visit.

Consequently, in this context, the mnemonic 'RARA' may be useful: 'Remove the risk; Avoid the risk; Reduce the risk and Accept the risk'.

The managers of construction sites have a 'Duty of Care' and are legally bound by Sects. 3 and 4 of the Health and Safety at Work Act, 1974. Further requirements under the Management of Health and Safety at Work Regulations 1992 and the Occupiers' Liability Acts, 1957 and 1984 (see Sect. 8.1, Legal Liability) exist and these insist that all staff and bona fide visitors (like the DOCO) are provided with the necessary safety equipment—hard hat, fluorescent jacket and approved footwear being the normal minimum requirements.

However, it naturally follows that the DOCO who arrives on-site already in possession of his or her own personal protective equipment (PPE), exudes a more professional demeanour. And in this context, who wants to borrow and wear a pair of manky wellington boots and hard hat that may have been worn by countless others?

Furthermore, since the UK police service came under the full scope of the Health and Safety at Work Act, 1974, they have been under a legal obligation to provide the following suggested list of equipment in respect of such site surveys:

- Safety Shoes (in lieu of public order boots where issued)
- Safety Wellingtons
- Safety Helmet

- Fluorescent Jacket
- Gloves
- Goggles/Protective Eyewear

In the UK, Health and Safety is addressed during initial training for the role of DOCO. Personal protective equipment (PPE) is covered, as are and also the concepts of Strategic and Dynamic Risk Assessments.

In essence, a Strategic Risk Assessment should be in place prior to any visit by the DOCO to construction sites, commercial premises and other locations where potentially dangerous activities take place or hazardous materials are stored or used. The Dynamic Risk Assessment is what we should all conduct immediately prior to entering such an environment—and during the course of that inspection process, if conditions change. For example, when conducting a survey of a construction site, should an incident occur that suggests conditions have changed (perhaps, suddenly and without warning scaffolding comes loose), this may cause the DOCO to remove himself and where necessary, others from the scene.

It is important to recognise that this is no simplistic risk-aversion exercise, as the lives of the individual *DOCO* and others might be under threat if no such action is taken. Indeed, there is also a responsibility (legal and moral) on the part of the *DOCO* for their own personal safety and training—including possession of the correct PPE. This issue is especially pertinent where the site in question has yet to be formally occupied by the construction company, or when they have left. We should never presume such a site is automatically safe. That said, the overwhelming majority of *Site Surveys and Environment Visual Audits (EVAs)* will be conducted in environments where there is little if any perceived risk. The skill is to routinely make this differentiation and undertake the inspection accordingly. As an example of the seriousness now afforded to Health and Safety by the construction industry, on the Queen Elizabeth Line across London workforce diversity meant that such notices were produced in 46 different languages!

Finally, lone worker and personal safety guidance should exist for when the *DOCO* is visiting such construction sites and other environments. The *DOCO* has a responsibility to ensure that her or his colleagues and manager know their intended destination and the proposed duration of their visit. They should also be in possession of a mobile phone and/or police radio.

Assets and/or function requiring protection

3.3 Site Surveys and Environmental Visual Audits (EVAs)

In the mists of time (that is to say, as recently as the mid-1990s), the following practice was not unknown in the UK. Namely, a distinct minority of Crime Prevention Design Advisers and Architectural Liaison Officers (CPDAs/ALOs) gave their recommendations and certified the development as meeting the criteria for a Secured by Design award (see Chapter 5 below), without ever visiting the site! In effect, they took the builder and developer at their word and consequently, became the ultimate hostages to fortune once the development was completed and occupied.

Animosities often begin when something subsequently goes wrong. Indeed, this author can testify to the blame game that ensues after a particular development is subject to criminal attack. When it is a police building where key recommendations made by the DOCO were never implemented, reports from around the world indicate that recriminations may fester for many years. Furthermore, the absence of site visits and surveys demonstrates that at best, the full professional DOC service was not being delivered.

Such inconsistencies aside, as previously indicated there is no great mystery about the relevance of, or the way in which such a survey or inspection should be conducted—and thereafter put into writing (see the following Sect. 3.4, Writing the Report/EVA: the 'Crown Jewels' of its Written Presentation).

As previously described, the easiest means of conducting the site survey or inspection is to follow the classic Layered Security approach (from the outside inwards). However, there is a fundamental advantage in doing this whilst already aware of what will be stored, the tasks undertaken and/or what needs to be protected within the premises. For example, all too obviously far higher levels of security will be necessary at a storage facility for drugs, firearms or cash, compared to ordinary residential premises.

To put the Layered Security ('Onion Peeling') approach in perspective (like statement taking—one of the foremost skills learnt by police officers, amongst many others), it in effect tells a story and therefore will have a beginning, middle and an end. Consequently, in ideal circumstances and as a general guide, the survey will most probably take the following order of assessment:

- Knowledge of what needs to be protected at the site
- Crime Pattern Analysis/Crime Impact Statement and prior research
- Approach to the site
- Site perimeter
- Grounds of the development
- Physical building
- Tasks to be undertaken therein and/or the property to be stored/protected (in effect, a return to the first issue albeit under greater scrutiny).

3.3.1 Knowledge of What Needs to be Protected at the Site in Question

Before considering the level(s) of security that will be required at a given site, it is imperative to know in advance what needs to be protected at that location, what will be stored there and what functions will take place. For the majority of instances, this will be little more than the domestic arrangements that exist in respect of the average home/dwelling. However, where the premises in question are commercial, or a school, hospital etc., it will be obvious in advance of conducting the site survey that increased levels of security will be necessary. Nevertheless, the DOCO must still ask about and confirm such expected practices—as well as establishing whether there are any additional, risk-associated functions.

3.3.2 Crime Pattern Analysis/Crime Profile and Crime Impact Statements

Crime Pattern Analysis, Crime Profile or the Crime Impact Statement should be perceived as the first part of the Layered Security approach. Wherever possible, this will be addressed before both the site survey and the meeting with the architect(s), developer etc. Its relevance to a development on a 'green field' site will be far less than with a refurbishment project—where the aim is to dramatically reduce existing crime, FOC and ASB, if taking place at that location. Moreover, in both instances, the possession of crime analysis in advance mirrors the old adage, being forewarned is forearmed.

In the Greater Manchester Police area, Design for Security Consultants (formerly known as Architectural Liaison Officers) systematically produce a Crime Impact Statement in respect of all such DOC projects, utilising existing crime data to compile this report. In addition, the Crime Impact Statement includes local crime trends and details regarding history of the site (where known). Similarly, where insufficient information has been supplied by the architect(s), developer etc., it may be necessary to conduct basic research regarding the proposed project, the location in question and the function of the building.

3.3.3 Approach to the Site

Very much the precursor to the site survey proper, the journey to the location of the proposed development will almost certainly provide Environmental Cues (see Sect. 2.8). Such cues will be particularly relevant in the immediate vicinity of the proposed development. For example: graffiti and other forms of criminal damage; the remains of 'torched' (burnt out) vehicles; evidence of unofficial rights of way in existence (broken down fencing, vegetation trampled into the dirt) and the potential for Natural Surveillance (see Sect. 4.4.2) from existing properties that overlook the development.

3.3.4 Site Perimeter

Is there a clearly defined perimeter to the proposed development, one that accords with the site plan already (hopefully) in your possession? What form does it take: wall, fencing, gates, hedge, trip-rail, no physical perimeter as such or a combination of these? If there is a physical perimeter, to what extent does it provide for Defensible Space (see Sect. 4.4.1) within the grounds of the proposed development? Will it prevent casual intrusion or does it need to be enhanced (raised in height, strengthened) or replaced? Are there special points of weakness (such as the aforementioned unofficial rights of way) that will almost certainly require specific measures/attention?

The relevance of this issue cannot be over-stressed, because it mirrors the thinking of the offender's mind—be he professional/career/travelling or opportunist (see Sect. 2.3 Crime Opportunity). His (and remember, world-wide these are crimes overwhelmingly committed by males) decision-making process, to target and break into (burgle) a particular building, will almost always **NOT** be made when he gets to the doors, windows or fabric of the building. Instead, he will most often make that decision before approaching the premises and will be guided by a number of key perceptions—Pascoe (1993a); Pascoe (1993b); Pascoe and Topping (1997):

- Is there something in the premises worth stealing?
- Does the perimeter security (see Sect. 4.2.3)—walls, fencing, hedges, gates etc. limit his potential Attack and Escape Routes?
- Are the premises occupied?
- Are there sufficient opportunities for anonymity: can he access and exit without being seen, or worse still, being apprehended? Incidentally, a strong body of academic opinion contends that the fear of being apprehended rarely features in a criminal's decision-making process. See Pascoe, 1993a; Nee and Meenaghan (2006); and Armitage (2017).

Key Issue

It is of fundamental importance that the DOCO understands the decision-making processes of virtually all offenders' intent on committing some form of property crime. For example, in the instance of residential burglary (house-breaking), Physical Security 'Target Hardening' Measures and Technological Innovations (see Sects. 4.2.1–4.3.3) of the building's structure are very much the last line of defence—be they stronger doors, better quality locks, laminated glass, intruder alarm, CCTV etc. Moreover, perhaps of greatest relevance, the offender will often be unable to identify such protection when he is 'casing the joint' (conducting his own pre-crime survey). See also Sect. 2.4, Rational Choice Theory.

3.3.5 Grounds of the Development

Once within the curtilage of the premises (in effect, the grounds between the perimeter and the proposed building, the DOCO should be looking for other special factors. For example: are trees and overgrown vegetation a barrier to Natural Surveillance? Is the development overlooked from surrounding properties or from the street? Does a change in gradients cause particular problems on this site (assisting trespassers and criminals to vault over walls, fences, hedges etc. and/or compromise potential Defensible Space.

All of the above issues could be said to rest with the subjective interpretation of the Environmental Elements of Crime Prevention Through Environmental Design (CPTED—see Sects. 4.4.1–4.4.6). Subjective, because they require the DOCO to apply his or her knowledge of issues including Defensible Space, Natural Surveillance, Territoriality and Permeability to the proposed development, whilst conscious of existing Crime Pattern Analysis/Crime Profile data and any other factors considered relevant.

3.3.6 The Physical Building

This is where the majority of the Physical Security 'Target Hardening' Measures and Technological Innovations (see Sects. 4.2.1–4.3.3) elements come into play. These often fall into the category of 'tick the box' requirements, but notice how by following the Layered Security process, they take place after consideration has first been given to the Environmental Elements.

In reality, an interplay will exist (especially in respect of the recommendations that are ultimately made), between the Environmental elements, Technological Innovations and Physical Security 'Target Hardening' Measures, mindful of the property being stored or protected and/or the activities being executed at the location. The important point is NOT to begin with the 'tick the box' physical security considerations, which may change depending upon your capacity to influence/receptiveness of the architect, planner, developer etc., in respect of the Environmental Elements. And taken together, they will influence your calculation of Commensurate Risk (see Sect. 1.5.15) which will in turn impact on the level of physical security measures required.

With refurbishment projects there will obviously be a building(s) to survey from the outset. In many ways this means it is far easier to make recommendations in respect of Physical Security 'Target Hardening' Measures and Technical Innovations—with evidence of past failings that caused crime, FOC and ASB often being all too apparent (visually, in the crime statistics, or through resident/occupier/user community consultation). However, the downside will almost certainly be a massive reduction in the capacity to incorporate the Environmental Elements of CPTED, being most usually hampered by the original design layout. That said, as Sect. 9.1, The Four Towers at Duddeston Manor demonstrates, such capacity may still exist.

3.3.7 The Meeting With the Architect(s), Developer, Planning Officer etc.

The first meeting with the architect, developer, planning officer etc. may take place in their offices or actually on-site. The latter has advantages in that from a time management perspective, you are able to undertake two tasks simultaneously. However, where the venue is a cold, wet and invariably muddy building site, lacking a shelter in which to spread out the plans and converse with the aforementioned parties, it can be both dispiriting and hardly conducive to the negotiation process (if necessary). In such circumstances and where key elements of the design are causing problems from the perspective of DOC, it may be advantageous to adjourn the meeting to a more comfortable and purposeful venue.

Indeed, the purpose of such a meeting with the architect, developer, planning officer etc., is first and foremost to discuss the project in question from the perspective of DOC and make necessary recommendations. Unfortunately, it is extremely rare to be consulted at the concept stage of a project. Outline planning approval will usually have been given, sometimes full planning approval and it is not unknown for the DOCO to be first consulted whilst the development is being built or even after the project is completed!

It is highly unlikely the architect, developer, planning officer etc. will accede to ALL the recommendations you make. As previously indicated (see Sect. 1.5 Role of the DOCO; Sect. 1.5.9. Diplomatic Skills; and 1.5.10., The Art of Compromise and Negotiation), the real skill is knowing which elements are mandatory requirements and those that are open to compromise, e.g. PAS 24 or equivalent doorsets are a mandatory minimum requirement for all SBD domestic premises and as a result, non-negotiable. Whereas, your desire to remove all rear access footpaths may be unacceptable for the developer/client, in which case the DOCO may have to compromise or lose the whole project. The DOCO is in a far stronger position where the developer must attain the Secured by Design award, as was formerly the case with the Homes and Communities Agency in respect of social housing projects in England. But in any situation SBD does not provide an guaranteed carte blanche wish-list.

Key Issue

Where full planning permission has already been given, or construction is taking place/complete, the decision rests with the DOCO as to what advice is appropriate in such circumstances. There will almost certainly be no capacity to incorporate the Environmental Elements of CPTED at such a late stage. Indeed, Good Housekeeping Measures (see Sect. 8.3) may be the only recommendations the developer can incorporate. Most importantly, retrospective applications for the Secured by Design award will undoubtedly be inappropriate once construction has begun—and certainly after completion! (see Chapter 5, Secured by Design).

3.3.8 Final Site Inspection

It is desirable (but by no means common practice, except with large, multi-stage developments) to conduct ongoing site surveys/inspections whilst the development is being built. Normally, the final site inspection will take place when the project is all but complete and ideally, a few weeks before it is handed over by the builder/construction company to the client. This margin enables the DOCO to perform what amounts to a 'snagging' exercise—checking that the recommendations she or he originally made and were agreed, now appear in the finished development. Moreover, this allows the developer time to correct any missing 'Target Hardening' elements.

Almost from the moment they enter the site, the experienced DOCO will instinctively know as to whether a development has been built according to her or his original recommendations. Nevertheless, such instinct alone is insufficient. As a cross-discipline parallel, in 2017 a rediscovered Leonardo da Vinci masterpiece, 'Salatore Mundi', sold at auction for a record US$450 million. Authenticated by Martin Kemp, Emeritus Research Professor in the History of Art at Oxford University, he observed how on first sight he instinctively knew it was a Leonardo, but then needed to conduct extensive research to prove it. Obviously, the responsibilities placed on the DOCO are not in the same league of expertise, but an enquiring mind is a necessary attribute.

Consequently, it is wise counsel for the DOCO to take with them the original file of correspondence and to check against the list of prescribed conditions, that each of the requirements originally made has been incorporated into the finished project. Features like doors, windows, glazing and locking mechanisms will have to comply with the relevant standards. Be especially aware that it is not unknown for a builder to substitute these with an inferior (and cheaper) alternative. Even when the relevant standard or symbol (such as the SBD logo or British Standard 'Kitemark') is engraved on the product in question, you must also ask to see the relevant paperwork and certification that confirms this provenance (see Sect. 4.2.1, Doors and Door Security). Indeed, it is a mandatory requirement that such certification is submitted with the Secured by Design application.

Minor failings can be addressed during this visit (enhancing perimeter protection by raising the height of fencing at key junctions or by the addition of wooden trellis, is a repeated necessity). But more serious concerns may require remedial action being taken by the builder/construction company and a subsequent re-inspection undertaken. Where the developer refuses or is unable to undertake such modifications, the full diplomatic and persuasive skills of the DOCO will be necessary.

Ultimately, the balance between refusing the award, or allowing it with reservations, rests with the DOCO. At such times liaison with more experienced colleagues and/or the former DOCOs (Development Managers etc.) at Secured by Design can be highly beneficial. And remember, if you confer the award only to later discover the crime, FOC and ASB you originally predicted/feared if certain measures were not incorporated does indeed take place, there will be plenty of time (many years) to repent at leisure!

As previously indicated, appeals regarding a failure to meet the necessary criteria for an SBD award, are addressed in Sect. 5.1.24. SBD—Appeals.

3.3.9 Conclusion

The fundamental conclusion that emanates from this chapter is the necessity for an in-depth, systematic and highly professional form of site survey and examination process, one that addresses the full spectrum of CPTED issues (see Chapter 5) and simultaneously draws down evidence from crime data, local consultation, published research and Crime Science—as appropriate.

This is worlds apart from the DIY 'tick the box' West Midlands Police Knowing Your Home (1995) or the former Home Office Crime Reduction College's Home Security: an introduction to domestic surveying (2001) style documents.

Such booklets (excellent and useful in many respects) owe at least part of their ancestry to Professor Gloria Laycock's (1989) An Evaluation of Domestic Security Surveys. She concluded that the efforts of specialist Crime Prevention Officers in producing such domestic security surveys could be better spent. And indeed, they have been—witness the evolution of Designing Out Crime and the DOCO role.

Key Issue
This final site inspection ably demonstrates the necessity for the DOCO to keep accurate records and to include the recommendations made and agreed in the original correspondence sent to the architect, developer, planning officer etc. Because, several years may have elapsed between making the recommendations and conducting the final site inspection (possibly by a different DOCO) and it is vitally important that an Administration System (see Sect. 1.5.12) is maintained for this purpose. Moreover, the absence of such accurate record keeping may mean that an appeal against the decision of the DOCO is successful—as has occasionally happened in the past and with consummate embarrassment (see Sect. 5.24).

3.4 Writing the EVA/Report: The 'Crown Jewels' of Its Written Presentation

The possession of a successful formula and style for producing concise yet detailed Environmental Visual Audits (EVAs), other surveys and recommendations, is regarded by many as the 'Crown Jewels' of an ability for the Designing Out Crime Officer (DOCO) and security reviewers in general—and guarded as such. Indeed, many are loathe to share such a formula/style, having honed this skill after many years of practice (although it might also demonstrate that a distinct minority do not want to advertise their skills shortage!).

In reality, there is no great mystique about the manner in which such a security survey should be conducted (see the previous chapter) and the way it is then constructed and put into writing. Many different formulas and styles exist—the report written from a Designing Out Crime perspective being a bespoke variant of the holistic security survey. Ultimately it is being presented for the benefit of the end user—the reader and often the individual who will have to take the recommendations made and authorise their inclusion.

Such a written report has distinct similarities with the art of taking a witness statement. And once again, there is no mystery to this process providing three overriding rules are obeyed:

- The report is laid out chronologically
- The Layered Security ('Onion Peeling') process is adhered to in the telling
- You explain your reasoning for the observations and recommendations made. Never presume that the reader is a fellow DOCO or 'Street Level Criminologist/Crime Scientist' (see Sect. 1.5.3) who will automatically understand your rationale—even though this necessity to explain will inevitably compromise to some extent the capacity to be concise.

This last point is especially relevant to public service buildings, where those with no Situational Crime Prevention training or experience often believe they know best. And incidentally, repeated anecdotal evidence suggests this is an international phenomenon. Furthermore, whilst detailing your thought processes may appear simplistic, this has repeatedly demonstrated benefits in the subsequent negotiation process where developers are prepared to adhere to some, but not necessarily all of the recommendations you have made—providing they have been fully explained.

3.4.1 Before Putting Pen to Paper/Finger to Keyboard

You should already be aware of the proposed function of the development in question (be it housing, a school, hospital or a project with inherently greater risk, such as an armoury); possess the plans and had time to examine them fully (see Sect. 3.1, Reading Plans). Thereafter, the next step is to obtain Crime Pattern Analysis/Crime Profile data and where required/local practice, complete a Crime Impact Statement.

Unfortunately, this last task is often much more difficult in practice. Where Crime Analysts are employed (usually but not exclusively by the police service), they are often weighed down with management requests and demand for analysis of the most serious types of crime. As is repeatedly the case, the preventive/reductive function tends to takes second place. Not that the DOCO most usually desires the forms of 'advanced blobology' (a description coined by an internationally respected Crime Scientist) which some crime analysts produce.

The advent of computer driven crime databases accessible by all police and often local government employees (not to mention increasingly, the public at large), has opened-up Crime Pattern Analysis to the DOCO who can often conduct their own research electronically. Such data may be of minor relevance where the proposed development will occupy a green field site. But it is imperative for brown field or in-fill projects ('pepper-potting') where existing crime patterns should provide an indication of any threats to the project—both during construction and following occupation. Consequently, it arms the DOCO with background data and necessary 'ammunition' to undertake the survey and conduct what is often the first meeting with the architect, developer or their agent. Therefore, the golden rule is that Crime Pattern Analysis should be a prerequisite before every survey is conducted and put into writing.

3.4.2 Constructing the Evaluation/Survey

As previously described and akin to taking a witness statement, there should be a beginning, middle and an end—with the ultimate aim of telling an accurate security story. If the main body of the survey sticks to that chronological formula, combined with the Layered Security approach (see Sect. 3.3), it is a simple framework upon which to hang additional sub-headings.

3.4.3 Executive Summary

Most importantly, you are strongly advised to begin with an Executive Summary. The value of this opening precis cannot be over-emphasised, if only because many end users will read nothing more in the report (often the person charged with authorising payment for the recommendations you are making!) Ideally, such an Executive Summary should be limited to no more than can be obtained on a single A4 sized sheet of paper, minimum size 11 font (or the electronic equivalent). From repeated personal experience, this author knows that anything of longer length will almost certainly not be read! And the more senior the intended recipient within a particular organisation, the more likely this will be the case.

The Executive Summary should include the most important recommendations contained in the evaluation, listed in descending order of priority and if there is space—a brief description of their rationale. Remember, you are attempting to communicate your line of reasoning and why it is imperative that your recommendations are executed—almost certainly because you aim to Design Out Crime, fear of crime and anti-social behaviour.

One additional point regarding the Executive Summary, it is advisable to include a paragraph describing how the recommendations contained within the body of the report should be treated as a whole. That is to say, they need to be implemented in full as a holistic package of measures. End users are prone to pick and choose from the list of recommendations (often on the basis of minimum cost) and need to be reminded how serious security/safety issues may result if they are not applied in their entirety.

3.4.4 Body of the Report

The Executive Summary should be followed by an opening introduction about the development in question and a paragraph devoted to Crime Pattern Analysis (CPA) at that location. For small-scale developments and where there is no great crime history in the area, this should be sufficient. However, large projects or those associated with specific crime problems will almost certainly require more detail. One note of caution: avoid attaching Excel or similar spreadsheets containing masses of data. These can easily confuse the end-user (and sometimes the author). It is far more advisable to describe and prioritise crime concerns in your own words.

Thereafter, suggested chapter headings can be expected to follow those contained previously in Site Surveys and EVAs (Sect. 3.3) which can be listed as:

- Approach to the site
- Geographical Juxtapostion
- Site perimeter
- Grounds of the development
- The physical building(s)
- The tasks undertaken therein and/or the property to be stored/protected.

And in addition:

- Good Housekeeping Measures. There is a powerful rationale for including a sub-section relating to Good Housekeeping Measures, (see Sect. 8.3) as a prelude to an Operations Manual.

3.4.5 Recommendations

As in the Executive Summary, list your recommendations in descending order of priority. Include a sentence to this effect, as it might otherwise confuse the reader on seeing the numbers are NOT listed sequentially, i.e. as 1,2,3,4 etc. and the way they appear in the body of the report in accordance with the Layered Security process.

There is, of course, a potential conflict here. Whilst listing the recommendations in descending order of priority, the Executive Summary will have stated that these measures are a holistic package and not a 'pick and mix' gallery of options. The argument for saying both is that experience demonstrates how time and again, despite your best efforts this highly selective tactic by the client will almost certainly take place. Consequently, prioritisation is an attempt to reduce the impact on security issues when this actually occurs.

In this last regard, be especially mindful that the client may decide that instead of your financially expensive (in her or his eyes) upfront proposed Situational Crime Prevention measures, security guards ('manned guarding') are a more expedient solution. Unfortunately, such a response is invariably unsustainable and at some point in the future, cost savings will be made and the security guards removed on economic grounds. Again, experience demonstrates how once the security guards disappear, that is exactly when the premises will come under attack. Whereas, had the developers invested in the long-term, sustainable Designing Out Crime features originally proposed by the DOCO...

All too obviously, prioritising the recommendations requires perhaps the greatest level of skill by the DOCO. As with the Environmental Elements (see Sects. 4.4.1–4.4.6), it necessitates considerable subjective application learnt from practical experience and knowledge of the key elements, Crime Science and other considerations that underpin our discipline. And don't be concerned that your opinion on such issues has a habit of changing over time. Apparently, that is how true wisdom evolves.

3.4.6 Academic References

The main body of the document is the obvious location for especially pertinent academic research you want to reference. For example, Professors Shane Johnson and Kate Bowers (2009) Permeability and Burglary Risk: Are Culs-de-Sac Safer? (see Sect. 6.2) in relation to Permeability, burglary and 'Leaking' Culs-de-Sac. Limit such referencing to one or just a few such articles (e.g. Professor

Rachel Armitage's, Crime Prevention Through Housing Design—2013). And if necessary, first allude to it in the Executive Summary.

3.4.7 Legal Responsibility Caveat (Disclaimer)

Historically, some DOCOs have included a paragraph disclaiming legal responsibility for any negative outcome emanating from the recommendations made. From a legal perspective, the value of such a paragraph is open to debate. Suffice to say, this author has never included such a disclaimer as it tends to suggest he does not have the courage of his convictions. However, you will need to be guided by practice and procedure in your own organisation as to whether such a legal caveat is included.

3.4.8 Photographs

For additional visual impact, the inclusion of digital photographs of the existing site (or bad examples you wish to avoid), can be highly impactive. These will hold the attention of the reader and in this increasingly multi-media driven age, their inclusion is likely to become all but mandatory in the future (see Sect. 1.5.8. Photograph now!). However, limit the use of such photographs to a minimum and where the greatest impact on the decision-making process is required.

3.4.9 Conclusion

This final element, 'Conclusion', is not strictly necessary as it can be argued it is interchangeable (at least in part) with the Executive Summary. However, it is more professional to include such an end of report summary, as it authoritatively completes the finished document.

3.4.10 Large-Scale Development Projects

Finally, with large-scale developments the skill is to break down the project into easily digestible portions, whilst simultaneously maintaining an overall perspective of the fully completed design. This should be covered in the body of the evaluation/survey, together with references to those parts of the overall development already completed and those that will follow—where available.

Indeed, one of the greatest challenges relating to large-scale development projects is often the sheer timescale of the construction process. It is not unusual for this to be of many years' duration and consequently, the DOCO originally charged with responsibility for Designing Out Crime may not see it through to completion—having retired or moved on to a new role. Consequently, there is a professional responsibility by the original DOCO to maintain accurate records of the recommendations made and ensure these are safely passed to their successor. The latter will in most circumstances adhere to the recommendations made by the original DOCO. However, where the uses of the development change or new threats appear, it is not unknown and entirely logical that the security requirements similarly evolve over time.

As an example of how this can be achieved in a real-life setting, see Sect. 9. 2, Brindleyplace: City Centre Commercial, Retail and Leisure Complex. This project took 17 years to complete and involved successive changes of DOCOs (known as Architectural Liaison Officers and Crime Prevention Design Advisers during that time period).

References

Armitage, R. (2013). *Crime prevention through housing design: Policy and practice.* Palgrave Macmillan.

Armitage, R. (2017). *Burglars' Take on Crime Prevention Through Environmental Design (CPTED): Reconsidering the Relevance from an Offender Perspective.* Macmillan.

Nee, C., & Meenaghan, A. (2006). Expert Decision Making in Burglars. *British Journal of Criminology, 46*, 935–949.

Newman, O. (1972). *Defensible space: Crime prevention through urban design.* Macmillan.

Newman, O. (1973). *Defensible space: People and design in the violent city.* Architectural Press.

Cozens, P., Love, T., & Davern, B. (2019). Geographical juxtaposition: A new direction in CPTED. *Social Sciences, 8*, 252. Available online at: https://www.semanticscholar.org/paper

Brantingham, P. and Brantingham, P. (1995) Criminality of place: Crime generators and crime attractors. *European Journal on Criminal Policy and Research 3,3*, special issue on Crime Environments and Situational Prevention, 5–21.

Eck, J., Chainey, S., Cameron, J. G., Leitner, M., & Wilson, R. E. (2005). *Mapping crime: Understanding hot-spots.* National Institute of Justice.

Beckford, C., Hankins, M., & Stagg, C. (1996). *Under lock and quay. Reducing criminal opportunities by design.* Metropolitan Police Service 2 Area CPDA Group.

UK Government. (1974). *Health and safety at work act*. The National Archives. Accessible at www.legislation.gov.uk/ukpga/1974/37

UK Government. (1992). *The management of health and safety at work regulations*. The National Archives. Accessible at www.legislation.gov.uk/uksi/1992/2051/contents/made

Pascoe, T. (1993a). *Domestic burglaries: The burglar's view*. Building Research Establishment Information Paper IP 19/93. Building Research Establishment.

Pascoe, T. (1993b). *Domestic burglaries: The police view*. Building Research Establishment Information Paper IP 20/93. Building Research Establishment.

Pascoe, T., & Topping, P. (1997). Secured by design: Assessing the basis of the scheme. *International Journal of Risk, Security and Crime Prevention, 2*(3), 161–173.

West Midlands Police. (1995). *Knowing your home: Straightforward advice and a simple DIY survey to help you manage the risk to your home and keep your home contents where they belong*. West Midlands Police.

Home Office Crime Reduction College (2001) *Home security: An introduction to domestic surveying*. HOCRC

Laycock, G. (1989) *An evaluation of domestic security surveys*. Crime Prevention Unit Paper 18. Home Office.

Crime Prevention Through Environmental Design

4

4.1 Introduction

CPTED (usually pronounced SEP-TED) and in full as Crime Prevention Through Environmental Design, was the description invented by the criminologist C. Ray Jeffery in his book of the same title (1971). It was originally described as a multi-disciplinary approach to deterring criminal behaviour, delivered through the implementation of changes to the built environment. In essence, CPTED strategies concentrate on the capacity to influence offender behaviour, before they reach the target in question and as a means to dissuade them from committing criminal acts.

According to Robinson (1996), Jeffery's CPTED concept emerged from his experiences with a rehabilitative project for juveniles in Washington D.C. By using the psychological learning theory of B.F. Skinner (1953), CPTED stressed the role of the physical environment in the development of pleasurable and painful experiences for the offender and capacity to alter behaviour. 'Jeffery 'emphasised' material rewards and the use of the physical environment to control behaviour' (Jeffery & Zahm, 1993, p. 330). The key theme was that by making changes to the physical environment (removing what we might describe as the Crime Opportunity—see Sect. 2.3), the crime itself would not take place.

Professor Tim Crowe (2000, p. 46) provided a succinct definition of CPTED: '*The proper design and effective use of the built environment can lead to a reduction in the fear and incidence of crime, and an improvement in quality of life. The goal of CPTED is to reduce opportunities for crime that may be inherent in the design of structures or in the design of neighbourhoods*'. However, like the previous definition, Crowe does not explain what CPTED involved or the mechanisms by which it might be achieved. See also Taylor (2002).

M. Stokes, *Designing Out Crime*, Crime Prevention and Security Management,
https://doi.org/10.1007/978-3-031-99840-9_4

In the UK with its history of police service delivered Situational Crime Prevention (SCP) dating back to the early 1960s, CPTED was embraced within police Crime Prevention training during the mid-1980s. However, in this author's experience the Environmental Elements of Defensible Space, Natural Surveillance, Territoriality, Image, Milieu etc., were generally perceived to be bolted-on to the Physical Security 'Target Hardening' Measures that had existed and evolved over time (especially since the Victorian era with the advent of new engineering and manufacturing techniques), together with more recent Technological Innovations. In this context, the all-embracing nature of CPTED has been a far more recent phenomenon—at least in the UK. Meanwhile, in the wider world there have been further developments.

Greg Saville and Gerry Cleveland (1997) conceived the concept of Second Generation CPTED. Through its four core strands of Social Cohesion, Connectivity, Community Culture and Threshold Capacity, Second Generation CPTED strives to promote community and neighbourliness in the cause of Crime Prevention—Community Interaction. And subsequently, the notion of Third Generation CPTED has been articulated by both the MIT (Massachusetts Institute of Technology) MIT SENSEable Lab (2013) and Professor Lorraine Gamman and Dr Adam Thorpe (2012) of the Design Against Crime Research Centre at the University of the Arts, London. In a presentation given by Mateja Mihinjac and Tarah Hodginson at the ICA (International CPTED Association) conference held in Mexico City, the former envisages utilising green technology in urban design. For example, living walls' that deter graffiti and crime absent eco-friendly urban furniture. Whilst Gamman and Thorpe, inspired by the Mayor of Seoul, Park Won-soon and his city's 'asset-oriented approach', suggest reframing crime problems through 'design thinking' that '…realises CPTED drivers without being CPTED led'.

Therefore, it was inevitable that the definition of CPTED would evolve over time. For example, Professor Paul Ekblom (2011b, p. 4) provides:

> Reducing the possibility, probability and harm from criminal and related events, and enhancing the quality of life through Community Safety; through the processes of planning and design of the environment; on a range of scales and types of place, from individual buildings and interiors to wider landscapes, neighbourhoods and cities; to produce designs that are "fit for purpose", contextually appropriate in all other respects and not "vulnerability led"; whilst achieving a balance between the efficiency of avoiding crime problems before construction and the adaptability of tackling them through subsequent management and maintenance.

Meanwhile, Professor Rachel Armitage's (2013, p. 23) definition of CPTED includes elements of social and environmental sustainability: *'...design, manipulation and management of the built environment...'. and '... to enhance sustainability through the process and application of measures at the micro (individual building/ structure) and macro (neighbourhood) level'.*

However, compare this to Professor Paul Cozens' (2014, p. 21) own updated definition of CPTED that includes *'...analyzing and assessing crime risks in order to guide the design, management and use of the built environment (and products)...'. Not only has his definition been expanded to include 'products' and 'fear of crime'. It now also encapsulates the promotion of '...public health, sustainability and quality of life'.*

Ultimately, we should remember the descriptions CPTED, SCP and DOC have existed for little more than half a century. Consequently, it is perhaps inevitable that their core ideas, influences and definitions often overlap with each other and are still being shaped. Don't be frightened by this debate and become involved if you feel the urge. Otherwise, simply enjoy the capacity to Design Out Crime on a long-term, sustainable basis.

Most commentators now agree that as a minimum CPTED includes: Physical Security 'Target Hardening' Measures (Newman, 1972, 1973, 1996) like Doors and Door Security, Windows and Glazing, Perimeter Security and even Defensive Planting; Technological Innovations such as Lighting, Intruder Alarms, CCTV, ANPR (Automatic Number Plate Recognition), Fogging Agents, Biometrics, etc. and Environmental Elements including Defensible Space, Natural Surveillance, Territoriality, Image and Milieu, Geographical Juxtaposition, Permeability (Access and Movement).

Meanwhile, in 2021 the International Organization for Standardization (ISO) based in Geneva, Switzerland, issued ISO 22341:2021. This was the first international CPTED standard to be published, with a remit that concentrates on process rather than prescription. Indeed, in the Introduction it states: 'This document is intended to promote a common understanding of crime prevention through environmental design (CPTED) in the field of security, law enforcement and related risks, and their preventive measures, through environmental design and management.' The standard includes seven principles, process orientation, contribution to urban development, inclusivity of stakeholders, scientific and evidence-based, pro-social focus, behavioural change an objective, and quality of life the outcome (see Pascoe, 2025). The next three sub-sections seek to examine these core concepts of CPTED: Physical Security 'Target Hardening Measures, Technological Innovations and Environmental Elements. Chapters are devoted to their constituent parts and briefly outline those of most relevance to the DOCO. Then in Sect. 4.4.6, attention turns to perhaps the most contentious issue of Permeability (Access and Movement), described by one observer as '*...the fault-line between the Designing Out Crime and Planning professions*' (personal witness, 2003).

Baddesley Clinton mediaeval moated manor house, Warwickshire

4.2 Physical Security 'Target Hardening' Measures

Physical security measures as a form of defence have a historical pedigree that can be traced back to the dawn of civilisation. For example: the stone doors and false tunnels within the pyramids of ancient Egypt; the Iron Age hill forts of the Ancient Britons; the Great Wall of China; the moat, drawbridge, portcullis and walls in the castles and forts of Arabia, Japan and Europe and the 'priest-holes' (hiding places) of Tudor England and Wales. Moreover, such techniques are also identifiable within the world of nature: spikes on plants such as Cacti and the likes of Berberis, Brambles, Pyracantha and Roses (see Sect. 4.2.4, Defensive Planting).

However, the evolution of purely defensive mechanisms into those identifiable as physical security measures, has been a subtle process, tied to perceptions of safety in society at large and new manufacturing processes. For example, the moated manor house of Tudor England and Wales can be seen as having developed from castle design. By the eighteenth century the moat had for the most part disappeared—unless the gentry could not afford to rebuild in accordance with (for example) Regency tastes and style. In this last respect the now National Trust owned moated manor house of Baddesley Clinton, Warwickshire, is a classic example of such financial distress.

Nevertheless, the walls and/or metal railings that formed the perimeter around the ancestral homestead persisted after the Napoleonic Wars, when returning soldiers were often treated as vagabonds (see the Vagrancy Act, 1824) and FOC, revolution and Disorder were of constant concern amongst the ruling classes. Meanwhile, those at the bottom of the social ladder did the best they could, often securing their homes after dark from the inside—a feature that can be identified in many countries around the world to this day.

What Oscar Newman (1972, 1973) described as Physical Security 'Target Hardening' Measures (and by default a key section of Crime Prevention Through Environmental Design (CPTED) in the opinion of the majority of observers), only really began to appear as the norm during the Victorian period, i.e. the second half of the nineteenth century. Moreover, such measures (primarily door locks and associated ironmongery) rapidly became a world-wide phenomenon, manufactured in the industrial heart-lands of countries like the UK (such as at the former Chubb lock factory in the City of Wolverhampton) for both domestic consumption and export to the colonial empires. The advent of the safe, strongroom, bars on windows, bolts on doors and high-quality key-operated locks can be traced back to this period.

Bringing this up to the present day, the full range of Physical Security 'Target Hardening' Measures can often be identified in new buildings. Indeed, Secured by Design (see Chapter 5) demands that a host of these are tested and certified before being fitted, as a minimum requirement to achieve the award.

However, a potentially worrying trend has begun to appear (largely in respect of non-residential buildings located in town and city centres) where physical security is concentrated in the building itself. The net result is that electronic door entry systems and CCTV are being used to protect the building, but with no secure perimeter beyond. This delivers uncontrolled Permeability (see Sect. 4.4.6) and denies the efficacy of the Layered Security approach (see Sects. 3.1 and 3.4). Moreover, the creation of such non-defensible public and semi-public space (see Sect. 4.4.1 and HOCPC Police Architectural Liaison Manual of Guidance (1987,

1994) provides Crime Opportunities, together with those for FOC and ASB to develop in the immediate vicinity of such buildings.

4.2.1 Doors and Door Security

As the easiest means of accessing the premises, the door will always be the most obvious entry point for legitimate callers and criminals alike. Consequently, over the centuries much effort, invention and innovation has been placed into making the door secure. According to Eras (1957) the first key-operated locks were introduced several thousand years ago by the Ancient Egyptians, Greeks and Romans, although they were predated by simple sliding bolts and wooden bars. From the Victorian period onwards, the quality of key-operated locks improved dramatically—especially in the adoption of multi-lever construction designed to prevent the lock from being 'picked', by means of a skeleton key, hair-pin or similar device.

Indeed, to this day in the UK Designing Out Crime Officers (DOCOs) recommend '5 lever mortise deadlocks to BS (British Standard) 3621' as a means of providing extra security to existing exterior doors. The reference to 'mortise' means that the lock is installed within the fabric of the door itself, whereas a 'rim-latch' is fitted to the surface of the door. Mortise locks are inherently more secure than the rim-latch, simply because they are in effect an integral part of the door fabric. However, even before the launch of Secured by Design (SBD) in 1989, it was realised that a single locking element would not spread the load should the door come under attack from a jemmying type device like a crowbar, spade or screwdriver—or the sheer force of a human shoulder or kick aimed at the appropriate point(s) of weakness (thus the original SBD requirement for two locking elements on exterior doors).

A large part of this chapter relates to exterior doors for residential premises—as the sub-headings dealing with spyholes, door chains or opening limiters and door viewers tend to suggest (see Sect. 5.4). However, the methodology (if not the specification) is equally applicable to non-residential doors (see Sect. 5.5).

Front Door

In the beginning, SBD required that two key-operated locks be fitted to the Front Door (main entrance door in a conventional dwelling), one a rim-latch one third from the top and the second a mortise deadlock one third from the bottom (or vice-versa) both thereby spreading the load if the door was attacked. Other mandatory requirements included: any glazing to be of Laminated Glass;

interior bolts fitted to the top and bottom of the door; a door chain or door limiter; door viewer (spyhole) and a restricting device on the letterbox (to prevent intrusion by mechanisms designed to release the locks) and more importantly, in terms of recent criminal history, incendiary items being pushed through same.

On this same theme, unfortunately the combination of key-operated security mechanisms (whilst ensuring the apparent safety from crime and security of residents) had the potential to contravene extremely justifiable Fire Safety concerns. This logic is sound. Because, having to manipulate two key-operated locking systems whilst attempting to escape from a smoke-filled dwelling, may cost precious seconds and ultimately, lives. Thus, in the UK and beyond, legally enforceable Fire Regulations prohibit the use of key-operated (from the inside) locking systems in all dwellings accessed from a corridor or landing area—for example, flats, apartments, sheltered housing schemes (for the elderly and other vulnerable groups) and HMOs (homes of multiple occupancy).

Furthermore, even before the original launch of the SBD award, it was obvious that whilst a high-quality door locking system might be installed, the quality of the door itself was often highly dubious. And even when the door was up to standard, much depended upon the professionalism of its installation within the door frame and that of the frame itself. When closed, too great a gap between the edge of the door and the frame would mean that locks would not fully engage inside the 'keep' and therefore, the door was weaker than imagined and more susceptible to a successful criminal attack.

In many ways, these safety and security dilemmas were resolved by the introduction (more than 25 years ago) of a single key-operated (on the outside), thumbturn (on the inside) multi-point locking system for such Front Doors to the individual dwellings. The door is secured by at least three 'espagnolette' type bolts (also known as 'shoot bolts') that emerge from the edge of the door when closed and often in combination with hinge bolts.

Instances of such doors being forced open are extremely rare. Indeed, on at least one refurbishment project there have been complaints from the Fire Service and specialist police departments (like the former Drugs Squad) that they were unable to effect entry through such doors, even when using hydraulic jacks (see Sect. 9.1, The Four Towers at Duddeston Manor). This is a classic example of 'over-egging the target hardening pudding' (over-engineering) and an instance of security wrongly outweighing safety concerns.

Key Issue

The important point to note (and the overwhelming golden rule in our Designing Out Crime (CPTED) discipline), is how security considerations should never be allowed to outweigh those of safety—and of Fire Safety in particular. This is probably the most important issue for a Designing Out Crime Officer (DOCO) if only because tragedy might follow and as secondary consequences, litigation and the credibility of our profession.

In the closing years of the twentieth century, the advent of new standards and testing regimes began to eliminate such problems in the UK. To comply, the door, frame, glazing (most often Laminated Glass due to its inherent strength) and door furniture (including the locking system) are factory constructed as a holistic entity and supplied in that form. Doorsets constructed to the standards set out in Chapter 5 and both independently tested and certificated, are a mandatory requirement for Secured by Design and detailed in a number of SBD Development Guides. Moreover, these standards apply to both the Front Door and all other exterior doors (with the exception of Fire Doors). Nevertheless, existing exterior doors (except Fire Doors) can continue to be made more secure by fitting the additional locks, bolts, door chain or door limiter,, door viewer (spyhole) and letterbox restrictor detailed previously—although this will almost certainly be insufficient to attain an SBD award.

As always, the Development Guides contained on the Secured by Design website, https://www.securedbydesign.com provide a list of the most up to date requirements in respect of Main Entrance Doors and all other Exterior Doors, for both residential and non-residential premises.

Side and Back Doorsets

Conventional housing design in much of the world will include Side and Back doorsets (rear access exterior doors). These are often especially problematic. '2XG' doors with a 4 mm thick plywood panel in the lower half of the door and a single sheet of glass in the top (or '2XGG' doors with glass top and bottom) were notorious across the UK from the 1950s through to the 1990s due to their inherent lack of strength and attraction to burglars. Two thirds of all residential burglaries were executed via the rear or side of the house, yet these secondary entrance doors almost always possessed poorer security standards than the Front Door (Main Entrance Door).

The aforementioned security standards sounded the death knell for such inherently insecure '2XG' and '2XGG' doors, as SBD now demands the same security requirements for Front, Side and Back doorsets. Moreover, this technical improvement has been instrumental in delivering sustainable reductions in residential burglary and other forms of property crime. It therefore accords with the *Domestic burglary drop and the security hypothesis*, Tseloni et al. (2017); and *The Crime Drop and the Security Hypothesis* of Farrell et al. (2011) the latter relating to vehicle crime. Furthermore, this testifies to the long-term effectiveness of SBD, witness the research findings of Brown (1999), Pascoe (1999), Armitage (1999), Teedon et al. (2009), Armitage and Monchuk (2009), Jones et al. (2016), and Stokes (2022).

However, in the UK context such enhanced security standards are now only mandatory in Scotland and Wales in respect of new and refurbished public housing projects by and for the RSLs (Residential Social Landlords). Thankfully, in recent years the private for sale sector has begun to emulate the RSLs. But the legacy of '2XG' and '2XGG' doors facilitating easy access by burglars may linger for many decades to come and until all such doors have been replaced.

Bolts

Surface–mounted, manually operated bolts at the top and bottom of the door have been a feature of exterior doors for at least a century. Moreover, these or key-operated mortise bolts were required by the original Secured by Design specification. Once again, the new security standards have made their use unnecessary.

Surface-mounted or key-operated mortice bolts continue to be installed by householders on existing doors—that do not comply with the new security standards or have no such multi-point locking system (wooden doors in particular). However, as such doors will almost always be a Designated Means of Escape (in the event of a fire or other emergency situation), it is strongly advised that the DOCO **NEVER** recommends their installation—purely on grounds of safety and the potential for civil litigation.

Door Chain or Opening Limiter

Again, a mandatory requirement for the Front Door since the Secured by Design award was launched in 1989, the objective of the Door Chain or Opening Limiter is to allow the householder to open the door a short distance so as to confirm the legitimacy of the caller. An identification card can be passed through the gap and conversation between householder and caller take place. However, on grounds of safety, the Door Chain or Restrictor should only be engaged when there are unknown callers at the door. There remains considerable difficulty in communicating this message to householders.

Indeed, there has been much debate regarding the efficacy of Door Chains and Opening Limiters. One argument contends that as the device has no real strength, its use inherently weakens the capacity of the door to repel potential intruders. However, the counter-argument perceives the Door Chain or Opening Limiter as a means of preventing a potential burglar posing as an official of one of the utility companies, or simply pushing past the householder once the door has been opened—and incidentally a much-repeated Modus Operandi for the insidious crime of 'Distraction Burglary'. Consequently, a Door Chain or Opening Limiter remains an SBD requirement.

Door Viewer

Another SBD feature since the very beginning, the Door Viewer (or Spyhole) allows the householder to view callers without opening the door. Such a device is especially appropriate for solid doors or ones that contain obscured/translucent glazing. Various types of Door Viewer exist, some incorporating an electronically enhanced audible element that permits householder and caller to speak to each other without opening the door.

Fire Doors

Fire Doors are covered by legally enforceable Fire Regulations. Consequently, no extra security devices can be installed on an existing Fire Door. Where a Fire Door is causing concern due to criminal attack, liaison with the local Fire Authority (or equivalent) is strongly advised in order to produce a mutually appropriate solution.

Communal Entrance Doors

These are often the first line of defence for dwellings contained within a low or high-rise block or HMOs. As such, communal entrance doors are a barrier to intrusion beyond that of the dwellings themselves. However, this may also mean they provide cover for criminal behaviour, in that the Front Door to the individual dwelling does not receive the degree of Natural Surveillance (see Sect. 4.4.2) that might otherwise be the case. Consequently, it is imperative that communal entrance doors are of sufficient strength to resist attack and that appropriate measures are in place to ensure they remain closed and locked when not being used for access and exit by residents and legitimate callers.

At one level, this may involve little more than Good Housekeeping Measures (see Sect. 8.3) being applied, whereby residents and callers understand that such doors should be closed and locked when not in use and that a brick or stone should never be used to prop the door open, whatever the desire for convenience.

At the opposite extreme, it may be necessary to install steel core doors of considerable strength with electromagnetic locks operated by a proximity access control (PAC) system linked to a 24/7 staffed concierge or security office (as detailed in Sect. 9.1, The Four Towers at Duddeston Manor). See also SBD—Residential Guide 2025, Sect. 5.4.29 and Sect. 5.4.51. re 'Communal and shared entrance doorset standards'.'And for Doorsets providing alternative access', Sect. 5.4.38 and Sect. 5.4.53.

External Doorset Apertures

Secured by Design Non-Residential Guide 2025, (see Sect. 5.5.22) details how external doorsets shall be certified to one of the minimum police-preferred standards set out.

Avoiding Door Recesses

Doors in recesses should be avoided, as they provide Crime Opportunity for offences like arson, burglary and graffiti—together with opportunities for ASB. Where they are unavoidable, mitigating measures such as doorsets of higher security standards should be installed.

Glazing Within Non-Residential Doorsets and Secure Vision Panels

All glazing in PAS 24 or STS 201 certified doorsets, including glazed panels/ sidelights adjacent to doors installed within an integral doorframe and windows adjacent to doorsets (within 400mm), must incorporate one pane of often Laminated Glass, certificated to the standard set out in in Sect. 5.5.16.

Such doors must comply with the requirements of the Equality Act, 2010 (formerly the Disability Discrimination Act, 1995—DDA) re glazing, level access and door width. A door viewer (spyhole) is recommended for circumstances such as a remote building with no surveillance, instead of an external, unglazed door.

Door Installation

Doorsets must be securely fixed into the fabric of the building in accordance with the manufacturer's specifications or in accordance with the standard set out.

Secure Vision Panels

Where privacy with a degree of security is required, a secure vision panel can be fitted. This will permit control of vision into a private area or room, or views of an outside area for the purpose of manual visual access control.

Access Control

Access control features across the wealth of SBD design and development guides.

For further information, see Sect. 5.4 and 5.5 of the Secured by Design Residential Guide—2025 and Non-Residential Guide—2025 at: https://www.securedbydesign.com

4.2.2 Windows and Glazing

After exterior doors, windows are unsurprisingly the second most popular means of access into premises by individuals with criminal intent—often because like doors, they have a tendency to be inherently insecure. Historically, windows began to be incorporated into buildings (and especially in cooler climates) following the manufacture of glass. The original prohibitive cost restricted its use to stained glass in churches and other religious buildings, manor houses and palaces. But as innovation grew and costs fell, so its application spread. Thus, the relatively small leaded-glass windows synonymous with mediaeval Europe (including Tudor England and Wales) gave way to the bull-nosed glass of the Georgian and early Victorian periods, before float-glass manufacturing techniques allowed far larger surface areas of glazing to be installed.

Indeed, school design is but one example of the manner in which glazing has become an increasingly important part of the overall design and with some consequent problems as a result (see Sect. 5.7 SBD—New Schools 2014; and Sect. 6.17, Counter Terrorism Design).

'Georgian Wired Glass' has been used primarily to provide visibility through Fire Doors and as a modest defence to attack.

Even at high altitude on Europe's tallest mountain, Mount Elbrus, Russian Federation, window security is necessary

It was the advent of first *Toughened Safety Glass* and in the late 1980s *Laminated Glass,* that demonstrated the value of these new forms of glazing in preventing criminal attack—and more especially, mitigating the effects of terrorist bomb explosion (IEDs Improvised Explosive Devices—see *Sect. 6.17*).

There has been an ongoing debate for many years as to whether in a double-glazed unit, the pane of Laminated Glass should be fitted on the inside or exterior of the building. This author believes it should almost always be fitted on the interior side (inner face)—the logic being that it affords maximum protection to those inside the building 2023. However, be guided by existing practice and procedure and especially the reference to thermal stress in SBD—Non-Residential Guide 2025 (see Sect. 5.5.17 and 5.5.27) Similarly, locating windows in gable end walls is believed to promote Natural Surveillance and deter crime and ASB.

Key Issue

Toughened Safety Glass should never be confused with Laminated Glass. Whilst the former has through its use in motor vehicles undoubtedly saved countless lives and helped to prevent far more serious facial injuries, it has also been defined as 'burglar-friendly glass'—simply because it can be shattered noiselessly and leaves no sharp edges. Laminated Glass can be best described as a 'glass sandwich' in which the PVB (Polyvinyl Butyral) or similar middle layer forms the 'meat in the sandwich' and holds the glass together when it comes under attack. Indeed, a successful assault will require considerable time and effort using a sizeable implement that produces an enormous amount of noise. In theory, this will attract attention and/or make the attacker desist. In other words, by its very nature it has inherent Crime Prevention characteristics.

The issue of the form of glazing is no mere academic debate akin to the number of angels that can dance on the head of a pin. In recent decades and throughout the world, architects have demonstrably sought to incorporate various forms of glazing in their designs. Such 'architectural glass' has increasingly been used in non-residential developments and will rarely be Laminated Glass. However, the DOCO's capacity to influence the choice of glass is extremely limited. In the UK context, with high profile public buildings of Crowded Places (see Sect. 6.17, Counter Terrorism Design), liaison with the local CTSA (Counter Terrorism Security Adviser) is a requirement in all such circumstances. It should also be noted how the use of Laminated Glass has saved lives in the instance of terrorist attack, as at the South Quay, Canary Wharf, London, bombing in 1996.

Once again, the Secured by Design website, https://www.securedbydesign.com provides comprehensive and regularly updated guidance for its requirements in respect of windows and glazing. Authoritative information is contained within the most recently published Development Guides.

Secured by Design—Residential Guide 2025: Windows, Roof Windows, Roof Lights and Roof Lanterns

Sect. 5.4.24 details how all Easily Accessible windows (including such roof lights, roof windows and roof lanterns) shall be certificated to one of the standards quoted. For the definition of 'Easily Accessible' see SBD Residential Guide 2025, Sect. 5.4.23 Note 23.4a. The classifications for window locking systems are stated, together with the relevant material standards for windows constructed of aluminium, PVCu, timber and steel, performance requirements and accredited certification bodies. Additional guidance is provided in respect of double-opening French doors/windows and bi-fold or sliding patio doorsets.

Attack-resistant glazing/Laminated Glass meeting the standard quoted is required in respect of any window located within 400 mm of a doorset (to protect the locking system) and Easily Accessible, emergency egress windows with non-lockable hardware.

Glazing in and Adjacent to Doorsets

Sect. 5.4.23 and 5.4.43 details how any glazing within doorsets, including glazed panels/side lights adjacent to doors installed within an integral door frame and windows adjacent to doorsets (within 400 mm), must incorporate one pane of Laminated Glass meeting or exceeding the standard quoted. Sect. 5.4, Note 23.18 includes a powerful line of reasoning:

> There is no specific requirement to install laminated glazing on the inner or outer face of a double-glazed unit. However, specifiers may wish to take into consideration the fact that toughened glass is usually more resistant to accidental damage by blunt objects such as a football and therefore may be best placed on the external face of the double glazed unit. It is recognised however that there are many other factors that may also need to be considered such as thermal efficiency, aesthetics and the requirement for privacy or obscured glazing, which will influence the specifier's decision.

New 'Bespoke' Windows, Roof Windows, Roof Lights and Roof Lanterns

Section 5.4.46 sets out how windows should be installed with multi-point espagnolette locking systems that meet the security requirements of the standards quoted. There should be locking points within 100 mm from the corner of the casement. Further requirements are detailed in respect of Listed Buildings, hinges and pivots, heritage hinges, tilted window pivots, top retaining bolts and sash fasteners (fitch catches).

Security of Existing Refurbished Windows

Section 5.4.47 describes how the DOCO should be consulted where the intention is to retain existing windows during any refurbishment scheme, before embarking on any improvements.

SBD—Non-Residential Guide 2025

Automatic Opening Window Systems and Vents

Section 5.5.28 details the requirements in respect of automatic opening window systems and vents, including the fitting of bars and grilles to protect the window 'aperture' (Its total area).

Glazed Curtain Walling and Window Walls

Section 5.5.38 describes the four distinct types of glazed wall systems recognised by Secured by Design; the necessity for a secure glazing retention system; 'Easily Accessible' and the British Standard 'Code of Practice'.

Laminated Glass prevents intrusion and burglary, even when broken

Roof Windows, Roof Lights, Roof Lanterns and Sun Tubes/Tunnels

Section 5.5.43 is devoted to the standards required in respect of roof lights and roof windows, with a distinction made between low crime, low risk and higher crime, higher risk area applications. Whilst in Sect. 5.5.44, sun tubes/tunnels and to prevent unlawful access, the maximum diameter of a sun tube/tunnel should be 350 mm.

Glazing Within and adjacent to Doorsets

Section 5.5.16 provides guidance in respect of glazing within or adjacent to non-residential doorsets.

Security Glazing

Section 5.5.17 and 5.5.27 set out the long-held requirement that all ground floor and Easily Accessible glazing (e.g. from a flat roof space) must incorporate one pane of Laminated Glass.

SBD—New Schools 2014

Following the publication of SBD—Non-Residential 2025, SBD—New Schools 2014 requirements in respect of windows and glazinghave now been superseded The current UK SBD standards for windows and glazing are detailed in Chapter 5. However, these are regularly updated and the specific Development Guide on the SBD website should always be interrogated to ensure the most recent correct standard and requirements are obtained.

4.2.3 Perimeter Security

> **Key Issue**
>
> In practical terms, Perimeter Security should be regarded as the first line of defence. Once again, we are attempting to influence the decision-making process of the criminal(s). Indeed, it naturally follows that if the perimeter is secure, this in itself suggests that care has similarly been taken (an Environmental Cue—see Sect. 2.8) in securing the physical elements of the building. At the very least, such layers of security will delay the criminal in his task—and such delay (the removal/restriction of Crime Opportunity—see Sect. 2.3) is generally understood to dissuade the individual(s) concerned.

Cape Town, South Africa, where Opuntia 'Prickly Pear' Cacti reinforces the role of electric pulse fencing on top of this 6 feet/1.8 metres high wall (Table Mountain in the background)

Throughout this handbook there is a continuous theme of how Designing Out Crime (DOC) places great emphasis on a secure boundary. Indeed, whilst a 'psychological boundary' or 'controlled boundary' may deter intrusion, a 'secure boundary' is a prerequisite for a safe and secure site and to define ownership. Moreover, it is a constituent part of the Layered Security ('Onion Peeling') approach to keep those intent on crime and/or ASB away from (for example) the target building in question. This may also explain why stand-alone buildings (with no fencing, walls or perimeter beyond that of the fabric of the building itself) are especially vulnerable and problematic. In such instances an even greater emphasis, deployment and financial investment in Physical Security 'Target Hardening' Measures and Technological Innovations will undoubtedly be necessary.

Ordinarily, a secure perimeter will consist of various types of fencing, walls, or hedges (see also Sect. 4.2.4, Defensive Planting). The height of these will be determined by a combination of different factors:

- Aesthetics. This is an especially important factor for the Designing Out Crime Officer (DOCO) in attempting to avoid the accusation of possessing a Fortress

Mentality. For example, in recent years DOCOs have moved away from recommending bare metal palisade fencing in favour of colour-coated (often dark green) welded-mesh style fencing, which is as strong yet perceived as 'easier on the eye'.

- Assessment of the local Crime Risk. As a general rule, the greater the risk, the higher the fence, wall or hedge. However, in the UK context heights above 1.8 metres (6 feet) will almost certainly require Planning Permission—and, for example, in front gardens, lower heights or a total prohibition on any form of physical barrier may be contained within the local byelaws, especially if it is located within a Conservation Area, requires Listed Building Consent, or a covenant relating to that property exists. In the instance of SBD applications, Perimeter Security is covered under specific guidelines which should be adhered to. Namely: SBD—Residential Guide 2025; SBD—Non-Residential Guide 2025; and SBD—New Schools 2015 (see Sects. 5.4, 5.5, and 5.7).

From the existing Secured by Design guidelines, the following recommendations in respect of Perimeter Security have general application.

Local conditions may require certain boundary treatments which should restrict unauthorised access and exit as far as possible, particularly where Natural Surveillance (see Sect. 4.4.2) of the boundary is difficult to achieve. Railings, expanded metal and welded-mesh fencing are good examples. However, chain link and similar low security fencing are only suitable as boundary demarcation and should not be used where a real security barrier is required.

Where the development is located in an area of High-Crime Risk and the grounds, parking areas etc. back onto open land, footpaths, railway property, tow paths etc., walls or fencing (minimum height 1.8 metres) certificated to the standards detailed in the above Development Guides are advised. Trellis topped fencing can be useful in such circumstances. Additional deterrent features, such as increasing the height of fencing or alternatively, planting thorny shrubs, may also be considered (once again, see Sect. 4.2.4, Defensive Planting).

Developers are expected to install fencing of a high standard, in order to ensure the security and longevity of the boundary. A high-quality fence that lasts for many years will provide security and reduce overall maintenance costs for occupants and landlords. Furthermore, a fence that has a long-predicted life is also more sustainable.

The method of fixing between panel/rails and posts should create a secure mechanical bond, so that panels/slats cannot be easily removed. In addition, it should provide a 'chain linking effect' where each panel and post acts in concert with the next to resist attack by pushing and pulling. The fixings employed in the panel/pale to rail construction should be of galvanised steel or stainless steel, with a design life to match the timber components. Posts should allow the construction of an unbroken panel to post chain and be of a non-brittle material. Wall-mounted fencing panels or railings should be located as close as possible, to eliminate climbing opportunities or use as informal seating. LPS 1175 includes a range of different qualities of fencing standards appropriate to the perceived risk, from which the fencing is designed to protect.

Where fence panels are of slatted design, they should be oriented vertically in order to avoid step-up points for climbing. They should also be flush across the attack face (so as to resist being prised off) no less than 15 mm thick and securely affixed to the frame/rails. Fence heights should be a minimum 1.8 metres overall and capable of raking/stepping to maintain height over different terrain.

Pedestrian gates should be the same height as adjacent fencing, of a framed design and employ galvanised adjustable hinges and fixings mounted behind the attack face. On outward opening gates where the hinges/brace is mounted on the attack face, fixings should be of a galvanised coach bolt design. Hinge systems must not allow the gate to be 'lifted off' and therefore, should employ a method to restrict the removal of the gate from the fence-post or wall (most often by brazing over the hinge-pins. Gates should be fitted with a galvanised latch and key-lockable from both sides.

Entrance/driveway gates should be inward opening, of substantial framed construction and employ galvanised adjustable hinges and fixings mounted behind the attack face. Once again, hinge systems must not allow the gate to be 'lifted off'. Gates should be fitted with galvanised drop bolts with the facility for padlocking (manual gates), or electro-mechanical locking (automated gates). Such mechanical/electro-mechanical devices should also have the capacity to hold gate leaves in the open position. The gate construction should have the same design and construction attributes as the fence.

The tops of fences should finish flush with their posts and a securely fixed capping rail run across the fence and posts to affect a continuous chain. The tops/top rail/capping of fencing and gates should be of a design able to accommodate a security topping to deter attempts to scale over the perimeter.

Perimeter security issues that fall under the heading of Counter Terrorism Design are addressed in Sect. 6.17.

4.2.4 Defensive Planting

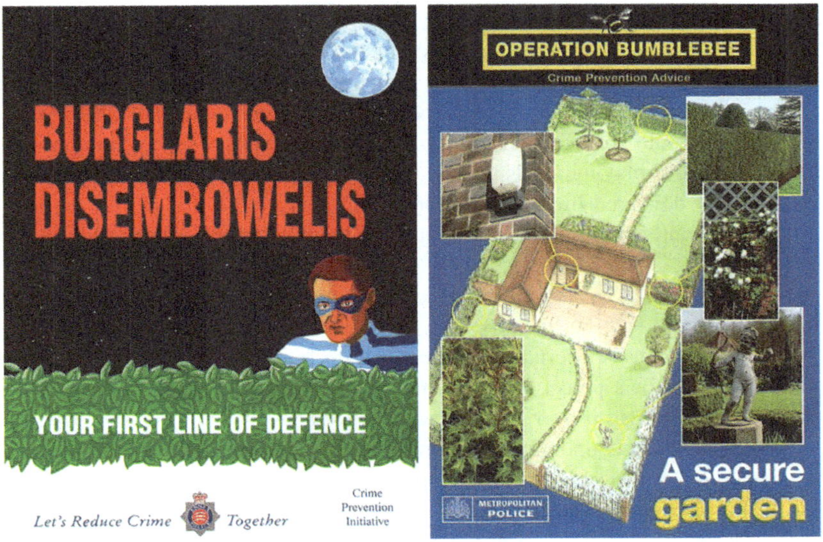

Burglaris Disembowelis: Your First Line of Defence; Stop Garden Thieves; and Operation Bumblebee: A Secure Garden, pamphlets

'Burglaris Disembowelis—Your First Line of Defence', half-joked a Crime Prevention initiative from Essex Police (1995) listing the prickly plants and shrubs suitable for the purpose of Defensive Planting. This caused certain elements in the media and some other commentators to raise an eyebrow, regarding the effectiveness of such measures. Yet as will be described in Sect. 4.4.2. Natural Surveillance, poorly laid out or maintained (overgrown) vegetation can act as a cover for crime or anti-social behaviour (ASB) and indeed, raise the fear of crime (FOC). In reverse, where the aim is to prevent intrusion the use of plants, shrubs and trees with spiny leaves or thorns can be highly effective. Or so we perceive, there being a distinct absence of what might be described as research and evidence emanating from Crime Science.

The use of such plants and shrubs to prevent intrusion is based on a double premise. First, that visually the thorns or prickly leaves inherent to these botanical specimens, will deter potential intruders (see Sect. 2.8, Environmental Cues). Second, that if they don't see such spiny elements or choose to ignore them, first contact with the plants or shrubs will be sufficient to make them desist from their anti-social and/or criminal activities.

Naturally, there is a history of practice to support the use of prickly shrubs and plants. For example, it is relatively common for householders in the UK to 'allow' (most often they have no choice in the matter as the things just simply appear!) a patch of stinging nettles or brambles to grow at the bottom of their garden, where it backs onto public open space or some other potential Attack and Escape Route for criminals—such as canal towpaths, railway lines, a golf course or school playing fields. No budding burglar wants to find himself stung by nettles or devoured by bramble thorns! But please note that nettles die back during the winter months and consequently their deterrent value is somewhat seasonal.

First, the mandatory and all too necessary health and safety warning. The use of prickly shrubs will be inappropriate in an environment where young children are expected to play or certain pets are likely to roam. As ever, the golden rule is that the cure for a perceived problem should not generate more issues for the innocent parties, than it ever will for potential offenders.

Prickly planting is undoubtedly more aesthetically pleasing than barbed wire, razor tape; or broken glass, carpet gripper and the like embedded into the top of a wall. Furthermore, whilst the use of the latter may be illegal (in contravention of the Occupiers' Liability Acts, 1957 and 1984 (see Sect. 8.1, Legal Liability) in England and Wales and equivalent legislation in Scotland and Northern Ireland), or require planning consent from local government (certainly true in much of the UK), it is difficult to see how this could ever be the case in the instance of planting prickly shrubs. The obvious exception is a 'high hedge' (more than 2 metres in height) covered by the Anti-Social Behaviour Act, 2003: Part 8. Similarly, unless used for agricultural purposes, electric 'pulse' fencing will often require planning consent, whereas prickly shrubs, e.g. rambling or climbing Roses, Pyracantha or Berberis, will not.

The use of prickly vegetation falls into two distinct categories. First, as ground cover—where the aim is to prevent trespass onto a particular area. Here the co-aim should be that such shrubs do not obstruct the aforementioned Natural Surveillance. Consequently, species that will grow to no more than one metre (3.25 feet) high should be used and where the aim is to prevent theft in car parks, this height should be reduced to 0.65 metres (2 feet) so as to provide no cover for crouching criminals attempting to break into parked and unattended vehicles. And on the negative side, such ground cover prickly planting is notorious for acting as a magnet to items of litter—which not surprisingly few people want to remove for fear of injuring their hands. And thus, the resort to litter picking equipment.

It is also vitally important that the correct species of prickly shrub is planted. This means that genuinely dwarf-growing species should be chosen if later problems with maintenance are to be avoided. There is abundant anecdotal evidence (to which this author can testify) that whilst dwarf-growing species were selected by the landscape architect or developer, full-growing versions costing far less financially were actually planted, the rampant results of which (acting as a barrier to Natural Surveillance—and often Formal Surveillance and Technological Surveillance), were not discovered until a few years later when these shrubs had become fully established. Maintenance issues (principally those of financial cost), are then incurred when such shrubs have to be pruned back—often as frequently as to be at least an annual exercise and sometimes even twice in a season.

The second and most common category of Defensive Planting concerns prickly shrubs that are used to protect the perimeter. Such shrubs can be used in isolation to produce a hedge (such as Hawthorn, Berberis or Hedging Rose—Rosa Rugosa), or to reinforce an existing perimeter structure such as a wall or fence. In the case of the latter, rambling or climbing Roses, Pyracantha and Blackthorn are especially appropriate.

Choice of species will almost certainly be determined by local growing conditions—extremes of temperature, drought etc. However, by their very nature, prickly shrubs tend to be extremely hardy and unlikely to wilt at the first shock to their system. That said, young shrubs will need to grow and become established. All too often prickly shrubs have been used in an attempt to close off an unofficial access point, only to find themselves trampled into the dust within weeks (if not days!) of being planted. In such locations, a combination of physical fencing (or at least post and wire) and prickly shrubs is recommended—and even this will not always be sufficient to prevent determined attempts to reopen such perceived (however inaccurately) rights of access.

One additional plea, over the years there have been some strange inclusions amongst the lists of prickly shrubs considered suitable for Defensive Planting. In particular, Gunera Manicata (otherwise known as Giant Rhubarb or Chilean Rhubarb), has repeatedly appeared. Whilst it is true that Gunera has prickly stems and can make for an impenetrable barrier during the summer months (as spectacularly demonstrated at the 'Lost Gardens of Heligan' in Cornwall), in climates where frosts are common during the winter (as in the UK), Gunera will die back completely and consequently has no deterrent value. And in addition, being a bog plant, it also requires copious amounts of water to survive.

Similarly, whilst at first glance Cacti such as the Prickly Pear (Opuntia) variety might appear to make an impenetrable barrier (and can be seen to be used for this purpose in Mediterranean countries), following a heavy frost these plants will turn to mush and almost certainly never recover. And to reiterate, Stinging Nettles (Urtica Dioica) also die back each winter!

The following list of trees, shrubs and prickly plants are considered suitable for the purpose of Defensive Planting. The Latin names by which they are internationally known, are given first, their more common UK name (where known) in brackets. The list is by no means exhaustive and other shrubs will undoubtedly be considered suitable for the purpose of providing a barrier to criminal attack—especially in hotter, more drought-prone and frost-free parts of the world. Local growing conditions will need to be considered, especially in the immediate months after planting when they are attempting to become established.

Special thanks are made to the aforementioned Essex Police (1995) and its Burglaris Disembowelis leaflet; to the Metropolitan Police for its pamphlets (distributed at the world-famous Royal Horticultural Society Chelsea Flower Show, Hampton Court Palace Flower Show and similar events), 'The garden strikes back: a Guide to plants that burglars hate' (1996); and Operation Bumblebee—A Secure Garden (1999). Appreciation is also given for the BBC Radio 4 programme 'Gardeners' Question Time' (GQT) devoted to the issue of Defensive Planting, broadcast in 2008. During that programme the GQT team members ably demonstrated how they were themselves highly experienced Street Level (or should that be garden level?) Crime Scientists. See also the below sub-section devoted to shed security.

List of suitable species:

Aralia Chinensis (Chinese Angelica Tree) shrub or small tree with stout spiny stems

Aralia Spinosa (Hercules' Club or Devil's Walking Stick) shrub with viciously spiny stems

Berberis (Barberry) different sized varieties such as Blenheim, Darwinii, Gagnepainni, Goldilocks, Hakenoides, Interposita, Julianae, Mucrifolia, Ottawensis, Stenophylla, Sublevis, Thunbergii Attropurpurea Rose Glow (ideal for window boxes), Verruculosa, Wilsoniae—and all with thorny stems

Chaenomeles (Quince) varieties such as Californica, Cathayensis, Japonica, Speciosa, Superba or Mooerloosei—a genus of deciduous thorny, spring flowering shrubs

Colletia (Anchor Plant, Jet-Plane Plant, Crucifixion Thorn, Thorn of the Cross) varieties such as Armata, Cruciata, Hystrix, Ferox, Paradoxa, Spinosa—a genus of spiny, slow-growing shrubs

Cortaderia (Pampas Grass) horticultural favourite of the 1970s, though currently 'out of fashion', with razor sharp leaves

Crataegus Monogyna (Hawthorn) the most common spiny agricultural hedging shrub/small tree in the UK producing white or pink flowers and red berries

Elaeagnus (Silverberry or Oleaster) varieties such as Angustifolia Oleaster, Ebingeii Gilt Edge or Limelight—large spiny shrub or small tree

Hippophae Rhamnoides (Sea Buckthorn) deciduous shrub with silver foliage and large thorns

Ilex (Holly) varieties such as Altaclerensis Atkinsonii, Balearica, Wilsonii; Aquifolium Handsworth New Silver, Ferox Argentea, Angustimarginata Aurea, Crassifolia, Madam Briot; Spinigera—the red (usually) berried and prickly leaved favourite of Christmas cards and the Festive Season

Maclura (Hedge Apple, Horse Apple, Mock-Orange) varieties such as Ponifera Osage Orange—small tree with thorny branches, often used in parts of the USA as an impenetrable hedge

Mahonia (Oregon Grape Holly) varieties Confuse, Fremontii, Japonica Bealei, Lindsayae Cantab, Media Charity, Pinnata—winter-flowering evergreen shrub with spiky leaves

Oplopanax (Devil's Club) varieties such as Horridus—deciduous shrubs with spiny foliage

Osmanthus (Fragrant Olive) varieties such as Armatus, Burkwoodii, Heterophyllus, Serrulatus—evergreen shrubs some of which have holly-like prickly leaves

Pernettya Mucronata (Prickly Heath) small compact shrub with prickly evergreen leaves and white, pink, red or purple berries

Poncirus Trifoliata (Japenese Bitter-Orange/Flying Dragon) a very spiny shrub, ideal as a protective hedge

Prunus Spinosa (Blackthorn) large family of flowering trees and shrubs, not all of which are spiny

Pyracantha (Firethorn) varieties such as Angustifolia, Golden Dome, Orange Glow, Shawnee, Cadrou also known as Saphyr Rouge—thorny shrubs bearing red, orange, or yellow berries in autumn and ideal for trellis or along fences

Rhamnus (Common Buckthorn) shrub or small tree, the branches of which end in a woody spine—varieties such as Cathartica

Ribes Uva-Crispa (Gooseberry) varieties such as Invicta—prickly shrub producing edible fruit

Roses—virtually all forms but especially:-

Rosa Eglanterias—almost all varieties are suitable for creating an impenetrable barrier

Rosa Rugosa (Hedging Rose) varieties such as Rosa Blanc Double de Coubert, Rosa Roseraie de l'Hay, and Rosa Fru Dagmar Hastrup—all ideal for creating an impenetrable hedge

Rosa Sericea f. Pteracantha (the author's personal favourite with 1/2 inch/12 mm shark's fin thorns!) Only to be used as a deterrent and never where it might cause harm to children, pets etc.

Rosa Spinosissima—again, virtually all varieties are ideal for creating an impenetrable barrier

Rosa (climbing roses) varieties such as Mermaid, Helenae, Paul's Himalayan Musk—ideal for growing through trellis and for the top of fencing

Rosa Kiftsgate (a particularly rampant climbing rose, to be used only where growth of potentially 15 metres/50 feet is required and will be expected!)

Rubus (Fruiting and Ornamental Brambles) varieties such as Cockburnianus Goldenvale, Bilorus, Lambertianus—all renowned for their barbed stems

Smilax (Sarsaparilla or Rough Bindweed) varieties such as Aspera—evergreen climbing shrub with sharp thorns

Ulex Europaeus (Common Gorse or Furze)—dense, prickly shrub that thrives in poor conditions and on acid soils in particular

Yucca Gloriosa (Spanish Dagger) clump forming shrub with long spiky leaves

Zanthoxylum (Hercules Club, Toothache Tree) varieties such as Clava-Herculis, Hirsutum—large shrub or small tree bearing pairs of sharp prickles along the branches.

See also: https://thecrimepreventionwebsite.com.

Garden Security

As one of the Physical Security 'Target Hardening' Measures of CPTED, the issue of prickly planting is most usually included under the heading of 'Garden Security'. In the UK, such Crime Prevention campaigns are rolled out each spring with the regularity of hearing the first cuckoo. And with good reason. Sheds and outbuildings often have the poorest forms of security (if any!) imaginable, yet are used to store items of considerable resale value. Lawn mowers and power tools (such as electric hedge trimmers) provide easy pickings for both opportunist thieves and their professional/career/travelling cousins (see Sect. 2.3, Crime Opportunity).

At the top end, high value items (such as antique garden furniture and ornaments), will often feature in the UK Art Loss Review (www.artlossreview.com) and/or Antiques Trade Gazette (www.antiquestradegazette.com) and on Trace (www.trace.com) as the proceeds of theft by such professional/career/travelling criminals. Standard Crime Prevention/Reduction advice should be applied, in that such valuable and most often highly identifiable items, need to be photographed and property marked by means of a system appropriate to the cost of replacing the object in question. In other words, a bronze statue worth several thousand pounds merits more than simply engraving the item with the householder's post-code (zip-code). At the very least, proprietary Property Marking identification and an alarm unit designed for such specialised applications, should be installed.

Furthermore, where gardens feature in the design process of either new or refurbished premises, it is strongly advised that the following security measures are incorporated—or advice provided as per Good Housekeeping Measures (see Sect. 8.3).

Boundary, Fencing, Walls, Railings, Gates and Hedges

Perimeter elements in a rear garden should be difficult to climb—and thus an impenetrable hedge and/or prickly planting (as detailed above) can be utilised for this purpose. Gates that provide access to the rear garden should be of the same height as adjacent fencing/wall/hedge (ideally a minimum of 1.8 metres—6 feet), but see also Sect. 4.2.3, Perimeter Security; and Sect. 5.4, Secured by Design—Residential Guide 2025). They should be bolted at the top and bottom and secured with a close-shackled padlock or similar locking device. Wooden trellis of no more than 650 mm (2 feet) high can be fixed to the top of perimeter fencing and walls, of such structure that it has no use as a climbing aid and if used as such will break, create noise and hopefully attract attention—or at least give the impression that it might do so. Such trellis can also assist the growth of the aforementioned prickly plants and shrubs, although once again raising the height of the fencing above 1.8 metres (6 feet) may require Planning Permission from the relevant local authority (local government).

Fencing, walls, railings or a hedge providing the perimeter to a front garden should be no more than 1 metre (3.25 feet) in height so as not to inhibit Natural Surveillance, see Sect. 4.4.2, unless constructed of railings or otherwise easy to see through. Yet again, see Sect. 5.4. Secured by Design—Residential Guide 2025. And remaining with the theme of front gardens, only trees with a high (above 1.8 metres/6 feet) canopy should be planted. Young conifer trees may be aesthetically pleasing (and beloved by show-home landscape designers), but once fully grown their low canopies obstruct Natural Surveillance—especially in the instance of fast-growing species like Leyland Cypress (Cupressocyparis Leylandii).

Garden Sheds and Other Small Outbuildings: 'Sheducation'

Where garden sheds are a necessary element of the design process, the doors should be secured with a good quality hasp held in place with concealed screws or coach bolts pinned through a steel plate, in order to prevent tampering. Hinges should be similarly secured. A close-shackled padlock, minimum width 64 mm (2.5 inches) made of hardened steel should be used to secure the hasp. Window grilles may be useful and consider advising the end user to hang net curtains, so that a potential thief cannot see what is inside. Ideally, valuable items (including motorised lawnmowers, power tools and bicycles) should not be stored in sheds where forced entry could be easily achieved. Nevertheless, it is better to secure such items (plus garden tools and ladders) inside, rather than outside where they could be put to illicit use by criminals. And whether indoors or outdoors, ladders and garden tools should be chained and padlocked to immovable fixtures/each other.

In addition, the DCLG's (Department for Communities and Local Government) Code for Sustainable Homes (2006—now repealed) insisted that such sheds on new housing developments must also be provided with a ground-anchored security arm to which bicycles, lawn mowers etc. can be locked for security purposes.

An edition of the aforementioned BBC Gardeners Question Time broadcast in December 2011, addressed the issue of allotment security. Highly practical advice was given to grow fruit and vegetables that are either difficult or take time to steal (e.g. prickly stemmed fruit). Whilst the panel's esteemed Bob Flowerdew recommended post-coding, securing together with steel cable and spraying/painting garden tools in lurid colours. The use of audible alarms and Property Marking systems were also discussed.

Security Lighting

The use of Lighting to deter crime, ASB and reduce the fear of crime (FOC) is dealt with in the following Sect. 4.3.1 and the same philosophy (as a proven deterrent to crime) applies in a garden setting. However, it is perhaps wise to repeat the warning regarding PIR (passive infra-red) controlled Exterior Lighting. Namely, that it is prone to flash on and off at the behest of small animals (birds and cats in particular) or when tree branches move in strong winds, thereby having the capacity to raise FOC. Moreover, when the property is unoccupied and not overlooked by neighbours (see Sect. 4.4.2, Natural Surveillance), there is little point in illuminating the environment if it will only serve to light the way for potential intruders. Nevertheless, an argument could be made for bathing in light a prickly hedge as someone approaches, so as to provide a warning and thereby reinforce the deterrent nature of such a barrier. As is often the case, the local 'site specifics' must be considered as they will inevitably influence the final decision(s) made by the Designing Out Crime Officer (DOCO).

Whilst on the subject of Garden Lighting, it is timely to reiterate the counsel against Bollard Lighting mentioned in Sect. 4.3.3, Closed Circuit Television (CCTV). These have been prized by architects since the age of Le Corbusier and Frank Lloyd Wright. However, as the Lighting unit is most usually situated on top of the 'bollard' 1 metre (3.25 feet) or less from the ground and with the light projecting downwards, they have an inherent tendency to illuminate only the lower half of someone's body. Consequently, whilst they are ideal for marking out pathways and intended directions of movement during the hours of darkness, they unwittingly raise the FOC—in that you cannot automatically identify someone who is walking towards you as their face is usually hidden in darkness.

Worse still, when used in conjunction with CCTV cameras (see Sect. 4.3.3) bollard lights have a tendency to seriously compromise the images produced, causing glare/flare and invariably resulting in the inability to recognise individual faces—the primary role of CCTV in any investigative (and indeed evidential) process.

Conclusion

Defensive Planting is a valuable constituent of CPTED and arguably best located within Physical Security 'Target Hardening' Measures. As a result, it is a key intervention for the Designing Out Crime Officer with a major role in preventing crime and ASB. And whilst evidential support from Crime Science is currently somewhat lacking, experienced gardeners (like this author) will testify to its effectiveness.

Moreover, when combined with Garden Security measures in general, Defensive Planting has the capacity to provide a holistic strategy for Crime Prevention/Reduction, especially in respect of the garden area—between the property and the boundary. And when perceived as part of the overall package of measures designed to protect the Target Enclosure (most often the dwelling containing the actual target—jewellery, cash, electrical goods etc.) its role as part of a Layered Security approach can be easily appreciated.

4.3 Technological Innovations

4.3.1 Lighting

The Police Crime Prevention Initiatives Guide to Lighting

Background

The value of light (and originally fire) in deterring attack has been appreciated by humanity for many thousands of years. Moreover, it can be similarly identified as both a defence mechanism and a lure within the animal kingdom (e.g. Glow

Worms, Electric Eels and certain types of Jellyfish). In early nineteenth century London, even Members of Parliament were forced to employ torchbearers to light their way home following late night sessions and to protect themselves from attack—one of the seminal reasons cited for the creation in 1829 of the Metropolitan Police in London. And throughout the Victorian era the advent of gas-powered street lighting to prevent crime (and equally if not more importantly, reduce the fear of crime (FOC) was rapidly practised across Britain and around the world. The invention of electric light accelerated this process, to the extent that light pollution is now considered a menace by many—and is certainly wasteful in terms of energy efficiency and its carbon cost (see Sect. 2.11).

Not recommended! A lethal combination: Buddleia Davidii ('Butterfly Bush'), electric pulse fencing and broken glass embedded atop a six feet (1.8 metres) high brick wall

From the perspective of Designing Out Crime, Lighting is important because of both its perceived and real effects in reducing crime, FOC and anti-social behaviour (ASB). Consequently, this chapter is divided into three parts. The first provides a summary from The Police Crime Prevention Initiatives Guide to Lighting (2021) and the standard BS 5489-1:2020. The second deals with the effects of Street Lighting on crime, in which arena there has been considerable academic research in recent decades and in the UK context, most especially over the past quarter century. The third part is devoted to Security Lighting, be it for domestic or commercial/non-residential environments.

The Police Crime Prevention Initiatives Guide to Lighting

This guide was produced in 2021 by Police CPI at Home Office request in support of the Safer Streets Fund. The latter encourages Police and Crime Commissioners (PCCs) and Local Authorities (LAs) to bid for funding in respect of initiatives proven to prevent crime, make residents safer and reduce demand on the police.

The guide aims to increase awareness of important aspects that must be considered before and during delivery of the intervention. This is vital to ensure the opportunities presented by Safer Streets to achieve sustainable reductions in crime are realised. In addition, accurately matching the Crime Prevention tactic to a specific Crime Risk evidenced by statistics and local knowledge, is critical to the success of Safer Streets. The guide is strongly recommended as a valuable tool for the DOCO.

What Is Lighting for Security?

All human beings want to feel safe and for this purpose good lighting conditions at night are vital. In certain circumstances, darkness can promote the fear of crime (FOC). Indeed, in a public place we may feel apprehensive unless able to see that this is not true.

Lighting is an effective security measure and a useful tool for public reassurance. It enables people to confirm they are safe, assess a developing threat and if necessary, take an alternative route. Recent events reported in the national news have raised concern over safety in public places. These reflect the value people place in being able to move around public places at night under high-quality lighting.

Good lighting is defined as an amalgam of qualities ranging from the appropriate amount of light and uniformity, high colour rendering qualities of lamps to energy efficient installations and the prevention of light pollution. The right type of light should be provided at the right time in and in the right places, with an appropriate control system. Secured by Design and the Institution of Lighting Professionals' Lighting Against Crime guide defines good Lighting as, a system designed to distribute an appropriate amount of light evenly with uniformity values between 0.25 Uo and 0.40 Uo, using lamps with a rating of at least 60 on the colour rendering index.

The Definition Explained

1. An appropriate amount of light: the level of illumination provided is appropriate for the visual task to be completed and suitable for the environmental conditions of the system.
2. Overall Uniformity (Uo): even distribution of light across the area being illuminated.
3. Colour Rendering: the ability of lighting to show the colours of objects as close as possible to their true tint, hue and tone.

4. Colour Rendering Index (CRI): the method used to indicate the colour rendering properties of a lamp with values ranging from CRI 0—CRI 100 (representing colour rendering qualities of conditions close to natural daylight).
5. Light Pollution: where the lighting installation causes 'nuisance' to others not directly involved.
6. Energy Efficiency: lighting equipment specified making efficient use of the electrical energy used.

Why Install Good Lighting?

Targeted Lighting improvements reduce crime and fear of crime, whilst simultaneously enhancing the general feeling of public pride in a location. In addition, unintentional benefits can radiate out from where Lighting improvements have been made and when allied to reductions in recorded crime, create a substantially positive picture. Whilst good lighting serves the public well, for criminals it can deter, delay and/or detect criminal activity:

> **Deter**—appropriate levels of lighting correctly switched at the right time, deters criminals as they cannot act under the cover of darkness.
> **Delay**—lighting forces criminals to use a time-consuming subterfuge as their criminal activities are able to be observed by responsible guardians.
> **Detect**—the correct levels of lighting provided will ensure criminal activity will be seen in commission by anyone who cares to look.

There may not be anyone there to harm us, but we will only be reassured if the lighting regime is so organised it allows us to confirm this is the case.

Where to Employ Lighting for Security?

Crime Pattern Analysis will identify where lighting improvements could be best employed to combat crime and fear of crime in areas of large public resort, high levels of pedestrian traffic, town centres or entertainment venues etc. In addition, it may be useful to assess the public's perception of their safety at night in local public spaces. Regardless of crime reports, it is the locations where people feel most vulnerable at night that targeted lighting improvements may benefit most.

When Is It Appropriate to Use Lighting for Security?

A Problem-Solving approach will help identify and analyse the crime problem of an area and determine whether improved lighting is an appropriate solution. This can be best described in the context of the four stages of the SARA model (Eck and Spelman, 1987)—a rational method to systematically identify and analyse problems, develop specific responses to those problems and subsequently assess whether the response has been successful:

Scanning—identification and prioritising of potential crime and disorder problems.
Analysis—analysis of a potential problem, by gathering information and intelligence to identify the underlying causes.
Response—development and implementation of tailored activities to address the causes of the problem, as identified in the analysis phase.
Assessment—measuring impact of the response to test if it had the desired effect and to make changes to the response if required.

As an example, Scanning may identify a neighbourhood is suffering a disproportionately high level of crime and/or fear of crime. Analysis will examine the data more closely and might reveal the active timeframe during which crime is being committed, environmental factors such as poor lighting and the absence of Natural Surveillance contributing to higher levels of crime. Assessment of the approach is used during and after the initiative to understand if the intervention was successful and how it could be improved. For the best outcomes, assessment should be an ongoing process, applied throughout an initiative so that adaptations can be made as necessary. In addition, how much lighting should be provided and at what times these levels should be applied, are critical for the success of a lighting strategy as this issue will, if correctly calculated, enhance the public's experience of an area whilst simultaneously increasing their safety.

To explain in more detail the issue of what time and how much lighting should be provided, consider a foot route heavily used by large numbers of people, perhaps in the morning or evening between specific times by commuters and/or children travelling to or from school—sometimes with parents but at other times alone. Appropriate lighting is necessary and especially during winter months. However, it may not be desirable to encourage the use of such a foot route late at night when, due to its remote location or low rate of use, it becomes a place of isolation for pedestrians with a potential for ambush or create fear of such an event. The provision of lighting in late night circumstances might work against the public's interest. Consequently, curfew closing times may be thought appropriate, following consultations with all interested and concerned parties. Similarly, it may not be desirable to encourage gatherings in settings such as a children's playground late at night, where street drinkers or other anti-social elements are attracted by the utility and convenience of the light. Careful consideration of such matters is advised when designing a lighting system to suit local conditions and therefore, those with knowledge of the area should be included in the design consultation process.

Local Agency and Public Commitment

A successful strategy depends on several factors, chief amongst which are PCCs, LAs, police and communities working in a coordinated and information-sharing manner to agree a target area where improved lighting could achieve maximum benefit for law-abiding people, in terms of reduced Crime Opportunity and increase in public reassurance.

How to Enhance Lighting for Security?

Installation guidance and compliance with BS 5489-1:2020 is required for any public lighting initiative. Inclusive participation by qualified lighting engineers and design professionals should be in place from the earliest stages of the initiative. Once the target area for lighting improvements has been identified by tactics such as local consultation and Crime Pattern Analysis, a number of key questions have to be addressed: lighting improvements can take several different forms including, increasing illumination levels of an existing installation through augmenting an extant system with extra luminaires, or installing an entirely new lighting system, with columns or fixings in a previously unlit area. Whatever lighting improvements are necessary, getting the engineering and design decisions right in the following key categories of specification will be critical factors in achieving the Crime Prevention success of a lighting regime.

The impact, appearance, quality and utility of a lighting system, together with its ability to fight crime and enhance public reassurance, will be facilitated and improved by compliance with advice and guidance for lighting engineers and designers contained within British Standard 5489-1:2020. This in turn supports the content of British Standard EN 13201-2:2015 Road Lighting. These standards represent essential guidance for lighting professionals to ensure designs and specifications are appropriate and justifiable.

The British Standards Institute state that their BS 5489-1:2020 includes:

- Recommendations on the general principles of Road Lighting and its aesthetic and technical aspects
- Guidance on operation and maintenance
- Guidance on means of minimising energy consumption and limiting the impact on the environment and adjacent property

- Recommendations for the design of lighting for all types of highway and public thoroughfares, including those specifically for pedestrians and cyclists and for pedestrian subways and bridges
- Recommendations for the design of lighting for urban centres and public amenity areas
- Recommendations and guidance for lighting relating to smart cities
- Additional recommendations for lighting around aerodromes, railways, coastal waters, harbours and navigable waterways to minimise the possibility of the lighting interfering with these modes of transport.

How Has BS 5489-1:2020 Changed Since the Last Update?

The principal changes compared to the 2013 edition are to align the standard with current best practice and with the BS EN 13201 series. New guidance is provided for lighting design when located within a 'smart city' environment or when co-located with electric vehicle (EV) charging points.

Amongst the many design and engineering principles that need to be considered when improving lighting to fight crime and fear of crime are the following:

- Creating a sense that the the lighting system complements the utility and distinguishing characteristics of an area
- Employing only time and circumstances sensitive levels of light for the task in hand, so that it is entirely sympathetic to the use that a given area is being put to and the local environment
- Lighting should be suitably controlled with the possibility of variable switching, remote management and monitoring considered to make the temporal/utility changes in light levels possible, that would better the potential for altering circumstances in a public place. Large public events would require high light levels, whereas the same areas when unoccupied might not
- Efficient maintenance regimen should be adopted to avoid premature fall-off in performance and efficacy of a system
- Careful positioning of lighting columns and fittings to avoid providing unnecessary climbing aids, or where they might cause obstruction, or contribute to any overall street clutter
- Through careful design and flexible management, a lighting design can promote public safety whilst minimising light pollution, sky glow and public nuisance, e.g. by using cut-off luminaires which prevent light being emitted in an upwards direction

- Care must be exercised when placing lighting on domestic property, within industrial parks or on privately owned places of public gathering because, subject to specified regulatory exemption, light being emitted from such premises can constitute a nuisance as defined under the Clean Neighbourhoods and Environment Act 2005
- Using lamps with high colour rendering qualities (e.g. 60 or above on the Colour Rendering Index) often improves visual performance and people's personal experience of an area. Colour rendering qualities of lamps refers to their ability to represent the colour of objects under illumination. To help with lamp selection, a measure is given from 100 (the colour rendering qualities of daylight) all the way down to 0 where no colour quality is available. Seeing true colours of objects under night-time lighting conditions, will make the surroundings appear more familiar and comfortable with attendant benefits of higher levels of public reassurance and satisfaction
- Overall lighting uniformity (Uo) levels of 0.4 or 40% are recommended where possible to ensure lighting does not create dark patches next to lighter ones where the human eye has difficulty in adjusting quickly to see it is safe to proceed along any route. If high levels of uniformity are neither achievable nor appropriate for technical or locally applying environmental reasons, the highest levels of uniformity possible should be achieved
- Lighting systems, either newly installed or as works undertaken to improve any existing lighting, should be specified with the requirements of existing or future CCTV systems. Different camera specifications will operate more efficiently under lighting matched to such specifications. It is important to match the Operational Requirements (OR) of the CCTV system with the Lighting's performance, as should the camera be expected to capture facial images for identification purposes, different levels of light may be required from those only needed to facilitate a camera capturing images of groups of people moving at a distance
- It is a failure of some CCTV systems to perform to their fullest potential, because the lighting under which they are expected to work at night is not compatible with their camera's specification and performance.

Who Should be Involved in Lighting for Security?

Partnerships formed between residents, LAs, police, local commercial interests and visitors, particularly those that drive the local economy such as in holiday destinations, should be canvassed to take account of their opinions. These opinions should be folded into how the campaign is designed. It is recommended representatives of these groups are consulted at the design stage of the initiative to improve the chances of a successful outcome.

Police Crime Prevention Initiatives

Police CPI is a police-owned organisation that works to deliver a wide range of innovative and groundbreaking Crime Prevention and demand reduction initiatives to support the wider UK police service, Government and the general public. Police CPI maintains close working links with the National Police Chiefs' Council National Leads, central government, manufacturers and companies involved in security products (within the UK and those in countries that supply the UK), standards authorities and key stakeholders such as academics, architects, developers, housing associations, LAs, planning officers and the public. Police CPI is a not-for-profit organisation, which raises its own money and delivers significant crime reductions at no cost to the police service or the public purse. Senior police officers from England, Scotland, Wales and Northern Ireland control and direct the work Police CPI carries out on behalf of the police service.

Secured by Design

Secured by Design is a police initiative operating throughout the UK, that encourages manufacturers of security products, to have them successfully tested and third-party certificated by a UK Accreditation Service (UKAS) body to police-preferred standards of security. SBD is one of several Crime Prevention schemes operated by PCPI, which is owned by the UK police service, supported by the Home Office.

SBD lists many approved Crime Prevention products and services. It is an invaluable resource for guidance on many things including security standards, Crime Prevention and product procurement. For more information go to: https://securedbydesign.com

Research into the Effectiveness of Street Lighting in Reducing Crime

Professors David Farrington and Brandon Welsh (2002) produced the first highly detailed systematic review of the accumulated evidence and meta-analysis of the effectiveness of Street Lighting in reducing crime. Indeed, their updated (Welsh & Farrington, 2008) study is believed to be sufficiently important to warrant its inclusion in Sect. 10.2, Suggested Reading.

Welsh and Farrington (2008) detail how studies evaluating the effectiveness of Street Lighting in reducing crime began in North America in the 1960s. Early results were positive (Wright et al., 1974). However, Tien et al. (1979) reviewed

15 out of an original 103 Street Lighting schemes (88 were rejected for failing to contain sufficiently rigorous evaluative information), concluding the results were mixed and generally inconclusive. Indeed, even these 15 were seriously flawed and cited issues with weak project design, misuse or absence of sound analytic techniques, inadequate measures of Street Lighting and failure to appreciate the effects of lighting on different types of crime. The research of Tien et al. (1979) suggested Street Lighting had no effect on crime and investigation in this field then came to a halt.

Very little research relating Street Lighting with crime took place in the UK until the late 1980s—Fleming and Burrows (1986). Atkins et al. (1991) and Ramsay and Newton (1991) asserted that improved Street Lighting had little if any effect on crime. However, Painter (1994) concluded that in three different areas of London (within Edmonton, Tower Hamlets and Hammersmith/Fulham), crime, disorder (ASB) and FOC reduced, whilst simultaneously pedestrian street use increased. Enter Crime Prevention 'Tsar', Professor Ken Pease who declared: 'the capacity of street lighting to influence Crime has now been satisfactorily settled…' and that the debate be moved from the sterile 'does it or doesn't it…' to the more positive 'how can I flexibly and imaginatively incorporate lighting in crime reduction strategy and tactics' (Pease, 1999, p. 72).

Rationale for the Effectiveness of Street Lighting in Reducing Crime

Welsh and Farrington (2008) detail how Street Lighting improvements can be viewed as a Situational Crime Prevention/Reduction measure that focuses on reducing Crime Opportunity (see Sect. 2.3) and increasing perceived risks, by changes to the built environment (Clarke, 1995). Other explanations contend increased street use will strengthen informal social control and community cohesion (Angel, 1968; Jacobs, 1961) and neighbourhood investment (Taub et al., 1984; Taylor & Gottfredson, 1987). Similarly, Kelling and Wilson's Broken Windows theory (1982) and Skogan (1990) stress the importance of neighbourhood investment in order to strengthen community confidence, cohesion and social control. As such, the introduction of increased Street Lighting should NOT be perceived as relevant solely to night-time crime, a repeated failing amongst those evaluations that witnessed no decrease.

Most importantly, the authors observe that whilst enhanced Street Lighting is a tangible alteration to the built environment, it is not a physical barrier to crime. Nevertheless, it can act as a catalyst to stimulate Crime Reduction by changing the perceptions and behaviour of residents and offenders.

Causal Links: Street Lighting and Crime

A number of sample hypotheses of how Street Lighting may reduce crime are described by Welsh and Farrington (2008):

- Improving visibility: deterring offenders by increasing the risk they will be recognised/interrupted, whilst authority figures become more visible
- Encouraging street usage which enhances Natural Surveillance and with it the number of potential Capable Guardians (Cohen & Felson, 1979)
- Enhanced visibility and increased street usage may interact to increase possibilities for Informal Surveillance and thereby influence potential offenders' perceptions of risk (Bennett & Wright, 1984; Newman, 1972, 1973, 1996)
- The renovation of the environment through increased Street Lighting may act as a psychological deterrent (see Sect. 2.12)
- Lighting may improve community confidence
- Improved Lighting may reduce the FOC.

Meta-Analysis of Existing Studies

Welsh and Farrington (2008) employed a systematic review of all the existing published research relating to improvements in Street Lighting and its effect on crime. The research criteria were especially severe and consequently, only eight (of the original 103 American studies) and five British studies (at least eight others from the UK were rejected, largely because there was no control area, or crime was not measured, or too small a sample size area), together with one from Australia and another from The Netherlands). Thereafter, a meta-analysis of the 13 studies that had made it past the final hurdle was conducted to produce the final results and conclusions.

Summary of Results

Four of the eight American studies demonstrated that improved Street Lighting led to reductions in crime, whilst the other four showed that it did not. However, a meta-analysis of all eight American studies disclosed that improved Street Lighting led to a 'near significant' (statistical definition) 7% decrease in crime. Meanwhile, the five British studies showed that improved Street Lighting produced reductions in crime and in the meta-analysis, a considerable 30% overall. A meta-analysis of all 13 American and British studies indicated a 20% reduction in experimental areas as compared with control areas. Or in reverse, crime increased by 25% in the control areas compared to the experimental areas.

Three other conclusions are also worthy of note. In two of the British studies the financial savings from reduced crime greatly exceeded financial costs (by

620% in Dudley, 540% in Stoke-on-Trent) of the improved Street Lighting; the hypothesis that increased Street Lighting causes more crime can be authoritatively rejected; and third, increased Lighting appears to be more effective in reducing crime in stable heterogeneous communities. In other words, '…increased community pride results in decreased crime' (Farrington & Painter, 2002, p. 37).

Conclusions

The evidence in support of the effectiveness of Street Lighting in reducing crime—and by extension FOC and ASB, is emphatic. However, at a time when many cash-strapped local authorities are considering and on occasion taking the decision to switch off their Street Lighting, counsel should be given that this may plunge their citizens into a new dark age—in a range of different forms.

Secured by Design and Street Lighting

Street Lighting features in a number of Development Guides produced by SBD, most especially in the aforementioned PCPI Guide to Lighting (2021); SBD—Residential Guide 2025 (see Sect. 5.4) and in SBD—New Schools 2014 (see Sect. 5.7). The key elements are repeated below. However, the actual Development Guide and subsequent revisions must be accessed online, to ensure the most up to date advice is being obtained.

All Street Lighting for both adopted highways and footpaths, private estate roads, internal roads, footpaths and car parks, must comply with BS 5489-1:2020. Where conflict with other statutory provisions occurs (such as developments within Conservation Areas), these requirements should be discussed with the DOCO and the LA lighting engineer.

It is recognised that a number of local authorities have 'dark sky' policies and deliberately light some of their rural, low crime areas to very low levels of illumination. Whilst others are currently experimenting with 'switching off' street lamps in low crime areas between certain hours of the night, in order to save energy costs and reduce CO_2 emissions.

Landscaping, tree planting and Lighting schemes should not be in conflict with each other. This is a recurring issue with developments, largely due to the number of different parties and local authority departments involved.

The Overall Uniformity (evenness) of light distribution is almost always more important than the levels of illumination being achieved by the system (the levels are determined by BS 5489-1:2020). The British Standards Institute have issued an advisory note stating that they recommend the Uo be at least 0.25 or 25%. A Uo value of 0.4 or 40% is recommended for an SBD Lighting system, but where technical reasons prevent this the very best levels possible are still required and under no circumstances should the rating fall below 0.25 Uo.

The Colour Rendering qualities of lamps used in an SBD development should achieve a minimum of at least 60 Ra (60%) on the Colour Rendering Index. This is scaled from 0 to 100 and indicates the colour rendering qualities of lamps. 0 is a non-existent ability to render colour under illumination (such as low-pressure sodium lamps) and 100 is the colour rendering qualities of daylight. The 'whiter' the light, the better the colour rendition qualities. Properly controlled white light will illuminate an area to higher satisfaction levels for people, whilst actually delivering less light than would be required for similar levels of satisfaction if non-white light sources were used.

The DOCO may request to be provided with a 'Lux Plan' in order that the lighting system can be assessed. The details on the plan must include the maximum, minimum and average lux levels proposed. The plan must also show the Uo and Ra values for the scheme.

SBD—Residential Guide 2025 (and previous versions) authoritatively advises that light pollution must be minimised. All things adjust their behaviour according to natural light. The application of artificial light has improved our experience of the night-time environment, but if this light is not properly controlled both physiological and ecological problems may occur. Minimising light emitted in directions where it is neither necessary nor desirable, is extremely important. Obtrusive Lighting is a statutory nuisance and illuminating areas unintentionally is wasteful. SBD requires that only luminaires with suitable photometry which reduce light spill and direct light only to where it is required, may be used.

In terms of sustainability, consideration must be given to the consequences of turning off street lights. Such a measure may be counter-productive, in terms of CO_2 emissions and lead to the greater use of motor vehicles, because residents are too afraid to use unlit streets. Crime levels and in particular, fear of crime, must also be carefully monitored to assess what impact such an action has made to the community. There are other possible technical alternatives to simply 'switching off', including the use of street lights that are sensitive to levels of moonlight; those that are switched on through the detection of pedestrians or vehicles; and emerging LED technology, which is 80% more efficient than contemporary Street Lighting. Indeed, new LED Street Lighting is already producing better quality and more appropriate levels of light, whilst simultaneously using less energy and therefore being more cost-effective.

Security Lighting

On a practical note, those of us involved in making recommendations in respect of Lighting over the past three decades (especially in relation to the Secured by Design award) have tended to refine our views during the intervening years. For example, as the previous section describes, Lighting is now a proven deterrent to crime, FOC and ASB. However, in the wrong location (isolated away from any hope of Natural Surveillance, Formal Surveillance or Technological Surveillance) such Lighting may simply illuminate the scene for the benefit of the criminal. This is especially important with the Repeat Victimisation 'Modus Operandi' (means of committing the crime).

Similarly, use of PIR (Passive Infra-Red) controlled Security Lighting on dwellings can all too easily prove to be counter-productive, because external PIRs are notorious for being subject to non-criminal conditions. For example, tree branches moving in the wind will activate the PIR and pets or flocks of birds often have the same effect. Thus, rather than alerting the householder or potential witnesses to a pending criminal attack, Security Lighting flashing on and off in such circumstances has the potential to raise the FOC. This is a particular concern in relation to housing for the elderly and vulnerable groups, especially where such householders already have difficulty sleeping.

Consequently, in the twenty-first century DOCOs tend to recommend dusk to dawn ambient level Security Lighting, controlled by either a time-switch or photo-electric cell. And again, it is vitally important than such Lighting is both appropriate (neither too weak or too over-powering) and directed at the key vulnerable points, in order to avoid light pollution.

Secured by Design and Security Lighting

Once again, Security Lighting is a constituent part of the SBD award system, and the following advice has been gleaned from across the Development Guides, most especially SBD—Residential Guide 2025; SBD Non-Residential Guide 2025; and SBD—New Schools 2014.

For homes (dwellings) Security Lighting is required to illuminate all external doors, car parking, garage areas and some footpaths leading to dwellings including blocks of flats. The use of low energy consumption lamps with an efficacy of greater than 40 lumens per circuit watt is required.

Note: The Code for Sustainable Homes (2006—now repealed) required Security Lighting to be switched by PIR and for the lamp not to exceed 150w. SBD has not specified this type of Security Lighting for a number of years following advice from the Institute of Lighting Engineers and police concern regarding the increase in the FOC (particularly amongst the elderly) due to the previously described repeated PIR lamp activations. Research has shown that a constant level of illumination is more effective at controlling the night-time environment.

External Security Lighting must be switched using a photo-electric cell (dusk to dawn) with a manual override. 24-hour lighting to communal parts of blocks of flats will be required (detailed in SBD—Residential Guide 2025). Internal Lighting can give the impression that a dwelling is occupied and possibly act as a deterrent to crime. However, SBD requires that at least 75% of the fixed internal light fittings utilise energy efficient lamps.

In Secured by Design—New Schools 2014, it states that in terms of security, Lighting should be provided to illuminate doors and windows that are overlooked from surrounding, occupied buildings or are surveyed by CCTV cameras (see Sect. 4.3.3). This requirement can be seen as having practical application to many forms of non-residential (commercial) development. The same SBD—New Schools 2014 Development Guide also describes that whilst bollard lights can be useful for way finding, they are generally insufficient for security. This is because they fail to properly model the facial features of pedestrians, hinder immediate identification and produce glare—all of which does little to promote a feeling of safety. SBD—New Schools 2014, provides specific advice in relation to the use of Internal Lighting (inside the building).

Well illuminated external environment—despite flare caused by the (not recommended) bollard lights

And Finally...

There are also forms and colours of light which provide the potential for some innovative applications. For example, ultra-violet (UV) light being used to discover identifying features, previously applied with a pen and whose ink will only show up under such UV reflection. Many proprietary Property Marking systems utilise the same or similar means of identification. And around the world, personal witness testifies to the fact that ultra-violet light is being used to dissuade intravenous drug injection—usually in toilet facilities. However, there are problems with the latter, discussed in Sect. 6.13, Public Conveniences (Toilets) where intravenous drug abusers mark the relevant vein in advance with a marker pen. Similarly, the use of pink light to show up acne and supposedly cause young people not to congregate in a given location, is unscientific, reeks of being yet another urban myth and is also morally suspect.

More seriously, use of blue light to dissuade those contemplating suicide to refrain from taking their lives on the railway lines in both Japan and the UK, has a well-intentioned if evidence-deficient pedigree. The theory is that blue light has both a calming effect and association with the emergency services. Sadly, this appears to be based more on wish-fulfilment rather than scientific evidence. Such blue Lighting could be identified (personal witness in 2017) at the platform ends of stations on the Yamanote commuter line in Tokyo—reportedly repeat suicide locations. However, more effective platform gates (that open simultaneously when the doors of a train stop at the platform) had been subsequently installed. This followed their installation at virtually all stations on the Shinkansen ('Bullet Train') lines and at stations across the Tokyo underground train system. In the UK, blue lighting has been used along platforms at railway stations such as those serving Gatwick Airport and Romford. Nevertheless, platform gates along the eastern section of the Jubilee Line and Queen Elizabeth Line in London have proved to be far more effective. The most important point to remember is that those wishing to take their lives are often both location and method specific—and thus the value of Physical Security 'Target Hardening Measures. See also Sect. 10.1, The Future of Designing Out Crime.

Conclusion

Lighting is now a proven deterrent to crime, fear of crime and anti-social behaviour. Indeed, for the DOCO it is one of the strongest mechanisms in the Crime Prevention/Reduction toolbox, there being a raft of academic evidence that proves its effectiveness. However, concerns regarding light pollution and both financial and carbon profligacy should be taken into account when recommendations are made, so as not to compromise the integrity of this powerfully effective intervention.

4.3.2 Intruder Alarms

Author's Anecdote

When Buckingham Palace was bombed by the Luftwaffe during the 'London Blitz' of 1940, the then Queen of England, Queen Elizabeth the Queen Mother, observed how she was now able to 'look the East End in the eye'. This was a reference to how the people of the East End of London had already been remorselessly bombed, but until that night Buckingham Palace had remained unscathed.

Scroll on almost 60 years to 1996 when the home of this author was burgled (broken into) whilst he was absent from the premises. The burglars forced entry by way of a ground floor kitchen window at the rear of the house, having gained access to the garden via that of a neighbour's whose occupied house they had ignored.

The fact that the premises appeared to be unoccupied, was undoubtedly the final act in their decision-making process (part of their Crime Script, see Sect. 2. 4, Rational Choice Theory) that caused them to decide to break in—using most probably little more than a screwdriver to force open the kitchen window. On which theme, the key-operated integral window lock proved to be totally ineffective (this burglary pre-dating the window security requirements of BS 7950:1997, PAS 24:1999 and subsequent updates).

Following a fairly tidy search the burglars escaped with a number of household electrical goods (TV, video recorder and laptop computer) and to their credit uniformed police officers attended the scene within a few minutes of being summoned. However, the criminals were never caught nor any of the property recovered. And incidentally, the plethora of 'Leaking' Culs-de-Sac that infest this modern village (designed in the late 1960s) undoubtedly acted as ideal Attack and Escape Routes for the criminals.

Furthermore (and the real point of this tale of embarrassment—for an experienced Architectural Liaison Officer (as the DOCO was known at that time), came from the observation made by the Scenes of Crime Officer (now known as

Forensic Scene Investigator—and elsewhere as Crime Scene Investigator—CSI), DC Dave Sheffield. He commented, 'Why didn't you have a proper alarm system fitted? I very rarely attend burglaries at homes that have even so little as a "bells only" (Audible Alarm) burglar alarm system installed'. DC Sheffield's remonstration was taken to heart and an Intruder Alarm system was installed a few days later.

When subsequently delivering the Crime Prevention/Reduction message to any number of residents' groups (Community Consultation) this author was at least able to empathise with the plight of some in the audience (those who had also been burgled)—to look them in the eye.

From a Crime Science perspective, perhaps the greatest relevance in this Crime Script was how the light left illuminated inside the premises (and the dummy alarm boxes fixed to the front and rear of the house just below roof level) had no deterrent effect. In other words, they had no value as Visual Environmental Cues (see Sect. 2.8). However, it could be that the burglars in question were more skilful than might have been supposed. Even if they had spotted the dummy alarm boxes, having approached the building under cover of darkness from the rear, they could have looked through the windows (the curtains/drapes were not drawn—a fundamental mistake) and noticed there were no PIRs (passive infrared detectors)—the primary means by which the majority of domestic Intruder Alarm systems are activated.

Intruder Alarms—Recent History

Despite the proven success of real Intruder Alarms as a deterrent to criminal attack, it may come as something of a surprise for the newly appointed Designing Out Crime Officer to discover that they are not a mandatory requirement of the Secured by Design (SBD) award system (see Chapter 5). Originally, they were to have been a constituent element of SBD guidance, but ongoing problems with the number of false activations in the UK (each necessitating various degrees of police response) meant that the requirement was removed.

Consequently, as a minimum standard for domestic premises, for many years SBD required no more than a 13-amp non-switched fused spur for the later installation of such an Intruder Alarm system. And whilst the number of false calls has been reduced, they still account for the overwhelming majority of all alarm activations that the police attend. This helps to explain why the installation of an Intruder Alarm does not feature as a mandatory requirement within the latest SBD—Residential Guide 2025 development guide (see Sect. 5.4)—or in any previous version. However, the same guide does reference the SBD Alarm Standard for Intruder Alarms.

Types of Intruder Alarm

In the UK, Intruder Alarms fall under two distinct headings:

- Type A—Remote Signalling Systems (registered with their home police force and linked to an alarm company's Alarm Receiving Centre (ARC)
- Type B—Security Systems including Audible Alarms ('bells only' systems).

The overwhelming majority of Intruder Alarms systems in use in the UK (most often in homes) fall into the category of Audible Alarms ('bells only'). However, the police will not normally attend Audible Alarm activations, unless there is corroborating evidence (e.g. from a neighbour who telephones the police to state they can see intruders on the premises).

Alarm Receiving Centre (ARC) Intruder Alarms also require a corroborative element, but this is usually supplied electronically (when the cable is cut, the alarm box 'foamed', or the alarm otherwise disabled), or/and by way of CCTV. Indeed, at the turn of the millennium CCTV corroboration heralded a whole new generation of ARC linked Intruder Alarm systems.

National Police Chiefs Council (2025) Security Systems Policy

This *document* sets out the most recent national (England, Wales and Northern Ireland—separate advice exists for Scotland) police guidance in respect of Security Systems. Indeed, Intruder Alarms are but one part of the rapid development of technology that these guidelines address. It also includes a link to the SBD Alarm Standard Technical Guide.

Type A—Remote Signalling Systems

Detailed guidance covers the monitoring of such systems; the issue of a URN (unique reference number) by police forces to each system; the relevant British Standards requiring compliance; and its application to stolen vehicle tracking systems and own company premises.

All premises with Type A systems shall have at least two keyholders—with replacements available to cover their unavailability.

Non-residential premises may be required (by the local Chief Constable or Commissioner) to have their warning devices (e.g. bells or sirens) delayed for a maximum of 10 minutes to maximise capacity for arrests to be made.

Additional directives are provided in respect of false alarms; police attendance ('Immediate' or 'Withdrawn'); CCTV systems; and Hold Up Alarms (HUA—commonly referred to as a 'Panic Attack Alarms').

Type B—Security Systems

Police attendance to all non-Type A systems 'will require evidence from a person at the scene that a criminal offence is in progress which indicates that a police response is required' (p. 9). There is no guarantee of a police response to Type B systems. And automatic dialling equipment must not be programmed to summon the police. https://www.securedbydesign.com/wp-content/uploads/2018/08/Security_Systems_Policy_2018.pdf.

In addition, the NPCC issued two consumer leaflets in respect of Intruder Alarms:

Commercially Monitored Remote Signalling Intruder Alarms for Police Response

This leaflet is aimed at both non-residential and residential premises. It reiterates the main points contained in the above guidelines, together with practical advice under the headings:

- Choosing the supplier
- Does the quotation…?
- Police policy on Alarms
- Installations
- Police response
- What happens if police remove response?
- Noise nuisance.

Intruder Alarm Response for Domestic Properties

Targeting solely residential installations and building on previous guidance, this leaflet provides easy to comprehend advice under the headings:

- Did You Know…
- Professional installers will…
- Audible Alarms
- DIY
- Remote Signalling Alarms
- Alarm Terminology
- Alarm Sounders
- Zone
- The Control Panel.

Burglars' Opinion of Intruder Alarms

To produce *The Burglars' take on Crime Prevention Through Environmental Design (CPTED): reconsidering the relevance from the offender perspective,* Armitage (2017) conducted semi-structured interviews of twenty-two prolific, convicted and incarcerated burglars. Moreover, the '*rich qualitative data*' was used to produce a more user-friendly summary of the burglar descriptions.

Each burglar was shown 16 photos during the course of each interview and their responses noted. The concept of Surveillance was mentioned 133 times; Physical Security 'Target Hardening' Measures on 103 occasions; Management and Maintenance, 40; Movement Control (Permeability), 39; and Defensible Space, 11.

On the subject of Intruder Alarms, with the exception of one particular brand of ARC monitored system, these were seen as an ineffective deterrent to burglary. Indeed, the offenders also described how to disable Intruder Alarms, by 'foaming' the external alarm box(es) to deaden the sound emanating from the bell(s) or siren. They also stated that after breaking into the house, removing the internal alarm panel would deactivate the majority of alarms. Furthermore, in most instances after breaking into the property, the alarm would fail to activate (had not been set?) And when it does activate, neighbours and passers-by rarely take any notice.

However, it should be remembered that these were 'prolific' burglars being interviewed, described elsewhere in this handbook as 'professional/career/ travelling criminals'. Consequently, a level of sophistication might be expected within their suggested Modus Operandi.

Using the 'international crime drop' hypothesis (Farell et al., 2011), Tseloni et al. (2017) concluded that rather than Intruder Alarms, 'combinations of the most effective devices (door and window locks plus security lighting' (2017, p. 1), led to a decline in forced entry burglaries. The authors drew on extensive research material: British Crime Surveys, Crime Surveys for England and Wales and Office for National Statistics. This includes asking victims about their crime experience. Unfortunately, expecting victims to know which security failing caused their burglary is no guaranteed route to accuracy. Certainly, amongst DOCOs the overwhelming experience is that retrospectively fitted additional door or window locks has negligible effect in preventing burglary by moderately determined offenders.

Intruder Alarms for the Designing Out Crime Officer

As previously described, there is no requirement by Secured by Design for Intruder Alarms to be installed in new or refurbished homes. However, this does not preclude their being advocated by the DOCO, where there is a proven Crime Risk in a given location. Meanwhile, in the instance of virtually all commercial (i.e. non-domestic) SBD developments, Remote Signalling Alarms have become the norm—largely at the behest of the insurance industry.

Conclusion

Despite the qualified findings of Armitage et al. (2017) and Tseloni et al. (2017) detailed above, Intruder Alarms are generally accepted amongst DOCOs as providing a deterrent to criminal attack. However, the add-on cost of such alarms together with ongoing issues in relation to false activation, necessitates that the DOCO thinks very carefully before insisting on their inclusion in any particular project—most especially in respect of new-build or refurbished homes. Indeed, it is the responsibility of the DOCO to decide, having assessed all the crime data in their possession (including local crime trends) and accumulated experience in the role, as to whether the Commensurate Risk (see Sect. 1.5.15) indicates an Intruder Alarm system is a necessary additional requirement which she or he wishes to stipulate at a particular development.

4.3.3 CCTV (Closed Circuit Television)

History and Background

Despite such literary references like those contained in George Orwell's (1949) legendary novel, *Nineteen Eighty-Four*, in a rudimentary form CCTV (Closed Circuit Television—now referred to as Video Surveillance Systems within the Secured by Design development guides, see Chapter 5) has been in practical existence for little more than 70 years. However, during that time literary, movie, TV and indeed journalist portrayal of CCTV have repeatedly over-emphasised its capabilities—all but imagining the camera alone would climb down from its position and secure the arrest, conviction and imprisonment of a guilty person.

CCTV cameras 'guard' the grave of composer Peter Tchaikovsky at the Tikhvin Cemetery, St Petersburg, Russian Federation

CCTV as a weapon in the fight against crime and anti-social behaviour (ASB/disorder) only truly began in the late 1980s. Indeed, to put this in perspective, in 1990 public area CCTV in the UK consisted of only three town/city centre schemes with approximately 100 cameras. Whereas, by 2008 it was estimated there were more than 500 such schemes with 50,000 cameras (Fry, 2008).

Figures that put the total number of CCTV cameras (public area and private) in the UK at between 4 and 6 million (some 20% of the supposed world total) should be treated with considerable scepticism, being based on highly dubious mathematical extrapolations.

Nevertheless, it is generally accepted that no other country in the world embraced so early the Technological Surveillance opportunities of CCTV to the same extent, as in the UK. This is illustrated by its annual value in terms of revenue, reaching a high point of £560 million in 2006, before contracting after the economic recession of 2009/10 (Key Note Ltd, 2014). More recently, growth of 1.8% per annum (2019–2023) puts the market value at £1.3 billion (IBISWorld, 2023). Admittedly, 28% of this figure relates to monitoring fire alarm systems. But in summary, the UK CCTV industry now generates approximately £1 billion annually with market growth anticipated throughout this decade. The exact cost of revenue spending (including the staffing costs of monitoring all these cameras) is unknown. Nevertheless, it can be confidently expected to exceed that of capital expenditure in any given year—other than those in which commissioning or a massive investment in technology takes place.

The reasons for the relative dominance of CCTV in the UK are complex, yet they help to explain why DOCOs have often been involved with public area schemes since the beginning. Indeed, the primary motivation for the surge of investment in CCTV that started in the early 1990s, can be identified in the massive rises of Reported Crime (especially Property Crime) at that time. The alleged effectiveness of CCTV in preventing such offences led directly to a series of five Home Office (central government) 'CCTV Challenges' that provided match-funding for public area CCTV schemes, both small and large in England and Wales. Separate government funding took place in Scotland and Northern Ireland.

A further motivation for this investment in *CCTV* can be attributed to its value in the investigative and evidential processes. Seminal instances in this regard include high profile cases like the conviction of the Harrods bombers and the abduction and murder of Merseyside toddler James Bulger, both in 1993; the investigation and conviction of the Brixton, Brick Lane and Soho bomber David Copeland in 1999; and perhaps most importantly from an international perspective, the investigations following the terrorist atrocities in London on 7 July 2005 and those attempted two weeks later. Such instances have illustrated the immense worth of *CCTV* in assisting the police senior investigating officer (SIO) during these highly charged enquiries, thereby ensuring the investigation steers a more direct course than might otherwise be the case. Tangential enquiries and an inordinate amount of misdirected effort can be avoided. However, the sheer volume of recorded material (often tens of thousands of hours of *CCTV* footage) also has the capacity to produce the opposite effect, especially in the absence of little other evidence.

In the British context, it is generally accepted that the poor-quality shopping centre CCTV images of James Bulger being led away by his killers, provided a watershed in terms of the acceptability of public area CCTV cameras. No matter that the CCTV images of him and his abductors were insufficient for identification. They were of instrumental importance as part of the investigative process—together with the effect they had on public perception of this technology.

Similarly, it now appears an international watershed took place after CCTV images of the London suicide bombers (and of the thwarted attacks a fortnight after 7 July 2005) were broadcast around the world. This seems to have convinced key decision-makers in nations formerly hostile to the medium (especially in Western Europe and North America) of its potential value. Certainly, Peter Fry (2008) the then chair of the UK CCTV User Group and many more believe this to be the case. Concerns relating to the intrusion of CCTV in matters of human rights and civil liberties rightly remain to this day, but after 1993 and 2005 they were effectively subsumed by the much-reported advantages delivered by Closed Circuit Television.

Instrumental in assuaging fears relating to civil liberties, since 1998, a series of legislative changes have impacted upon the practical use of CCTV in the UK. The most notable of these have been the Human Rights Act, 1998; Regulation of Investigatory Powers Act, 2000; Freedom of Information Act, 2000; Protection of Freedoms Act, 2012; and a succession of Data Protection Acts—most recently in 2018. Whilst the Biometrics and Surveillance Camera Commissioner has amended and reissued the Surveillance Camera Code of Practice (2021) and updated in 2022. https://www.gov.uk/government/publications/update-to-surveillance-camera-code.

Of greatest importance, it is imperative that CCTV systems are used in a manner which stands up to scrutiny and are accountable to the very people they aim to protect. This translates as meaning full support for the individual's right to privacy, with a strong recommendation this remains paramount for all agencies involved in public surveillance systems. If users, owners and managers of CCTV schemes are to command the respect and support of the public, there must be standards and checks to ensure that systems are used with the utmost probity at all times.

Evidence of the Effectiveness of CCTV

To date, research into the effectiveness of public area CCTV as a deterrent to crime, fear of crime (FOC) and anti-social behaviour (ASB) has been at best, varied witness: Poyner (1992), Brown (1995), Squires and Measor (1996), Squires (1998), Armitage (2000, 2002), and Gill et al. (2005).

Internationally, within police services (personal witness) there is near universal acceptance that CCTV has had a positive impact in respect of crime, FOC and ASB. Moreover, CCTV has demonstrated its worth in terms of: controlling incidents of crime and ASB; monitoring the movement of people in emergency situations; and the more effective deployment of resources. Most especially, CCTV has proved itself to be effective in assisting in the arrest of offenders and (to a lesser extent) providing evidence that has subsequently been used in court.

However, criticism of the effectiveness of CCTV invariably concerns its ability to prevent and deter crime and ASB, together with its role in reducing FOC—often the much-heralded motivations for its original installation. Such evidence to date has been insubstantive and few studies have been methodologically valid (Armitage, 2002). For example, Brown (1995), Mahalingham (1996) and Sarno (1996) all found evidence that CCTV reduced levels of FOC. However, Ditton and Short (1998) challenged the manner in which this questioning was conducted.

Similarly, Brown (1995) reported Property Crime was reduced in three town and city centres (Birmingham, Kings Lynn and Newcastle-upon-Tyne), although some offence categories increased. This pattern of inconsistency in respect of Property Crime was similarly identified by Tilley (1993), Squires (1998), Skinns (1998), and Armitage (2000).

In crimes where alcohol can be a key factor (e.g. assaults and woundings, public order offences), evidence of the effectiveness of CCTV in preventing these is far weaker (Brown, 1995; Skinns, 1998; Squires, 1998). However, this both accords with and powerfully lends support to Rational Choice Theory (Cohen & Felson 1979; Cornish & Clarke, 1986), where the ability to apply rational decision-making is reduced by the effects of alcohol (see Sect. 2.4, Rational Choice Theory).

Evidence relating to the Displacement or Diffusion of Benefits (prevention/deterrent effect spreading to a neighbouring area not under CCTV surveillance) effects of CCTV is similarly inconsistent—see also, Sect. 2.1, Displacement of Crime and Diffusion of Benefits. Brown (1995) found evidence of a Diffusion of Benefits in respect of Property Crime, but Geographical Displacement for the more serious offences of robbery and Theft from the Person.

Skinns (1998) also discovered evidence of Geographical Displacement and whilst Armitage et al. (1999) did not, both identified some degree of Diffusion of Benefits. Poyner (1992) reported Diffusion of Benefits to a neighbouring car park and in his second study, across a fleet of buses when CCTV cameras had only been installed on a small proportion of their number.

At the more local level, anecdotal evidence of Displacement of Crime (see Sect. 2.1) is legion, but hardly scientific. This also pertains to the concept of 'micro-displacement' (this author's description, albeit something of a 'Higgs boson' in terms of real evidence for the existence of such a phenomenon!), e.g. where drug-dealers will move a short distance so as to be outside the field of view of a CCTV camera(s).

As a consequence of this often contradictory evidence, the Home Office commissioned Professor Martin Gill and his team (2005) of the University of Leicester's Department of Criminology to conduct a £1.5 million, three-year study into the effectiveness of CCTV. They examined 13 schemes from the Round 2 CCTV bids for Crime Reduction Programme funding, with particular regard to: the impact of CCTV on crime, fear of crime and disorder (anti-social behaviour); its cost-effectiveness; novel uses of targeted CCTV; evidence of Crime Displacement or a Diffusion of Benefits to surrounding areas; impact on arrest, detection and conviction; and those features which make a CCTV scheme a success or failure.

Gill et al. (2005) concluded that 6 of the 13 schemes showed a 'relatively substantial reduction in crime' whereas in the other seven, crime increased. Systems installed in a mixture of areas (e.g. car parks, a hospital and other areas covered by one CCTV system) 'displayed the most encouraging results' in terms of Crime Reduction—especially in car parks. Impulsive crimes (especially those committed under the influence of alcohol) were less likely to be reduced than premeditated ones—which again accords with Rational Choice Theory (see Sect. 2.4). Spatial Displacement was not common but did occur, whilst feelings of safety increased in all but one of the areas. Eight schemes led to reductions in the feelings of victimisation. In all but one of the schemes, public support for CCTV was maintained at more than 70%.

They conclude, 'The use of CCTV needs to be supported by a strategy outlining the objectives of the system and how these will be fulfilled. This needs to take account of local crime problems and crime prevention measures already in place'.

To summarise, there is perhaps an unpalatable truth—namely that despite haphazard and anecdotal evidence to the contrary, the effectiveness of CCTV in preventing crime and ASB is at best, unproven. However, this has to be balanced against the repeatedly proven worth of CCTV in the investigative and to a lesser extent, evidential processes. Furthermore, a well-designed and efficiently managed CCTV system has the capacity to be a powerful security intervention. This is especially the case when dovetailed within the overall framework of security management and at its most effective when forming part of an overall security strategy, e.g. within the Designing Out Crime process.

Role of the DOCO in Respect of CCTV

Despite the relative lack of evidence regarding the effectiveness of public area CCTV in preventing crime (detailed above), historically a whole half day was given over to this subject on the Designing Out Crime Officer's Course (previously known as the Architectural Liaison Officer's or Crime Prevention Design Adviser's Course).

As such and certainly in the UK context, the police-employed Designing Out Crime Officer is repeatedly designated as the default CCTV expert. In one sense this automatic role allocation could be interpreted as 'mission creep'. Nevertheless, it can be powerfully argued how anything that strengthens the role and application of the DOCO should be applauded.

When Police managers do consider it wise to have a DOCO input in respect of CCTV (and the advice of this handbook is that they should), this can be perceived as both another string to the bow of the Designing Out Crime Officer whilst simultaneously providing an additional medium against which to assess the success of Crime Prevention/Reduction interventions. Nevertheless, there are some warnings that first need to be given.

In particular, the Designing Out Crime Officer is strongly advised NOT to become involved in the technical specification of the CCTV scheme in question. That should be left to the consultant and system experts who are advising the scheme managers. Similarly, DOCOs (and police managers) are counselled to be extremely wary of biometric applications of CCTV, whose track record of success has been somewhat limited in certain areas. There are undoubtedly more commercial companies seeking police (or other perceived authority) approval in the world of CCTV, than just about anywhere else in the vast arena of security products.

Instead, the DOCO should limit their advice to the following subject areas:

1. Ensuring a detailed Operational Requirement (OR) exists prior to the CCTV system being commissioned, maintained during the lifetime of the scheme and detailing in easily understood plain English (at least in the English-speaking world!) what the CCTV scheme is expected to do. For example:
   - The proposed purpose of the scheme
   - That the cameras should be able to provide images of sufficient high quality to enable both general view (monitoring activity) and close-up (clear facial identification of the persons depicted and of vehicle plate recognition)

- That such high-quality images are available during periods of poor visibility (such as during the hours of darkness)
- That the cameras' field of view overlaps so as to allow a particular suspect to be 'tracked' and thus provide Continuity of Evidence
- That such images are recorded in real time (ideally) and maintained for ideally 30 days
- That such images can be easily downloaded onto a WORM (Write Once Read Many times) medium, such as CD-ROM or DVD, for both investigative and evidential purposes.

2. Ensuring a Code of Practice (a 'living document' updated regularly and at least once a year) exists for the operational use of the CCTV scheme. This will include the necessity for:
   - Conducting spot-checks of the quality of the images being produced and recorded
   - Checking that the CCTV scheme meets the requirements of the Biometrics and Surveillance Camera Commissioner's Camera Code of Practice (2021).
3. Advising on camera location, based on analysis of the existing recorded crime data and the operational desire to view particular communications, e.g. Hot-Spot areas: late night queues alongside bus stops, taxi ranks, or outside nightclubs. Full details regarding how to produce a CCTV Code of Practice can be found in the former Home Office Scientific Development Branch (HOSDB) Operational Requirements Manual (2009) originally written by Jim Aldridge (1994).

The HOSDB became the Centre for Applied Science and Technology in 2012 and has since been absorbed into the Defence Science and Technology Laboratory (2018).

In addition, the Technological Innovation of digital CCTV (cameras, transmission and especially recording systems), has grown dramatically since the beginning of the twenty-first century and is now the dominant form of recording for both non-residential (commercial) and domestic applications (see Aldridge, 2000). Such innovation—including Internet Protocol (IP) technology, has provided largely positive improvements. For example, high-quality colour images in conditions of poor light quality.

However, there have been problems with some digital recording systems. A few standalone CCTV cameras/combined recorders possess a very limited recording period. For example, it was not unknown for images to be recorded over after 72 hours—far too short a period to recover evidence of criminal events that took place during, for example, a public holiday weekend. And some recording systems 'steal' memory from those pictures recorded several weeks beforehand, producing heavily pixelated images as a result. Clauses to prevent such problems occurring should be contained within the Operational Requirement (OR), detailed above.

Historically, there has been considerable debate regarding the appropriateness of including CCTV within the DOC arena. Indeed, this was a repeated subject area for the former UK Designing Out Crime Association (DOCA). However, at worst it could be argued that CCTV is a Technological Innovation of CPTED (thus its inclusion within this section of the handbook) and we ignore its relevance within the built environment at our peril. Moreover, whilst CCTV does not have the inherent sustainability characteristics of the Environmental Elements of CPTED (see Sects. 4.4.1–4.4.6), as demonstrated in Sect. 2.15, Intervention Sustainability Spectrum, it is increasingly being used as either a bolt-on security measure or as an alternative (proxy) to creating Defensible Space, Natural Surveillance and Territoriality. Consequently, a powerful argument can be made for the DOCO to have an in-depth knowledge of both the scope and limitations of CCTV.

National CCTV Strategy

Launched in 2007, the United Kingdom National CCTV Strategy, aimed to ensure a common standard of application and use of CCTV across the whole of the UK. Drafted by a team headed by the then ACPO lead on CCTV issues, the Deputy Chief Constable of Cheshire Police, Graham Gerrard, they proposed to standardise procedures and requirements in respect of the management of recorded images derived from public area CCTV systems.

The strategy was based around the following themes:

- The necessity for standards in all aspects of CCTV
- Training of all personnel
- Police use of CCTV footage and evidence
- Storage, volume, archiving and retention issues
- The need for CCTV networks—of live and stored images
- Equipping, resourcing and standardisation within the Criminal Justice System
- Emerging technologies, changing threats, new and changing priorities
- Partnership Working
- Financial and resource management.

CCTV User Group

Established in 1996, the CCTV User Group has a wide membership of service providers (including the police); is a highly respected authority across a wide range of CCTV issues; and a bastion of knowledge and information which it will readily share with interested parties. Its secretary.

The aims and objectives of the CCTV User Group are to:

- Provide help, assistance, professional advice and the opportunities for networking for those involved in the deployment, management and operation of public area CCTV systems
- Encourage best practice amongst all those involved in the planning, installation, implementation or management of CCTV systems
- Encourage the introduction of universal standards for the implementation and use of CCTV
- Promote the effective and efficient operation and management of CCTV systems
- Formulate nationally recognised standards by which all end users of CCTV will wish to abide
- Promote the need for the professional training of CCTV managers and operators and encourage the development of appropriate courses
- Discourage the unnecessary or inappropriate proliferation of CCTV in public places proposed without justification or by an Operational Requirement
- Discourage the use of public area CCTV surveillance methods by any groups or organisations which are not publicly accountable.

The CCTV User Group can be contacted at: https://www.cctvusergroup.com

A single CCTV guards 'Bully' from vandalism

4.4 Environmental Elements

4.4.1 Defensible Space

California Fire Services interpretation of Defensible Space

C. Ray Jefery and Oscar Newman might be described as the two founding fathers of CPTED and Designing Out Crime. Both men traced their cross-discipline academic ancestry to the seminal works of Jane Jacobs (1961, 'Urbanist'—her self-description), Elizabeth Wood (1967, Housing Specialist) and Schlomo Angel (1968, Planner). However, coming from separate academic disciplines, Jeffery and Newman produced a two-edged response to the ills of the 1950s/60s built environment that was arguably, twice as powerful.

Simultaneous to Jeffery's development of the CPTED concept, the New York based Architect, Oscar Newman, was working in the same subject area—albeit one that stressed the importance of design elements rather than psychology. Newman's Defensible Space: Crime Prevention Through Urban Design (1972) might be seen as developing the themes of Jacobs (1961) and Jeffery (1971), with CPTED deconstructed into a number of core elements including: Defensible Space, Natural Surveillance, Territoriality, Image and Milieu and Geographical Juxtaposition.

In *Defensible Space: People and Design in the Violent City* (1973, p. 84) Newman describes how 'Defensible Space is a model for residential environments which inhibits Crime by creating the physical expression of a social fabric that defends itself'. He asserts how the Defensible Space elements have a common purpose in providing a safe, productive and well-maintained living space. The aim is that a potential criminal perceives such an environment to be under the control of its residents and as a result, he will be easily recognised.

Newman observes this is partly '*...target hardening—the traditional aim of security design as provided by locksmiths*' (1972, p. 73). However, he contends how rather than delegating all security issues to the police, people must also protect themselves as a community (the Corporate Approach to Crime Prevention and Community Safety—see the *Morgan Report*, 1991). For him, Defensible Space is a collective term for a series of mechanisms: real and Symbolic Barriers; strongly defined areas of influence; and improved opportunities for surveillance. These are expected to empower the residents by giving them control over their environments.

Newman's four characteristics of Defensible Space are:

1. Natural Surveillance—capacity of physical design to provide surveillance opportunities for residents/agents;
2. Territoriality—capacity of the physical environment to create perceived zones of territorial influence;
3. Image and Milieu—the ability of design to influence perceptions of uniqueness, isolation and stigma;
4. Geographical Juxtaposition—with 'safe zones' and the security of adjacent areas.

Over the subsequent two decades he continued to develop the concept of Defensible Space, first adding (1976) the 'Five Principles' for creating same:

1. Assignment to different resident groups of specific environments they are best able to utilise and control, as determined by their ages, lifestyles, socialising activities, backgrounds, incomes and family structures;
2. Territorial definition of space in residential developments to reflect the zone of influence of specific inhabitants. Residential environments should be sub-divided into zones towards which adjacent residents can easily adopt proprietary attitudes;

3. Juxtaposition of dwelling interiors with exterior spaces and placement of windows to allow residents to naturally survey the exterior and interior public areas of their living environments and the areas assigned for their use;
4. Juxtaposition of dwellings (their entries and amenities) with city streets so as to incorporate the streets within the sphere of influence of the residential environment;
5. The adoption of building forms and idioms that avoid the stigma of peculiarity so as to allow others to perceive the vulnerability and isolation of a particular group of inhabitants.

Newman argued that there was a hierarchy of living and community spaces. Housing developments that incorporate Territoriality are said to have the strongest deterrents to criminal activity. This impacts upon the way that housing is grouped and space laid out to prevent Crime. Crucially, according to Newman in such an environment people should not only challenge unusual behaviour, but feel it is their duty to do so. And the fear of this should be a powerful deterrent to criminals.

For the DOCO this all-embracing definition of Defensible Space may appear over-reaching. Certainly, in the UK Defensible Space is generally understood to be a far more limited entity where the words 'defensible' and 'space' have all but literal meanings. Indeed and with distinct parallels to the Public Order Act 1986, Defensible Space is usually described in terms of the following forms: public space (e.g. a road and its pavements/sidewalks); semi-public space (enclosed front garden); semi-private space (secured rear garden/yard); and private space (inside the premises): see HOCPC Police Architectural Liaison Manual of Guidance (1994). Consequently, in practical terms and at its most simple, Defensible Space can be reduced to the creation of a building, Target Enclosure or space that assists legitimate occupiers to keep offenders out.

Newman's critics have concentrated on his methodological weaknesses. Rubenstein et al. (1980, p. 6) described Defensible Space as a '*rats nest of intertwining hypotheses*'. Cozens (2014) contends Defensible Space is difficult to measure and define. And there is also the evidence deficiency of much of the early work claiming to be CPTED. Nevertheless, Newman's ideas can still be perceived as revolutionary and are more than simply a catalyst to further research of distinct value to Crime Prevention Practitioners—especially DOCOs/CPTED Officers.

4.4.2 Natural Surveillance

Historic telephone kiosk in Gamla Stan, Stockholm, Sweden repeatedly vandalised due (at least in part) to an absence of Natural Surveillance

Unlike Defensible Space, there is an overwhelming consensus regarding the veracity of Natural Surveillance (Informal Surveillance)—most probably because it is a benign and largely non-contentious phenomenon. For Oscar Newman it is the second of his four characteristics of Defensible Space and he defines Natural Surveillance as '*The capacity of physical design to provide surveillance opportunities for residents and their agents*' (1973, p. 78).

Warren and Stollard (1988, p. 32) provide their own definition '*…the impression that residents keep, or are able to keep watch on their neighbours. As a result, potential offenders feel conspicuous and are deterred from committing crime*'. Moreover, it has increasingly become acceptable to apply CPTED concepts like

Natural Surveillance and Defensible Space to the whole of the built environment (and beyond) so as to include the occupants of non-residential buildings and those walking, cycling or driving between them.

In essence, Natural Surveillance encapsulates the notion that offending behaviour (crime and anti-social behaviour/disorder) can be prevented if those committing these acts believe they are being watched. 'Natural', means the potential for observation by members of the public, as opposed to the Formal Surveillance opportunities provided by police officers, police community support officers, security guards etc.; or the Technological Surveillance delivered by CCTV (Closed Circuit Television—see Sect. 4.3.3) cameras, Alarm Receiving Centre control rooms, ANPR (Automatic Number Plate Recognition) etc.

The aim is to maximise the opportunities for all such forms of *Surveillance*. For example: by pruning back vegetation that is obscuring vision—a concept Kuo and Sullivan (2001) report can be traced back to King Edward I and the *Statute of Winchester 1285*; (see below); ensuring main entrance doors and windows in normally *Active Rooms* (like kitchens and living rooms) face the street; putting window(s) in gable end walls; locating fire escape stairs externally; and improving Street Lighting (see Sect. 4.3.1) in an effort to aid Surveillance during the hours of darkness. As with so many issues in CPTED and Designing Out Crime, these examples raise further questions and debates.

> The highway from one merchant town to another shall be cleared so that no cover for malefactors should be allowed for a width of two hundred feet on either side; landlords who do not affect this clearance will be answerable for robberies committed in consequence of their default and in the case of murder they will be in the King's mercy. Statute of Winchester 1285, Chapter V, King Edward I.

For Newman (1973), surveillance has a demonstrable effect in reducing irrational fears and anxieties amongst inhabitants. He was especially interested in the Natural Surveillance that takes place when residents conduct their daily activities. However, Newman accurately warns how there is no guarantee that having witnessed a particular criminal act the resident will necessarily react—either by directly challenging the offender or calling the police or other agency.

In recent decades, criminological research has produced a varied picture regarding the value of surveillance in all its forms. For example, Jane Jacobs (1961, p. 35) invented the concept of Eyes Upon the Street—and concluded that increased street activity led to reduced crime. Repetto (1974) observed that offenders avoid properties where there is visual surveillance to and from neighbouring properties; and select those with the opposite characteristic—also the conclusion of Nee and Meenaghan (2006).

Brown and Bentley (1993) investigated offender target selection and deduced that burglars avoid dwellings which are occupied, or they believe are occupied. Similarly, Brown and Altman's (1983) research indicated that those properties with fewer signs of residents being present experience more crime.

Mayhew (1979) examined the effects of surveillance upon vandalism in public areas and against 217 telephone kiosks in particular. She concluded that surveillance had a small impact on crime and that consequentially, action taken by such witnesses is low and depends largely on their role (e.g. police officers and security guards are far more likely to act against such behaviour than ordinary residents). She also found that Formal Surveillance by such official agents is the most effective form of surveillance. See also Armitage et al. (2010) regarding the practical experiences (not necessarily positive!) of residents living in 'Award-Winning Design' developments.

Winchester and Jackson (1982) researched residential burglary offences in the Kent Police area of England and reported that occupancy, surveillance and access opportunities were important determinants when criminals were selecting targets. They concluded that properties overlooked by neighbouring ones experience less crime (and more crime where the reverse was true)—also the verdict of Van der Voordt and Van Wegen (1990) and Armitage (2006, 2013).

Macdonald and Gifford (1989) interviewed 44 convicted male burglars and used 50 photographs of different forms of housing, displaying a range of Environmental Cues (see Sect. 2.8) involving surveillance. The burglars were then asked to arrange the photos in terms of poor, moderate, or good burglary targets. The Natural Surveillance opportunities emanating from the properties proved to be a good indication of their attractiveness to burglars. Secluded homes were especially favoured, whereas corner plots and those with good visibility or overlooked by neighbours, far less so.

Cromwell et al. (1991) discovered that properties located in close proximity to a stop sign, traffic lights, commercial business establishment, park, church or busy road are more attractive to offenders. This apparent contradiction with Eyes Upon The Street is also evidenced by Winchester and Jackson (1982) and Groff and La Vigne (2001) who found that properties located on a main road experience more crime.

Groff and La Vigne (2001), Van der Voort and Van Wegen (1990), Welsh and Farrington (2008) all conclude that properties located in areas with poor street lighting experience increased crime.

Meanwhile, Hearnden and McGill (2004) interviewed convicted burglars and discovered how belief of goods within the property worth stealing, was the most important consideration. Out of 72 respondents, only 35 thought the 'type of property' important. Moreover, the fear of being seen came very low on the list of offenders' concerns—perhaps of no great surprise when their primary motivation was to steal goods to sell and finance their drug addiction—once again a finding that supports Rational Choice Theory (see Sect. 2.4).

Armitage (2006) also concluded that properties visible from nearby footpaths experience more Crime; as do those located within viewing distance of traffic lights. Whilst Reynald (2009) reported how increased levels of Surveillance leads to enhanced levels of Guardianship activity.

At the micro level, Atlas (2008) logically contends windows are of little value if residents do not use them, or there is nothing to see. In practice, the expression '*located in Active Rooms*' (see Sect. 5.4. Secured by Design—Residential Guide 2025) is increasingly used to describe windows that overlook areas of risk, such as car parking areas adjacent to dwellings. Atlas also notes how criminal gangs have utilised Natural Surveillance principles in reverse when locating drug dens, cannabis farms (Commercial Cannabis Cultivation) etc., as a means of identifying intruders such as their rivals and the police, well before such law enforcement officers break down the doors.

In conclusion, there is no desire to dismiss Natural Surveillance as an unimportant commodity. Jacobs, Jeffery and Newman stressed its importance and even the most critical observer would concede that a street that is not overlooked by premises with windows, has the perception of being one that instinctively feels less safe than one that does. Extending this theme, barriers to Natural Surveillance such as rampant vegetation (see Sect. 4.2.4) can have the opposite effect. And similarly, the argument that increased levels of Street Lighting can reduce crime (by improving surveillance opportunities during the hours of darkness) now carries with it considerable academically proven research in its support (see especially Welsh and Farrington (2008).

4.4.3 Territoriality

For Oscar Newman (1972, 1973, 1996) '*Territorial Reinforcement*' (Territoriality) is a characteristic of Defensible Space. He defined it as '*The capacity of the physical environment to create perceived zones of territorial influences*' (Newman, 1973, p. 51).

Newman maintained *Territoriality* promotes social control through increased designation, ownership and influence over space. Strangers and intruders stand out; and owners more likely to challenge their presence. Walls/fencing/gates;

trees/shrubs/plants; lighting; paving/surface treatments; signage and notices can reinforce Territoriality.

The objectives of *Territoriality* can also be achieved in existing locations by assigning space to designated users. Moreover:

- A well-maintained environment exudes an aura of care and activity—as supported by the Broken Windows theory of Kelling and Wilson (1982)
- The provision of soft landscaping, e.g. trees, shrubs and plants will assist in reinforcing Territoriality
- Real and Symbolic (Psychological) Barriers can similarly enhance *Territoriality*
- Activities restricted to well-defined private areas
- Signage and notices displayed at location entrances
- External communal areas, e.g. parking bays, drying areas and play spaces, should be close to the entrances they serve, or accessed directly from them
- Outdoor seating to be located under the territorial control of specific residents or businesses willing to accept that role and not where it will act as a generator of ASB. In the UK this philosophy can often be observed in the fenced-off seating areas outside pubs.

Territoriality is crucial in creating *Defensible Space* with few entry and exit points. *Natural Surveillance* reinforces *Territoriality* and thereby reduces fear by suggesting intruders are under surveillance from residents. Moreover, *Territoriality, Natural Surveillance, Image and Milieu* combine to deliver an environment under the residents' control.

A number of criticisms have been made of *Territoriality* for example, Newman's original overarching presumption that design alone (Architectural Determinism) can generate territorial control. Merry (1981) and Atlas (1991) point to the necessity for positive social conditions if Newman's concepts are to work. Newman (1976) subsequently accepted the relevance of social factors necessary for Territoriality (Newman & Franck, 1982). However, Reynald and Elffers (2009) contend that at its heart lies an unspecified level of 'space'.

Territoriality is instrumental in making legitimate users feel safer, potential offenders at risk. Moreover, each *CPTED* element is reinforcing the others—be it as a secure environment, the number of *Eyes Upon the Street* (Jacobs, 1961, p. 35) or perceived ownership and control. Brown and Altman (1983) included the social factors involved in creating *Territoriality,* but concentrated more on boundary-regulation. Indeed, they argued that burglars assess Territoriality at the neighbourhood, street and target levels. They also combined Newman's *Territoriality* with Altman's (1975) theory of territorial behaviour and privacy.

Most importantly, Newman (1972, 1973, 1996), Brown and Altman (1983), Brown and Bentley (1993) all assert that increasing *Territoriality* delivers a Crime Reduction benefit. Nevertheless, as Reynald and Elffers (2009) observe, further investigation is required—most especially in relation to the effect that an area's accessibility has on residents' ability to exercise *Territoriality* over their space.

4.4.4 Image and Milieu

An architect by profession, Newman highlighted two CPTED elements: *Image and Milieu* (or Image Management and Maintenance). He believed residents' perceptions of areas and facilities were linked to urban safety. Likewise, the design and appearance of a development can influence its '*uniqueness, isolation and stigma*' (Newman, 1972, p. 102) by those who reside, work, commit crime, or hold an opinion. Newman's contention was of a connection between poor design and victimisation and compares '*upper-middle-income high-rise housing*' with '*publicly supported housing*', where '*frills are strictly forbidden*' (1973, p. 105). Moreover, this is equally true if public housing costs more than private to build—a statement with relevance on both sides of the Atlantic and indeed worldwide. However, Newman is far weaker in evidencing how 'keeping the poor in their place' brings their vulnerability to the attention of others.

Oscar Newman was influenced by the 33 heavily vandalised high-rise blocks at Pruitt-Igoe, St Louis, Missouri and whose design he criticised. The 31, 11 storey tower blocks contained 2,870 flats/apartments and were built in 1954. Occupancy peaked at 91% (Jencks, 1984) before Pruitt-Igoe began its descent into worldwide notoriety. With occupancy below 70%, all 31 blocks were demolished between 1972 and1976 having become synonymous with crime, poverty and racial segregation (Ramroth, 2007).

Newman also argues against designing an institutional atmosphere, e.g. using glazed tiles (as in hospitals and prisons); corridor lights in vandal-proof housings; and external mercury vapour Lighting that exhibits a strong purplish hue. He does not advocate aesthetic treatment purely for beautification, but cites President Lyndon B. Johnson's Commission on Law Enforcement and Criminal Justice (1967) that recognised the debilitating effect on the soul of a deteriorated living environment—repeated by Kelling and Wilson (1982) in their Broken Windows theory. This author's personal witness when growing up on a series of council estates (public housing) half a century ago, suggests this was also the norm with public housing on this side of the Atlantic. In the UK, only since 'Estate Action' and similar refurbishment and demolition programmes began in the late 1980s, has the institutionalised appearance of much high-rise and low-rise social housing started to change.

In the UK and beyond, a growing hostility towards high-rise housing was attributed to alienation, vandalism and crime (Newman, 1996; Power, 1997). This reached its zenith in 1968, with a gas explosion in an eighteenth floor flat at Ronan Point, London. 'Progressive collapse' meant every room beneath it fell to the ground. Four people died and 17 were injured. The subsequent inquiry concluded inadequately researched national policy guidelines led to flawed construction methods. As Power (1997, p. 58) describes: '*Concrete became the main building medium, partly because it was vaunted as being long-lasting, if not indestructible*'. Yet, worldwide its use revealed early signs of decay including '*spalling*' (AKA '*concrete cancer*'—chemical reaction with internal steel reinforcements). Concrete also turned grey, losing the original pristine white appearance extolled by architects like Le Corbusier (1947). Eventually, this caused recourse to the use of cladding with potentially lethal effects—as evidenced at Grenfell Tower, London in 2017.

Another perspective contends that high-rise housing cannot be blamed as 'the product of architectural egotism or of an artistic and technological dream'. Nuttgens (1989, p. 95). The original proponents believed it answered a housing crisis and delivered massive improvements compared to the slums from which many were rehoused. Indeed, tenants were '*...grateful and excited at the break it offered from previous poor conditions*'. (Power, 1997, p. 59). Problems came from: building quickly; failing to research construction methods; delivering the amenities originally promised, and residents' aspirations.

4.4.5 Geographical Juxtaposition

One of Newman's concepts (1972), Geographical Juxtaposition (GJ) originally refers to adjacent or proximal land use. Agreeing with Armitage (2018) that *CPTED* needs to continually evolve, Cozens et al. (2019) cite an absence of discussion about GJ., despite its value in understanding '*Crime issues stemming from influences at a variety of levels of remoteness from the crime location*' (2019, p. 2). They contend GJ has four elements: micro (crime location); proximal (close to a crime location); meso (proximal to more distant areas); and macro (remote influences). Existing literature concentrates solely on the Crime location. Whilst

CPTED suffers in believing busy places are always safe: '*Eyes upon the street*' (Jacobs, 1961, p. 35). Indeed, Cozens (2011) suggests planning policy for high density, mixed-use developments in the USA, UK, Australia and beyond may exacerbate crime.

Cozens et al. (2019) found 14 GJ items (3.31%) in 423 works published 1968–2019, compared to 127 (30%) for Defensible Space. Yet, GJ challenges: *CPTED* in mixed-use development, Permeability, urban consolidation, New Urbanism and Image Management. Bowers (2014, p. 390) asserts how '*land use and crime are inextricably linked is both intuitively plausible and well evidenced*'. GJ is an '*important, if currently vaguely conceptualized, design issue*' (Cozens & Hillier, 2012; Cozens et al. 1999, p. 256).

The criminogenic capacity of the environment is described by Cozens (2011). Violent Crime and assaults are concentrated within 400 feet of bars (Ratcliffe, 2011); subway stations, up to 1200 feet away (Groff & Lockwood, 2014). Similarly, public use facilities on/near major roads are crime-related (Wilcox & Eck, 2011). More specifically: transport hubs—increased Crime Risk and geographical distribution of crime; subway stations, nearby hotels and motels—street robberies; fast food restaurants, convenience stores, pawn shops—greater opportunities for offenders to congregate; schools—more Violent Crime and Property Crime; banks—Risky Facilities (Clarke & Eck, 2007); playgrounds and parks—Crime Attractors; public housing projects—higher crime. Similarly, non-use/dereliction is associated with higher crime and '*contagion*' (Skogan, 1990).

Bernasco and Block (2011) assert how blocks with Crime Generators/Attractors experienced higher risk of Robbery which decayed over distance—likewise bars, clubs, fast food restaurants, off-licenses, grocers, petrol stations, laundromats, pawn shops, general stores. Such locations satisfy the 80:20 rule: 20% of facilities will generate 80% Crime Risks. Anderson et al. (2013) believe a limited understanding of microlevel details of crime and land use dictate the necessity for CPTED audits, as new developments may impact crime and be influenced by it.

GJ may influence local Crime Risks: concentration and dilution; malign or benign Crime Displacement; behavioural modification; motivation/demotivation; distribution of Crime Opportunities; nodes acting as Crime Attractors/Detractors/Facilitators/Enablers/Precipitators/Absorbers/Radiators/Reducers; density of offenders; paths and accessibility; edges; presence of Capable Guardians. Assessing crime before implementing CPTED solutions has always been part of the CPTED process (Crowe, 1991) and restated by Atlas (2008), Cozens (2016), Cozens and Love (2017). Yet most guidelines make no such reference. Crime and Crime Prevention operate as a complex system, but using GJ when conducting

Crime Risk Assessments helps to better comprehend the crime taking place and which CPTED methods will be most effective and where necessary, implement these in surrounding areas.

Positive risk feedback from CPTED may cause lower rates of crime and a Diffusion of Benefits—AKA Benign Displacement/Halo Effect see Sect. 2.1). Whilst negative risk feedback suggests crime is lower and requires increased CPTED resources to deliver the outcomes. Feedback effects may indicate the necessity to extend specific CPTED interventions beyond the targeted development.

GJ may also improve knowledge of CPTED concepts, e.g. Territoriality—the GJ of psychological signs of potential defenders between potential criminal and target. Surveillance (and Sousveillance)—the GJ of potential guardians between a potential criminal and target. Image Management—levels of maintenance and repair indicate how such space is cared for and how crime/ASB will not be tolerated. It provides a GJ of the owners/managers of space and potential offenders. Access Control—existence and separation of the GJ of spaces: safe and secure with legitimate, privately owned activities and resources; or higher motivations for crime and risks that threaten to exploit legitimate activities and resources.

Natural access control is the GJ of a psychological/habitual barrier between a potential criminal and the target. Activity Support—encouraging increased legitimate use and GJ having the potential to reduce crime rates, e.g. police stations located in areas with higher crime to reduce such activities. Target Hardening is based on the GJ of two kinds of space: safe and secure with legitimately owned resources and activities; or spaces with higher motivations for crime and risks that threaten to exploit legitimate activities and resources. It can be perceived as the GJ of a physical barrier between these two types of space '*that has high costs to cross*' (2019, p. 14). For example, high security doors between the potential criminal and Target Enclosure.

GJ is increasingly relevant due to urban consolidation and densification (Cohen & Felson, 1979) and linked to Crime Pattern Theory (Brantingham & Brantingham, 1981). Ackerman and Rossmo (2015) contend density and crime location determine offenders' interaction with their environment and travel behaviour—especially short distances.

Cozens et al. (2019) suggest an extended spectrum of GJ to identify: which elements impact on crime in a given area; new theories of geographical/spatial effects on crime; enhanced understanding of how CPTED operates; challenge planning archetypes that compromise Crime Prevention; and enhance links between urban planning, CPTED and Environmental Criminology. They propose Four New Principles of CPTED (2019, p. 14): GJ to explain crime and Crime Prevention issues; how at a distance GJ is inversely proportional to the CPTED measures applied; all Crime Risk issues are GJ Crime Risk factors; the benefits of distance can be maximised by obscuring Crime Opportunities, or raising visibility, e.g. 'active' rooms (like kitchens and lounges) to overlook public space.

Understanding GJ helps to design, manage and use spaces more effectively and avoid creating anti-social design. It can assist CPTED in reducing the propensity for offending and enhance Prosocial Behaviour, where the offender is a member of that community. Raising their emotional/moral attachment to the area may lessen their desire to commit Crime (Armitage, 2018). Ultimately, Cozens et al. (2019) conclude that GJ is an essential element of CPTED theory and may also improve understanding of the other CPTED concepts.

Cozens et al. (2001) also developed a 'new integrated model of defensible space'. Following which, Cozens and Hillier (2012) and Cozens (2014, 2016) restated the importance of GJ. Cozens and Linde (2015) convincingly argue that CPTED surveys should include measurement of GJ land uses (many UK-based DOCOs would claim they already do so); and as additional criteria in conducting CPTED street audits (Cozens & Babb, 2018). Finally, Cozens (2016) perceives GJ as a powerful unifying idea of how crime and environment are related.

4.4.6 Permeability

Permeability providing the same potential Crime Opportunities in Abu Dhabi and the UK

There is a major fault-line between the worlds of Designing Out Crime and Planning (especially amongst those who articulate a New Urbanist/Urban Villagers' viewpoint) and it can be encapsulated in the word Permeability—or Movement as it is often known within the Planning profession. The great irony is that both positions aspire to a safe and secure community for all those who reside, are educated, work, socialise or travel through the built environment. However and to be succinct, practitioners in the field of Designing Out Crime (DOC) can now provide repeated evidence from Crime Science of how Permeability directly facilitates crime—most especially Property Crime, although it also has the capacity to influence various forms of ASB, Violent Crime and other forms of serious crime when people are put at risk by being placed in less secure environments.

For the DOCO, Permeability is a highly contentious issue, as any trawl of the internet will demonstrate. Critics are also prone to use it as the primary justification for their Fortress Mentality and Architectural Determinism criticisms of Crime Prevention Through Environmental Design (CPTED) and Designing Out Crime. However, within the DOC community evidence of the criminogenic capacity of Permeability has become increasingly powerful.

Permeability was not originally included as a specific element within CPTED. However, all too obviously it is not a Physical Security 'Target Hardening' Measure or Technological Innovation and consequently, readily fits within the category of and as an Environmental Element of CPTED. Indeed, it shares features with Geographical Juxtaposition and the fourth of Newman's characteristics of Defensible Space (see Cozens & Love, 2015; Cozens, Love and Davern 2019). And a case might be made for it having an influence on Image and Milieu - another of Newman's characteristics of Defensible Space. Most importantly, Permeability can now be criminologically perceived as possessing the capacity to compromise the other key CPTED elements as previously described in this section, and by default, the safety and security of the development in question and the people who use it.

Historically, it is only since the beginning of this twenty-first century that Permeability has risen to seminal prominence within Designing Out Crime. Before that time there was more than a suspicion amongst DOCOs of an incompatibility, when Jane Jacobs concept of '*Eyes Upon the Street*' (1961, p. 35) was used to justify a highly permeable built environment. For example, offenders (like 'pick-pockets') who for centuries past deliberately target busy locations. But since 2000 the evidence from Crime Science that testifies to the negative consequences of excessive Permeability has risen to prominence. Moreover, and as Professor Rachel Armitage (2007) contends, the debate about the criminogenic features of permeable design has been a distraction to the core desire to reduce crime.

Supporters of increased Permeability often cite Jacobs, yet she wrote: '*I hope no reader will try to transfer my observations into guides as to what goes on in towns or little cities, or in suburbs which still are suburban*'. As this statement appears early in The Death and Life of Great American Cities (1961, p. 26) it begs the question as to just how many people have actually read the seminal work of their guru?

From '*Eyes Upon the Street*' to the 1992 Rio Earth Summit, Permeability has been a constant design ideal, with little if any scientific evidence that such Movement will encourage Informal Surveillance and Guardianship. In the UK, *Places, Streets and Movement: A Companion Guide to Design Bulletin 32—residential roads and footpaths* (1998) advocates a '*walkable neighbourhood*' as the '*key to creating a sociable, sustainable community*'. The DOCO might be persuaded about the latter, providing measures are incorporated to Design Out Crime (e.g. the elimination of 'Leaking' Culs-de-Sac—see Sect. 6.2) and nebulous footpaths. Not so *Towards an Urban Renaissance* (1999) and its statement: '*A user friendly public realm should make walking and cycling easy, pleasant and convenient by keeping the size of urban blocks small with frequent pedestrian cut-throughs to make a new development permeable and accessible to the existing neighbourhoods*'. Unfortunately, before and since publication, repeated evidence demonstrates how such a policy will design in Crime Opportunity (see Sect. 2.3).

From a DOC perspective, the major problem with these design desires is how they had previously been applied and catastrophically failed. Indeed, the cities and towns of the UK are peppered with 'sink estates', largely the product of the post-Second World War building boom. By the 1970s many of these were rife with Property Crime and excessive Permeability was a common denominator, identifiable in the Modus Operandi of the crimes committed. Moreover, a host of professionals from the built environment possess first-hand experience of the criminogenic capacity of Permeability.

For example, the 'Leaking' Culs-de-Sac and Radburn rear access footpaths rarely used by the residents they were designed to serve, but the perfect Attack and Escape Routes for burglars with their stolen goods; rear parking courtyards in which residents are terrified to park their cars due to thefts, criminal damage, arson, drug taking or illicit sexual activity; the vandalised 'residents garden' or public seating, located as 'amenity features' but without consulting residents (or the DOCO) and now the blatant generators of ASB and worse.

As a young police officer during the 1980s, this author recalls repeated attendance at burglary and other crime locations where 'Leaking' Culs-de-Sac had provided Attack and Escape Routes for offenders. Stolen goods were regularly discovered hidden in these footpaths (awaiting later collection). And there was the serial experience of abandoning the patrol car and chasing criminals through such footpath networks, after they had been disturbed during the commission of their offences. Furthermore, what such examples fail to communicate is the constant fear in which many residents now live following Repeat Victimisation (see Farrell et al., 2000).

It is truly unforgiveable when this terror emanates from consummate poor design, ideology and/or a failure of Community Consultation with those who will have to live in that built environment, often for decades thereafter. And whilst the negative aspects of Permeability were apparently not known half a century ago, advances in Situational Crime Prevention, CPTED and Designing Out Crime mean that excuse is no longer valid.

Armitage (2007) details the two protagonists and their key principles in this Permeability debate. New Urbanism: encouraging walkable neighbourhoods, mixed land use, rear access footpaths, hiding garages and permeable street networks replacing Culs-de-Sac. Whilst the Designing Out Crime /Situational Crime Prevention/CPTED, position might be best demonstrated by Secured by Design (SBD) with its repeated track record of independent, academic assessment: maximising Defensible Space, Natural Surveillance, Territoriality, Geographical Juxtaposition, Image and Milieu; ensuring properties and their boundaries are physically secure and minimising access and egress.

Evidence supporting the long-term, sustainable effectiveness of SBD is set out in Chapter 5. But there is also a raft of powerful research data that points to the efficacy of minimum entry/exit and limited Permeability: Brantingham and Brantingham (1975, 1993, 2008), Bevis and Nutter (1977), Brown and Altman (1983), Newlands (1983), Greenberg and Rohe (1983), Beavon (1984), Taylor and Gottfredson (1987), Cromwell et al. (1991), Poyner and Webb (1991), Armitage (1999, 2013), Rengert and Wasilchick (2000), Wiles and Costello (2000), SURF Centre (2002), Poyner (2006), Armitage and Monchuck (2009), and Stokes (2022).

On occasion, evidence appears to support maximum Permeability, e.g. Rudlin and Falk (1995). In the context of Space Syntax (see Sect. 2.10), when Hillier and Shu (2000) and Shu (2000) published their research findings, UK newspapers ran headlines proclaiming the '*End of the road for the Cul-de-Sac*'. However, it is vital to distinguish between 'Leaking' and true Culs-de-Sac and Hillier (2004) indicated it was 'leaky' Culs-de-Sac that were problematic in terms of security.

The DOCO received statutory support when *Department of the Environment Circular 5/94, 'Planning Out Crime'* (1994) was published. However, it did not mandate planners to liaise with the police. This situation was reinforced when 5/94 was superseded by Safer Places: The Planning System and Crime Prevention

(2004—see Sect. 6.4). The latter (repealed in 2014) set out 'seven attributes of safer places', yet failed to address the 'fault-line' of Access and Movement—Permeability.

On a related theme, the findings of Wiles and Costello (2000—see Sect. 2.3) describe the relatively short travelling distances of opportunist offenders, both in England and North America. Moreover, Al-Kahtani (1996) discovered similarly short travelling distances (Crime Travel Footprint/Opportunist Radius) amongst offenders in Saudi Arabia—thereby suggesting an international pattern of behaviour. It might therefore be concluded that overwhelmingly, offenders are travelling on foot to commit crime. In this context, Permeability is of crucial importance to the searching patterns and behaviour of these offenders, as it facilitates more Crime Opportunities in a shorter distance and space of time.

Despite Permeability now being a (if not THE) dominant issue amongst UK DOCOs, reading the accounts of CPTED Officers/Advisors elsewhere in the world, it is rarely articulated. But then in the UK, Second and Third Generation CPTED are scarcely mentioned. Perhaps this demonstrates the differing priorities confronting practitioners across the globe and how in the UK, DOCOs must increasingly consider offending behaviour in developments with far higher densities. Furthermore, the success of Secured by Design (see Chapter 5) is now mainstreamed into Designing Out Crime and used in the construction and refurbishment of thousands of homes and other non-residential properties.

References

Ackerman, J., & Rossmo, K. (2015). How Far to travel? A Multilevel Analysis of the Residence-to-Crime Distance. *Journal of Quantitative Criminology, 31*, 237–262.

Aldridge, J. (1994). *Operational Requirement Manual for CCTV.* Police Scientific Development Branch.

Aldridge, J. (2000). *Digital Imaging in the Criminal Justice System—Policing Issues.* Police Scientific Development Branch.

Al-Kahtani, S. (1996). *Travel to Crime Distances of Burglars in Saudi Arabia.* University of Manchester.

Anderson, J. M., MacDonald, J. M., Bluthenthal, R., & Scott Ashwood, J. (2013). Reducing Crime by Shaping the Built Environment with Zoning: An Empirical Study of Los Angeles. *University of Pennsylvania Law Review, 161*, 699–756.

Angel, S. (1968). *Discouraging Crime Through City Planning* (Working paper no. 75). University of California, Berkeley: Institute of Urban & Regional Development.

Armitage, R. (1999). *An Evaluation of Secured by Design Housing Schemes Throughout the West Yorkshire Area.* The Applied Criminology Group, The University of Huddersfield.

Armitage, R. (2002). *To CCTV or Not to CCTV? Community Safety Practice Briefing.* NACRO.

Armitage, R. (2006). Predicting and Preventing: Developing a Risk Assessment Mechanism for Residential Housing. *Crime Prevention and Community Safety: An International Journal, 8*(3), 137–149.

Armitage, R. (2007). Sustainability Versus Safety: Confusion, Conflict and Contradiction in Designing Out Crime. In G. Farrell, K. Bowers, S. Johnson, & M. Townsley (Eds.), *Imagination for Crime Prevention: Essays in Honour of Ken Pease. Crime Prevention Studies* (Vol. 21, pp. 81–110). Willan.

Armitage, R. (2013). *Crime Prevention Through Housing Design: Policy and Practice.* Palgrave Macmillan.

Armitage, R. (2018). *Urban Design and Crime—Pro-Social Design Would Reduce Crime.* Available online: https://www.hud.ac.uk/news/2018/july/pro-social-design-reduces-crime

Armitage, R., & Monchuk, L. (2009). *Re-evaluating Secured by Design (SBD) Housing in West Yorkshire.* University of Huddersfield.

Armitage, R., Monchuk, L., & Rogerson, M. (2010). It Looks Good, But What Is It Like to Live There? Assessing the Impact of Award Winning Design on Crime. *Special Volume of European Journal on Criminal Policy and Research, 17*(1), 29–54.

Armitage, R., Smyth, G., & Pease, K. (1999). Burnley CCTV Evaluation. In K. Painter & N. Tilley (Eds.), *Crime Prevention Studies* (Vol. 10). Criminal Justice Press.

Atkins, S., Hussain, S., & Storey, A. (1991). *The Influence of Street Lighting on Crime and Fear of Crime* (Crime Prevention Unit Paper No. 28). Home Office.

Atlas, R. (1991). The Other Side of Defensible Space. *Security Management March*, 63–66.

Atlas, R. (2008). *21st Century Security and CPTED.* Taylor & Francis Group.

Beavon, D. J. (1984). *Crime and the Environmental Opportunity Structure: The Influence of Street Networks on the Patterning of Property Offences* (Masters thesis). Simon Fraser University, Burnaby, BC.

Bennett, T., & Wright, R. (1984). *Burglars on Burglary.* Gower.

Bernasco, W., & Block, R. (2011). Robberies in Chicago: A Block-Level Analysis of the Influence of Crime Generators, Crime Attractors, and Offender Anchor Points. *Journal of Research in Crime Delinquency, 48*, 33–57.

Bevis, C., & Nutter, J. B. (1977). *Changing Street Layouts to Reduce Residential Burglary.* Paper Presented at the Annual Meeting of the American Society of Criminology, Atlanta, Georgia.

Bowers, K. (2014). Risky Facilities: Crime Radiators or Crime Absorbers? A Comparison of Internal and External Levels of Theft. *Journal of Quantitative Criminology, 30*, 389–414.

Brantingham, P. J., & Brantingham, P. L. (1975). The Spatial Patterning of Burglary. *The Harvard Journal, 14*(2), 11–23.

Brantingham, P. J., & Brantingham, P. L. (1981). Notes on the Geometry of Crime. In P. J. Brantingham & P. L. Brantingham (Eds.), *Environmental Criminology*. Sage.

Brantingham, P., & Brantingham, P. (1993). Nodes, Paths and Edges: Considerations on the Complexity of Crime and the Physical Environment. *Journal of Environmental Psychology, 13*, 3–28.

Brantingham, P., & Brantingham, P. (2008). Offender Mobility and Crime Pattern Formation from First Principles. In L. Liu & J. Eck (Eds.),*Artificial Crime Analysis Systems: Using Computer Simulations and Geographical Information Systems* (pp. 193–208). Idea Press.

Brown, B. (1995). *CCTV in Town Centres: Three Case Studies, Police Research Group, Crime Detection and Prevention Series, Paper 68.* Home Office.

Brown, J. (1999). *An Evaluation of the Secured by Design Initiative in Gwent, South Wales* (Unpublished MSc dissertation). Scarman Centre for the Study of Public Order. Leicester: University of Leicester.

Brown, B. B., & Altman, I. (1983). Territoriality, Defensible Space and Residential Burglary: An Environmental Analysis. *Journal of Environmental Psychology, 3*, 203–220.

Brown, B., & Bentley, D. (1993). Residential Burglars Judge Risk: The Role of Territoriality. *Journal of Environmental Psychology, 13*, 51–61.

Clarke, R. V. (1995). Situational Crime Prevention. In Crime and Justice Vol. 19, *Building a Safer Society: Strategic Approaches to Crime Prevention* (pp. 91–150). University of Chicago Press.

Clarke, R. V., & Eck, J. (2007). *Understanding Risky Facilities.* Center for Problem-Oriented Policing. Available online at: http://www.popcenter.org/tools/PDFs/risky_facilities.pdf

Cohen, L. E., & Felson, M. (1979). Social Change and Crime Rate Trends: A Routine Activity Approach. *American Sociological Review, 44*, 488–608.

Cornish, D., & Clarke, R. (1986). *The Reasoning Criminal.* Springer-Verlag.

Cozens, P.M. (2011). Urban Planning and Environmental Criminology: Towards a New Perspective for Safer Cities. *Planning Practice & Research, 26*(4), 481–508. https://doi.org/10.1080/02697459.2011.582357

Cozens, P. (2014). *Think Crime! Using Evidence, Theory and Crime Prevention Through Environmental Design (CPTED) for Planning Safer Cities* (1st ed.). Praxis Education.

Cozens, P. M. (2016a). *Think Crime! Using Evidence, Theory and Crime Prevention Through Environmental Design (CPTED) for Planning Safer Cities.* Praxis Education.

Cozens, P. (2016b). *Think Crime: Using Evidence, Theory and Crime Prevention Through Environmental Design (CPTED) for Planning Safer Cities* (2nd ed.). Praxis Education.

Cozens, P., & Babb, C. (2018).*What's Up on Beaufort Street? Exploring Crime, Traffic Risks and Crime Prevention Through Environmental Design (CPTED).* Available online at: https://inglewoodonbeaufort.com/2018/12/03/inglewood-cpted-report

Cozens, P., & Hillier, D. (2012). Revisiting Jane Jacobs 'Eyes on the Street' for the 21st Century: Evidence from Environmental Criminology. In S. Hirt & D. Zahm (Eds.), *The Urban Wisdom of Jane Jacobs* (pp. 196–214). Routledge.

Cozens, P., Hillier, D., & Prescott, G. (1999). The Sustainable and the Criminogenic. *The Case for New-Build Housing Projects in Britain. Property Management, 17*, 252–261.

Cozens, P., Hillier, D., & Prescott, G. (2001). Crime and the Design of Residential Property. Exploring the Perceptions of Planning Professionals, Burglars and other Users. *Property Management, 19*, 222–248.

Cozens, P., & Love, T. (2015). A Review and Current Status of Crime Prevention Through Environmental Design (CPTED). *Journal of Planning Literature*, 1–20.

Cozens, P., & Love, T. (2017). Crime Prevention Through Environmental Design (CPTED) Introducing and Learning from the "Dark Side". *Oxford Research Encyclopaedia of Criminology and Criminal Justice.* Available online at: http://criminology.oxfordre.com/view/10.1093/acrefore/9780190264079.001.0001/acrefore9780190264079-e-2

Cozens, P., Love, T., & Davern, B. (2019). Geographical Juxtaposition: A New Direction in CPTED. *Social Sciences, 8*, 252. Available online at: https://www.semanticscholar.org/paper

Cozens, P., & Van der Linde, T. (2015). Perceptions of Crime Prevention Through Environmental Design (CPTED) at Australian Railway Stations. *Journal of Public Transportation, 18*, 73–92.

Cromwell, P., Olson, J., & Avary, D. (1991). *Breaking and Entering: An Ethnographic Analysis of Burglary*. Sage.

Crowe, T. (1991). *Crime Prevention Through Environmental Design: Applications of Architectural Design and Space Management Concepts*. Butterworth-Heinemann.

Crowe, T. (2000). *Crime Prevention Through Environmental Design*. Butterworth-Heinemann.

Department for Communities and Local Government. (2006). *Code for Sustainable Homes*. DCLG.

Department of the Environment. (1994). *Circular 5/94 Planning Out Crime*. DOE.

Ditton, J., & Short, E. (1998). Evaluating Scotland's First Town Centre CCTV Scheme: Airdrie's Crime Data and Airdrie's Offenders. In C. Norris & G. Armstrong (Eds.), *Surveillance, CCTV and Social Control—The Rise of Closed Circuit Televisual Surveillance in Britain*. Gower.

Eck, J. E., & Spelman, W. (1987). *Solving Problems: Problem- Oriented Policing in Newport News*. Police Executive Research Forum.

Ekblom, P. (2011a). *Crime Prevention, Security and Community Safety Using the 5Is Framework*. Palgrave Macmillan.

Ekblom, P. (2011b). Deconstructing CPTED… And Reconstructing It for Practice. Knowledge Management and Research. *European Journal on Criminal Policy and Research, 17*(1), 7–28.

Eras, V. J. M. (1957). *Locks & Keys Throughout the Ages*. Bailey Bros and Swinfen Ltd.

Essex Police. (1995). *Bugularis Disembowlis: Your Garden Helping You to Bite Back Against Burglars*. Essex Police.

Farrell, G. (2013). Progress and prospects in the prevention of repeat victimisation. Tilley, N. (Ed.). *Handbook of Crime Prevention and Community Safety* (1st ed.). Willan. https://doi.org/10.4324/9781843926146

Farrell, G., Tilley, N., Tseloni, A., & Mailley, J. (2011). The Crime Drop and the Security Hypothesis. *Journal of Research in Crime and Delinquency, 48*, 147–175.

Farrington, D., & Painter, K. (2002, November). *How to Evaluate the Impact of CCTV on Crime*. Paper given at Conference on Evaluating the Real Evidence on CCTV in Leicester.

Farrington, D., & Welsh, B. (2002). *Effects of Improved Street Lighting on Crime: A Systematic Review* (Home Office Research Study No. 251). Home Office.

Fleming, R., & Burrows, J. (1986) *The Case for Lighting as a Means of Preventing Crime* (Home Office Research Bulletin, 22) (pp. 14–17). Home Office.

Fry, P. (2008) Paper presented at *CCTV User Group Conference*, Manchester, April 2008.

Gamman, L., & Thorpe, A. (2012). *The Korean Experience—3rd Generation CPTED?* Lecture given to the Designing Out Crime Association at Central St Martins, London, on 21 March 2013.

Gill, M., Spriggs, A., Allen, J., Argomaniz, J., Bryan, J., Hemming, M., Jessiman, P., Kara, D., Kilworth, J., Little, R., Swain, D., & Waples, S. (2005). *The Impact of CCTV: Fourteen*

Case Studies. Home Office Online Report. Home Office www.homeoffice.gov.uk/rds/pdfs05/rdsolrl505.pdf

Greenburg, S., & Rohe, W. M. (1983). External Solutions to Neighborhood-Based Problems: A Critical Review of the Relationships Between Fear of Crime, Bystander Surveillance, and Changes in the Crime Rate. *Victimology, 5*, 133–151.

Groff, E., & Lockwood, B. (2014). Criminogenic Facilities and Crime across Street Segments in Philadelphia. *Journal of Criminal Justice, 51*(3), 315–330.

Groff, E. R., & La Vigne, N. G. (2001a). Forecasting the Future of Predictive Crime Mapping. *Crime Prevention Studies, 13*, 29–57.

Hearnden, I., & McGill, C. (2004). *Decision-Making by House Burglars: Offenders' Perspectives* (Home Office Research Findings 249). Home Office.

Hillier, B. (2004). Can Streets Be Made Safer? *Urban Design International, 9*, 31–45.

Home Office Crime Prevention Centre. (1987). *Police Architectural Liaison Manual of Guidance*. HOCPC.

Home Office Crime Prevention Centre. (1994). *Police Architectural Liaison Manual of Guidance*. HOCPC.

Home Office Scientific Development Branch. (2009). *CCTV Operational Requirements Manual*. Home Office Scientific Development Branch.

IBISWorld. (2023). *Security Systems Services Industry Research Report* Accessed on 28 September 2023 at my.ibisworld.com/uk/en/industry/n80.200/products-and-markets

International Organization for Standardization. (2021). ISO 22341 Security and resilience—Protective Security—Guidelines for crime prevention through environmental design (CPTED). ISO: Geneva.

Jacobs, J. (1961). *The Death and Life of Great American Cities*. Random House.

Jeffery, C., & Zahm, D. (1993). Crime Prevention Through Environmental Design, Opportunity Theory and Rational Choice Models. In R.V. Clarke & M. Felson (Eds.), *Routine Activity Theory and Rational Choice. Advances in Criminological Theory* (Vol. 5). Transaction Publishers.

Jencks, C. (1984) *The Language of Post-Modern Architecture*. Rizzoli.

Jones, A., Valero-Silva, N., & Lucas, D. (2016). *The Effects of 'Secure Warm Modern' Homes in Nottingham: Decent Homes Impact Study*. Nottingham City Homes.

Kelling, G., & Wilson, J. (1982). Broken windows: The police and neighbourhood safety. *The Atlantic Monthly* (pp. 29–38). Washington DC.

Key Note Market Intelligence. (2014). *Market Report 2014 Closed-Circuit Television*. Key Note Ltd.

Kuo, F., & Sullivan, W. (2001). Environment and Crime in the City. Does Vegetation Reduce Crime? *Environment and Behaviour, 33*(3), 343–367.

Mahalingham, V. (1996). Sutton Town Centre Public Perception Survey. In M. Bulos & D. Grant (Eds.), *Towards a Safer Sutton? CCTV One Year On*. London Borough of Sutton.

Macdonald, J. E., & Gifford, R. (1989). Territorial Cues and Defensible Space Theory: The Burglars Point of View. *Journal of Environmental Psychology, 9*, 193–205.

Mayhew, P. (1979). Defensible Space: The Current Status of a Crime Prevention Theory, article published in *The Howard Journal of Criminal Justice* (Vol. XVIII, pp. 150–159). Howard League for Penal Reform.

Merry, S. (1981). 'Defensible Space Undefended: Social Factors in Crime Prevention through Environmental Design.' *Urban Affairs Quarterly 16* (3): 397–422.

Metropolitan Police. (1996). *The Garden Strikes Back—A Guide to Plants That Burglars Hate*. Metropolitan Police Service.

Metropolitan Police. (1999). *Operation Bumblebee—A Secure Garden*. Metropolitan Police Service.

Nee, C., & Meenaghan, A. (2006). Expert Decision Making in Burglars. *British Journal of Criminology, 46*, 935–949.

Newlands, M. (1983). *Residential Burglary Patterns in a Vancouver Neighbourhood* (Unpublished Honor's thesis, Department of Criminology). Simon Fraser University, Burnaby, BC.

Newman, O. (1972). *Defensible Space: Crime Prevention Through Urban Design*. Macmillan.

Newman, O. (1973). *Defensible Space: People and Design in the Violent City*. Architectural Press.

Newman, O. (1976). *Design Guidelines for Creating Defensible Space*. National Institute of Law Enforcement and Criminal Justice, Law Enforcement Assistance Administration. U.S. Dept. of Justice.

Newman, O. (1996). *Creating Defensible Space*. U.S. Department of Housing and Urban Development.

Newman, O., and Franck, K. (1982). The Effects of Building Size on Personal Crime and Fear of Crime. *Population and Development 5*: 203–20.

Nuttgens, P. (1989). *The Home Front*. BBC Books.

Office of the Deputy Prime Minister. (1998). *Places, Streets and Movement: A Companion Guide to Design Bulletin 32 Residential Roads and Footpaths*. ODPM.

Office of the Deputy Prime Minister. (ODPM) and the Home Office. (2004). *Safer Places: The Planning System and Crime Prevention*. Thomas Telford.

Office of the Surveillance Camera Commissioner. (2021). *Surveillance Camera Code of Practice*. https://www.gov.uk/government/publications/update-to-surveillance-camera-code

Orwell, G. (1949). *Nineteen Eighty-Four*. Secker and Warburg.

Painter, K. (1994). The Impact of Street Lighting on Crime, Fear and Pedestrian Street Use. *Security Journal, 5*, 116–124.

Pascoe, T. (1999). *Evaluation of Secured by Design in Public Sector Housing—Final Report*. Building Research Establishment.

Pascoe, T. (2025). ISO 22341 *Security and resilience—Protective Security—Guidelines for crime prevention through environmental design (CPTED)*. International CPTED Association (ICA) Masterclass given online on 19 October 2025.

Pease, K. (1999). A Review of Street Lighting Evaluations: Crime Reduction Effects. In K. Painter & N. Tilley (Eds.), *Surveillance of Public Space: CCTV, Street Lighting and Crime Prevention. Crime Prevention Studies* (Vol. 10). Criminal Justice Press.

Power, A. (1997). *Estates on the Edge: The social Consequences of Mass Housing in Northern Europe*. Macmillan.

Poyner, B. (1992). Situational Crime Prevention in Two Parking Facilities. In R. Clarke (Ed.), *Situational Crime Prevention: Successful Case Studies* (pp. 42–57). Harrow and Heston.

Poyner, B. (2006). *Crime Free Housing in the 21st Century*. Jill Dando Institute of Crime Science.

Poyner, B., & Webb, B. (1991). What Works in Crime Prevention: An Overview of Evaluations, presented at *the British Criminology Conference*. York.

President's Commission on Law Enforcement and Administration of Justice. (1967). *The Challenge of Crime in a Free Society.* Dutton.

Ramroth, W.G. (2007). *Planning for Disaster: How Natural and Man-made Disasters Shape the Built Environment.* Kaplan Publishing.

Ramsay, M., & Newton, R. (1991). *The Effect of Better Street Lighting on Crime and Fear: A Review* (Crime Prevention Unit, Paper 29). Home Office.

Ratcliffe, J. (2011). How Near Is Near? Quantifying the Spatial Influence of Crime Attractors and Generators. In *Patterns, Prevention, and Geometry of Crime.* Edited by Martin Andreson and Bryan J. Kinney. Monsey: Criminal Justice Press.

Rengert, G., & Wasilchick. (2000). *Suburban Burglary: A Tale of Two Suburbs* (2nd ed.). Charles C. Thomas.

Repetto, T. A. (1974). *Residential Crime.* Ballinger.

Reynald, D. M. (2009). Guardianship in Action: Developing a New Tool for Measurement. *Crime Prevention and Community Safety: An International Journal, 11*(1), 1–20.

Reynald, D., & Elffers, H. (2009). The Future of Newman's Defensible Space Theory: Linking Defensible Space and the Routine Activities of Place. *European Journal of Criminology, 6*(1), 25–46.

Robinson, M. B. (1996). The Theoretical Development of CPTED: 25 Years of Responses to C. Ray Jeffery. *Advances in Criminological* Theory, 8.

Rubenstein, H., Murray, C., Motoyama, T., & Rowse, W. (1980). *The Link Between Crime and the Built Environment. The Current State of Knowledge.* National Institute of Justice.

Rudlin, D., & Falk, N. (1995). *Sustainable Urban Neighbourhoods: Building the 21st Century Home.* Architectural Press.

Sarno, C. (1996). The Impact of Closed Circuit Television on Crime in Sutton Town Centre. In M. Bulos & D. Grant (Eds.), *Towards a Safer Sutton? CCTV One Year On.* London Borough of Sutton.

Saville, G., & Cleveland, G. (1997, December 3–5). 2nd Generation CPTED: An Antidote to the Social Y2 K Virus of Urban Design. Paper presented at the *2nd Annual International CPTED Conference,* Orlando, FL. Accessed 24 December 2024, http://www.veilig-ontwerp-beheer.nl/publicaties/2nd-generation-cpted-antidote-to-the-social-y2k-virus-of-urban-design

Secured by Design. (2021). *The Police Crime Prevention Initiatives Guide to Lighting.* PCPI.

Secured by Design. (2023). *Security Systems Policy.* Secured by Design. Accessible at www.securedbydesign.com/security-systems-policy/

Shu, C. F. (2000). Housing Layout and Crime Vulnerability. *Urban Design International, 4,* 177–188.

Skinner, B. F. (1953). *Science and Human Behavior.* Macmillan.

Skinns, D. (1998). Crime Reduction, Diffusion and Displacement: Evaluating the Effectiveness of CCTV. In C. Norris, J. Moran, & G. Armstrong (Eds.), *Surveillance, Closed Circuit Television and Social Control* (pp. 116–127). Ashgate.

Skogan, W. (1990). *Disorder and Decline: Crime and the Spiral of Decay in American Neighborhoods.* Free Press.

Squires, P. (1998). *An Evaluation of the Ilford Town Centre CCTV System.* Health and Social Policy Research Centre, University of Brighton.

Squires, P., & Measor, L. (1996). *Closed Circuit TV Surveillance and Crime Prevention in Brighton: Half Yearly Report*. Health and Social Policy Research Centre, University of Brighton.

Stokes, M. (2022). *An Investigation of the Sustainability of Crime Prevention in the Built Environment: Impact and Implementation Factors* (Published Doctoral Thesis). University of Huddersfield.

SURF Centre. (2002). *Hulme, Ten Years On*. University of Salford.

Taub, R., Taylor, D., & Dunham, J. (1984). *Paths of Neighborhood Change: Race and Crime in Urban America*. University of Chicago Press.

Taylor, R. B. (2002). Crime Prevention Through Environmental Design (CPTED): Yes, No, Maybe, Unknowable and All of the Above. In R. B. Bechtel (Ed.), *Handbook of Environmental Psychology* (pp. 413–426). John Willey.

Taylor, R., & Gottfredson, S. D. (1987). Environmental Design, Crime and Prevention: An Examination of Community Dynamics. In A. Reiss & M. Tonry (Eds.), *Communities and Crime* (pp. 387–416). University of Chicago Press.

Teedon, P., Reid, T., Griffiths, P., Lindsay, K., Glen, S., McFayden, A., & Cruz, P. (2009). *Secured by Design Impact Evaluation Final Report*. Glasgow Caledonian University.

Tien, J., O'Donnell, V., Barnett, A., & Mirchandani, P. (1979). *Street Lighting Projects: National Evaluation Program, Phase 1 Report*. National Institute of Law Enforcement and Criminal Justice, US Department of Justice.

Tilley, N. (1993). Understanding Car Parks, Crime and CCTV: *Evaluation Lessons from Safer Cities* (Police Research Group Crime Prevention Series Paper 42). Home Office.

Tseloni, A., Farrell, G., Thompson, R., Evans, E., & Tilley, N. (2017). *Domestic Burglary Drop and the Security Hypothesis*. Springer Open.

Urban Task Force. (1999). *Towards an Urban Renaissance*. Urban Task Force.

Van der Voordt, T. J. M., & Van Wegen, H. B. R. (1990). Testing Building Plans for Public Safety: Usefulness of the Delft Checklist. *Netherlands Journal of Housing and Environmental Research, 5*(2), 129–154.

Warren, F., & Stollard, P. (1988). *Safe as Houses: A Working Paper Reviewing the Importance of Housing Design and Layout in Achieving Security*. Institute of Advanced Architectural Studies.

Welsh, B. P., & Farrington, D. C. (2008). Effects of Improved Street Lighting on Crime. *Campbell Systematic Reviews, 2008*, 13. https://doi.org/10.4073/csr.2008.13

Wilcox, P., & Eck, J. (2011). Criminology of the Unpopular Implications for Policy Aimed at Payday Lending Facilities. *Criminal Public Policy, 10*, 473–482.

Wiles, P., & Costello, A. (2000). *The 'Road to Nowhere': The Evidence for Travelling Criminals* (Home Office Research Study 207). Home Office Research, Development and Statistics Directorate.

Winchester, S., & Jackson, H. (1982). *Residential Burglary: The Limits of Prevention* (Home Office Research Study No. 74). Home Office Research and Planning Unit.

Wood, E. (1967). *Social Aspects of Housing in Urban Development*. Department of Economic and Social Affairs. United Nations.

Wright, R., Heilweil, M., Pelletier, P., & Dickinson, K. (1974). *The Impact of Street Lighting on Crime*. University of Michigan.

5 Secured by Design

5.1 SBD—Introduction

Original Secured by Design marketing material from 1989

The historical significance and importance of Secured by Design (SBD) in the growth and support of Designing Out Crime (in both the UK and further afield), cannot be over-emphasised. Devised during 1988 and originally launched in south-east England the following year, SBD was primarily conceived as a promotional campaign/marketing tool for the then Crime Prevention Design Advisers and Architectural Liaison Officers (CPDAs/ALOs) as they had recently become known in the UK—aimed at the private (for sale) housing market. Moreover, by 1992 it had been rolled out across the whole of the UK, i.e. England, Wales, Scotland, Northern Ireland, Channel Islands and the Isle of Man.

M. Stokes, *Designing Out Crime*, Crime Prevention and Security Management,
https://doi.org/10.1007/978-3-031-99840-9_5

More than three decades later, Secured by Design has grown into an internationally recognised brand. Furthermore, it can now be identified as the catalyst that engineered the adoption of Designing Out Crime measures within the built environment across the UK and indeed beyond.

Most importantly, the longevity and success of Secured by Design have made it the most effective and successful (in terms of both Crime Reduction and sustainability) Crime Prevention initiative in the history of policing. And not just in the UK. SBD has subsequently been used as a model for similar initiatives in countries like Ireland, Chile, Abu Dhabi in the United Arab Emirates, South Korea and The Netherlands. Indeed, the Dutch version of SBD, 'Police Label Secure Housing' has processed more than one million homes (2025) to 'Secured Dwelling' standard. Consequently, it can be stated that there is nothing (and it is not through a lack of searching) as long running and both independently and academically proven to be effective as Secured by Design anywhere in the world—an accolade which in itself points to the efficacy of such Crime Prevention programmes.

Furthermore, because Secured by Design is universally accessible by way of the internet at https://www.securedbydesign.com (and continually updated), its philosophy, attributes and requirements are available to a potential worldwide audience. Indeed, this interactive website includes a wide variety of subject headings, publications and Development Guides.

5.2 SBD—History to Date

The new discipline of Architectural Liaison/Crime Prevention Design—referred to as Designing Out Crime (DOC) throughout this handbook, had become established across the UK by the late 1980s. It is now delivered by Designing Out Crime Officers (DOCOs) previously and on occasion still known as Crime Prevention Design Advisers (CPDAs), Architectural Liaison Officers (ALOs) and Design for Security Consultants (DfSCs). In the beginning they used BS 8220: Part 1: 1986 Security of buildings against crime Part 1. Dwellings (British Standards Institution, 1986) to establish their service as a necessary prerequisite for new homes within the built environment. However, the desire for more holistic guidance led to the birth of the SBD award system. As the original advertising material describes, *'Secured by Designhas been created by Senior Crime PreventionOfficers...'* and *'The Association of Chief Police Officers and the Home Office Crime PreventionUnit are fully backing the campaign'* (SBD, The Concept, 1989).

By combining traditional Physical Security 'Target hardening' Measures and Technological Innovations (see Sect. 4.2.1– 4.2.1–4.3.3) with the Environmental Elements of Crime Prevention Through Environmental Design (CPTED, see Sect. 4.4.1–4.4.6), the SBD award system systematically provides the necessary incentivisation to attract partners from a wide spectrum of providers within the built environment. They were encouraged to incorporate such CPTED measures and thus add a sustainability element, in terms of greatly reduced levels of Crime, fear of crime (FOC) and anti-social behaviour (ASB). Moreover, examples like The Four Towers at Duddeston Manor (see Sect. 9.1) demonstrate that Community Consultation has always been intrinsic to the SBD award system (and by default Designing Out Crime), thereby delivering sustainable Crime Prevention. Nor was this restricted to the short-term, but also over the medium and long-term time periods—proving how the initiative is extremely cost-effective in terms of both capital investment and reduced crime, FOC and ASB. Furthermore, it can now be confidently estimated that since 1989, in excess of 2 million homes in the UK have been built to the standard of and achieved a Secured by Design award.

From its initial launch in 1989, SBD has been reliant upon sponsorship from external companies (most usually manufacturers of building and security products), for the funding of its marketing. In 1999 this was placed on a more professional footing with the creation of ACPO CPI Limited. This metamorphosis allowed for the expansion of the number of commercial sponsors and the recruitment of more full-time staff (many former CPDAs/ALOs). It also permitted an enhanced quality of service to be delivered to individual DOCOs across the UK. In 2016 management of Secured by Design was passed to the Mayor's Office for London acting on behalf of all UK police forces.

One of the key spurs to the adoption of Secured by Design came in 1994 when the then Housing Corporation (Homes England since 2018) mandated Housing Associations to seek and attain the SBD award for their new developments—a situation that now exists only in Scotland and Wales. Nevertheless, this does not prevent the local authority (local government) from making attainment of the SBD award a mandatory planning condition—at the request of the DOCO.

Perhaps of greatest significance, there now exists a series of independent, academic evaluations of SBD which underscore its long-term, sustainable Crime Prevention benefits: Brown (1999), Pascoe (1999), Armitage (1999), Teedon et al. (2009), Armitage and Monchuk (2009), Jones et al. (2016), Stokes (2022) with little Crime Displacement and more likely a Diffusion of Benefits or 'Halo effect' (see Sect. 2.1).

Police label secure housing in The Netherlands

5.3 SBD—Development Guides and Publications

Full details of SBD requirements can be found at: https://www.securedbydesign.com/guidance/design-guides Most importantly, SBD applicants are asked to familiarise themselves with the appropriate Development Guide before contacting the Designing Out Crime Officer (DOCO), Crime Prevention Design Adviser (CPDA), Architectural Liaison Officer (ALO) or Design for Security Consultant (DfSC) in whose area the development is located.

To reiterate, hereafter and to avoid any confusion, all these titles will be referred to as Designing Out Crime Officer (DOCO). Contact details for the relevant DOCO in any part of the UK are provided on the Secured by Design website, courtesy of clicking onto an interactive map of Great Britain and Northern Ireland which identifies the police force boundaries within the countries and islands that make up the UK.

The below descriptions are a mere synopsis of the current Development Guides and the full, continually revised and updated versions **MUST** be accessed by way of the above website—where the relevant quoted standards will also be found. It should also be noted that these same Development Guides are copyrighted by and to Secured by Design.

5.4 SBD—Residential Guide 2025

Background and History

For the majority of UK DOCOs, the *Secured by DesignResidential Guide 2025* (or its successors) is the guide to which they will most often refer. Indeed, Designing Out Crime (originally Property Crime from dwellings) was one of the original key motivations for creating the SBD award system.

SBD was conceived with the specific aim of influencing private sector developers building homes for sale. Those who (like this author) were young police officers during the 1980s can well recall repeatedly attending Residential Burglaries where forced entry had been very easily achieved as a result of poor-quality Physical Security 'Target Hardening' Measures, often at brand new homes. In particular: useless window locks, patio doors with no 'anti-lift' device and the most pathetic of locking systems, plus that 'piece de no resistance'—the notorious 2XG doors containing a wafer-thin plywood panel in the lower half. One swift kick and the burglar easily and quickly gained access. This caused considerable distress for those tens of thousands of victims of residential burglary who were burgled (many we know repeatedly) by this form of Modus Operandi.

Unfortunately, 1989 (the year SBD was launched) and the early 1990s marked a depressed house building market across much of the UK and with some notable exceptions, adoption of SBD by the private, for sale sector was at best patchy. However, since the turn of the century, there has been a growing willingness by private sector developers to build new homes (on occasion at the very highest end of the market) in accordance with SBD standards. Nevertheless, for much of the first three decades public sector housing development (including refurbishment) has proved to be the mainstay of the workload for the UK DOCO.

As a partial explanation, it might be argued the public sector has a vested interest in Sustainable Crime Prevention, because they remain in ownership of the dwellings in the long-term (historically at least 30 years) and will almost always pay for the ongoing maintenance costs. Anything that can reduce such costs is a bonus and thus the attraction of the Designing Out Crime service, delivered by DOCOs.

The SBD Development Guides have increasingly evolved from the original largely 'tick the box' exercises and now require the DOCO and applicant to proactively evaluate the security measures being incorporated. Nevertheless, such checking exercises (Gawande, 2011; Sidebottom et al., , 2012) are necessary during the SBD final site inspection process. Indeed, the more than 2 million dwellings that have now achieved an SBD award testify to the efficacy of such 'tick the box' checking. Looking forward, the holy grail in terms of Designing Out Crime is for all homes to be built or refurbished to Secured by Design standards.

Readers of 'Secured by Design Residential Guide 2025' should be aware of major changes since 'SBD Homes 2024' was published. Most significant amongst these has been the change of titles from 'Homes' to 'Residential Guide'—see Sect 5.4. This is also replicated with 'SBD Commercial 2023' being updated and replaced by 'SBD Non-Residential (Commercial) Guide 2025'—see Sect. 5.5. Further changes are detailed within each development guide. Perhaps of greatest importance, a minute examination of these two new development guides reveals them to be the most detailed and in-depth publications to date—testament to the consummate professionalism of SBD and their authors.

Most important! What follows in this chapter amounts to a synopsis of the SBD development guides and other design guidance. the SBD website must always be accessed online to ensure the most up to date and accurate requirements and standards are obtained.

Preface

The Secured by Design Residential Guide 2025 (formerly known as the Homes Guide) has been written by PCPI in consultation with DOCOs and industry partners, such as the NPSA and UKAS.

The preface goes on to describe changes to the English Planning and Building Control regulations; the importance of police advice over the past 35 years—specifically the SBD initiative; the National Planning Policy Framework (NPPF), National Planning Practice Guidance (NPPG), Scottish Planning Policy (PAN

77) and Planning Policy Wales (TAN 12) reinforce the importance of a safe and secure external environment, as does the https://police.uk website.

Further guidance includes: advice by HMG's Chief Planning Officer; how the MHCLG has followed Scotland's lead and introduced physical security standards for new homes for the first time within Building Regulations); and the Welsh Government having decided to include a new Building Regulation to address the security of new homes.

This edition of the guidance for residential properties has been designed to cater for the security of all new and refurbished homes including those for disabled and older people.

Additional paragraphs relate to: constructing well-designed places—including the Home Office 'Violence Against Women and Girls' agenda; that SBD should not create additional barriers for disabled and/or older people; how accessible and inclusive design should be a material consideration; that the National Design Guide promotes active frontages; the National Model Design Code's considerations; how the police service places great importance upon the need to build sustainable and inclusive communities; that the guide's requirements and recommendations are based on academically sound research findings which prove SBD delivers significant crime reductions and cost efficiency savings; contributions to any guide can be made via sbdconsultations@police-cpi.co.uk and how the SBD Residential Guide 2025 applies to all SBD applications made after 1st March 2025.

1. Introduction

SBD is a police initiative to guide and encourage specifiers to design and build new homes (and those undertaking refurbishment) to adopt Crime Prevention measures through environmental design. The advice provided has been proven to reduce Crime Opportunity and fear of crime, creating safer, more secure and sustainable environments. SBD is owned by the UK police service and supported by the Home Office. Building Control Departments in England, Scotland and Wales all reference SBD as a means of compliance.

Additional paragraphs are devoted to compliance; role of the DOCO in conducting a crime risk analysis and providing advice; withdrawal of the SBD award; carbon cost of crime (see Chapter 16); how the environmental benefits of SBD are supported by independent academic research; that research documentation can be found on the SBD website at: https://www.securedbydesign.com Applications for an SBD Award can be made via the same SBD website.

The harm caused to victims and society by violence against women and girls (VAWG) in all its forms, including but not limited to, harassment, stalking, rape, sexual assault, murder, honour-based abuse and coercive control is incalculable. Whilst men and boys also suffer from many of these forms of abuse, they disproportionately affect women. In spring 2023, the Home Secretary announced Violence Against Women and Girls as a national threat and included it in the Strategic Policing Requirement (SPR) with terrorism, serious and organised crime and child sexual abuse.

SBD can assist with ensuring that the built environment in all its forms is designed to reduce the opportunity of VAWG and contribute to such places feeling safe, to live, work and socialise, for all. Under this VAWG sub-heading, further advice relates to NPCC guidance; liaison with the DOCO; hyperlinks to the Safer Parks and Make Space For Girls Resources Library and handbook Creating Places that Work for Women and Girls available at: https://wwwqueenelizabetholympicpark.co.uk/about-us/how-we-work/handbook-creating-places-work-women-and-girls

2. Scope of SBD Residential

This edition of the SBD Residential Guide addresses the community safety, environmental factors and physical security requirements for all types of dwellings including individual houses, housing estates, low and high-rise apartment blocks. The design, layout and physical security parts can be applied to both new and refurbished homes. Further guidance relates to any risk of terrorism, espionage, sabotage or other malicious threats and the Security Overlay to the RIBA Plan of Work (see Sect. 5.14).

3. SBD Residential Explained

Who Should Read This Document?

The SBD—Residential Guide can now fulfil the requirements of: Planning Authorities, Building Control, Architects, Designers and Project Teams, Developers, Social Housing providers, Private rented sector and Homeowners or occupiers.

SBD Residential Guide Format in Detail

Part 1: Development Design and Layout

Provides guidance on design and layout that impact on creation of a safe and secure environment, e.g. road layout, footpath design, communal areas, dwelling boundaries, car parking and lighting.

Part 2: Physical Security of the Home
Provides the Police Preferred Specification for all physical security requirements for new or refurbished homes. Part 2A reflects the requirements of the GB Building Regulations. Part 2B addresses bespoke new homes and existing refurbished homes.

Part 3: Additional Features for the SBD Gold Award
The essential security requirements in Part 2, are further enhanced by the requirements set out in this part. Part 3 addresses a range of additional or optional features, such as enhanced glazing, external garage doorsets, private external and dwelling lighting, intruder alarms etc. If a development contains any of the features within Part 3, the physical security requirements should be adhered to in order to achieve full SBD Gold Award compliance.

Compliance with any of these awards satisfies the GB Building Regulations.

SBD Graded Security Levels
SBD Gold combines external environment security with physical security specification of the home. SBD Silver offers new developments, major refurbishment and the individual to gain an award for the level of physical security provided. SBD Bronze provides a route to achieve a reasonable level of physical security for bespoke or refurbished properties, where a traditional enhanced security product is not available, or cannot be used due to the Listed Building or other conservation status.

SBD Gold
The SBD Gold Award is conferred on new developments or refurbishment schemes that achieve compliance with all the security features within Part 1, Part 2A and Part 3 of this document.

SBD Silver
The SBD Silver Award has two routes: new developments or refurbishment schemes that meet the security features contained within Part 2A. This is also the minimum qualifying criteria for SBD NBA (see Sect. 5.4.4). Or, new bespoke developments or refurbishment schemes that meet the required security features particular to the development contained within Part 1, Part 2B and Part 3.

SBD Bronze
SBD Bronze can be awarded to new bespoke developments or refurbished properties that do not meet the required security features contained within Part 2B.

4. Secured by Design National Building Approval
Secured by Design National Building Approval (SBD NBA) provides a structured means of discharging UK Building Regulations. Further guidance covers: advantages of SBD NBA; due diligence; certificate of conformity with UK Building Regulations and SBD Silver Award; increased confidence; reduced.

bureaucracy and UK police certificate of compliance. More information at: https://www.sbdnba@police-cpi.co.uk.

5. How to Apply for the SBD Award
Applicants should make themselves familiar with the relevant sections of the SBD guidance contained within the Development Guide and strongly advised to consult their local DOCO at the earliest opportunity details of which can be obtained at: https://www.securedbydesign.com/contact-us

6. Construction Phase Security—Advisory Note
SBD recommends that security should be in place before and during construction including: robust perimeter fencing; monitored Intruder Alarm system; signage; Video Surveillance Systems (VSS) and DOCO consultation.

7. UK Planning and Strategic Policies in Support of SBD
Crime is a Material Planning Consideration. The police service has worked in partnership with the governments, assemblies and local authorities throughout the UK, to incorporate DOC principles within strategic policy and planning guidance documents. Separate headings then describe the strategic guidance in support of SBD across England, Wales, Scotland and Northern Ireland.

Part 1

Development Design and Layout

8. Roads and Footpaths

Vehicular and pedestrian routes should be visually open, direct, well used and support the Defensible Space of neighbourhoods. Such routes should be safe, secure and accessible for all, in support of measures to combat Violence Against Women and Girls. See also the Active Travel England—Planning Active Places Webpage Link.

Through-Roads and Culs-de-Sac

Some road layout patterns have inherent advantages in frustrating criminals' searching behaviour and need to escape, e.g. denying excessive Permeability; unnecessary segregated footpaths; intuitive wayfinding and passive surveillance; review of available research; and short/non 'Leaking' Culs-de-Sac.

Footpath and Bicycle Route Design

Routes for pedestrians, cyclists and vehicles should be integrated to assist easy, intuitive wayfinding by use of inclusive design to increase activity and Natural Surveillance. Additional guidance covers: NPPF guidance; rear access public footpaths; unavoidable segregated/isolated footpaths; physical barriers to defeat 'desire' lines; footpath width; emergency access routes and pedestrian subways.

Planting Next to a Footpath

Planting next to a footpath should have lowest-growing adjacent to the path and larger shrubs and trees planted towards the rear. Additional advice relates to: careful selection of tree and plant species; footpaths running next to or close to roads; creation of Defensible Space and the prevention of graffiti and ASB.

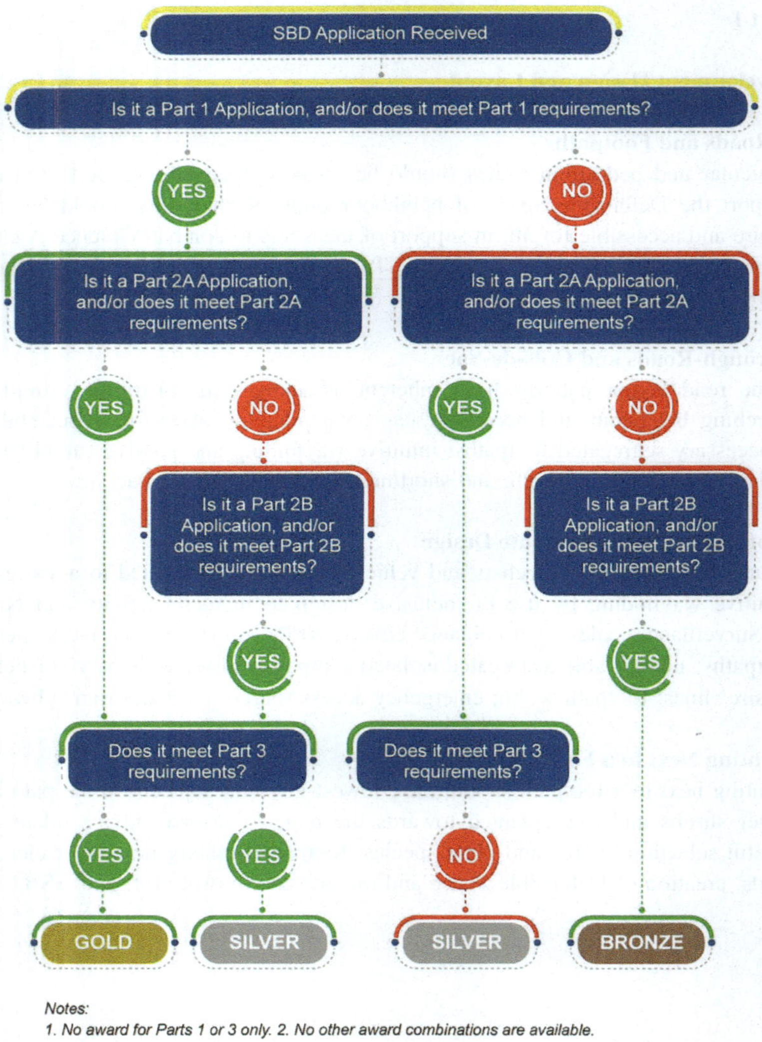

Award structure

Seating Next to a Footpath
Seating can be a valuable amenity, or a focus for ASB. Advice covers: maintenance; clear ownership cues; appropriate Lighting; position; sight lines; construction; security; design to avoid ASB; minimising opportunities to conceal drugs or weapons and positioned to prevent groups from blocking footpaths or gathering that intimidates passers-by.

Lighting of Footpaths
Footpaths in constant use to be lit as per BS 5489–1:2020. Otherwise, no Lighting to discourage use at night. Guidance covers: landscape architects, lighting designers and ecologists coordinating their plans; trees; technical requirements; 'dark sky' policies; light pollution; environmentally friendly (LED) Lighting.

Footpaths on Phased Developments
Advice covers delaying footpath completion due to phased development/long-term planning policy.

9. Communal Areas and Play Spaces
Poorly designed public realm and communal areas, e.g. playgrounds, seating facilities, may generate crime, fear of crime and ASB. Guidance includes: Natural Surveillance; safe and accessible routes; boundaries; Lighting; vehicular access; landscaping and maintenance; noise and nuisance; security at night; fencing; a dedicated entry and exit point; informal association spaces; drying spaces; fully inclusive design and early engagement with the DOCO.

10. Dwelling Boundaries
The boundary between public and private areas should be clearly indicated. For most housing developments, dwelling frontages should be open to view. Guidance covers front garden: wall, fence and hedge height; planting of feature shrubs/suitable trees (open-branched or light foliage etc.); layered Defensive Planting and both growth rate and maintenance issues.

Access Gates to Rear Gardens
Gates providing access to the side of the dwelling, rear gardens or yards must be robustly constructed, of the same height as the fence (minimum 1.8 m) and capable of being key-lockable from both sides. Such gates must be located on or as near to the front of the building line as is possible.

Side and Rear Boundaries
Vulnerable areas, such as exposed side and rear gardens, need more robust defensive barriers.

by using walls or fencing to a minimum height of 1.8 m. Guidance includes: Defensive Planting, fencing and gates quality/installation and automated gates.

Fencing in High Crime/Vulnerable Areas
Defines a vulnerable location and describes Defensive Planting to protect boundary fencing. Such areas may require time to become established, necessitating additional temporary protection. Any fencing should be 1.8 m minimum height and certified to a standard listed.

Sub-divisional Boundaries
Sub-divisional fencing design should be agreed with the DOCO and local planning authority and is dependent upon location and Crime Risks. Further guidance relates to specific features designed to achieve security, demarcation and privacy.

Exterior Balconies, Roofs and Terraces
The minimum height for balustrades is 1.1 metres, to provide adequate protection against children climbing and falling from height. DOCO advice is included.

11. Layout and Orientation of Dwellings
Dwellings should be positioned to face each other, to allow neighbours to easily view their surroundings, leading to increased Natural Surveillance, community interaction and engagement, thereby making potential offenders feel vulnerable to detection.

12. Gable End Walls
The creation of windowless elevations and blank walls adjacent to public space, should be avoided. End of terrace walls can attract graffiti, inappropriate loitering and ASB. Guidance covers: first floor windows to provide surveillance; buffer zones; hedges; climbing plants and anti-graffiti finishes.

13. Rear Access Footpaths
It is preferable that footpaths are not located at the rear of properties. Where essential, rear access footpaths must be gated and serve the fewest number of homes. Advice includes: position and locking of such gates; gate construction and standards as required by the DOCO; the inability to climb over, crawl under,

force open or be removed from their hinges and design of the gate to prevent and allow good Natural Surveillance from the street.

14. Dwelling Identification

Clear signage (naming and/or numbering) of properties is essential to assist residents, postal workers and attendance of emergency services. Such signage should be present before an award is granted.

15. Climbing Aids

Boundary walls, bins, fuel stores, metre boxes, street furniture, trees, low flat roofs, car ports or balconies should be designed and located so that they cannot act as climbing aids into the property.

16. Vehicle Parking

Vehicles should be parked in locked garages or on hard standing within the dwelling boundary. Guidance covers: DOCO advice re gates and bollards; parking bays outside 'active rooms' to be clearly marked; discouraging rear parking courtyards; fence height and use of trellis; position and security of garages; gable end windows; Lighting of communal parking facilities; how parking bays should benefit from good Natural Surveillance; that Street Lighting should comply with Sect. 5.4.19; restrictions on encroaching or uncontrolled undergrowth and use of GDPR compliant video surveillance systems (VSSs).

17. Bicycle Parking

Visitor Bicycle Parking—Short Term

Short-term visitor bicycle parking, e.g. delivery riders, shall be located next to the primary building entrance and in view of occupants. Advice covers VSSs; bicycle parking stands and their standards; requirements by DOCOs; charging and storage of Lithium-ion powered vehicles; guidance from the NFCC (e-bikes and e-scooters) and the Bicycle Association, Cycle Rail Working Group and SBD (re public cycle parking).

18. Planting in New Developments

The planting of trees and shrubs, should create attractive residential environments. Guidance covers: maintenance requirements; Crime Opportunity; careful design and choice of species; Natural and Formal Surveillance; enhancing health and wellbeing of residents; defensive planting; shrub, tree and canopy heights; building frontages; capacity to obscure doors, windows, lights and VSS cameras.

19. Street Lighting

All Street Lighting for adopted highways and footpaths, private estate roads, unadopted roads and car parks must comply with BS 5489–1:2020. Any conflicts with other statutory provisions should be discussed with the DOCO and local authority Lighting designers. Additional guidance relates to 'dark sky' policies; discouraging 'switch off'; bollard lights; Street Lighting blocked by trees; written confirmation from a competent lighting design professional that the Street Lighting scheme conforms to BS 5489–1:2020 and how SBD encourages use of the most environmentally friendly light sources. See: https://www.securedbydesign.com/initiatives/safer-streets

Part 2

Physical Security

20. Introduction

Planning policy in England and Wales has shown that security forms part of a sustainable and vibrant development. Approved Document Q in both countries that 'Reasonable provision must be made to restrict unauthorised access…' This was predated in Scotland since 2010 with the relevant document being Scottish Building Standard 4.13.

The experience gained by the UK police service over the past 35 years in the subject area of Designing Out Crime has led to the provision of a security requirement considered to be more consistent than that set out within GB Building Regulations. This is the result of testing products to the relevant security standards and independent certification.

Further guidance covers specifiers notifying DOCOs of the products they intend to purchase and documents thereafter containing the necessary certification details.

Part 2A

Physical Security for New Homes

21. Introduction

Provides technical guidance on the 'Police Preferred Specification' for new dwellings, including those formed by a material change of use and extensions to existing homes. It also covers: GB Building Regulations; Building Regulations (England and Wales); Scottish Building Standard 4.13; consultation with housing occupational therapists and the RiDC re locking systems, doors and door thresholds (Approved Document (M4) Volume 1: Access to and use of dwellings, and compartmentation; specifiers obligation to ensure compliance; achievement of SBD Silver Award or SBD Gold Award; major and minor refurbishments of one or more buildings.

22. Doorsets and Windows and Their Fitness for Purpose

From an SBD perspective, doorsets and windows must meet the security requirements of one of the standards set out. The design team, in consultation with the DOCO, shall specify the class of BS 6375–2:2009 for doorsets and windows commensurate with the location and estimated number of cycles. Further guidance covers: performance criteria; security aspects of primary doorsets and windows; cycles of repeated use; specifiers supplying proof to DOCOs; classified according to their intended use; minimum requirements of 'normal' and 'frequent' use.

23. Dwelling Entrance Doorsets

The term 'doorset' refers to a door, frame, locks, fittings and glazing as one combined unit. It refers to examples of fenestration through which people would walk to access or egress a building rather than openings that are traditionally identified as windows. Additional guidance relates to: door frames; DOCO examination; doorsets allowing direct access into to the home; certification and standards; 'Easily Accessible'—as defined within Approved Document Q Appendix A and the two routes for compliance (standards and certification).

Security, Fire and Smoke Rated Doorsets Including Those With Adjacent Glazing

This sub-section details the requirement of security, fire, smoke and security rated doorsets, e.g. flat or apartment entrance doorsets, interconnecting garage doorsets, some doorsets aiding security compartmentation and/or meet a building regulation requirement.

Garage Doorsets, Vehicular and Pedestrian

Approved Document Q, Section 1 (General), clause 1.1, states, '…where access to the dwelling can be gained via an interconnecting doorset from the garage, then either the garage doorset/s (vehicular and pedestrian) or the interconnecting doorset can be designated as the secure doorset'. The requirements of pedestrian doorsets (interconnecting or garage access doorsets) are also provided. And the relevant standards are also set out in extensive detail.

Further Requirements for All Pedestrian Doorsets
Doorsets shall also be certificated to the relevant material standards for: PVCu, Aluminium, Steel, Timber, Composite. Further guidance covers: third-party certification authorities; SBD NBA; proof of certification; SBD Licensing Scheme and product marking.

Locking Systems
To prevent burglary by ensuring the full locking system has been engaged, clear operating instructions must be attached to the inner face of the door for the information of the end-user. Further advice covers the removal of such a notice.

Apartment Entrance Doorsets That Are Designated Emergency Exit Routes
Where an apartment entrance doorset is designated an emergency exit route and with no alternative means of escape, an SBD compliant solid spindle locking mechanism will be considered. Occupants must be able to unlock the door from the inside without a key.

Glazing in and Adjacent to Doorsets
Glazing within doorsets, including glazed panels/side lights adjacent to doors installed in an integral door frame and adjacent windows (within 400 mm), must incorporate one pane of Laminated Glass meeting or exceeding the standard set out. Further advice covers: the faces of Laminated and Toughened Glass; thermal efficiency; aesthetics; privacy/obscured glazing; exceptions and installation.

Outward Opening Doorsets
Outward opening doorsets within SBD developments must form part of a certificated product range.

Door Limitation and Caller Identification
A door chain or opening limiter meeting the Door and Hardware Federation Technical Specification must be installed on a doorset to which callers can be expected (usually front doors (see Approved Document Q, Section 1: Doors, para 1.4). Guidance covers suitability; installation; door viewer Technical Specification; height and exceptions (clear glazing or side panel with clear glazing).

Doorset Installation
Door frames must be securely fixed to the building fabric as per the manufacturer's instructions and specifications and made available to the DOCO upon request if visual confirmation is felt necessary.

Doorsets hidden from public view (side or back doors) should not be recessed more than 600 mm unless: in wide recesses within public view from the street and never recessed more than 1000 mm.

Secure mail delivery to houses, bungalows, flats, apartments or maisonettes accessed via a private dedicated entrance doorset.

Increasing crime problems are associated with letter plate apertures, e.g. Identity Theft, Arson, Hate crime, lock manipulation and 'fishing' for personal items. SBD strongly recommends, where possible, mail delivery via a secure external letter box meeting the Door and Hardware Federation technical standard or delivery 'through-the-wall'. These should be Easily Accessible: at a suitable height for users.

Letter Plate Apertures in Doorsets

Where a letter plate aperture is required within a doorset it must form part of a certificated doorset range. Further guidance relates to: invalidation of the certificated doorset; validation where both doorset and letter plate are tested together within the Scope of Certification; Arson attack concern where the letter plate is replaced by an external letter box or installation of 'Anti-Arson' container; installation; standards and test methods; height of mailbox aperture. Doorsets with non-key-operated lockable internal hardware and letter plates installed, should comply with Sect. 5.4.39

External Surface-Mounted Letter Boxes

Surface-mounted letter boxes must be robust in construction. The specification offers reassurance these attributes have been met. In High-Crime Areas this is the safest way mail can be delivered. The letter box must be securely fixed to the face of the building in accordance with the manufacturer's specifications and located in a position that benefits from Natural Surveillance.

Through-the-Wall Mail Delivery

Where design prevents a letter plate with security cowl installed within a door, e.g. narrow hallway, or surface-mounted secure mail boxes are unsuitable, e.g. in a corridor, 'through-the-wall' mail delivery into a secure internal letter box may be preferred. These must include same design features as surface-mounted boxes; anti-arson features and meet the specifications set out (see Sect. 5.4.39).

24. Windows, Roof Windows, Roof Lights and Roof Lanterns

Window frames must be securely fixed in accordance with the manufacturer's instructions and specifications, made available to the DOCO if visual confirmation is necessary. Easily Accessible windows (including roof windows, roof lights and roof lanterns) shall be certificated to a standard listed. Additional guidance covers: methods of entry; emergency egress windows without locking hardware; key-lockable hardware; non-key locking hardware; fire resistant glass; opening restrictors for child safety; fit for purpose and certificated to the relevant material standard; alternative compliance; proof of certification; SBD Licensing Scheme; UKAS accreditation; Laminated Glass; trickle vents; automatic opening window and venting systems; Smoke Control Contractor or Fire Safety Professional; Easily Accessible windows, roof windows and roof lights in buildings containing multiple dwellings/bedrooms; independent third-party certification; adequate description; specification and location of security, fire and smoke rated windows.

25. Conservatories and Sunrooms

Where a conservatory or sunroom is installed, doors and windows must meet same physical security standards as Sect. 5.4.23 (doors) and Sect. 5.4.24 (windows). Further advice covers polycarbonate glazing and doorsets separating the conservatory from the rest of the dwelling.

26. Lightweight Framed Walls in Houses and Buildings Containing Multiple Dwellings or Bedrooms

Security can be compromised if lightweight framed walls do not offer sufficient resilience to prevent criminal attack. Advice to DOCOs covers: Approved Document Q; SBD requirements; doorsets; standards; 'robust details' alternatives and correct installation.

27. Private External Lighting

Where possible the Lighting requirements within BS 5489–1:2020 should be applied (see Sect. 5.4.19).

Further advice covers: additional guidance from the Chartered Institute of Building Services Engineers (CIBSE) and Society of Light and Lighting (SLL); areas to be illuminated; bollard lights; luminaires reducing light spill and light pollution; lamps with high colour rendering; overall Lighting uniformity (Uo)—levels of 0.4 or 40%; Public Lighting; dawn to dusk Lighting; SBD encouragement; use of LED light sources; the arguments against the use of both fluorescent and PIR-activated Security Lighting; and blue light reduction.

Lighting in Communal Areas Within Flats/Apartments

24-hour Lighting (switched using a daylight sensor formally called photoelectric cells) to communal parts of blocks of flats will be required. Further guidance covers dimming, locations and the use of environmentally friendly LED light sources—see https://www.securedbydesign.com/initiatives/safer-streets

28. Utility Metres
There is no requirement if 'smart metres' are installed (remote signalling). Otherwise, utility metres should be located outside the dwelling, at the front or as close as possible to the front building line and to the front of any fencing or gates. They should be incapable of being used as a climbing aid. Further advice covers buildings containing a large number of residences such as flats, apartments or maisonettes.

29. Access Control and Additional Security Requirements for Buildings Containing Multiple Dwellings or Bedrooms
Dwellings designed to provide accessible housing must consider disabled and older residents, e.g. access such as full automation via remote key fob for independent entry through all doors, car parks, communal areas, lifts, operation of heavy doors, reach controls and wall-mounted fobs.

Multiple Dwellings
Defined as a building containing flats, apartments, bedsits/individual bedrooms accessed from a semi-private area served by a shared or communal entrance doorset including Houses in Multiple Occupation (HMOs) and student accommodation.

Visitor Door Entry Systems
Covers visitor systems that call a dwelling—individual or served from a communal entrance. Further elements relate to: two-way conversation; see and identify visitor/location; locking device remote operation from dwelling; easy to operate, understand and display caller's image before answering; lens type; image recording/retention; communal entrance/landings if compartmentation is required.

Access Control Systems
Proximity access control systems provide electronic access through communal entrance doorsets.

Guidance includes: card or key fobs; access restrictions; recording every system event; memory storage; being fully programmable; no common code radio transmitters and encrypted 'key' security.

Physical Security Standards for Communal and Shared Entrance Doorsets
Defined as an external doorset leading from the street or otherwise public area to an internal semi-private communal area providing access to segregated flats, bedsit or individual bedroom. See Figure 2. Communal and shared entrance doorsets in the guide.

Luxury single dwelling easily achieving a SBD award. Photo credit: © Berkeley Homes (the developer)

Physical Security Requirements for Communal Entrance Doorsets That Serve 4 Dwellings or Less in Buildings That Are No More Than Two Storeys in Height
Instead of visitor door entry and access control systems, non-electronic keys requiring residents to meet and greet visitors at the communal door are encouraged—with exceptions. Guidance covers: Sect. 5.4.23 physical security standards; doorsets; meet and greet and tradesperson/timed-release.

Physical Security Requirements for Communal Entrance Doorsets That Serve 4 Dwellings or Less in Buildings That Are More Than Two Storeys in Height
These are required to have a visitor door entry system and access control system (regardless of the number of flats/apartments, bedsits or bedrooms). Further guidance covers: Sect. 5.4.23 physical security standards; lift controls and how meet and greet and tradesperson/timed-release are unsuitable.

Physical Security Requirements for Communal Entrance Doorsets With an Electronic Visitor Door Entry System Serving 5 to 10 Dwellings (Inclusive)
Sets out the relevant standards. Once again, such dwellings are required to have both a visitor door entry system and access control system. Further guidance covers: electronic release hardware; power failures and the unsuitability of both meet and greet and tradesperson/timed-release.

Physical Security Requirements for Communal Entrance Doorsets With an Electronic Visitor Door Entry System Serving 11 Dwellings or More
Communal entrance doorsets that serve 11 dwellings or more are required to have both a visitor door entry system and access control system. Specifiers shall satisfy the DOCO that the doorset is fit for its intended purpose and environment as in Sect. 5.4.22. The full range of appropriate standards are set out.

Additional guidance covers: the full range of appropriate standards and explanations; doorset failure; electronic release hardware; how meet and greet is not SBD compliant (see Sect. 5.4.29); tradesperson/timed-release are not permitted; and that visitor door entry systems utilising VSS must comply with Sect. 5.4.36.

Developments With 25 or Less, Flats, Apartments, BEDSITS or Bedrooms
Smaller developments of 25 or less apartments, bedsits or bedrooms are required to have both a visitor door entry system and access control system. The technology involved should include the attributes detailed. The full range of standards are set out. Meet and greet is not SBD compliant. Whilst tradesperson or timed-release mechanisms are similarly not permitted,

Developments With More Than 26 Flats, Apartments, Bedsits or Bedrooms
Larger developments containing more than 26 flats, apartments, bedsits or bedrooms shall have both visitor door entry and access control systems. The access control technology is outlined within UL 293, however it must also provide a number of attributes that are set out under this heading.

Security compartmentation of developments incorporating 26 or more flats, apartments, bedsits or bedrooms Developments of 26 or more flats, apartments, bedsits or bedrooms can suffer adversely from ASB, criminality and reduced sense of safety due to unrestricted access throughout the building. Extensive guidance covers: lift, stairwell and fire rescue service access; DOCO liaison; power failure; use of ACBs/PIBs; how meet and greet and tradesperson release are inappropriate; standards and the creation of compartmentation.

Locking Mechanisms for Security Compartmented Doorsets

Where security compartmentation is recommended, locking mechanisms must be capable of being overridden by people attempting to escape without the use of a key, or requiring more than one mechanism to be manipulated. Further guidance covers a range of issues under this heading.

This table sets out SBD Residential Guide 2025 compliance standards and requirements under the headings of: 'Doorset', 'Visitor Door Entry System', 'Access Control' and 'Compartmentalisation'.

30. Fire and Rescue Service Key Systems

Restricted key systems for an Access Control Box or, where fitted in respect of access doors, lifts, smoke ventilation or any other type of emergency override mechanism, must be the subject of agreement with the appropriate fire and rescue service and for SBD certification to take place. Failure to comply with Sect. 5.4.29.53 will result in failure of the development to receive SBD accreditation.

31. Lift Security

BS EN 81 details safety rules for construction and installation of passenger lifts and passenger goods lifts. BS EN 81-72:2022 refers to the vandal resistance qualities of such lifts. Three categories of testing parameters in terms of vandal resistance are then detailed.

32. Secure Bicycle Parking

Internal Bicycle Parking for Residents

Internal bicycle parking will be accessed by a fire, smoke and security rated doorset compliant with Sect. 5.4.23 (addressing e-bike storage). Further advice covers: construction; access doors; doorsets; access control and monitoring systems; bicycle parking equipment and standards; storage capacity; DOCO involvement and developer's responsibility. Hyperlinks connect to the COP for locking hardware on fire and escape doors; the NFCC's fire safety guidance for e-bikes and e-scooters and the Bicycle Association, Cycle Rail Working Group and SBD's Public Cycle Parking guide, see Sect. 5.17. of this handbook.

33. Integral Bin Stores and Plant Rooms

This section relates to bin stores, plant rooms and sub-stations situated within buildings, which have both external doorsets and internal doorsets. Such doorsets should comply with Sect. 5.4.23. If higher crime risks can be evidenced, higher security specifications can be employed. Additional guidance includes: access control systems; emergency egress, e.g. thumbturn or crash bar; automatic closers and ventilation.

34. Emergency Door Release Devices

Break glass emergency door exit release devices on communal external doors providing an important aid to egress in the event of an emergency have been abused, rendering some buildings insecure for long periods. SBD recommends vandal-resistant, stainless steel self-resetting emergency exit systems. Installation and system type must fully comply with Building Regulations and achieve final 'sign-off' by local Building Control or Approved Inspector. Further advice covers emergency egress route additional security.

35. Telephone and Internet Protocol (IP) Based Access Control/Visitor Door Entry Systems With or Without Remote Unlocking

To ensure viewed images are of appropriate quality, the systems shall be demonstrated to the DOCO on equipment similar to that used by residents (e.g. TV smart phone or tablet), prior to receiving the SBD award. Further advice covers UL293 (dedicated standard for access control) and Internet Protocol security; system capacity; device removal; SBD's 'Secure Connected Device' accreditation scheme; remote unlocking; system abuse; in-dwelling provision of audio and visual communication and no usage charges for residents.

36. Video Surveillance Systems (VSS)

VSS is used to describe all video surveillance systems capable of recording moving images or sound, including those with on-site or remote recording facilities and video doorbells that begin recording when a doorbell is activated. Further advice covers its purpose and Operational Requirement (OR); how VSS is not a universal solution but can help deter crime and criminal behaviour, assist with the identification of offenders, promote Personal Safety and provide reassurance; Data Protection Act 2018 and GDPR compliance; images of people and other identifying features; the importance of signage; VSS requirements; standards;

Operational Use and Requirements Table and the quality of the system together with the imagery produced.

Additional guidance covers: VSS cameras and door entry systems; monitoring; the ideal of 24/7 monitoring; image storage and retention; early discussions with an independent expert covering a range of issues that are detailed including Intruder Alarm requirements, facial recognition/identification and Automatic Number Plate Recognition (ANPR); areas and field of view activities to be monitored; use of recorded images; maintenance of equipment and management of recordings; training of operatives; fully operational system demonstrated to the DOCO on final inspection, before receiving the SBD award; and system compliance.

37. VSS Management and Maintenance Protocols

Describes how the data controller must facilitate image availability at all times on reasonable request from the police with their contact details clearly displayed in accordance with the requirements of the GDPR and Information Commissioner's Office. Further guidance covers: 'As Installed' system specification and schematic; drawing and logbook; maintenance inspections; cleaning and test recording; certificate of operational safety and security; system-operating manuals; certificate of compliance and requirements of the GDPR; NSI, SSAIB or equivalent certificate of compliance; and certificate of compliance with SBD guidelines. Further guidance is available at: https://www.securedbydesign.com/guidance/design-guides and: https://ico.org.uk/for-gdpr-guidance-and-resources/accountability-and-government/data-protection-impact-assessments-dpias/what-is-a-dpia/.

38. Doorsets Providing Alternative Access to Communal Areas (Other Than the Primary Shared or Communal Access Doorset) Including Emergency Egress Doorsets

Alternative access doorsets, emergency egress and Fire Doorsets that may be used by residents to access communal parts of the building are also required to be 'secure doorsets' as per Sect. 5.4.23.

Emergency Egress Doorsets From Underground Car Parks

Doorsets providing access to and emergency egress from underground car parking areas, must meet the requirements of the regulations set out. The dilemma of different performance requirements is described, and its solution after detailed discussions with the fire service—see Sect. 5.4.38.

Glazing in Communal Entrance Doorsets
Where a glazed vision panel is installed, it must form part of the manufacturers certificated doorset range. Additional guidance relates to security performance criteria, thicknesses of such glazing and standards.

Mail Delivery in Buildings Containing Multiple Dwellings or Bedrooms
There are increasing crime problems with the delivery of post to buildings with multiple dwellings or bedrooms. Mail delivery compromising security of such residential areas to deliver to each residence is not permitted. Facilities should enable mail to be delivered to safe and secure areas.

Communal Mail and Parcel Delivery
Communal mail delivery facilities serving multiple flats or rooms, e.g. student accommodation should incorporate: external or internal delivery facilities; access control with data logging; airlock entrance; access control strategy; internal and external video surveillance; secure system of depositing parcels; mail and parcel delivery boxes not subject to master key access, robust construction, incorporate anti-fishing design and fire resistant. Individual letter boxes shall have the maximum aperture size set out. All delivery boxes must be installed in accordance with the manufacturer's specification. And a secure system of delivering parcels shall be considered.

40. Loft Hatches in Communal Areas
Loft hatches located in communal areas, e.g. over landings must be locked to prevent access into dwellings via the loft space, to prevent spread of fire and smoke). No 'hinged' or 'lift out' loft hatches complying with recognised security standards currently exist. But where padlocks, hasps and staples are used to secure the hatch, they must be certificated to one of the standards listed. Responsibility for specification and location of fire rated security products lies with the developer or their agent.

41. Electrical Sockets and USB Charging Points in Communal Areas
To discourage mobile phone charging and ASB in communal areas, all 13-amp outlets or dedicated USB charging points for use by cleaners and maintenance staff should be capable of being key locked to prevent misuse. Installation and product types should comply with the standard set out.

Part 2B

Physical Security for Bespoke Homes and Existing Refurbished Homes

42. Introduction

This section provides technical guidance for bespoke new homes and upgrading of existing homes. To gain SBD approval, such homes with standard doorsets and windows are expected (as a minimum) to achieve the SBD Silver Award. However, bespoke developments subject to planning or specification restrictions due to their Listed Building status, locality within a Conservation Area or of national importance, special architectural or historical interest, may not be able to achieve the SBD Silver Award. Nevertheless, every effort should be made to achieve compliance with as many elements as possible.

Usually, SBD certification is only available if the provisions in Part 2A are complied with. However, Approved Document Q (English and Wales Building Regulations) and Scottish Building Regulation 4.13 both allow an alternative route using a door or window specification incorporating components tested to published security standards and SBD has responded by providing additional guidance.

SBD defines a bespoke doorset or window as: 'a single or small number of doorsets or windows installed within a development (normally no more than 4 homes) of unique design with non-standard features which preclude the use of conventional enhanced security door and window products. Doorsets or windows installed within buildings of specific architectural value, constrained by Listed Building or other conservation status may also be considered to be bespoke'. A client led requirement for SBD accreditation and compliance will lead to an SBD Bronze Award. Where combined with Part 1 compliance and (where applicable) relevant security recommendations in Part 3, an SBD Silver Award may also be achieved. Major refurbishment schemes should meet the requirements within Part 2A.

The guide also recommends that when introducing new security measures, the procurement process involves consultation with local authority access professionals in respect of four detailed accessibility issues.

43. New Bespoke Houses, Bungalows, Flats, Apartments or Maisonettes Accessed Via a Private Dedicated Entrance Doorset

Defines 'doorset' and how all new bespoke doorsets allowing direct access into the home, e.g. front and rear doors, interconnecting garage doorsets, French doors, Bi-fold or sliding patio doorsets, dedicated private flat or apartment entrance doorsets, Easily Accessible balcony doorsets, are required to be secure doorsets within GB Building Regulations. A definition of 'Easily Accessible' is provided courtesy of Approved Document Q. Where there is a requirement for a doorset to be fire, smoke and security rated, e.g. flat or apartment entrance doorsets, interconnecting garage doorsets and some doorsets aiding security compartmentation, it is the developer or their agent's responsibility to ensure compliance with all applicable Building Regulations.

Door and Window Materials
All bespoke window and doorsets constructed from materials commonly utilised for such purposes, shall meet the minimum material specific requirement and standards as described under this heading.

Doorset Hardware and Locking Systems
The primary entrance doorset should be fitted with a multi-point locking system that meets the requirements of one of the standards quoted. Alternative lock configuration for primary dwelling doorsets (usually the Front Doorset) are set out in great detail, as are those for non-primary doorsets and bespoke doorsets.

Glazing in and Adjacent to Doorsets
Any glazing within bespoke doorsets, including glazed panels/side lights adjacent to doors installed within an integral door frame and windows adjacent to doorsets (within 400 mm), must incorporate one pane of Laminated Glass meeting or exceeding the requirements of the standard set out.

Door Limitation and Caller Identification
Repeats the information given under this heading in Sect. 5.4.22.

Doorset Installation
Door frames must be securely fixed to the building fabric as per the manufacturer's specifications. Additional information is provided in the guide.

44. Security of Existing Refurbished Doorsets
Repeats the definition of 'doorset' and how it is difficult to provide a definitive requirement for each doorset type and material. If the existing doorsets are to be retained during a refurbishment scheme, the DOCO should be first consulted. Some areas for improvement may include: thorough inspection to ensure it is a good state of repair; how replacing material, dimensional and locking systems are referenced in Sect. 5.4.6.5; Euro profile cylinders replaced with products

certificated to the standards listed; sliding patio doorsets with a minimum of three locking points and anti-lift hardware; ‘slave’ door leaf of French or double doorsets; timber doorsets, including deadlock guard, or anti-thrust plate and frame strengthening; hinge bolts or specialist interlocking hinges; doorsets incorporating reinforcing panels; upgrading the glazing to meet the requirements of Sect. 5.4.3.3 and use of glazing security clips or tape.

45. Secure Mail Delivery to Bespoke Houses, Bungalows, Flats, Apartments or Maisonettes Accessed via a Private Dedicated Entrance Doorset
A letter plate complying with the Door and Hardware Federation’s technical standard quoted, will provide reassurance that the letter plate aperture being used to gain access to the home will be substantially reduced. Additional guidance relates to: Approved Document Q; height of letter plate and aperture dimensions; alternative compliance; fixing, testing and relevant standards; internal security deflector plates; dimensions and NHBC requirement. Alternative compliance is also detailed.

External Surface-Mounted Letter Boxes
Due to increasing crime problems with letter plate apertures, SBD strongly recommends mail delivery via a secure external letter box meeting the Door and Hardware Federations’ Technical Specification stated. Or, delivery ‘through-the-wall’ into a secure area of the dwelling. The letter box must be securely fixed to the face of the building in accordance with the manufacturer’s specifications and located in a position that benefits from Natural Surveillance.

Through-the-Wall Delivery
A surface-mounted secure mail box may be used when design constraints prevent a letter plate with a security cowl being installed within a door. Anti-arson design features may also be advised for such a box if such Crime Risks are present. The relevant technical specifications are also detailed.

46. New Bespoke Windows, Roof Windows, Roof Lights and Roof Lanterns
All new bespoke windows should comply with the material and dimensions requirements within Sect. 5.4.43.

Window Hardware
Windows should be installed with multi-point espagnolette locking systems that have been shown by test to meet the security requirements of PAS 24 for a component part of a window of the same material. There should be locking points

within 100 mm from the corner of the casement. Where a multi-point espagnolette locking system is not compatible or desirable (e.g. Listed Building application) then there should be a minimum of two locking points per opening light. Further advice relates to hinges and pivots installed within bespoke windows; heritage hinges; tilting window pivots and top retaining bolts and sash fasteners (fitch catches).

Glazing in Windows

All glazing in bespoke windows within 400 mm of an adjacent doorset shall incorporate one pane of Laminated Glass to the relevant standard. Guidance also covers: absence of inner- or outer-face requirement to install Laminated glazing; thermal efficiency, aesthetics; privacy or obscured glazing; PAS standards; Building Regulations/Standards; Laminated/Toughened Glass and emergency egress.

Window Installation

Advises how windows must be securely fixed to the building fabric in accordance with the manufacturer's specifications. Further guidance relates to vertical Sliding sash windows.

47. Security of Existing Refurbished Windows

It is difficult to provide a definitive requirement for each window type and material. Therefore, if the existing windows are to be retained during a refurbishment scheme, the DOCO should be consulted before embarking on any improvements. Some areas for improvement are also described.

48. Conservatories and Sunrooms

Where a conservatory or sun room is installed in a bespoke home and it is not possible to utilise PAS 24 doorsets and windows due to the reasons set out, the doors and windows must meet the same material, dimensional applicable and physical security standards within Part 2b. Further guidance relates to untested roofing systems and the consequences for doorsets.

49. Private External Lighting

Where possible the Lighting requirements within BS 5489–1:2020 should be applied (see Sect. 5.4.2.14). Further advice covers: Chartered Institute of Building Services Engineers (CIBSE) and Society of Light and Lighting (SLL); areas to be illuminated; bollard Lighting; luminaires reducing light spill and light pollution; lamps with high colour rendering; overall Lighting uniformity (Uo)—levels of

0.4 or 40%; external public Lighting; SBD encouragement and fluorescent versus PIR-activated Security Lighting.

Lighting in Communal Areas Within Flats/Apartments

Repeats the guidance given in Sect. 5.4.27.

50. Utility Metres
Repeats the requirements set out in Sect. 5.4.28.

51. Additional or Alternative Requirements for New Bespoke Buildings Containing Multiple Dwellings or Bedrooms
Relates to buildings containing multiple dwellings such as flats, bedsits or individual bedrooms accessed from a semi-private area and served by a shared or communal entrance doorset (including HMOs and student accommodation) there might be a requirement for doorsets to be fire, smoke and security rated.

Guidance covers: responsibility of the developer or developer's agent to ensure compliance; testing and certification to the standard stated; Listed Building or conservation status; DOCO liaison; glazing, glazed panels/side lights adjacent to doors and standards; mechanical locking systems to meet the same physical security standards as in Sect. 5.4.43; magnetic doorset locking systems; bespoke doorsets to be installed with hinge bolts or specialist interlocking hinges; enhanced security measures for timber doorsets; and door entry and access control systems to comply with the same standards as in Sect. 5.4.23.

52. Mail delivery in Bespoke Buildings Containing Multiple Dwellings or Bedrooms
To combat crime associated with delivery of post to buildings containing multiple dwellings or bedrooms, the measures detailed in Sect. 5.4.23 should be implemented.

53. Bespoke Doorsets Providing Alternative Access to Communal Areas (Other Than the Primary Shared or Communal Access Doorset) Including Emergency Egress Doorsets
Alternative access doorsets, emergency egress and Fire doorsets used to access communal parts of the building must also be 'secure doorsets' (Building Regulations, England and Wales). Bespoke doorsets shall meet the requirements within Sect. 5.4.43.

54. New Windows, Roof Windows and Roof Lights in Bespoke Buildings Containing Multiple Dwellings or Bedrooms
All Easily Accessible bespoke windows, roof windows and roof lights in buildings containing multiple dwellings or bedrooms shall, where possible, meet the requirements within Sect. 5.4.46.

55. Lightweight Framed Walls in Bespoke Dwellings
Identifies how the security of a development can be severely compromised if lightweight framed walls do not offer sufficient resilience to withstand a criminal attack and SBD requirements.

56. Loft Hatches in Communal Areas
Expands on the requirements given in Sect. 5.4.40.

57. Secure Bicycle Parking

Internal Bicycle Parking for Residents
These will be accessed via a fire, smoke and security rated doorsets compliant with Sect. 5.4.23; (addressing concerns over storage of e-bikes). Guidance covers dividing walls; thumbturns or other emergency furniture on access doors; emergency egress; access control and monitoring systems; relevant standards and maximum storage. Hyperlinks to National Fire Chiefs' Council (NFFC) and Standards for Public Cycle Parking (see Sect. 5.17) are also provided.

58. Internal Communal Drying Rooms
Where dedicated communal internal drying rooms are located in blocks of flats, they must be fitted with doorsets that meet the same physical specification as a front door in Sect. 5.4.23—to ensure they are only accessible to the residents. The locking system must be operable from the inner face by use of a thumbturn to ensure residents are not accidentally locked in by another person.

Part 3

Additional Features for an SBD Gold Award or a Silver Award for a Bespoke Development

59. Introduction
This section relates to those seeking the full SBD Gold Award or SBD Silver Award for a bespoke development. Ancillary requirements are not compulsory

features, e.g. underground car parking etc., but where installed they should meet the requirements within this section to ensure the full award is achieved. Bespoke developments cannot achieve a full SBD Gold Award because either/or both doors and windows have not been proven to resist an attack. This part of the SBD Residential Guide may be also used to ensure the security of supplementary or ancillary requirements are also catered for.

60. Doorsets Providing Access/Egress From Communal Areas, Houses and Buildings Containing Multiple Dwellings or Bedrooms

All doorsets providing access to communal areas of a building containing multiple dwellings (e.g. flats) bedrooms (student accommodation) and communal facility areas such as bicycle stores, bin stores (with external access), underground car parks (including fire egress doorsets) shall meet the security requirements of Sect. 5.4.23

61. Additional Window Requirements for an SBD Gold Award

In certain High-Crime locations (to ensure security is commensurate with the risk) the DOCO may require Laminated Glass of the standard quoted to be installed in all ground floor and basement windows and those Easily Accessible above the ground floor. Developers are advised that a late SBD Gold Award application for a development in a High-Crime area may require glazing to be replaced if it does not meet the standard required.

62. Vehicle Access Doorsets

Details the requirements if the developer wishes to gain the full SBD Gold Award in respect of external pedestrian and vehicle access doors to the garage and the necessary standards. Such products must be certificated to Sold Secure Bronze level or above.

63. Car Parking

Communal Car Parking Areas

Details the requirements where communal car parking areas are necessary. Defines 'active' rooms and how Street Lighting must be at the levels recommended by BS 5489-1:2020. The DOCO shall be provided with written confirmation of conformity to BS 5489-1:2020 by a competent lighting design professional and a CDM risk assessment completed.

Underground Car Parking

Where a development incorporates an underground car parking facility, security enhancement measures and standards are listed in detail.

64. External Bicycle Parking for Residents in Public Areas

External bicycle parking facilities will be designed for secure storage using bicycle lockers, hangars or dedicated storage devices certified to a standard listed. Further guidance covers positio*n, Natural Surveillance*, Video Surveillance, Lighting by energy efficient lamps, Lithium-ion powered vehicles and the National Fire Chiefs' Council fire safety guidance for e-bikes and e-scooters available at: https://www.nfcc.org.uk/our-services/campaigns/e-bikes-and-e-scooters-fire-safety-guidance Similarly, the Bicycle Association, Cycle Rail Working Group and SBD have published a UK quality and security standard for public cycle parking available at: https://www.securedbydesign.com/images05132-Cycle-Parking-and-Security-Standards-June2021-REV-6.pdf

Sheds Used for Bicycle Parking

Where sheds are provided for bicycle parking, they shall be certified to Sold Secure SS301 Bronze or above and securely fixed as described.

65. Intruder Alarms

Intruder Alarm systems shall meet the requirements of the standard quoted and shall be in accordance with current electrical regulations. If an immediate police response is required, installers must meet the requirements of the National Police Chiefs' Council (NPCC) policy Guidelines on Police Requirements & Response to Security Systems available at: www.policesecuritysystems.com/national-police-chiefs-council-security-systems-policy, www.policesecuritysystems.com/secured-by-design-alarm-standard.

Further information is available at: www.policesecuritysystems.com/secured-by-design-alarm-standard/sbd-technical-guide.

66. Future Homes Standard

The Future Homes Standard is a key component of the UK government's ten-point plan for a green industrial revolution and an important part of enabling the UK to become net carbon zero by 2050.

Advice relates to Mechanical Ventilation with Heat Recovery (MVHR), energy efficiency, excess heat, current and acceptable standards.

5.5 SBD—Non-Residential (Commercial) Guide 2025

This Guide Was Updated and Published Under Its New Name in 2025. Major Changes Have Taken Place and Readers of This Handbook Are Strongly Advised to Check Online and Ensure the Correct Guidance Is Obtained.

Preface

The Secured by Design Non-Residential Guide for 2025 (formerly known as the Commercial Guide) has been written by Police CPI in consultation with DOCOs and industry partners, such as the National Protective Security Authority and UKAS accredited test houses. The intention is to periodically update this guide as a result of further consultations with partners, signalling new iterations of standards and improved styles of applying Designing Out Crime methods. The police service places great importance upon the need to build sustainable and inclusive communities and to raise awareness of the significant impact that low crime has on the ongoing and long-term sustainability of good business and commerce.

Constructing well-designed places, buildings and environments that promote both sustainable communities and health and well-being, is an objective the SBD Initiative widely supports. However, it is imperative that they must also be safe, secure and accessible. Mitigating the opportunities for crime is not only about reducing and preventing injury and crime, but also building strong, cohesive, vibrant and participatory communities.

Crime, fear of crime and the unhappiness and disorder they create, significantly damages communities. There is an expectation that all those with responsibility for the built environment from local authorities and the police to architects, builders and developers, include crime prevention measures and practices within their design; creating environments that reduce opportunities to commit crime with design-led solutions is a cost-effective, resource efficient and highly impactive means of improving the health and well-being and quality of life for all.

Secured by Design works to deter criminal and anti-social behaviour through the design, layout and specification of buildings and the spaces around and between them. Academic research shows that crimes are more likely to be committed where a criminal feels more comfortable committing them, for instance, where a physical environment offers easy unrestricted or at least uncontrolled access, where clear messages of ownership are absent, where either natural and formal surveillance are absent or where wrong doers feel free to move within an area assured of their anonymity.

Design-led solutions to prevent crime include (but are not limited to) natural surveillance, high-quality lighting, clearly defined and well-overlooked defensible spaces, the eradication of unnecessary pedestrian routes and to see these concentrated together to form busier traffic routes, the sensible and sensitive positioning of play and leisure areas, provision of clear and appropriate boundary treatments and careful removal of chances to hide unseen and create ambush points. For example, an obvious advantage of creating spaces with clear sight lines across and around them that do not provide places where anyone can hide, is the fact that people can see for themselves that they're safe moving through an area that reassures them.

Police CPI is aware that due consideration should be given to not creating additional barriers for disabled and/or older users. Taking an inclusive design approach which aims to remove barriers that create undue effort and separation will ensure that buildings, places and surrounding spaces can be easily and comfortably accessed and used by everyone. This guide incorporates the latest security standards, developed to address emerging criminal methods of attack and includes references to the Building Regulations and other statutory requirements across the UK.

The National Design Guide sets out the characteristics of well-designed places and demonstrates what good design means in practice. Detailed guidance is provided by the National Model Design Codes Parts 1 and 2. Its purpose is to provide detailed guidance on the production of design codes, guides and policies to promote successful design. It expands on the ten characteristics of good design set out in the National Design Guide, which reflects the government's priorities and provides a common overarching framework for design. The National Design Guide says that 'the long-standing, fundamental principles for good design are that it is: fit for purpose; durable and brings delight'. This concept embodies the places where we live, work and spend our leisure time.

Accessible and inclusive design should be seen as a material consideration specifically for occupation by disabled and/or older users so as to ensure that these members of our community are not subsequently made less secure. For example, if door locking systems are complex and difficult to operate, some people may leave their door insecure due to their inability to operate such a locking system. The standards contained within this document are based upon those developed by SBD with various standards owners and trade associations. Should you wish to contribute to this or any other SBD guide, please contact SBD at: sbdconsultations@police-cpi.co.uk This iteration of the Secured by Design Non-Residential Guide is applicable to all SBD applications made after 1st March 2025.

1. Introduction

Secured by Design (SBD) is a crime prevention initiative operated by the police services of the UK. Secured by Design Non-Residential Design 2025 is one of several SBD Design Guides which aim to reduce crime and fear of crime in the built environment. SBD continues to evaluate the effectiveness of its guidance and periodically amends the guides to take account of new research findings and standards. This guidance incorporates new security standards that have been developed to address new and emerging methods of criminal attack. SBD Non-Residential provides a practical level of risk commensurate and sustainable security measures compatible and sympathetic to successful business. Further guidance covers: the types of crime committed against non-residential properties; National Planning Policy Framework (NPPF); how the police service places great importance on the need to build sustainable developments; the carbon cost of crime; management and maintenance via the SBD partner initiative 'Secured Environments' https://securedenvironments.com and mitigating the impact of terrorist attacks at: https://www.npsa.gov.uk/system/files/documents/responding-to-terrorist-incident-effective-command-and-control-npsa-march-23.pdf

Violence Against Women and Girls (VAWG)

The harm caused to victims and society by violence against women and girls (VAWG) in all its forms, including but not limited to, harassment, stalking, rape, sexual assault, murder, honour-based abuse and coercive control is incalculable. Whilst men and boys also suffer from many of these forms of abuse, they disproportionately affect women. In spring 2023, the Home Secretary announced Violence Against Women and Girls as a national threat and included it within the Strategic Policing Requirement (SPR) alongside terrorism, serious and organised crime and child sexual abuse.

Additional guidance under this VAWG heading includes: how Secured by Design can assist with ensuring that the built environment is designed to reduce the opportunity of Violence Against Women and Girls; the National Police Chiefs' Council's (NPCC) 'Policing Violence Against Women and Girls—The National Framework for Delivery: 2024–2027' outlining the need for building safer spaces and places: https://www.npcc.police.uk/SysSiteAssets/media/downloads/our-work/vawg/vawg-framework-for-delivery.pdf; DOCO liaison; emphasising the importance of collaboration and joined-up thinking; consult the handbook—'Creating Places that Work for Women and Girls', developed by the London Legacy Development Corporation see: https://www.queenelizabetholympicpark.co.uk/about-us/how-we-work/handbook-creating-places-work-women-and-girls; the 'Green Flag—Safer Park

Improving access for women and girls' documentation produced in collaboration with the University of Leeds and the 'Make Space for Girls' campaign resources library. These are available at: Safer Parks Pdf Link and Make Space For Girls Resources Library Webpage Link.

2. The Designing Out Crime Process

This edition of the Non-Residential Guide is effective from 1st March 2025. It applies to all types of non-residential premises from small storage units to multi-use business parks. Since the crime risks associated with non-residential premises can range from low to high value crimes, there can be no 'one size fits all' solution applicable to all development. The advice given by the police DOCO will be provided directly from the content of this guide and will be dependent upon a crime risk analysis (CRA) and local crime trends. Where justified by the results of the CRA, some sections of this guide allow for the adoption of enhanced measures commensurate with an increased risk and specified by the DOCO. The National Police Chiefs' Council (NPCC) designated the DOCO role as a specialist role within policing in October 2021.

If SBD approval is required, applicants should follow the application process by familiarising themselves with the relevant parts of this guide and consulting the DOCO for site-specific information at the earliest opportunity—ideally at the pre-planning stage, and always before a planning approval submission on a new development or refurbishment project.

3. How to Apply for an SBD Award

SBD award applicants should complete the checklist and application form which can be downloaded by visiting: https://www.securedbydesign.com Further guidance includes: how the application form must be read in conjunction with the full SBD Non-Residential Guide; a new application is required if the development has not started on-site within 3 years; how the guide will now be published in 'Editions' to allow for updates throughout the year; and developers wishing to apply for SBD approval should contact their local DOCO whose contact details can be found on the SBD website at: https://www.securedbydesign.com/contactus

4. Scope of SBD Non-Residential

The SBD Non-Residential Award ranges from new schemes to refurbishments of existing buildings. The scheme encompasses both non-residential use where the public have no formal access, and those where public access is integral to non-residential use, such as retail premises. Advice covers: SBD design guidance and specification requirements to reduce crimes against the person or property crime, fear of crime and ASB; Counter Terrorism State Threat security measures; CTSA referral and the National Protective Security Authority (NPSA), Royal Institute of British Architects (RIBA) and Police CPI *Security Overlay to the RIBA Plan of Work*: https://www.architecture.com/knowledge-and-resources-landing-page/security-overlay-to-the-plan-of-work

5. Phased or Speculative Developments
Some non-residential or mixed-use developments can be phased (developed when the market allows) or speculative (a developer provides the infrastructure but no specific building designs. SBD certification can be achieved in either situation provided the relevant DOCO's approval is obtained. Where a mixed-use development has both residential and non-residential elements, SBD certification can only be gained for the residential development once the non-residential development has been secured in compliance with the SBD Non-Residential Guide. SBD residential developments is available at: https://www.securedbydesign.com/images/HOMES%20GUIDE%20May%202024.pdf Mixed-use developments will require separate application forms to be completed for the residential and non-residential elements.

6. Explanation of Graded Security Levels

SBD Non-Residential BRONZE Award
To achieve the Bronze Award, applicants must have complied with as many elements of the Silver Award as is possible and all the relevant recommendations detailed in the Bronze Award section.

SBD Non-Residential SILVER Award

Applicants must have complied with all the relevant recommendations detailed in the Silver Award section.

SBD Non-Residential GOLD Award

To achieve approval of the highest award for security in development design, applicants must have complied with all the relevant recommendations detailed in both the Silver and Gold Award sections.

7. Construction Phase Security—Advisory Note
Unfortunately, many crimes which occur during the construction phase (including enabling works) of a development. The most significant include theft of plant equipment, materials, tools and diesel fuel. Guidance includes: SBD recommends

security should be considered throughout the life-cycle of the development, prior to and during construction; signage; video surveillance systems (VSS); DOCO consultation re perimeter fencing or hoarding; nature and *surveillance* of adjoining footpaths and/or roads and how additional advice on such security can be obtained from the SBD website: https://www.securedbydesign.com/images/CONSTRUCTION_SITE_SECURITY_GUIDE_A4_8pp.pdf.

8. Planning Policy, Design Codes, Guidance

Policy and strategic guidance in support of SBD reflects established principles of Designing Out Crime The application of these principles, design details and specifications for the particular development, must be agreed between the developer/developer's agent and police DOCO. Guidance covers: local planning conditions, crime risk assessment and other statutory provisions; specific design guidance for England, Wales, Scotland and Northern Ireland and the National Design Guide (England and Wales).

9. Design and Access Statements

Proposed non-residential developments may require a Design and Access Statement (DAS)—a report accompanying and supporting a planning application that enables those responsible for the design of new developments to demonstrate their approach to designing out opportunities for crime within their planning proposals. Compliance with SBD award criteria can be a major indication that a proposal adequately addresses the crime prevention component in the DAS. Guidance includes: how 'crime' has a potentially adverse economic, social and environmental impact; the National Planning Policy Framework requires that crime and fear of crime should be afforded due consideration within the DAS; insufficient or inadequate crime prevention information; DOCO advice dependent on the issues detailed; specifically targeted crime preventative design measures and how SBD guidelines are minimum requirements and in higher risk areas greater crime resistance will be required.

10. Levels of Security Standards

Physical security should be integral to the design and commensurate with the risk. Where justified by a crime risk analysis, some sections of this guide allow for commensurate enhanced measures to be specified by the DOCO, the details of which are contained within each relevant section. Further guidance covers SBD's Police Preferred Specification' and website link.

11. Unknown Occupiers and Risk

This guide details the minimum level of police preferred security for non-residential developments. If during the design process the intended end use changes, the DOCO must be advised of this fact to ensure that the security measures remain commensurate with the risk.

Bronze Award

12. Introduction

If doors or windows are being replaced, new and refurbished non-residential developments will be expected to achieve the SBD Silver Award, as a minimum. However, some bespoke developments may not be able to achieve SBD Silver compliance due to planning or specification restrictions owing to listed building status, position within a designated conservation area or determined to be of national importance, or special architectural or historical interest. This said, every effort shall be made to achieve compliance with as many elements of the Silver Award as possible, following consultation with the DOCO. Further guidance covers: how the SBD Bronze Award offers bespoke non-residential developments a route to compliance with door and window components tested to published security standards; definition of 'doorset' and fire and/or smoke rated doorset compliance in relation to developer or developer's agent and Building Regulations.

13. Doorset and Window Materials

All bespoke window and doorsets constructed from common materials such as timber, PVCu, aluminium, steel and composite shall meet the following minimum material specific requirements:

Timber Products

Doorsets and windows should be manufactured from solid or laminated timber with a minimum density of 600 kg/m^3. Guidance covers: doorset and window dimensions; maximum length and height dimensions by window type and constructing good quality timber windows and doorsets from the standard stated.

Aluminium Products

All windows and doorsets should be constructed from aluminium profile fabricated from designated alloys and conforming to the standard set out or alternative.

Guidance covers: the thickness of aluminium profiles; bespoke aluminium products and construction of good quality windows and doorsets conforming to the standard stated.

Composite Products
The Association of Composite Door Manufacturers has advised SBD it is unwise to produce a specification for a 'bespoke' application. Therefore, it is not possible to create a bespoke composite non-residential doorset that complies with the SBD Bronze Award.

Steel Products
Guidance for the construction of good quality bespoke steel windows and doorsets is detailed in the standard set out: 'Steel-framed windows and glazed doors. Specification'.

PVCu Products
All PVCu windows and doorsets should be constructed from profile meeting the requirements of the standard set out. Guidance covers: bespoke PVCu products; reinforcement; the standard for good quality bespoke PVCu windows and doorsets.

Doorsets and windows must also be fit for purpose and certified to one of the relevant material standards listed.

14. Dorset Hardware and Locking Systems
The primary entrance doorset shall be fitted with a multi-point locking system that meets the requirements of one of the standards set out. Guidance includes: alternative lock configuration for primary doorsets (usually the front doorset) and standards; additional surface-mounted rim locks; non-primary doorsets; bespoke doorsets; how electronic locking systems that form part of the access control system and controlled by access codes, key cards etc. shall not prevent emergency egress; electronic locks for a primary doorsets should be certified to one of the minimum standards set out; Sect. 5.5.36 for information about Internet of Things devices and how further advice on doorset hardware and locking systems from a suitably qualified locksmith, is available at: https://www.locksmiths.co.uk

15. Window Hardware and Locking Systems
Windows shall be installed with multi-point espagnolette locking systems shown by test to meet the security requirements of PAS 24, as a component part of a window of the same material. There shall be locking points within 100 mm

from the corner of the casement. Guidance covers: where multi-point espagnolette locking systems are not compatible or desirable; hinges and pivots within bespoke windows; heritage hinges; tilting window pivots and top retaining bolts; sash fasteners (fitch catches) and how further advice on window hardware and locking systems from a suitably qualified locksmith, is available at: https://www.locksmiths.co.uk

16. Glazing Within and Adjacent to Doorsets
Any glazing within PAS 24 or STS 201 certified doorsets, including glazed panels/side lights adjacent to doors installed within an integral door frame and windows adjacent to doorsets (within 400 mm), must incorporate one pane of laminated glass certificated to the standard stated. Additional guidance relates to: laminated glazing on the inner or outer face of a double-glazed unit; exemptions; shown to be part of the manufacturer's certified range of doorsets and be specifically referenced within the Scope of Certification; any vision panels forming part of the doorset testing and circumstances attracting the recommendation that a door viewer is installed.

17. Security Glazing
All ground floor and easily accessible glazing must incorporate one pane of laminated glass or glass tested to the standard set out, unless protected by a roller shutter or grille. Further.

guidance relates to glazing failure, or cracking, due to thermal stress.

18. Secondary Glazing
Where existing windows cannot be replaced, owing to their listed building status, location within a designated conservation area, national importance, special architectural or historical interest and sufficient internal reveal exists, secondary glazing can be installed to meet as a minimum, one of the standards listed.

19. Roof Design and Lightweight Wall Construction

Roof Construction
Roofs are potentially vulnerable to criminal intrusion and damage through vandalism, if criminals gain access to them. Therefore, careful consideration must be given to their construction. Guidance covers: the relevant minimum standards; installation in accordance with the manufacturer's approved fixing system; DOCO requirements such as a monitored intruder and hold-up alarm system

(see Sect. 5.5.54) where traditional roof installations cannot meet these security standards or to protect a vulnerable ceiling void.

Roof Access and Aids to Climbing

Unauthorised access to roofs can make a range of crimes from burglary to arson easier to commit. Some buildings have a common roof covering that extends across different rooms, areas or even occupying businesses. Therefore, from a single breach point all of these spaces are potentially at risk from intruders. Guidance covers: the construction and flush-fitting of rainwater pipes; flat roofs (especially at low level); relevant legislation (including the Occupiers' Liability Acts 1957 and 1984) and signage. Designers should also avoid creating climbing aids to upper windows and flat roofs and incorporate escape routes that avoid the installation or deployment of external staircases.

Roof Windows, Roof Lights and Roof Lanterns

The security requirements for easily accessible roof lights and roof windows differ depending on the crime risk. Low crime risks require roof windows, roof lights and roof lanterns to be certified to one of the minimum standards listed. High crime risks require roof windows, roof lights and roof lanterns to be certified to one of the minimum standards listed. Alternatively, an easily accessible roof light or roof window that does not meet the above requirements must be protected by an internal grille certified to one of the standards listed. Guidance includes: explanations for the above; how the DOCO must be provided with proof of certification; fitting in accordance with the manufacturer's specifications or, in accordance with one of the standards set out.

Sun Tubes/Tunnels

To prevent a sun tube being used as an access point into a non-residential building, the maximum diameter allowed is 350 mm. Where larger diameters of sun tube are required, the specification of alternative means of protection shall be agreed with the DOCO.

Lightweight Wall Construction

Where lightweight wall construction is being considered and crime risk analysis identifies this may be vulnerable to attack, use of insulated sheet cladding and a reinforced lining such as welded steel mesh, can enhance the security of the building fabric. These should be installed by approved contractors, in compliance with the manufacturer's instructions and meet one of the standards listed.

Silver Award

20. Introduction

The SBD Silver Non-Residential Award is available for new and refurbished developments that include the relevant security features contained in this section. All doorset and window component parts must form part of the manufacturer's certified range to comply with the requirements of the SBD Silver and Gold Awards.

21. Doorsets and Windows and Their Fitness for Purpose

The standard set out defines the resilience and operational characteristics of both doorsets and windows which, in simple terms, means to test a doorset or window to ensure that it will withstand the demands of repeatedly being opened and closed over a significant period of time. Guidance includes: SBD perspective of security requirements; Class 4 (moderate use); importance of SBD developments having the correct level of security specification and durability; doorsets and windows classified according to intended use; minimum requirement of 'moderate' for classification; 'normal' and 'frequent' classification; secondary doorsets and specifiers to supply proof to the DOCO that doorsets and windows meet the duty of rating.

22. External Doorsets

External doorsets shall be certified to one of the minimum standards listed (and the notes thereafter).

Where the DOCO or developer does not know who the occupier will be or the nature of their business, the preferred level of SBD doorset security will be required. This may be higher than the minimum standard and is influenced by the DOCO's knowledge of the area, local crime trends and a crime risk analysis. Further guidance covers: an expectation that security exceeds the minimum standards when the building is unoccupied; extra security measures to be agreed with the DOCO; doorsets must be certified to one of the UKAS accredited certification bodies and websites listed; further information about certification bodies and test houses weblinks; continuous assessment; alternative third-party accreditation and DOCO referral; how doorsets must also be fit for purpose and certified to the relevant material standard listed; DOCO to be provided with proof and scope of certification via one of the bodies listed (and the exemption); how doorsets must be securely fixed and installed in accordance with the manufacturer's specifications (or the standard set out) with a copy given to the DOCO.

23. Security, Fire and/or Smoke-Rated Doorsets Including Those With Adjacent Glazing

It is important that doorsets are installed by a UKAS third-party certificated installer. There is a requirement for manufacturers to demonstrate that their fire doors and associated products such as fixings, side and over panels etc. are able to resist the passage and spread of fire and smoke for a specified time, depending on the certification requirements. Guidance covers: how correctly installed, appropriate doorsets will assist building management—and weblink; third-party certification from a single UKAS accredited certification body for doorsets required to be security, fire and/or smoke rated and/or meet a building regulation requirement; changes to doorset component parts; doorsets incorporating multiple performance characteristics; glazed panels fitted adjacent to a doorset and certification; responsibility for specification and location of security, fire and/or smoke rated doorsets; how it is imperative fire and smoke resistance is professionally assessed/measured; installation by a competent person who possesses appropriate qualifications and fire doors to be alarmed.

24. Avoiding Door Recesses

Recessed doorways should, where possible, be avoided as they provide venues where crimes such as graffiti, arson and burglary are committed. If recessed doorways are unavoidable, other mitigating measures will be required and are detailed within this Section.

25. Locking Systems for Doorsets and Gates

Locking mechanisms incorporating cylinders (Euro or oval profile) must be certified to the standard set out. Guidance covers: minimum requirements of the standard; cylinder certification scheme and standard; certification schemes for lock cylinders recognised for use in SBD developments; British Standard references developed from BS EN 12,209; the difference between standards; doorsets designated as 'emergency' or 'panic' exits fitted with hardware appropriate to the specific use; doorsets supplied with suitable (easily removed) label and leaflet, outlining the operating instructions for the occupier; how emergency exit doors can be fitted with a localised alarm if a crime risk assessment justifies such and a weblink to the Master Locksmiths Association.

26. Glazed Vision Panels, Top Lights, Side Panels and Fan Lights

Where a doorset incorporates a number of performance characteristics, i.e. enhanced security, fire performance and smoke control and its construction includes the addition of, but is not limited to, a vision panel, top light, side panel and/or fan light, these must be fitted in accordance with the manufacturer's instructions and be referenced within the approved scope of 3rd party certification of the doorset. Guidance includes: glazed panels fitted adjacent to a doorset (within 400 mm) and matching the standard stated; and those circumstances where it is recommended a door viewer be installed.

27. Security Glazing
All ground floor and easily accessible glazing must incorporate one pane of laminated glass or glass tested to the standard set out, unless it is protected by a roller shutter or grille that meets the security standards within Sect. 5.5.39. Guidance includes: the definition of easily accessible; how on occasion, large panels of laminated glass on south facing elevations have cracked due to thermal stress; the 'laminated toughened' recommendation; and a reminder to specifiers that the standard stated is the minimum SBD requirement.

28. Windows
All easily accessible windows shall be certificated to one of the minimum standards set out.

Note 28.1: Easily accessible is defined as:

A window or doorset, any part of which is within 2 metres vertically of an accessible level surface such as a ground or basement level, or
An access balcony, or a window within 2 metres vertically of a flat roof or sloping roof (with a pitch of less than 30°) that is within 3.5 metres of ground level'.

Guidance includes: roof lights and roof windows and roof lanterns (see Sect. 5.5.43); minimum level of SBD window security where eventual occupier not known; certification to a higher standard of security; windows certified by a UKAS accredited certification body; weblinks to the SBD Certification Bodies webpage and SBD Test Bodies webpage; how certificated products undergo continuous assessment; alternative third-party accreditation; requirements of easily accessible glazing; that windows must be fit for purpose and certificated to the relevant material standard stated; DOCO provided with proof of certification and how the DOCO may require installation specification. Windows must meet the requirements of Building Regulations regarding safety glazing and emergency egress. In addition, SBD requires security glazing in windows below 800 mm (from floor level) or 1500 mm, if within 300 mm of a doorframe and non-key locking hardware on designated emergency egress windows together with security glazing. Finally, there is a weblink for advice on window hardware and locking systems from a suitably qualified locksmith.

Automatic Opening Window Systems and Vents

Many sustainable industrial and warehouse buildings being proposed require an automatic building management control system, in order to ventilate the premises and regulate temperatures, including at night. This can cause security problems. Guidance includes: automatic opening window systems, vents, pressure relief panels, bars or grilles used to secure a building aperture associated with the mechanisms; automatic opening window systems; and DOCO consultation at earliest opportunity to ensure security is not compromised by the proposed automatic building control management system.

29. Hostile Vehicle Mitigation

Where there is a risk of motor vehicles being used to attack premises or an ATM etc., the standards for vehicle security bollards are set out. Other measures that may be considered to address vehicles used to commit crimes in non-residential settings are described. Further information on impact rated vehicle security barriers (VSBs) and current standards, can be found on the National Protective Security Authority website at: https://www.npsa.gov.uk/blog/hvm-impact-rated-barriers

30. Underground Car Parking

Every effort must be made to prevent unauthorised access into the underground car park. Therefore, an electronic access control system is required for all pedestrian and vehicular entrances, operated by access codes or key cards and with strict management of same. Further guidance includes: egress requiring the same code be inputted; location of inward opening automatic gates, roller shutters or grilles and certification to one of the minimum standards detailed; safety detection systems; automated gates meeting statutory safety standards; lighting scheme certified to the standard stated; use of pale colours on walls and ceilings to produce higher levels of illumination; crime risk analysis indicating a higher level of security is required; the importance of design stage consultation with the Fire and Rescue Service and the 'Park Mark' initiative.

31. Emergency Egress Doorsets From Underground Car Parks

Doorsets providing access to and emergency egress from underground car parking areas, must meet the relevant building regulations in relation to fire safety and security requirements in this guide. These can be diametrically opposed to one another and problematic if not dealt with appropriately at the design stage. As a result of detailed discussions with the Fire and Rescue Service, a possible alternative design solution has been agreed. This includes: doorsets allowing emergency egress access to comply with Sect. 5.5.22; doorsets equipped with audible warning (or where this is not possible due to design constraints); how it is vital these recommendations are discussed in full and agreed with the Fire and Rescue Service at the design stage; and automated access control systems on all doors meeting the requirements within Sect. 5.5.48.

32. Internal Bicycle Stores

Internal bicycle parking facilities will be accessed via a security, fire and/or smoke rated doorset compliant with Sect. 5.5.23 (addressing e-bikes). Guidance includes: bicycle stores being well lit; access doors fitted with crash bars or other emergency furniture; doorsets; access control and monitoring systems; bicycle parking stands, anchor points, rack systems and dedicated lockers; standards and certification; consideration for the less able-bodied; separation into discreet units for more than 70 bikes; security measures agreed with the DOCO; Standards for Public Cycle Parking and website link; DOCO requiring extra security measures; storage of Lithium-ion powered vehicles and devices; National Fire Chiefs' Council guidance re electric bicycles, electric scooters and website link and The Bicycle Association/Cycle Rail Working Group guidance and the SBD website link.

33. Telecommunications and Utility Access Covers

Telecommunication cables shall enter buildings below ground and be protected by secure access covers, certificated to the standard set out and positioned in highly visible locations. Increased security measures may be required if a crime risk analysis identifies these are necessary and commensurate with the risk. This will help to delay or prevent the occurrence of burglaries, where the perpetrators cut the VSS or alarm signal prior to committing an offence.

34. Internal Lighting

It is recommended that most internal office lighting is operated by detection devices that will automatically switch lights on and off, due to movement or the absence of movement in each room. This will reduce energy consumption and identify the presence and progress of intruders when the building is closed. Guidance covers: security of controlling mechanisms; two-stage lighting in critical movement areas and emergency lighting to ensure continued on-site safety and security following a power failure—together with the relevant standard.

35. Video Surveillance Systems (VSS)

Although video surveillance systems (VSS) are not a solution to all security problems, they can help deter criminals and assist with the identification of offenders after a crime has been committed. If a VSS has only a recording facility, it can be a useful investigative tool. Whereas a monitored system allows a real-time reaction to criminal activity. VSS are also often very useful in mitigating against risk where other forms of security are not feasible.

Guidance includes: mandatory compliance with the Product Security and Telecommunications Infrastructure Act 2022 (PTSI Act); and the 'Secure Connected Device' accreditation scheme—which is fully detailed. Further advice covers: necessity for an operational requirement covers: (OR); evidential quality recordings; staff trained to make recordings for the police; the ideal of live monitoring; post-event investigation and image retention; site-specific camera location; early discussions with an independent expert and potential installers; VSS installation in accordance with the standard stated; compatibility with the existing or planned lighting system; vandal-resistant housings; registration with the Information Commissioner's Office (ICO) and compliance with Data Protection and Human Rights legislation; specifiers and data controller ensuring compliance with the GDPR; guidance as per the standards stated; Biometrics and Surveillance Camera Commissioner's Code of Practice; passport to compliance and relevant website; plus the recommended standards for the surveillance camera industry and website.

36. Cybercrime and the Internet of Things

Internet of Things (IoT) and smart devices that enable data to be exchanged, combined and analysed must have achieved SBD's Secure Connected Device accreditation and demonstrate they have met the appropriate and relevant IoT standards and certification conducted by an SBD recognised body. Guidance includes: mandatory compliance with the Product Security and Telecommunications Infrastructure Act 2022 (PTSI Act) and the 'Secure Connected Device' accreditation scheme—which is fully detailed.

37. Building Shell

Lightweight Wall Construction

Repeats Sect. 19.15 to 19.16 of this guide.

38. Glazed Curtain Walling and Window Walls
Specifiers should be aware that due to their size and configuration, it is common for glazed curtain walling and window walls to commonly fall outside the scope of testing or certification procedures. Consequently, in some circumstances an alternative method may be required to demonstrate that their security is commensurate with the risk. Guidance covers: how SBD recognises four distinct types of glazed wall systems; their installation and glass retention standards; 'easily accessible' glazing; British Standards 'Codes of Practice'; and guidance produced by the National Protective Security Authority (NPSA) together with their web link.

39. Roller Shutters and Grilles
Roller shutters and grilles can provide additional protection to both internal and external doors and windows. Guidance covers: minimum standards; how in new-build developments roller shutters should be integrated into the fabric of the building; higher security ratings may be required and dictated by type and level of crime risk, building location and security level of the door or window being protected.

40. Doors Providing Vehicular Access
Doors providing vehicular access for deliveries and other entrances where no other door is present, shall be certificated to one of the standards set out or above. Additional guidance.

covers: the DOCO requiring a higher a higher level of security due to crime risks; how in new-build developments, roller shutter doors should be integrated into the fabric of the building and where minimum free air movement as calculated by designers in respect of smoke etc. is not compatible with minimum-security requirements, applicants must seek to achieve the required free air movement by alternative means, such as forced ventilation and other open apertures..

41. Roof Construction
Repeats Sect. 19.1 to 19.4 of the guide.

42. Roof Access and Aids to Climbing
Repeats Sect. 19.5 to 19.9 of the guide.

43. Roof Windows, Roof Lights and Roof Lanterns
The security requirements for easily accessible roof lights, roof windows and roof lanterns differ depending on the crime risk. Guidance includes: a definition of

'easily accessible'; low crime risk standards; high crime risk standards; alternative protection by way of internal grilles; the DOCO being provided with proof of certification; fitting in accordance with the manufacturer's specifications or, in accordance with the standard set out.

44. Sun Tubes/Tunnels
Repeats Sect. 19.14 of the guide.

45. Internal Layout

Entrances into the Building—Main Public/Visitor Entrance
Unless a non-residential development intends to allow large numbers of the public to gather inside, it would be better that public access is restricted to one main visitor entrance. Guidance covers: high-risk premises with electronic access control linked to a reception desk/security office; audio-visual links; illumination of adjacent internal and external areas; clear view of entrance approaches; disabled access including suitably located vandal-resistant proximity reader and where the crime risk is high, an airlock door system.

46. Staff or Additional Entrances
There may be circumstances where multiple entrance doors are required for staff convenience or building functionality. All such doorsets must be SBD compliant. Guidance includes: reducing criminal opportunity by use of access codes, key cards etc.; integrated access control systems; biometric and voice recognition systems; and how such measures should not be allowed to prevent emergency egress and Internet of Things (IOT) devices.

47. Reception Area and Visitor Control
Where a business manages large visitor numbers, access beyond the reception area to other parts of the building, including stairs and lifts, shall be controlled by staff preferably on a raised dais using automatic locking doors or barriers and/or by the use of proximity reader technology. Guidance covers: how doors and barriers must allow emergency egress and be integrated into the fire/smoke alarm system; the relevant fire authority's advice; social distancing recommendations; providing staff with clear views of entrance doors, approaches and public waiting areas; reception desk construction; staff or visitors using wheelchairs; escape routes to a place of safety and its security; audible personal attack alarms and staff training in their use and where the risk dictates, video surveillance.

48. Automated Access Control Systems
Automated access control systems (AACS) are likely to be required at the main entrance to non-residential developments and may also be required on some internal doorsets. Therefore, an AACS is required for all pedestrian and vehicular entrances. Guidance includes: access codes or key cards; security levels; lone worker or vulnerable person protection; preventing access and trespass into parts of the building; types of business or business practices; local crime risk factors; a common entrance serving two or more businesses; an intrusion detection system (IDS); data logging facility; compliance with the standard stated; codes of practice produced by the National Security Inspectorate and Door and Hardware Federation.

49. Automated Access Control Standards
Specifiers are advised to make reference to guidance published by the British Security Industry Association (BSIA) 'A specifiers guide to the security classification of access control systems': https://www.bsia.co.uk/zappfiles/bsia-front/pdfs/132-specifiers-guide-access-control-systems[1].pdf.

All systems shall meet the standard set out and if applicable the Internet Protocol security shall meet the requirements of the SBD 'Secure Connected Device' accreditation scheme, by being tested and certificated by an SBD approved certification body: https://www.securedbydesign.com/internet-of-things

50. Access Control Systems and Security Staff
Where security officers are to be employed at the entrance, an automated access control system will still be required and should therefore be discussed with the DOCO at the earliest opportunity.

51. Secure Internal Doorsets
Specifiers should contact the DOCO if there are any internal doorsets requiring additional security, e.g. computer server room. Guidance covers: security standards; high crime risk doorsets; risks where end users are not known; and post-occupation/change of use issue of an SBD Silver or Gold Award.

52. Lift Security
BS EN 81 is a multi-part document detailing safety rules for construction and installation of passenger and passenger goods lifts. Further guidance includes: vandal resistance qualities and standard testing parameters; and DOCO requirements in high crime areas.

53. Mail Delivery

For the majority of non-residential buildings, mail delivery will take place during business hours and handed in at reception. Guidance includes: parcel drop and dedicated mail rooms; communal mail and parcel delivery; external and internal delivery facilities; access control; airlocks, VSS; delivery drop boxes; letter plates; letter boxes; and security standards.

54. Intruder and Hold-Up Alarm Systems

Intruder and hold-up alarm systems shall meet the requirements of the standards detailed. If immediate police response is required, installers must comply with the NPCC policy document available at the web addresses detailed. The SBD Alarm Standard is also detailed and weblink provided.

55. Security Fogging Devices

Security fogging devices can be incorporated within the intruder alarm system to disorientate the intruder on activation. Such devices must conform to the standard detailed.

56. Public Address Systems

In large non-residential buildings, a public address system is recommended to provide instant, effective communication to all staff members especially in emergency situations, initiating a prearranged and rehearsed response to particular situations.

57. Physical Security Standards for Computers and Server Rooms

Consideration must be given to the structure of internal walls, floors and ceilings of computer server rooms to provide appropriate security and prevent damage by fire, smoke or from firefighting (water) from other parts of the building. Where impractical to secure a room housing a computer server, two alternative solutions are recommended and the relevant standards are detailed.

58. Safes and Strongrooms

SBD recommends commercial safes and strongrooms certificated to one of two standards quoted. Guidance covers: resistance grades determined by the value of the safe's contents; liaison with insurers to select a safe or strongroom and variations with safes certificated in European countries.

Self-storage Centres

Self-storage centres shall comply with the relevant security measures in the guide. Guidance includes: individual storage units; integral locks certified to one of the security standards listed; cylinder certification; and padlock security standards.

59. Security Guards and Guard Houses

The provision of security guards may require additional design features to be built into the buildings and grounds, to reduce overall costs. Guidance covers: electronic control monitoring systems; BS 7499:2020 compliance; security control rooms or guard houses and their standards; doors, windows and glazing and certification to the LPS standard set out.

60 Arson and Fire Protection

If arson is identified as a crime risk factor in an area, non-residential developments must be designed with arson prevention in mind. Architects must consult with a Fire Risk Assessment Consultant Service to ensure that appropriate fire suppression measures are in place.

61. Buildings with Undercrofts

Depending upon the nature of the building, its location and use, undercroft car parking can assist in the commission of a terrorist attack and therefore it is important to recall advice given in the section entitled *'Scope of the development'* as follows:

> If the development is at risk of terrorism, espionage, sabotage or other malicious threats, this matter should be brought to the attention of the DOCO at the earliest opportunity to ensure that the proposed design, layout and physical security measures appropriately mitigate the risk. If the police service identifies vulnerable buildings or activities that require detailed counter terrorism advice, a referral will be made to the Counter Terrorism Security Advisor (CTSA).

Gold Award

62. Introduction

The Secured by Design Gold Non-Residential Award is available for new and refurbished developments that include the relevant security features contained in the Gold and Silver Award sections of this guide.

63. Urban Design and Planning Policy

Creating a Sense of Place

Creating an easily legible sense of place is a key element of SBD developments. A sense of place.

describes a relationship between people and the physical space around them. Crime and fear of crime are reduced where spaces are clearly defined, obviously owned and easily overlooked.

64. Location and Adjoining Land Use

A major factor influencing the security of a development, is the nature of surrounding land use and its location. Other factors to consider include crime risk, business type, business hours, business size, number of employees, visitor numbers, transport links etc. Good examples of such Geographical Juxtaposition (Newman, 1972, see Sect. 4.4.5) are described in respect of both non-residential and residential developments. A crime risk analysis is described as 'decisive' in the consultation process.

65. Building and Site Layout

A prudent site layout will maximise natural surveillance and deter criminals. However, outside normal hours, with fewer users and lower levels of natural surveillance, formal methods of security and surveillance may be required e.g. onsite security guards, locked gates or remote VSS monitoring. Site conditions that adversely affect natural surveillance must be discussed with the DOCO and security solutions agreed.

66. External amenity Spaces

External amenity spaces should be carefully planned to benefit from natural surveillance. Additional security measures such as fencing may be needed and agreed with the DOCO. Guidance includes: promoting sustainable communities, health and well-being; play spaces and communal areas.

67. Roads and Footpaths

Vehicular and Pedestrian Routes

Poorly designed vehicle and pedestrian routes allow criminals to tour an area anonymously and make their escape after a committing a crime. To combat this, routes must: take the most direct route from origin to destination and should not undermine the defensible space and natural surveillance of non-residential units. Guidance covers: symbolic barriers; secure emergency vehicle access points; access for large emergency vehicles; consultation with the fire and rescue service; the Violence Against Women and Girls (VAWG) agenda and how further information is available at: https://activetravelengland.gov.uk/planning-active-spaces

68. Through-Roads and Culs-de-Sacs
Road layouts that frustrate the searching patterns of criminals will deny reasonable excuses for their presence, whilst limiting potential access and escape routes. If a through route must be incorporated, limiting permeability will deny criminals excuses for trespass. To facilitate natural surveillance by responsible guardians, routes for pedestrians, cyclists and vehicles should not be segregated. Movement frameworks based on primary routes and shared space remove the need for little used alleyways, shortcuts, footpaths and large numbers of minor access points that may become ambush points and environmental factors in promoting the fear of crime.

69. Footpath Design
Pedestrian, cycling and vehicle routes should not be segregated, have a rational purpose, follow natural 'desire lines', be functional, integrated and subject to natural surveillance. Guidance covers: unrestricted movement along quiet alleyways; informal shortcuts; gated private footpaths; lighting; VSS; cycle restrictors; maximising surveillance; subways, footbridges and underpasses; dimensions; defensive planting and liaison between the designer, DOCO and fire authority.

70. Planting Next to Footpaths
Ideally, low maturing shrubs should be planted at footpath edges with taller plants positioned towards the rear, to avoid overgrown vegetation creating pinch points and areas of concealment. Where possible, public footpaths should be routed away from building facades but if unavoidable, a buffer zone created using defensive planting to keep pedestrians away from windows and to protect walls from graffiti. Further guidance relates to correctly specifying shrubs and maintenance.

71. Lighting of Roads and Footpaths
All lighting of roads and footpaths within SBD developments must comply with BS 5489–1:2020. However, if this conflicts with local conditions such as in a conservation area or 'dark sky' policy, this should be discussed with the DOCO and local lighting authority. Guidance includes: variable lighting systems; bollard

lighting; landscaping and lighting designs; completing a CDM risk assessment; and a competent lighting design professional providing the DOCO with written confirmation.

72. Phased Developments and Footpaths

Where completion of a footpath is delayed by phased introduction of elements or other long-term planning policy considerations, it is recommended footpaths are not constructed until interaction with the local environment can be properly understood. This avoids creating underused footpaths that lead to undesirable areas, if only in the short to medium term.

73. Perimeter Security and Site Access

Enclosed Non-Residential Developments—Industrial and Office Use

SBD recommends single and multiple unit non-residential developments contained within secure perimeters, with access-controlled entrances. Use of one main entrance serving vehicles and pedestrians, is advised. Guidance includes: mutual supervision; clear demarcation between roadway and footpath; additional entrances for emergency access and egress controlled by electronic access and video surveillance systems (VSS); high risk considerations; staffed gatehouses; crime risk analysis and DOCO liaison.

74. Open Non-Residential Developments—Office and Industrial Use

Ideally, public footpaths, roads and other public areas should be kept at a distance from non-residential buildings to prevent vehicles being used to ram and breach the ground floor or, drawing a vehicle close to use it is as a climbing platform to access higher windows and flat roofs. Guidance covers: creating defensible space in open-plan developments; design elements to keep vehicles away from buildings; emergency access and escape; soft landscaping; hard paving/uneven surfaces and avoiding use of landscaping aggregates.

75. Perimeter and External Areas

Boundary Types

Three categories of boundary—Secured, Controlled and Psychological are described, as is protection by fences, gates or barriers and certification to one of the standards listed.

76. Security Fencing
Security fencing is effective at delaying or deterring intrusion, due to the need to penetrate or climb over. Elimination of climbing aids is important e.g. trees, lamp columns or buildings. SBD recommends effective fencing, without creating a 'fortress' impression. Guidance covers: security fencing materials; visually permeable fencing; being commensurate with the local crime risk; separate inner security fences; height and minimum standards; public footpaths and Visual Security Systems (VSS).

77. Gates
The height and length of gates shall be consistent with the dimensions of the boundary fence into which they are set and shall close any gap beneath them to ground level. Guidance includes: secure access control; minimum height; standards and automated gates (see Sect. 5.5.30).

78. Defensive Hedging
It may be a planning requirement to use or retain a defensive hedge, to protect the site perimeter or bolster an existing or proposed fence. As it can often take considerable time for a new hedge to mature and become an effective barrier, a fence that meets the minimum-security requirements of Sect. 5.5.72 would be required, to ensure security integrity is in place from the date of commission.

79. Signage and Unit Identification
The building's reception area and car park shall be clearly signposted from the entrances and further on into the site. Clear signage will assist to dismissing the excuse by intruders they could not find reception and help security staff and police establish their legitimacy. Signage identifying areas not open to public access can act as a reminder to challenge unauthorised persons. Guidance covers: industrial estate site maps; anti-graffiti coatings and the clear numbering of individual units.

80. Parking and Access
The routes from site entrance to reception and to car parks and delivery points, should be clearly defined and maximise natural surveillance from reception and occupied offices. Guidance includes: new industrial estate road systems; identifiable parking for staff and visitors; the design to allow secure delivery and collection of goods and discussing such issues with the DOCO.

81. Vehicle Parking and Access

Local infrastructure will often determine the location of vehicular and pedestrian routes and entrances. This must be considered when designing the site layout, to ensure security is easily and effectively controlled. Guidance covers: main site and other entrances; secure emergency vehicle entry; Fire and Rescue consultation; delivery vehicle roadways; retractable bollards and gates.

82. Vehicle Parking and Deliveries

Routes from site entrance to reception, visitor car parks and delivery points, should be as short as possible, open to view from reception and secured from the site. Guidance covers: separate parking for staff and appropriate fencing; automatic access-controlled gates; parking provision for emergency vehicles; site design for secure deliveries; how parking facilities should comply with 'Park Mark' criteria for safer car parks; DOCO advice and how further information can be found at: https://www.saferparking.com

83. Electric Vehicle Charging in Secure Non-Residential Parking Areas for Employees Only

To avoid potential conflict, electric vehicle charging points (EVCPs—employee use only), should be located in parking bays subject to *Natural Surveillance*. A management process should ensure charging bays are used for their intended purpose and steps taken to prevent vehicles overstaying. Security could be compromised where non-employees use pay-as-you-go EVCPs within secured car parks and is therefore not advised.

84. Electric Vehicle Charging in Covered Car Parks

When considering the location of EVCPs, any relevant Building Regulations must be complied with and consultation undertaken with the relevant Fire Strategy Authority. This is particularly the case when locating EVCPs in 'covered car parks'.

85. External Bicycle Parking for Employees and Visitors in Public Areas

External bicycle parking facilities will be designed for secure storage using bicycle lockers, hangers or dedicated storage devices and certified to one of the standards set out. Parking facilities will be located as close as possible to the primary entrance (within 50 metres) and subject to *Natural Surveillance* by building occupants. Guidance covers: video surveillance systems (VSS); facilities to be well lit during the hours of darkness; DOCOs requiring extra security measures;

developer's responsibilities; National Fire Chiefs' Council fire safety guidance and Bicycle Association/Cycle Rail Working Group quality and security standard.

85. Motorcycle, Scooter and Moped Parking
Theft of motorcycles, scooters and mopeds is a major problem, especially in urban areas. They are often broken up for their parts or used in further *crimes*, e.g. snatch-theft or robbery.

Dedicated Motorcycle Parking Facilities
Parking is often in small, on street bays. These attract high rates of theft owing to the Crime Opportunities they provide. Guidance covers: ground anchors/robust metal support stands; their construction and standards; dedicated parking facilities; bike covers and signage.

British Parking Association—Park Mark Safer Parking Award Scheme
A paragraph is devoted to the Park Mark Safer Parking Award Scheme (see Sect. 5.5.19) including the role of the DOCO and the website: https://www.britishparking.co.uk

87. External Security Issues

Landscaping
The planting of trees and shrubs in new developments to create attractive environments is encouraged and supported by *SBD*, provided the following matters are addressed:

- Landscaping layouts must allow sufficient space to accommodate the planting choice
- Future maintenance requirements should be considered at the design stage
- Landscaping must be maintained to ensure it does not blind *VSS* cameras or obscure *Lighting*
- Planting designs must mitigate *crime* rather than contribute towards *Crime Opportunities*.

88. External Furniture and Litterbins
Benches and planters should be of robust, vandal and graffiti-resistant design and to prevent street sleeping. They should be fixed into the ground to prevent theft, use as climbing aids or as a tool to assist burglary. Litterbin location should be

carefully considered, to prevent use as climbing aids or to start fires. Guidance covers: design of benches and planters; fixing of furniture to prevent its theft; and security of litterbins.

89. Natural Surveillance and Recessed Doorways
It is important to avoid creating areas or building features that cannot be overlooked from occupied parts of a development. Recessed doorways will obstruct *Natural Surveillance* and collect wind-blown litter that can be used to start fires. If the recessing is unavoidable, it would be prudent to use a secure doorset with emergency exit hardware and integral secure vision panel.

90. Temporary Buildings
Temporary buildings, e.g. portable buildings, are notoriously difficult to secure due to their construction and location outside the secure envelope of permanent building structures. 'Temporary' means up to 6 months occupation or use and after this point will be considered in terms of security requirements as permanent. Temporary buildings used as classrooms long past their original intended lifespan, are a good example. Guidance covers: locating temporary buildings together; discontinuation of use; not being used to store high value equipment; inclusion in the main buildings' intruder and fire alarm regimen; how the location of temporary buildings must be discussed with the local fire authority to ensure spread of fire to other buildings is minimised and advice from a suitably qualified locksmith at: https://www.locksmiths.co.uk

91. Wind Turbines and Photovoltaic Installations
Consideration must be given to protecting wind turbines, photovoltaic installations (PVs) and biomass boilers from vandalism or theft, by using access control, appropriate fencing and removal of climbing aids. PVs should be located on difficult to access roofs and secured with theft-resistant fastenings. Landscape design should never include loose pebbles which could be used to damage the panels.

92. Storage Facilities

Equipment Storage
Heavy plant and building equipment should be protected from theft and criminal damage in secure storage areas, within the main building if possible or inside a compound with high security fencing. VSS coverage may be required if the value or nature of the items suggests a High-Crime Risk. The DOCO can advise on the best way to protect this equipment.

93. External Waste Storage
Waste and rubbish containers, particularly wheelie bins, assist climbing and use in arson attacks. They should be fitted with lockable lids and kept in lockable roofed stores away from main buildings. Local fire authorities can advise regarding stand-off distances, fire and smoke detection devices, sprinklers, and storage of combustible materials. Additional guidance covers recyclable material and bin stores for kitchen waste.

94. Fuel Storage
Central heating oil and cans of fuel used for grounds maintenance and plant machinery, should be stored in a secured location away from the main building. Where fuel has to be stored inside the main building, advice should be sought from the relevant fire authority.

95. Cleaning Equipment Storage
A secure store(s) inside the building must be provided for safe storage of cleaning equipment/fluids, as many will be flammable and possibly toxic. Advice should be sought from the relevant fire authority.

96. Utility Services

Telecommunications Access Covers, Ducting and Utility Metres
Utility access covers, protecting access to drains, sewers, telephone cables, electricity cables and other services, must be secured to prevent access and damage. Further guidance covers: business continuity and how security provided by the Intruder or Fire Alarm, demands sufficient secure ducting into the site and an appropriate number of secure access covers; utility metres sited in a secure building, e.g. a plant room; and the desirability of using smart meters.

97. Windowless Building Elevations
A common feature of many industrial units and warehouses are long runs of elevations, often to the rear and side, that have no windows for observation of the site. Yet these often feature windowless emergency exit doorway(s) with *Crime Opportunities* such as graffiti, burglary, arson and loitering. Further guidance covers: windows in blank elevations; keeping trespassers away from buildings in the interests of preventing unauthorised entry, graffiti, arson or other criminal damage; pathways between buildings intended to service emergency exits; gates; fencing; separating verges; 'defensive' hedging; anti-graffiti coatings; VSS; DOCO liaison and professional or community led artwork.

98. Walls—Facades, Apertures, Recesses and Graffiti
Facades of buildings should be designed to minimise the opportunities to hide along them or climb up them—especially important when access to vulnerable windows or flat roofs is possible. Further guidance relates to: reducing opportunities for burglary; unauthorised entry to goods lifts, fuel delivery points or ventilation ducts; grilles, air ventilation apertures and shutters; and graffiti prevention.

99. Security Lighting
The need for *Lighting* will be determined by local circumstances. Lighting an inner-city footpath will only reduce *crime* if matched by *Natural Surveillance* from nearby premises and reaction expected. Illuminating isolated and little used footpaths may give a false sense of security. Further guidance covers: closing the path at night with the aim of deterring or detecting an intruder; co-ordination with any *VSS* system; landscape design to avoid conflicts; protecting light fittings from vandalism; uniform lighting and good colour rendition; avoiding glare and light pollution; external illumination when a building is unoccupied and bollard lights.

5.6 SBD—Self Build

1. Introduction and Scope

The SBD Self Build Guide provides advice on how to create a safe and physically secure home. Most self-build homes will be single detached properties, but the guide also applies to community or joint venture self-build initiatives for affordable terraced homes (be they new or refurbishment schemes) creating a safer, more secure and sustainable environment.

Adoption of SBD guidance has been proven to reduce crime by up to 87%, achieved by focussing on issues of design and layout and supported by using effective physically secure products. Clients are advised to source products that meet Police Preferred Specification via the SBD Member Companies pages of the SBD website. Additional benefits include meeting the needs of planning authorities and building control professionals. And generating an SBD—Self Build Award certificate.

SBD is owned by the UK police service, supported by the Home Office and refers to the English, Scottish and Welsh building regulations collectively as UK Building Regulations. (Northern Ireland has no Building Regulation for the physical security of dwellings). Contact for the free of charge DOCO service should be made via the SBD website and SBD Self Build Application Form. It is intended for those creating a home for their own occupation rather than the sale or to-let market. Further detailed advice can be found in SBD—Residential Guide 2025 where the topics discussed are significantly expanded.

2. How to Apply for the SBD Self Build Award
The guide consists of: Section 1: Development layout and design. Section 2: Physical security of the home. Section 3: Advice re additional features that may or may not exist in the original home design.

3. Your Carbon Footprint and Planning Policies in Support of Secured by Design
Academic research shows that crime generates significant levels of carbon dioxide. Therefore, by preventing crime applicants will create a more welcoming and pleasant place to live, as well as a cleaner, healthier and greener environment. Crime is a material consideration in planning and therefore a determining factor in gaining planning consent.

Section 1: Plot Layout and Design

4. General Layout
In terms of crime, there exists the public face of any home—designed to engage with surrounding properties and receive guests and the more secure private side. Positioning the home within the plot can maximise its self-policing qualities. Homes facing one another (allowing neighbours to see each other) are preferred to those hidden by a substantial wall or large hedge, denying surveillance and assisting criminal behaviour. Further advice relates to vehicular and pedestrian routes, use of Symbolic Barriers, Defensible Space, public footpaths, landscaping, planting and tree species.

5. Dwelling Boundaries Front Boundaries
The boundary between public and private areas should be clearly indicated. Dwelling frontages may be open to view with walls, fences and hedges kept low. Or a combination of wall (maximum height 1 metre) and railings or timber picket fence, can create a more substantial front boundary. Planting of feature shrubs and suitable trees should be set back from paths, to avoid obstructing visibility of doors, windows, access gates and allow clear line of sight to parked vehicles, pavement and road.

Access gates to rear gardens must be robustly constructed, of same height as the fence (minimum 1.8 m), capable of being locked and located as near as possible to the front building line. Exposed side and rear gardens require more robust barriers, using minimum height 1.8 m walls or fencing, deterrent features, e.g. increasing the height or planting thorny shrubs. High-quality fencing will ensure longevity, security, reduce maintenance costs and be more sustainable. Further guidance relates to: inward opening entrance or driveway gates; substantial framed construction employing galvanised adjustable hinges and fixings. Automated gates must meet the relevant statutory safety standards and be CE marked accordingly. Organisations providing guidance and training are set out.

Sub-divisional boundaries fencing installed correctly will prevent the possibility of future neighbourly disagreements over ownership. Its style will depend upon local circumstances, Crime Risk and the local planning authority. The addition of a trellis topping can help to deter climbing—especially useful on exposed and vulnerable rear boundaries. Close liaison with the DOCO from the outset is advised.

6. Gable End Walls

It is important to avoid creating windowless elevations and blank walls adjacent to public spaces, which can attract graffiti, inappropriate loitering and ball games. If unavoidable, possible solutions include: a window above ground floor level offering additional surveillance; a buffer zone using railings; or a 1-metre-high mature thorny hedge. Where there is insufficient room to create Defensible Space, use an appropriate climbing plant or a finish applied and that allows easy removal of graffiti.

7. Dwelling Identification

Clear Signage (naming and/or numbering) of the home is essential to assist visitors, postal workers and the attendance of emergency services. This should be in place prior to first occupation.

8. Climbing Aids

Avoid accidental creation of climbing aids around the walls, providing access to first floor windows or along any boundary line, designed to keep offenders out. The siting and design of bin and fuel stores, fixed outdoor furniture, trees, low flat roofs, car ports or balconies to ensure such access has not been provided, requires special care. Easily accessible windows will need additional security.

9. Vehicle Parking
Ideally, vehicles should be parked in locked garages. If not, they should be on hard standing within the dwelling boundary. Where dedicated garages are provided within the curtilage of the dwelling, vehicle or pedestrian entrance doors should be easily observed from the street and neighbouring homes. Locating garages forward of the building line can obscure views and should be avoided.

10. Street Lighting
If the new home is on an adopted highway, it is likely the Lighting provided will be BS 5489-1:2020 compliant. Further guidance covers: conflict with other statutory provisions, e.g. Conservation Areas; trees blocking Street Lighting; location; seasonal variations on planting and bollard lights.

Section 2: Physical Section

11. Introduction
This section provides physical security details necessary for an SBD award to be made. It includes technical guidance on 'Police Preferred Specification' for dwellings, to ensure compliance with UK Building Regulations. Design and specification of products like doors and windows will impact on other Building Regulations. Such regulations in all four UK nations are considered equal to one another. Where specified, products must be 'Certificated' to relevant security standards—and confirmed before purchase. SBD member companies manufacture a range of products that will meet Building Regulation compliance and a very effective level of physical security protection.

12. Dwelling Entrance Doorsets
The term 'doorset' refers to a door, frame, locks, fittings and glazing as one combined unit. It includes front and rear doors and numerous other doorsets set out in the guide. Sub-paragraphs relate to: definition of Easily Accessible doors and windows; fire and security rated doorsets or interconnecting garage doorsets; glazed or non-glazed adjacent side panels; responsibility for specification; pedestrian doorsets; certification of vehicular access doorsets; primary security provided by the vehicular doorset; suitably qualified and recognised third-party certification authorities; glazing in and adjacent to doorsets; necessity as part of the manufacturer's certificated range of doorsets; outward opening doorsets; door limitation and caller identification; door viewers; door frames to be securely fixed; external doorsets not be hidden from public view; secure mail delivery; letter plate apertures in doorsets; arson attacks necessitating the

omission of a letter plate within a doorset; surface-mounted letter boxes; and through-the-wall mail delivery.

13. Windows, Roof Windows and Roof Lights

Window frames must be securely fixed in accordance with the manufacturer's instructions and specifications. Further guidance relates to: 'Easily Accessible' windows certificated to one of the standards listed; key-lockable hardware unless designated as emergency egress routes within the Building Regulations; windows as part of a designated fire escape route; windows not Easily Accessible; Laminated Glass; fire rated-windows and Technical Schedule.

14. Conservatories and Sun Rooms

Should a conservatory or sun room be installed, the doors and windows must meet the same physical security standards as described at Sect. 5.6.12. If a conservatory is installed with a polycarbonate glazing system, a doorset shall be installed separating the conservatory from the rest of the dwelling. The doorset shall again comply with the requirements of Sect. 5.6.12.

15. External Lighting for Dwellings

Lighting is required to each exterior dwelling elevation containing a doorset and can assist in identifying the door and locking mechanisms. It should operate dusk till dawn and use energy efficient lamps.

16. Utility metres

If 'smart metres' are used (remote signalling) there is no location requirement. Otherwise, utility metres should be located outside the dwelling at the front or as close to the front building line as possible (to ensure they are visible and deter vandalism) and to the front on any fencing or gates.

17. Visitor Door or Gate Entry Systems, With or Without Remote Unlocking, Utilising Telephone-Based Internet Protocol (IP)

Used to access homes with operation via the Internet, the Internet Protocol for security shall be tested and certificated to the standard stated for Internet of Things (IoT) devices. Remote unlocking (those operated outside the home boundary using mobile equipment, e.g. smart phones and tablets) should only be permitted where there is both live audio and visual feed. Systems should not permit remote release where there is audio only communication, e.g. poor signal area, loss of signal, etc.

13. CCTV and Recording

CCTV is not a universal solution to security problems but may form part of an overall security plan. It can help deter crime, assist the identification of offenders, promote personal safety and provide reassurance.

Even small developments will benefit from a good quality, inexpensive CCTV system. Cameras should be: contained in vandal-resistant housings; sited to avoid glare from concentrated white light sources; record images in colour HD quality and stored for 31 days. Further advice covers: camera location with time and date stamping and suitable methods of export, incorporating the required software to view the exported footage.

If CCTV cameras cover areas beyond the boundaries of the home and record images of people and/or vehicle registration numbers, the CCTV system may need to be registered with the Information Commissioner's Office and signs displayed. Further advice, including the ICO CCTV Code of Practice can be found at: https://www.ico.org.uk

Section 3: Additional Security Requirements

19. Additional Security Features (If Present in Your Design)

In order to receive an SBD—Self Build Award, it is important the additional detail provided in this section is met if the feature exists within the self-build home.

20. Additional Window Requirements

In certain High-Crime locations and to ensure security is commensurate with risk, the DOCO may require Laminated Glass meeting the standard quoted is installed to all ground floor and basement windows and those 'Easily Accessible' (as previously defined in this guide) above the ground floor.

21. External Garage Doorsets

If the home incorporates a garage(s), the security of both dwelling and garage must be considered. If a garage is not secured as part of the dwelling or it is detached from the dwelling, then external pedestrian access doors must meet the same physical, locking and fixing specification as described at Sect. 12. Vehicle access doorsets shall be certificated to one of the standards listed.

Alternatively, a vehicle access door not certificated to one of the standards and not subject to the English and Welsh Building Regulations, may be satisfactory if an external 'garage door defender' security product is also fitted. These must be certificated to Sold Secure Bronze level or above.

22. Secure External Storage Facilities and Bicycle Security (Including Mopeds, Scooters and Motorcycles)
Any external containers specifically designed for the secure storage of bicycles and other property must be certificated to one of the security standards listed. Further guidance relates to: bicycle storage in a wooden shed (SBD—Residential Guide 2025 details such requirements); security anchors for mopeds, scooters, motorcycles or bicycles stored within that same shed and how such security anchors must be securely fixed in accordance with the manufacturer's specifications.

23. Intruder Alarms
Where an Intruder Alarm system is installed, it shall meet the requirements of BS EN 50,131 (wired and wire free systems). If immediate police response is required the system must comply with the National Police Chiefs' Council (NPCC) policy document—Guidelines on Police Requirements & Response to Security Systems, available at: https://www.securedbydesign.com

5.7 SBD—New Schools 2014

This SBD Development Guide (first produced in 2004 and fully revised in 2010 and 2014) provides guidance to establish and maintain a safe and secure environment within school premises. It also details what is required for a school building and campus to achieve the SBD award. Until the late 1990s, several decades had passed before school building began in earnest in the UK. Indeed, Cottesbrooke Infant School (see Chapter 67) was a rare exception). However, beginning before the turn of the millennium an intensive school building programme began across the UK—much of it financed under the Private Finance Initiative (PFI) and thereafter, the Building Schools for the Future programme.

Section 1. Introduction
A sensible and practical level of security (that will not adversely affect the efficient running of the school) is essential to a successful teaching and learning environment. The majority of criminal incidents in schools relate to Property Crime, due to the availability of portable and desirable goods with a ready market, e.g. personal computers, laptops, digital projectors etc. Other crimes that occur include acts of vandalism (such as graffiti), arson and assaults. The latter range from staff being physically assaulted by parents and students, to bullying by one or more students against another. More recently, cyber bullying has

become a noticeable problem in schools. The victims of school crime can also extend beyond the staff and students, as many schools open into the evenings and at weekends for use by the local community.

The police service places great importance upon the need to build sustainable developments. This not only includes the need to use environmental-friendly products, materials and construction methods, but also the necessity to raise awareness that Crime Prevention is a positive sustainability issue.

Academic research conducted on behalf of SBD has confirmed that Crime alone is responsible for the release of at least 6,000,000 tonnes of CO_2 into the atmosphere each year. It therefore follows that the achievement of an SBD award for the school not only indicates that the designer has made a significant effort to create a secure teaching environment (recognised by the police), but by doing so has also reduced the school's carbon footprint. To that end, applicants are also encouraged to build to the Building Research Establishment's environmental and sustainability standard BES 5051: Issue 2. The standard's requirements can be found in the BREEAM Education 2008 Assessor Manual.

It is also important that the benefits of a new secure school are complemented with a clear management and maintenance programme, to ensure a safe and secure teaching and learning environment. Further information about risk management in both new and existing schools can be obtained from the SBD partner initiative Secured Environments at https://www.securedenvironments.com (see Sect. 5.12)

Scope of This Document

For the purposes of this document a school is defined as a building or collection of buildings located on one or more sites and used for the purposes of full and part time education of pupils between the ages of 2 and 19 and other community uses, in line with the 'Extended Schools' agenda. However, the scope of this document does not extend to universities or other tertiary colleges.

This document provides design guidance and specification requirements for reducing Crime Risk against people and property in all schools and school grounds such as burglary, theft, arson, vehicle crime and assault. The same advice is also intended to reduce the fear of crime and the incidence of ASB. Section 1 concentrates on generic environmental Crime Prevention issues, whilst Sect. 2 deals with the specific physical Crime Prevention requirements. Compliance with both sections is required to achieve SBD certification.

Applying for SBD Certification
In order to stand the best chance of achieving a SBD award, it is of the utmost importance to consult the Designing Out Crime Officer (DOCO) prior to a planning application being made. In some areas SBD is administered by a Crime Prevention Design Adviser (CPDA), Architectural Liaison Officer (ALO) or Design for Security Consultant (DfSC). However, all subsequent references within the guide refer to the process being administered by the DOCO. In the Development Guide, the term CPDA is used as it predates adoption by the UK Crime Prevention Design Group (CPDG) of the role title DOCO in 2016 and by the National Police Chiefs' Council (NPCC) in 2021.

Local planning conditions, Crime Risk Assessment and other statutory provisions may influence the measures to be adopted and the applicant is referred to specific guidance for England, Scotland, Wales, Northern Ireland, the Channel Islands and Isle of Man. Further information relates to acts of terrorism and other extremist activity. As is guidance on how the advice given by the DOCO will be dependent upon the outcome of Crime Risk analysis and an understanding of local crime occurrence.

2014 Section 1: The Development—Layout and Design (Planning Issues)

Urban Design and Planning Policy

1. Creating a Sense of Place
Creating an easily legible sense of place where staff, pupils and legitimate visitors are able to go about their daily routine without undue fear of crime (FOC), is a key element of the SBD—New Schools 2014.

2. Phased or Speculative Developments
It is recognised that some educational developments may be phased, developed when funds are available or when a school expands. This will not necessarily preclude achievement of SBD certification, provided there is continuous dialogue with the DOCO.

3. Location and Adjoining Land Use
Security requirements will be influenced by the location of a new or existing school, hours of operation, numbers of staff and visitors and by transport links.

Similarly, adjacent land use, sustainable design principles and capacity for Natural Surveillance will be important considerations. However, Natural Surveillance must work with Defensible Space.

4. Multi-site Schools
It is recommended new schools are planned on a single site. Multi-site schools generate movement between sites, with potential for unauthorised access and difficulty in managing Perimeter Security.

5. Configuration of Buildings
Whilst security is more easily managed in a single building, it is accepted this may not be possible due to site constraints, energy efficiency, daylighting, sound insulation, environmental considerations and community use. Future school expansion and linking this to existing buildings so as to appropriately control access, are all issues where the DOCO will work closely with the designers to ensure the best outcome.

6. Location of Buildings
Whilst site constraints and solar heating may ultimately dictate the location of a new school building or buildings, there are Crime Prevention advantages in locating them close to the site boundary and surrounding community and to orient them to offer surveillance to as many elevations as possible.

7. Location of Outdoor Activities
Provision for different activities and their relationship to the school should be considered carefully. Informal hard play areas and external social spaces should be within view of occupied rooms. Formal recreation spaces, e.g. multi-use games areas and external all-weather sport facilities can encourage trespass outside normal school hours and may require fencing.

8. Caretaker or Warden Accommodation
The permanent presence of a warden or caretaker who lives on-site is less common these days, yet such a person can be regarded as a Capable Guardian, i.e. a deterrent or witness to crime, and play an important role in Crime Prevention. On-site caretaker housing must be built to the latest SBD Residential Guide standard.

Perimeter Security

9. Site Boundaries

A clearly defined boundary using a fence, wall or other effective barrier against intrusion is a prerequisite for a secure site and to define ownership. A secure boundary will help staff manage the school site, by limiting trespass and by channelling visitors to the site through appropriate entrances. And it will also frustrate the intruder intent on breaking into the school. A secure boundary can also provide for the security and safety of very young children, keeping them in and the 'uninvited' out. Public footpaths immediately outside boundary fencing can affect security. If the footpath already exists and cannot be re-routed, Defensive Planting should be considered to reinforce fencing.

10. Security Fencing

Fencing is effective at delaying or deterring intrusion because of the need to climb over or penetrate the fence. It is therefore important that there are no structures close to or over the fence that will aid climbing, e.g. trees, lamp columns or buildings. Such fencing should be effective, but not create a 'fortress' impression. Appropriate types may include welded-mesh fencing, railings or expanded metal fencing, ideally colour-coated to soften appearance. It is accepted that some security rated fencing systems can be both costly and aesthetically unpleasing. However, the type of fencing required to gain SBD recognition must ultimately be determined by local Crime Risks.

11. Gates

The design, height and construction of any gates within a perimeter fencing system should match that of the adjoining fence and not compromise boundary security (see also Sect. 5.7.50).

12. Defensive Hedging

In some locations the use of defensive hedging (such as Hawthorn) may be effective in protecting a site perimeter, or to further bolster the security of an existing or proposed fence. A hedge can take several years to develop into an effective barrier and therefore a temporary fence will be required in the short to medium term. When the hedge has matured to provide an effective barrier, the fence can be left in place ('lost') or removed. Species selection is important, as hedging will require continuous maintenance.

Entrances at the Perimeter

13. Pedestrian and Vehicle Entrances

SBD prefers one entrance area at the perimeter of the site serving both vehicles and pedestrians, or separate entrances located in close proximity to one another so as to aid mutual supervision. There should be clear demarcation between the roadway and footway and a safety barrier between the two will be necessary. Additional guidance covers: community facility buildings requiring their own secured entrance; additional entrances at the perimeter of the site; gatehouses; physical barriers, such as 'anti-ram' bollards; access for emergency and service/ delivery vehicles; restricting vehicle and pedestrian access beyond the entrance area; and locking entrance gates when the school is closed and unoccupied.

14. Drop-Off and Collection Facilities

Whilst SBD promotes walking to school along safe routes, it is recognised some students will still be driven to school by car. This can result in traffic delays outside the school caused by parked vehicles and damage to footpaths. The Development Guide recommends that school security procedures, road safety and traffic regulations must be considered from the outset to minimise these problems. Parking provision for school buses will also need to be considered. SBD recognises that school security procedures normally require parents/guardians of very young children to pick up their children directly from the school, in a secure area.

15. Signage

The school's reception entrance should be clearly signposted from the entrances onto the site. Colour-coding of signs can be helpful in identifying the routes to various school departments. Site maps, if required, should be correctly oriented for the visitor and located in places where they cannot be attacked or treated with anti-graffiti coatings to allow easy cleaning.

16. Community Access

Co-location of facilities on school sites is being encouraged by central government, particularly when new or replacement schools are being planned, or when major refurbishments take place. This should be taken into account in the security planning.

Vehicle Parking and Access

17. Access by Pedestrians and Vehicles

Local infrastructure, vehicular and pedestrian routes (together with an appropriate site layout) will determine the location of the car parks, delivery points and bus stops and can help to identify the appropriate location of security-controlled entrances. The main site entrance is the first line of defence against trespass and should be the focus for directing and controlling all access. Other entrances should be kept to a minimum and all must be controlled.

18. Vehicle Parking and Deliveries

The area forming the route from the public site entrance to the reception and visitor car parks and delivery points should be as short as possible, open to view from the reception area and secured from the rest of the site. This will help to deter trespass into other parts of the school grounds and channel all visitors through reception. Additional guidance relates to: separate parking for staff; parking provision for emergency vehicles and secure deliveries to the school. This may include a secure external compound or easy access to delivery doors; and vehicle parking facilities should comply with the police service's 'Park·Mark' criteria for safer car parks (although attaining the award is not a requirement of SBD New Schools 2014).

19. Motorcycle, Moped and Bicycle Parking

Secure motorcycle, moped and scooter parking should be made available for staff and older students (if permitted to bring vehicles onto the site). Such parking provision should benefit from: surveillance from occupied school offices and/or classrooms; be provided with secure ground anchors and lit after dark when in use. Secure bicycle parking should be located in view of occupied school offices and/or classrooms, with stands to which the bicycles can be secured. To encourage cycling to school, it is recommended cycle parking is contained within a securable, roofed building. The building should be lit and secured during school hours. Separate parking for staff should be considered.

Access Within the School

20. Routes Within Grounds

Footpaths between school buildings and the public entrance, must be direct and follow known or anticipated desire lines. They should be well lit, unobstructed and at least 3 m wide to allow passing without conflict.

21. Public Use of Grounds

Trespass onto school grounds by the local community for dog walking, golf practice etc., should be discouraged by using signage and fencing. Some schools open the school grounds for appropriate supervised public use during school holidays and outside school hours. This is to be encouraged, since lawful occupation can help prevent crime.

22. Public Footpaths Through Grounds

Unless local circumstances dictate otherwise, public footpaths through the school grounds should not exist. Paragraphs 8 and 12 in Schedule 6 of the Countryside and Rights of Way Act 2000 make provision to extinguish or divert Public Rights of Way through school grounds that create Crime Opportunities.

External Issues

23. Landscaping

The use of plants in landscape design is encouraged to soften boundary treatments and introduce visual interest, although plant growth above 1 metre and below 2 metres should be absent to provide a window of surveillance. Plant growth below 500 mm may be required in respect of car parks to deter vehicle interference. The planting of new trees should be considered in tandem with the installation of any CCTV system, Lighting columns and the denial of aids to climbing. Whilst, species selection of trees and shrubs should take account of their future maintenance.

24. External Furniture, Litterbins and Sports Equipment

External furniture such as benches, structures and planters (together with sports and play equipment) should be of robust vandal and graffiti-resistant design. Furniture should be fixed into the ground and should not be located at or close to a building line or boundary fences, where it can be used to climb onto roofs. Litterbins can also be used to assist climbing and the contents used to start a fire. It is preferable that litterbins are locked onto a fixed base and located away from buildings, not mounted beneath windows or on walls covered in combustible material.

25. Troublesome Meeting Places

Students of all ages like to meet together within the school grounds and certain locations are preferred. If they are meeting to indulge in behaviour contrary to school rules (such as smoking or bullying), these congregation points tend to be in places that are out of sight of the school buildings or at a blind corner. Recommendations are set out in pursuit of minimising out of site locations including: removal or repositioning of a building (such as a shed or other lightweight storage unit); the trimming of a hedge; CCTV surveillance; eliminating recessed doorways; the careful design of soft and hard landscaping and consulting students and staff about known troublesome congregation points, so that these matters can be considered during the design process.

26. Natural Surveillance and Recessed Doorways

For new buildings, it is important to avoid the creation of areas and building features (such as recesses) that cannot be overlooked from another occupied building or room. Recessed doorways can obstruct surveillance and collect windblown litter that can be used to start fires. Where a recessed doorway is unavoidable because of site constraints, it would be prudent to use a secure doorset (see Paragraph 63) with emergency exit hardware and in-built secure vision panel. Additional advice is provided in respect of inner courtyards and isolated teaching areas.

27. Temporary Classrooms

Temporary classrooms (such as portable buildings) are notoriously difficult to secure, due to their construction and location outside the secure envelope of permanent building structures. Moreover, voids under these buildings must be secured to prevent litter collecting underneath, which may be used to start a fire. They should not be used for the storage of high value equipment, e.g. computers and projectors; must be included within the school's Intruder Alarm system; constructed of non-combustible materials and their location must be discussed with the local fire authority to ensure spread of fire to other buildings is minimised and the fire service's access is not hampered.

28. Special Sports Facilities

Multi-use games areas and artificial playing surfaces (usually with Lighting for night-time use) are expensive facilities that are often targets for intrusion, vandalism and misuse. They need to be carefully planned, managed and protected using all appropriate SBD guidelines and specifications. Lighting needs to be coordinated with actual occupation and use of the particular facility. Whilst spectator stands and sports pavilions should be constructed of non-combustible materials and sited appropriately, to avoid visual obstruction of other buildings.

29. Wind Turbines, Photovoltaic Installations and Biomass Boilers
Consideration must be given to protecting wind turbines, photovoltaic installations (PVs) and biomass boilers from vandalism by using access control, fencing and removal of climbing aids. PV panels are susceptible to criminal damage from thrown missiles and are likely to be the subject of theft, as their installation becomes more commonplace. Therefore, PVs should be located on roofs and secured with theft-resistant fastenings. School landscape design should never include the use of loose pebbles.

Storage Facilities

30. Equipment Storage
Secure storage for play and sports equipment (including seasonal storage for goalposts etc.) should if possible, be provided within the main building, with secured access from the outside. Generous storage provision should avoid the future need for additional outbuildings that are more vulnerable to attack.

31. External Waste Storage
Waste containers (particularly those with wheels) can be used for climbing and the contents used to start fires. Consideration should be given to using waste containers with lockable lids. Advice in respect of safe stand-off distances, fire and smoke detection devices and fire sprinkler systems should be sought from the relevant fire authority. Adequate secure provision for storage of materials to be recycled should be included. Locating bin stores for kitchen waste is discussed.

32. Fuel Storage
Fuel (such as central heating oil and cans of fuel used for grounds maintenance machinery), should be stored in a secured building located well away from the school. Where there is insufficient space and fuel has to be stored inside a main building, advice should be sought from the appropriate fire authority.

33. Cleaning Equipment Storage
A secure store, or stores inside the school building on each floor, must be provided for the safe storage of cleaning equipment and cleaning fluids, given they may be flammable and possibly toxic. Advice should be sought from the appropriate fire authority.

Utility Services and Mail Delivery

34. Metre Reading and Security of Utility Services

Utility access covers protecting access to drains, sewers, telephone cables, electricity cables and other services, must be secured to prevent access and damage by unauthorised persons. Utility metres must be sited in a secure building and if possible, should allow for metre reading without having to enter the main building. Alternatively, instructions should be given to utility providers to carry out their readings during hours of occupancy or by prior appointment.

35. Mail Delivery Arrangements

For the majority of schools, it is expected that mail delivery will take place during school opening hours with mail handed in at reception. For out of hours deliveries refer to Sect. 5.7.61.

Building Shell

36. Automatic Opening Window Systems and Vents

Many sustainable school buildings require an automatic building management control system to ventilate the building and regulate the temperature. This can cause security problems and a number of factors are set out in the Development Guide.

37. Walls—Facades, Apertures and Graffiti

Facades of buildings should minimise the opportunity for hiding and climbing up to windows or onto roofs. The potential for unauthorised entry using facilities such as goods lifts, fuel delivery points or ventilation ducts should be considered and are detailed. Further recommendations relate to graffiti removal and prevention. Designers should consider wall finishes that make this task easier to perform, particularly when the risk of graffiti is high. Various graffiti-resistant glazes and sacrificial coatings are available for treating surfaces already subject to 'tagging'.

38. Roof Design and Access

Preventing easy access to roofs should be considered at the building design stage. External rainwater pipes can be used for climbing and physical barriers should be used to prevent access to an existing roof. Similarly, low level eaves/roofs can sometimes pose security problems and should be avoided. Flat roofs, particularly those at a low level, may be more easily accessed and vulnerable to intrusion.

They may also create a risk of falling by a person who has climbed onto the roof recklessly (see Paragraphs 59 and 60 regarding roof construction and roof lights).

Attention is drawn to relevant legislation (including the *Occupiers' Liability Acts 1957 and 1984*) concerning the responsibilities and liabilities of building owners/occupiers for the safe use of roof areas and signage.

Internal Layout Issues

39. Entrances into the School Building Main Entrance

There should be one main entrance into the school building, which may incorporate more than one set of doors. For example, one set for volume access by students and possibly staff; and another set for parents, visitors and staff. These should be access controlled for all or part of the day, depending on local Crime Risk. Additional advice covers: relevant opening times; Lighting the entrance; reception staff having a clear view of the approaches to the school entrance; disabled access and use of an 'airlock' door system.

Additional Entrances

There will be instances, especially within multi-building schools, where further entrance doors are required for the convenient movement of students between classes. Electric door lock solutions for such external doorsets can be operated in numerous ways—detailed in the Development Guide. Additional guidance covers: ID cards; integrated access control systems; biometric access control and voice recognition systems and separate access-controlled entrances for staff and sixth formers.

40. Reception Area, Visitor Control and Interview Room

A visitor's first impression of a school and its security is often gained in the reception area. The reception area should therefore be welcoming and create a positive and interesting impression of the school. Colour schemes and textures of wall finishes and furniture should be carefully considered to create a calming environment for the waiting visitor. It is recommended that the reception is staffed or supervised at all times and access beyond the reception area is controlled using automatic locking doors, turnstiles or another barrier operated by the receptionist or by proximity reader technology with fobs/ID cards issued to staff and older students.

Further guidance relates to: relevant Health and Safety Regulations to minimise the risk of assault; reception desks providing the receptionist with a clear view; reception desk design; an escape route to a place of safety; audible personal attack alarms; visitor toilets and shared community facilities.

41. Toilet Facilities
Badly designed and located toilets in schools can provide opportunities for bullying, assault, arson and ASB. Consequently, guidance includes toilet block location. If it is likely the community will use part of the school building(s), consideration should be given to providing additional toilet facilities close to the rooms or section of the school used, to prevent unnecessary movement around the school. SBD recommends full ceiling to floor cubicle enclosures, to provide increased student privacy and reduce ASB. Space within the toilet facility should be kept to a minimum, with hand-washing facilities visible and potentially unisex—by being located outside the cubicle area.

42. Corridors, Circulation Areas and Stairways
Modern school design often includes very wide circulation corridors that are similar to 'streets'. They have many uses beyond circulation and are used to display students' work, for 'break-out' study groups and meeting parents. Consequently, it is recognised recesses along its length may be a deliberate design feature. Lines of sight can be maximised through the imaginative use of glazed screens and reflective surfaces. Improved surveillance of the corridor from adjacent classrooms and offices (with due regard to unwanted interruption) may be useful. Stairways are often the places that attract misbehaviour and bullying. A series of design considerations are described that can help control this type of behaviour.

43. Internal Doorsets
As a general rule, all internal doorsets should be locked when the building is left unoccupied with specific attention paid to locking storerooms, staff restrooms, head teachers' offices and general office areas, at the end of the school working day.

44. Changing Rooms—Field Sports, Swimming and Physical Education
Students and staff should be provided with individual secure storage facilities to hold personal belongings during a lesson.

45. Kitchens and School Restaurant
Areas of the school that may sometimes be available for community use (including serving of food and beverages) should be planned to allow secure separation

from the rest of the premises. External courtyards as part of restaurant areas, should have secure boundaries with access-controlled exits. Card reader systems can be programmed to include cashless catering and electronic registration.

46. Medical Room
The medical room should be located on the ground floor close to the main entrance door and emergency vehicle parking, to allow easy access to medical professionals and other emergency services. It should be equipped with a secure storage facility for first aid equipment and materials.

Security Lighting

47. External Lighting
In terms of security, the objective of Lighting a school site after dark is to improve the chances of detecting an intruder and to enhance an intruder's fear of detection. Lighting design should be coordinated with any CCTV system and landscape design. A Lighting scheme should be sufficient to cater for lawful after dark activity around the school site.

External illumination when the building(s) is/are unoccupied is recommended for entrance gates and routes to the main entrance and doors, car parks and observable building elevations. However, whilst the use of bollard lights can be useful for way finding, they are generally insufficient for security. Lights placed at lower levels can fail to properly model the facial features of pedestrians, hinder ready identification and might produce a glare effect and raise the fear of crime. To be effective, Lighting should ensure a realistic chance that there will be witnesses to an intrusion. Consequently, the DOCO will place greater emphasis on physical and electronic security for vulnerable areas of the building, which are not adequately monitored or overlooked at night.

48. Internal Lighting
It is recommended that most internal school Lighting is operated by detection devices that automatically switch lights on and off due to movement activity or its absence in each room. Apart from being a considerably more efficient method for reducing energy consumption (in that it does not require a deliberate and remembered action by the user) such a system will identify the presence and progress of intruders in the building when it is closed. It therefore follows that controlling mechanisms for the system should be securely contained.

Section 2

Physical Security Specifications Introduction

49. Security Standards

It is important that an effective and realistic level of physical security (commensurate with the risk) is incorporated into building construction. The physical security standards outlined within this Section, together with those of Section 1, indicate the minimum requirements needed in order for a development to be awarded an SBD certificate. It should be noted, however, that in some higher risk locations, additional or alternative measures may be required.

Perimeter and External Areas

50. Fencing, Gates and Security Bollards

The demarcation between public space and the school grounds is important. However, in many circumstances there is also a need for fencing that offers greater security in order to protect a particular risk. It is therefore important that the boundary treatment is discussed in detail with the DOCO at the earliest possible opportunity. The guide details how there are at least five reasons for providing a perimeter boundary fence. The height of the fence will be determined by local circumstances, Crime Risk and the system chosen. In most circumstances heights between 1.2 m (demarcation) and 2.4 m (higher security) will be appropriate. Lower heights of fencing (1.2 m to 1.6 m) are only suitable for boundary demarcation and controlling movement—not for security.

The height of security fencing will generally start at 1.8 m and above. It is normally preferable that the perimeter fencing allows clear views over the educational buildings and the grounds from the surrounding land and buildings. Dark colour finishes to fencing reduce the reflection of light and therefore makes it easier for passers-by to observe activity within the grounds/premises. Depending on Crime Risk, the SBD requirements for fencing are set out, including materials, construction, gates, high security fencing, security standards and security bollards.

51. Telecommunications and Utility Access Covers

Telecommunication lines and cables should enter buildings below ground, be protected by secure access covers certificated to one of the standards quoted and positioned in highly visible locations. For high-risk areas or vulnerable businesses, access to other utilities should also be secured to such a standard.

52. External Lighting Standard Requirements
All Street Lighting for adopted highways, estate roads, footpaths and car parks must comply with the standard quoted. Where conflict with other statutory provisions occurs, e.g. developments within Conservation Areas, requirements should be discussed with the DOCO and local authority Lighting Engineer. Additional guidance relates to: landscaping; Overall Uniformity of light for an SBD development; evenness of light distribution; Colour Rendering qualities of lamps; Colour Rendering Index and how light pollution must be minimised. The DOCO must be provided with a 'Lux Plan' in order to assess the Lighting system. The plan must detail the maximum, minimum and average lux levels proposed. The plan must also show Uo and Ra values for the scheme.

53. Closed Circuit Television (CCTV)
CCTV (Video Surveillance System—VSS) is not a universal solution to security problems. It can help deter vandalism or burglary and assist with the identification of culprits once a crime has been committed, but unless it is monitored continuously and appropriately recorded, CCTV will be of limited value in relation to the personal security of students, staff and visitors. Nevertheless, the provision and effective use of CCTV fits well within the overall framework of security management and is most effective when it forms part of an overall security plan. Further recommendations are made concerning the: CCTV Operational Requirement (OR); recording capability producing images of evidential quality; location of cameras; installation standard; co-ordination with the existing or planned Lighting system; registration with the Information Commissioner's Office (ICO); compliance with guidelines in respect to Data Protection and Human Rights legislation and remotely monitored detector activated CCTV systems.

54. Securing of Unattended Bicycles, Mopeds, Scooters and Motorcycles

Bicycles
The securing of bicycles left unattended must be considered within school design. Research by the Design Against Crime Research Centre suggests cyclists should be encouraged to lock both wheels and crossbar to a stand, rather than the crossbar alone. Therefore, a design of cycle stand that enables this method of locking is recommended. Further guidance covers: minimum requirements for cycle stands; external and preferably roofed bicycle stores; external containers designed for the secure storage of 2 or 3 bicycles; ventilated bicycle stores within a school building and their illumination.

Motorcycles, Mopeds and Scooters

External parking stores for mopeds, scooters and motorcycles should be covered and located close to, in view of the school building and provided with secure anchor points certificated to Sold Secure Silver Standard.

Building Envelope

55. Wall Construction

School walls in remote locations and in elevations that are out of sight, are attacked on occasion and should therefore be constructed to resist such an attack. Materials resistant to manual attack or damage should be used to ensure the initial provision of security.

56. Glazed Curtain Walling and Window Walls

SBD recognises four distinct types of glazed wall systems. These are detailed, together with the standards relating to glazing retention, attack-resistant glazing, patent glazing and sloping glazing.

57. Roller Shutters and Grilles

Grilles and shutters can provide additional protection to both internal and external doors and windows. The minimum standards for such products are set out, together with the appropriate circumstances and necessity. Security ratings higher than the minimum will be dictated by the location of the building and/or security level of the door or window being protected.

58. Roller Shutter Doors

Roller shutter doors providing access for deliveries and other apertures without a door, must be certificated to one of the standards listed.

59. Roof Construction

Roofs are vulnerable to criminal intrusion and damage from vandalism. Careful consideration must be given to their construction. Lightweight roofing systems are becoming common in school buildings. They must be certificated to one of the standards listed. Where traditional roofing systems are used, the DOCO must be consulted to discuss alternative security measures. Vulnerable ceiling voids should be protected by a monitored Intruder Alarm system. There must be full compliance with the 'Duty of Care' obligations under the Occupiers' Liability Acts 1957 and 1984.

60. Roof Lights and Sun Tubes

Roof Lights

Based on a site-specific risk assessment (which will take into account contributing factors such as accessibility and visibility), a roof light aperture must be protected by either a roof light or roof light and internal grille, certificated to the standards quoted for low or High-Crime Risk applications.

Sun Tubes

To prevent a sun tube being used as an access point into a school, the maximum diameter is 350 mm. If larger diameters are required, the DOCO should be contacted to discuss alternative protection.

61. Mail Delivery Letter Plates

There are increasing crime problems associated with letter plate apertures, such as arson, hate crime, lock manipulation and 'fishing'. In order to address such problems SBD strongly recommends (where possible) mail delivery via a secure external letter box or delivery 'through-the-wall' into a secure area of the building. If a letter plate is utilised it must form part of the certificated doorset range in order to be accepted within SBD developments. Dimensions and standards are set out, as is advice in respect of internal security deflector plates and anti-arson containers.

Surface-Mounted Letter Boxes or Through-the-Wall Delivery

Surface-mounted or 'through-the-wall' letter boxes greatly reduce Crime Risk problems associated with letter plates and also heat loss through the door. Surface-mounted letter boxes, through-the-wall delivery and the standard required, are specified.

62. Access for Persons with Disabilities

In accordance with the Equality Act, 2010 (Disability Discrimination Act, 1995 in Northern Ireland), security features should not compromise guidance found in relevant Building Regulations. Security features (e.g. locking devices, door closers etc.) should be designed for ease of use by those with disabilities. Features introduced to meet building regulation access, e.g. external handrails, should be considered alongside the security risks that these required features might present, such as climbing from a handrail onto a flat roof. Measures to counteract such risks must be discussed with the DOCO.

63. External Doorset Apertures
It is important the doorset aperture is protected. Due to the nature of educational establishments and their location, there is an expectation the security will meet one of the minimum standards detailed for when the building is unoccupied. It is expected that all doorset products are fit for purpose and will require certification to the material specific standards for doorsets. In addition, such certification must be conducted by one of the UKAS accredited bodies listed.

Avoiding Door Recesses
Recessed doorways should, where possible, be avoided as they provide opportunities for crime and ASB, i.e. graffiti, arson and burglary. Doorsets designated as 'emergency' or 'panic' exits must be fitted with the hardware appropriate to the specific use and to the standards quoted.

64. Glazing Within Doorsets and Secure Vision Panels
All glazing in and adjacent to doors must include one pane of attack-resistant glass, securely fixed in accordance with the manufacturer's instructions. Glazed panels adjacent to the doorset and an integral part of the doorframe, should be tested as part of the manufacturer's certificated range of door assemblies. Alternatively, where manufactured separately from the doorframe, they shall be certificated to one of the standards quoted. Access for the disabled should comply with the Equality Act, 2010. Doorset attributes which may require closer attention include: opening mechanisms, glazing, level access and door width. Further guidance is provided in respect of door viewers.

65. Door Installation
Doorsets must be securely fixed into the fabric of the building in accordance with the manufacturer's instructions. The DOCO may require a copy of the manufacturer's installation specification.

66. Security Glazing
All ground floor and Easily Accessible glazing must incorporate glass to the standard quoted. This may not be required if windows that are not immediately adjacent to doors and protected by a roller shutter or grille as described in Paragraphs 57 and 58 above. When large Laminated Glass panels are used on south facing elevations, there have been occasional incidents of glazing failure (cracking) due to thermal stress. Use of Toughened Safety Glass appears to be a simple solution to this problem, except it offers no security resistance. Therefore, it is recommended the inner pane of glass used in a double-glazed unit is 'Laminated

Toughened'. This combination of the two sheets of Toughened Glass and Laminated Glass, offers both resistance to intrusion and thermal stress associated with large, glazed areas.

67. Electronic Access Control

Electronic access control may be required at the main entrance to a school and on some additional external and internal doors. These requirements will be influenced by many factors in the Development Guide. The doors must incorporate electronic access control systems, with electronic lock release and (for main entrance and possibly other entrances) reception-linked entry phone. In some cases, visual verification by CCTV camera incorporated into the call-panel or separately located, may be required.

68. Electronic Access Control Standards

In the absence of dedicated UK security standards for access control systems (i.e. the interface between the user and the lock control mechanism) a small number of access control systems are currently licensed by SBD following Government evaluation.

69. Access Control and Security Staff

Although it is very unusual for schools to employ formal security staff, there are some that do. Where security officers are employed at the school entrance, electronic access control will still be required.

70. Protection of Window Apertures

Window apertures may require differing levels of security protection, dependent upon Crime Risk and crime levels. They must also meet the local Building Regulations in respect to safety and thermal performance. The minimum SBD standard for protecting window apertures at ground floor, basement and Easily Accessible locations, is they must be certificated to one of the standards quoted.

All glazing in Easily Accessible windows must include one pane of attack-resistant glass (see Paragraph 66), securely fixed in accordance with the manufacturer's instructions. Windows must be fit for purpose and certificated to the relevant material standard. Windows installed in SBD schools must be certificated by one of the UKAS accredited certification bodies or demonstrate alternative compliance. A higher level of security can be attained by either installing windows to the security standards detailed together with the relevant material standard or by additional protection of the minimum standard windows by the use of roller shutters or grilles as described in Sect. 5.7.57.

Internal Security Considerations

71. Intruder Alarm Systems

A suitably designed, fit for purpose, monitored Intruder Alarm system must be installed. For police response, the system must comply with the Police Security Systems Policy. Security Fogging and offender marking devices can be included within the Intruder Alarm system, to disorientate and mark an intruder as the alarm system is activated. They must conform to the standard quoted.

72. Public Address Systems

A public address system is recommended to provide instant, effective communication to the whole school, particularly in emergency situations where a prearranged and rehearsed response to particular situations can be initiated, e.g. fire escape where the PA system may be linked to the fire alarm system.

However, this aspect must be discussed with the appropriate fire authority.

73. Physical Security Standards for Computers and Server Rooms

Consideration must be given to the structure of internal walls, floors and ceilings of computer server rooms to provide appropriate security and prevent damage by fire, smoke or fire extinguishment (water) from other parts of the building. Numerous materials are recommended. Computer files should be backed-up to an off-site location. The service provider should be certificated to the standards quoted, once reference has been made to the body described.

74. Internal Doorsets

Where an internal doorset must be secure, e.g. computer server room, this should be certificated to a standard quoted. New building presents an ideal opportunity for an integrated access control system-operating most doorsets, with proximity reading technology least susceptible to tamper/damage.

75. Staff Room

Staff rooms should provide each staff member with secure storage for clothing and personal effects.

76. Student Lockers

Students should be provided with secure storage for clothing and personal belongings. If the lockers are provided with a power supply for charging of equipment, e.g. mobile phones, the locker system must form part of the manufacturer's certificated range. Lockers should be located in areas with high circulation and Passive Surveillance. If located along corridors the lockers must be non-combustible.

77. Metal Detectors
Metal detectors used responsibly can be effective in reducing violence in schools. Metal detection using arches or wands is more effective if used as part of an overall Anti-Violence, Anti-Bullying programme, involving numerous remedies for example: local police, CCTV monitoring, periodic locker searches and reviewing use of the school's internet or intranet service.

Other Security Considerations

78. Security Staff, Manned Guarding and Guard Houses
Although it is highly unusual for schools to employ security staff, advice includes: provision of manned guarding; electronic patrol monitoring systems; security control room operation; key holding facilities; construction and standards. Manned guarding may require additional design features at the outset in order to reduce overall costs. The standards relevant to manned guarding are set out.

79. Arson and Fire Protection
Due to the high incidence of arson in schools, premises should be designed with arson prevention in mind. Architects should consult with the Fire Service, insurance specialists and a full cost–benefit analysis undertaken on the installation of a sprinkler system for all new building projects/substantial refits and renovations. See also Department for Education and Arson Prevention Bureau advice.

80. Other Useful Security Equipment to Recognised Standards
SBD licences security products useful in the school environment, yet many are not referenced within this document. These products are independently certificated to Police Preferred Specifications or standards or assessed as fit for purpose by SBD's Independently Chaired Product Assessment Panel.

81. Safes and Strongrooms
For commercial safes and strongrooms, SBD recommends certification to the standard quoted. The required resistance grade is determined by the value of the items to be kept in the safe overnight. The ratings in the table should be used as a guide, as insurers will define their own ratings depending on the performance of the safes and the situation in which the safes and strongrooms are located. It is therefore vitally important that insurers are consulted prior to selecting a safe or strongroom.

5.8 SBD—Hospitals

Hospitals can be especially problematic in terms of Designing Out Crime, many occupying large and sprawling sites with poor perimeter and interior security. Moreover, as the primary objective is to treat those who are ill, historically the necessity for adequate security was repeatedly overlooked.

1. SBD and Other Design Guidance

Good design must be the aim and encouraged by all those involved in the development process. SBD strives to achieve a good standard of security for buildings and the immediate environment. It attempts to deter crime and ASB, using appropriate design features to create a sense of ownership and responsibility. Incorporating security measures in a hospital can reduce crime, fear of crime (FOC) and ASB. The aim of the DOCO is to assist the design process to achieve a safe and secure location for patients, staff and visitors, without creating a 'fortress environment'.

2. Principles

The primary role of hospitals is to provide medical support for the community. Crime costs the NHS £600 million p.a. (in 2005) and open accessibility makes hospitals difficult to secure. However, the vulnerability of people and property to crime can be massively reduced, if the following measures are incorporated.

3. Role of the DOCO

Police forces throughout the UK employ DOCOs (Architectural Liaison Officer and Crime Prevention Design Adviser are used in the Development Guide). Their main mechanism for delivery is the SBD initiative and award scheme, which uses CPTED principles.

4. The Concept of Secured by Design

A key government planning objective is to secure quality, sustainable environments for where people live, work and play. Designing for Community Safety is a key element. SBD encourages the building industry to adopt Crime Prevention measures in development design, to reduce Crime Opportunity and the Fear of Crime, creating a safer and more secure environment. Endorsed by the NPCC and with Home Office backing, it was drawn up in consultation with central government. Once a development is complete, the main opportunity to incorporate Crime Prevention measures is gone. It is the responsibility of NHS Trusts etc. to instruct designers to liaise with the DOCO at the concept stage. This allows advice to be included at the earliest opportunity, rather than at the planning application stage, which often causes delay.

5. Crime Pattern Analysis
The DOCO will assess the Crime Risk and associated problems likely to feature in hospitals, soliciting Crime Pattern Analysis to provide an accurate picture of local crime trends. By combining police and hospital records, site-specific advice commensurate with perceived risk, will be provided. Careful design need not cost more when considered at the outset and will quickly show sustainable benefits.

6. Introduction
Provides advice on how to establish and maintain a safe and secure hospital site. It aims to help those involved in the design, development, procurement and management of hospitals. The overall objective is to reduce Crime Opportunity, FOC and ASB. A further aim is to provide a system of self-assessment of new proposals and the changes necessary to achieve SBD status. This guide includes the process by which accreditation can be achieved.

7. Assessment Guidance
The following features are the minimum for SBD accreditation. Additional security features may be required in medium or High-Crime Areas.

8. General Comment
To control access by pedestrians and vehicles, the perimeter should be enclosed. Fragmented/remote hospital sites pose security management problems and must be avoided where possible. Buildings should be arranged to avoid long distances, difficult to overlook hidden areas and isolated buildings.

9. Boundary Treatment
Details construction and minimum height of boundary fencing and gates, anti-climb toppings, unobstructed Natural Surveillance, thorny shrubs and signage.

10 Public Entrances to Hospital Site
The number of entrances to hospital sites should be reduced to the minimum practicable, preferably one main entrance with an adjacent security gatehouse controlling public movement onto the site.

11. Signs
Signage from the site entrance through to the destination ward/department, should be clear and multi-lingual, where appropriate.

12. Vehicular Access
Vehicular access within the hospital site should be restricted to a minimum and to areas where it is necessary, e.g. parking and service areas. Physical features to enforce this restriction are detailed.

13. Pedestrian Access
Casual intrusion onto the hospital campus should be discouraged. The footpath layout should serve the development, with the number kept to a minimum.

14. Parking for Cars, Motorcycles and Bicycles
Car parking should be restricted to designated areas where vehicles are more readily supervised. Secured and managed motorcycle and bicycle parking must be provided. All parking areas shall incorporate the measures required by the Safer Parking 'Park Mark' Award detailed at https://www.parkmark.co.uk

15. Landscaping
Landscaping should not prevent Natural Surveillance or reduce the effectiveness of any CCTV system. Future landscape growth and maintenance must be taken into account at the design stage.

16. Surveillance
Accessible areas should benefit from Passive Surveillance, i.e. overlooked by windows from Active Rooms or CCTV. Areas providing concealment (Crime Opportunities) shall be fenced-off.

17. CCTV Systems
Hospital CCTV systems (Video Surveillance Systems—VSS) should be registered with the Information Commissioner and comply with its CCTV Code of Practice available at: https://www.informationcommissioner.gov.uk An Operational Requirement (OR) should include design and performance specification, image standards and the areas covered.

18. Lighting
Good, but not oppressive levels of light should be evenly distributed to allow clear colour rendition. Well-positioned Lighting deters and reveals potential intruders and reduces the fear of crime.

19. Access to Drainage/Services
Access to telephone junction points and manhole covers should be secured, in order to prevent interference and/or removal.

20. Site and Building Layout
Building site lines should be kept as simple as possible in order to avoid the creation of hiding places, recesses and congregation points. Where possible, buildings should be oriented to maximise Natural Surveillance and Formal Surveillance opportunities.

21. Building Shell Security
The DOCO should be consulted regarding the building shell, to minimise climbing/scaling aids and high-risk elements, e.g. drug stores and pharmacies.

22. Windows
Details the security specification for all window types and how ground floor and other vulnerable windows should use glazing that complies with BS EN 356 P1A.

23. Perimeter Doors
Sets out the security standard for external hospital doors, including links to any alarm/warning system and glazing that complies with BS EN 356 P1A.

24. Roller Shutters
Provides the security standard for roller shutters and connection with the Intruder Alarm system.

25. Internal Layout
Advises grouping Accident and Emergency (A&E), intensive care, outpatients, radiography, operating theatres and MRI scanner functions to assist patient access. Locating obstetrics, gynaecology, special care baby unit, anti-natal clinic, labour ward and maternity beds together can enhance security.

26. Hospital Entrance and Visitor Control

Advises reducing the number of public entrances into the hospital, all within full view of the reception area. Further guidance covers: automated doors, Laminated Glazing; and rising/retractable bollards in front of high-risk perimeter doors.

27. Accident and Emergency

The A&E ward is often a flash-point for crime and its entrance should have an access control system for use at night. Using security staff to grant entry, can be highly effective in preventing trouble within the hospital. CCTV should cover ambulance entry points.

28. Reception Area

The design of the reception area can enhance staff safety. Further advice relates to: Laminated Glass screens; personal attack alarms with direct link to the police; surveillance by reception staff; CCTV; triage area; furniture; fire hoses; public telephones; food and drink dispensing machines; calming colour schemes; background music; securely fixed TV screens and segregating arrested or intoxicated patients.

29. W.C. Facilities

Suggests measures required to prevent vandalism of such facilities, including access control at night.

30. Treatment Rooms

Patients should not be positioned between staff and the main entrance door. Doors should be installed between treatment rooms, together with panic alarms.

31. Maternity and Paediatrics

Maternity and paediatric wards should not be located on the ground floor. Further advice covers: standalone facilities, baby tagging, alarm sensors, double doors, Defensive Planting, visitor control, glazing and CCTV.

32. Pharmacy

A vulnerable part of the hospital, the pharmacy can attract drug-dependent people and should possess a separate alarm zone. Located within easy reach of the main entrance, it should not form any part of the external structure. Further advice covers: wall construction, staff entrance, pharmacy counter, CCTV and controlled drugs storage cabinets.

33. Laboratories
Details regulative and registration arrangements for hospital laboratories and advises consultation with the local CTSA and (NaCTSO) at https://www.nactso.gov.uk

34. Cash Office
Provides security advice for the cash office, including wall construction, floor, ceiling and separate zone within the main Intruder Alarm system.

35. Retail Outlets
Retail outlets—including ATMs (Cashpoint Machines) benefit patients, staff and visitors. But they also attract thieves and remedies are suggested.

36. Police Office
The local police should be contacted regarding a police office within the hospital and best location—ideally close to the main entrance and A&E ward.

37. Security Office
Recommends all hospitals employ dedicated security staff providing 24/7, 365 days coverage, patrol patterns and the security office preferred location.

38. Hospital Patient Records Office
Suggests the physical security measures necessary for the patient records office, including the transit of records to and from such facilities.

39. General Wards/Access Control
Recommends the level of security required on general wards and staff locker/changing areas.

40 General Office Security
Details the physical security measures necessary for general offices and store-rooms.

41. Mortuary
Provides advice regarding the minimum-security requirements for the hospital mortuary.

42. Corridors and Circulation
Describes corridor design, access control and signage.

43. Security Management
Details advice re key security, access control, public address systems and wearing identity badges.

44. Computers
Provides advice in respect of the security of computers and office equipment.

45. Staff Accommodation
Secured by Design—Residential Guide 2025 (see Sect. 5.4) guidelines apply to on-site staff accommodation. Safe routes between such accommodation and the hospital must be designed.

46. Needle Exchange
The design guide suggests advice should be obtained from the DOCO.

47. Secured by Design
If the advice given by the DOCO is incorporated, an SBD certificate will be awarded. An SBD application form must be completed by the agent or client prior to siteworks commencing.

Secured by Design Checklist—Hospitals
SBD—Hospitals 2005 provides a checklist covering all issues from campus layout to the maintenance of a Crime Log and Police Contact Records.

Appendix A—Security Measures
Details how selecting the most appropriate measures will depend on a number of risk and cost factors.

SBD—Hospitals 2005 repeats the logical mantra that the standards it sets are a minimum level required to obtain SBD accreditation. High-risk sites may require additional measures.

5.9 SBD—Safer Tram Stop Award

1. Introduction

Grand Central Tram Stop outside New Street railway station, Birmingham

The Safer Tram Stop Award is aimed at tram operated fixed rail systems that operate throughout the UK. Its aim is to reduce Crime Opportunities and ASB from taking place at tram stops. Operators are required to take recognizance of Crime Patterns, passengers' perceptions of safety and management practices.

The award is aligned with the Secure Stations Award, (see Sect. 5.15—a DfT and British Transport Police initiative for heavy rail, London Underground and Docklands Light Railway) and the Secured by Design award.

It aims to: reduce crime and fear of crime at tram stops; provide guidance to operators on how to create and maintain safe environments and raise awareness amongst designers and architects by providing them with a design framework.

2. Responsibilities

Key responsibilities rest with: the operator of the tram system; Secured by Design; the police service with responsibility for policing the system; Passenger Transport Executives. Any disagreement as to whether a tram stop is suitable for the STSA will be decided by the assessor—a Designing Out Crime Officer (DOCO).

3. How to Apply (see also Appendix A)

The operator should approach the relevant Force Crime Prevention/Reduction Unit who will appoint an assessor. The assessor will conduct a Risk Assessment based on the annual passenger count, passenger perception survey and crime committed at the tram stop and in the surrounding area.

The award will be conferred providing the assessor is satisfied: the tram stop is located in a low crime environment; has good management practices; a favourable passenger perception survey and its design is conducive to Crime Prevention. Remedial work may be necessary and if still refused, the operator may appeal to Secured by Design.

4. The Safer Tram Stop Award

The award has been created for fixed rail transport systems not covered by the DfT/BTP Secure Stations Award, to gain recognition for good practice. Relevant tram systems are named. The Award is initially granted for one year, after which the assessor may increase this to two/three-year periods.

5. Crime Recording and Statistics

There should be regular liaison meetings between the police and the tram operator to discuss current and emerging crime trends. In addition, regular meetings with the local Crime Reduction Officer can implement measures designed to combat any increase in crime. The Crime Profile will determine what measures are appropriate at a particular tram stop, with the overarching objective of reducing the likelihood of members of the public becoming the victims of crime.

6. Passenger Perception Survey

It is important that tram operators are fully aware of passengers' perceptions of safety when using the tram network. A minimum of 100 respondents is recommended, 25% of whom should use the tram stop after 1900 hours. Less than 15% of those surveyed should describe themselves as 'feeling unsafe' or 'very unsafe' when using the tram stop.

7. Management Practices (see also Appendix B)
The operator should have in place an overarching Security Policy, which sets out the parameters for managing security, network wide. Operating Procedures for the safe operation and control of the tram stops should be regularly reviewed. Role specific Risk Assessments should be in place for all front-line staff, together with the dissemination of information about recent incidents.

At stops where staff are located, they should be contactable and readily available to any member of the public requiring assistance. At unstaffed stops, the public should be able to contact staff via help points, direct dial telephone or constantly monitored CCTV.

Staff induction and refresher training should include: conflict and dealing with aggression, to reduce the risk of being assaulted; dealing with victims of crime; involvement with cash handling and response to being a target for robbery and Planned General Inspections (PGIs) at the tram stop. A list of items that would be checked is provided.

The operator should conform with the Security Industry Authority's (SIA) regulations or be accredited under the Community Safety Accreditation Scheme (CSAS) or the Railway Safety Accreditation Scheme (RSAS).

8. Design Issues (see Appendix C)
The design of a tram stop can affect the perceptions of passengers using it; either deter or encourage crime; and therefore the management's response to such incidents. Design and the use of anti-vandal fixtures can prevent a deterioration in quality at the tram stop and any increase in fear of crime.

9. Parking Facilities Including Cycles and Motorcycles
Where the tram stop incorporates parking facilities, it is recommended that an application is made for the 'Park Mark' Safer Parking Award (see Sect. 5.19) Failing such an application, it is good practice to emulate the theory set out in the Safer Parking Scheme assessment guidelines.

Cycle storage facilities should provide secure parking for passengers' bicycles and motorbikes. This may be achieved through use of the different forms of surveillance (as detailed below).

10. Boundaries and Perimeters
The tram stop should have a defined boundary or perimeter that signifies the limit of ownership. This may include physical or Symbolic (Psychological) Barriers, which should not obstruct surveillance. Fencing standards and recommended heights (including those for vegetation) are listed. All boundary treatments should

be graffiti resistant. The assessor will decide whether the tram stop is sufficiently distinguishable from the surrounding footways for an award.

11. Surveillance

Descriptions of Natural Surveillance, Staff (Formal) Surveillance and Technological Surveillance (CCTV or Video Surveillance Systems) are provided—together with the potential obstructions to each of these elements.

Any CCTV system must comply with the Data Protection Act 2018 and be registered with the Information Commissioner, as it covers a public area. An Operational Requirement (OR) outlining all the stakeholders' needs should be completed. The potential issues for consideration and requirements of the OR and by extension, those of the CCTV system, are then set out.

12. Signage

Signage in a corporate style should be clearly visible and capable of being read under natural or artificial Lighting. Design of the signage strategy should include: way finding signs; contact details for the operator of the system; other local transport information; a map of the local area; timetables or frequency of service details; live and up to date tram running information; Data Protection Act 2018 compliant CCTV signage and any emergency and information help points clearly indicated.

13. Lighting

Ideally, Lighting at the tram stop should conform to the relevant British Standard and the minimum levels of Lighting are stipulated. In addition: all Lighting fixtures and fittings should be protected from damage; Lighting columns should not be able to be used as climbing aids or used to mount CCTV cameras and the Lighting at the tram stop should be sustainable.

14. The Whole Journey

This Development Guide recommends consultation and working closely with various organisations and committees, in order to prevent instances of Anti-Social Behaviour and drunkenness, throughout the whole journey by tram.

15. Special Considerations

These include: instances where adjacent train stations already possess the DfT/BTP Secure Stations Award; co-operation between relevant police forces; consultation with the management of the facility in question; whether the tram stop is of sufficient design to receive the award and conducting a Risk Assessment in respect of cash collection from any ticket vending machines.

16. References
A list of references is provided in the guidance.

Appendix A—Application Process
Sets out a flow chart for the application process.

Appendix B—Checklist for the Award—Management
Provides a checklist of management considerations necessary to achieve the Safer Tram Stop.

Award.

Appendix C—Checklist for the Award—Design
Provides a checklist of the design considerations necessary to achieve the Safer Tram Stop Award.

5.10 Specialist and Supported Housing Schemes

Preface

Changes to the English Planning and Building Control regulations by the Ministry for Housing, Communities and Local Government (MHCLG) underline the importance of police advice, especially the SBD initiative. The National Planning Policy Framework (NPPF), National Planning Practice Guidance (NPPG), Scottish Planning Policy (PAN 77) and Planning Policy Wales (TAN 12) reinforce the importance of safe and secure external environments. This guide sits alongside SBD—Residential Guide 2025.

Introduction
Details the opportunistic behaviour of criminals and how physical measures can strengthen security. A minimum level of physical security is recommended—to be enhanced should the risks of crime increase.

Academic research demonstrates the appropriate combination of physical security measures can dramatically reduce Property Crime. A synopsis is provided of Police Crime Prevention Initiatives (Police CPI) and how SBD is a police initiative operating across the UK. https://www.securedbydesign.com

Doorsets and Windows and Their Fitness for Purpose
Discusses the guiding principle of fitness for purpose and the various security certification ratings. A table sets out the testing parameters in terms of repetitious use. Doorsets and windows should meet the duty rating that is commensurate with their intended location and use. A number of scenarios are set out and extensive guidance is provided in respect of fire, smoke and security rated doorsets.

1: Timber Front Door With Single-Point Locking
If the property has been awarded SBD certification, the timber front door will be an enhanced security doorset. Guidance covers: DOCO confirmation; replacement doors meeting Police Preferred Standard; thumbturn release if front door only means of escape; single-point locking systems and standards; additional mortice or Euro cylinder deadlock; deadlock guard; additional automatic dead-locking rim-latch lock or escape night latch lock (Fire Safety Regulations); frame reinforcement; hinge bolts; door viewer; door limiter; letterbox security cowl; glass panels in or adjacent to doors; Toughened and Laminated Glazing; BS Kitemarks/ standards; Anti-Shatter Security Film and standard.

2: Timber Front Door With Multi-Point Locking
If the property has been awarded SBD certification, the timber front door will be an enhanced security doorset. Guidance covers: DOCO confirmation; replacement doors and police-preferred standard; thumbturn release where front door is the only means of escape; multi-point locking systems and standards; Euro profile lock cylinders; frame reinforcement; hinge bolts; door viewer; door limiter; letterbox security cowl; glass panels in or adjacent to doors; Toughened and Laminated Glazing; BS Kitemarks/standards; Anti-Shatter Security Film and standard.

3: PVC-U, Aluminium or Composite Front Door With Single-Point Locking
Composite doors usually comprise two moulded surface skins, inner door framing material and foam insulating core. Guidance covers: three categories of composite door; SBD certification; DOCO confirmation; replacement doors and police preferred standards; thumbturn unlocking if the front door is the only means of escape; single-point locking Euro profile lock cylinders; additional locking measures if front door the only means of escape; interlocking security hinges; door viewer; door limiter; letterbox security cowl; glass panels in/adjacent to doors; Toughened and Laminated Glass; BS Kitemarks and standards; Anti-Shatter Security Film and standard.

4: PVC-U, Aluminium or Composite Front Door With Multi-Point Locking
Composite doors usually comprise two moulded surface skins, inner door framing material and foam insulating core, producing a door that is strong and secure. Guidance covers: three categories of composite door; SBD certification; DOCO confirmation; replacement doors and police preferred standard; thumbturn unlocking if front door the only means of escape; Euro profile lock cylinders; interlocking security hinges; door viewer; door limiter; letterbox security cowl; glass panels in or adjacent to doors; Toughened and Laminated Glass; BS Kitemarks and standards; Anti-Shatter Security Film; Easily Accessible doorsets and windows; SBD and Approved Document Q Appendix A.

5: Timber Side/Rear Door With Single-Point Locking
If the property has been awarded SBD certification, timber side/rear doors will be enhanced security doorsets. Guidance covers: DOCO confirmation; replacement doors and police preferred standard; single-point locking; mortice deadlock/sashlock; deadlock guard; mortice rack bolts; surface-mounted throw-bolts; frame reinforcement; hinge bolts; side/rear doors with panels; glass panels in/adjacent to doors; Toughened and Laminated Glass; BS Kitemarks and standards; Anti-Shatter Security Film and its standard.

6: Timber Side/Rear Door With Multi-Point Locking
If the property has been awarded SBD certification, timber side/rear doors will be enhanced security doorsets. Guidance covers: DOCO confirmation; timber side/rear doors; replacement doors and police preferred standard; Euro profile lock cylinders; multi-point locking systems and standard; frame reinforcement; hinge bolts; side/rear doors with panels; glass panels in/adjacent to doors; Toughened and laminated glazing; BS Kitemarks/standards; Anti-Shatter Security Film and its standard.

7: PVC-u, Aluminium or Composite Side/Rear Door With Single-Point Locking
Composite doors usually comprise two moulded surface skins, an inner door framing material and foam insulating core. Guidance covers: the three categories of composite door; SBD certification; DOCO confirmation; replacement doors and police preferred standard; single-point locking; Euro profile lock cylinders; additional locking systems; surface-mounted throw-bolts; interlocking security hinges; side/rear doors with panels; glass panels in/adjacent to doors; Toughened and Laminated Glass; BS Kitemarks and standards; Anti-Shatter Security Film and its standard.

8: PVC-U Aluminium or Composite Side/Rear Door With Multi-Point Locking.

Composite doors generally comprise of two moulded surface skins, an inner door framing material and a foam insulating core, resulting in a door that is strong and secure. Guidance covers: three categories of composite door; SBD certification; DOCO confirmation; replacement doors and police preferred standard; Euro profile lock cylinders; multi-point locking systems and standards; interlocking security hinges; side/rear doors with panels; glass panels in/adjacent to doors; Toughened and Laminated Glass; BS Kitemarks and standards; Anti-Shatter Security Film and its standard.

9: Sliding Patio Doors With Single-Point Locking
If the property has been awarded SBD certification, sliding patio doors will be enhanced security doorsets. Guidance covers: DOCO confirmation; replacement/existing sliding patio doors and standards; single-point locking systems; Euro profile lock cylinders and standards; a minimum of three surface-mounted, key-operated patio locks; anti-lift devices; Toughened and Laminated Glass; BS Kitemarks and standards; Anti-Shatter Security Film and its standard.

10: PVC-U or Aluminium Sliding Patio Doors With Multi-Point Locking
If the property has been awarded SBD certification, sliding patio doors will be an enhanced security doorsets. Additional advice covers: DOCO confirmation, replacement/existing sliding patio doors meeting standards; multi-point locking; Euro profile lock cylinders and standards; anti-lift devices;

Toughened and Laminated Glass; BS Kitemarks and standards; Anti-Shatter Security Film and its standard.

11: Timber Glazed Double/French Doors With Single-Point Locking
If the property has been awarded SBD certification, timber glazed double/French doors will be enhanced security doorsets. Guidance covers: DOCO confirmation; timber glazed double/French doors; replacement/existing windows and standards; single-point locking; deadlock guards; Euro profile lock cylinders and standards; interlocking security hinges; mortice rack bolts/surface-mounted locking throw bolts; hinge bolts; additional locking points; glass panels in/adjacent to doors; Toughened and Laminated Glass; BS Kitemarks and standards; Anti-Shatter Security Film and its standard.

12: Timber Glazed Double/French Doors With Multi-Point Locking

If the property has been awarded SBD certification, timber glazed double/French doors will be enhanced security doorsets. Guidance covers: DOCO confirmation; timber glazed double/French doors; replacement/existing windows and standards; multi-point locking systems; Euro profile lock cylinders and their standards; interlocking security hinges; glass panels in/adjacent to doors; Toughened and Laminated Glass; BS Kitemarks and standards; Anti-Shatter Security Film and its standard.

13: PVC-U or Aluminium Glazed Double French Doors With Single-Point Locking

If the property has been awarded SBD certification, PVC-U or aluminium glazed double/French doors will be enhanced security doorsets. Guidance covers: DOCO confirmation; replacement/existing windows and standards; single-point locking; Euro profile lock cylinders and standards; additional locking points; interlocking security hinges; glass panels in/adjacent to doors; Toughened and Laminated Glass; BS Kitemarks and standards; Anti-Shatter Security Film and its standard.

14: PVC-U or Aluminium Glazed Double/French Doors With Multi-Point Locking

If the property has been awarded SBD certification, PVC-U or aluminium glazed double/French doors will be enhanced security doorsets. Guidance covers: DOCO confirmation; PVC-U or aluminium glazed double/French doors; replacement/existing windows and standards; multi-point locking systems; Euro profile lock cylinders and standards; interlocking security hinges; glass panels in/ adjacent to doors; Toughened and Laminated Glass; BS Kitemarks and standards; Anti-Shatter Security Film and its standard.

15: Glazing

Glazed panels in and adjacent to doors, including adjacent windows within 400 mm, must incorporate one pane of Laminated Glass meeting, or exceeding, the standard quoted. Further guidance covers: the inner or outer face; thermal efficiency; aesthetics; privacy; obscured glazing; doorset exceptions; adjacent windows; doors manufactured separately from the door frame and secure fixing.

16: Door Chains/Limiters

Door chains or opening limiters must meet the requirements of the Technical Specification stated; be suitable for the door material and installed in accordance with the manufacturer's instructions.

17: Door Viewers
A door viewer must meet the requirements of the Technical Specification quoted and be fitted according to the measurements set out—unless clear glazing is immediately present.

18: Secure Mail Delivery
New doorsets should have no letterbox apertures due to associated crime problems. Further advice covers: blanking plates/letterbox security cowls; secure external letter boxes and standard; letterplates within certificated doorset range; Scope of Certification; Technical Standard; test methods and requirements; concern for arson/repeat arson attacks and 'anti-arson' containers.

19: Compartmentation
Structural division of a building, usually by access controlled doorsets, can enhance security. Further advice covers: suitability; creation of clearly identified spaces; auditable access control data; tracking building life; Crime Prevention tactic; residents' assigned access and methods; installation of secure doorsets; security encrypted electronic keys; unrestricted emergency egress; lift security protocol.

20: Fire and Rescue Service Access
The fire and rescue service should have unrestricted access to all floors in the event of an emergency. Therefore, the internal access control system utilised should incorporate: control of the lift(s); electronic release of dedicated external doorsets; access control box (ACB) switch and standards; installation of a secure information box (SIB), or premises information box (PIB), its standard and location.

21: Locking Mechanisms for Security Compartmented Doorsets
Details the security compartmentation strategy of controlling movement through stairwells and landing doorsets; fire strategy; doorset locking mechanism emergency escape override; unsuitability of multi-point locking or solenoid door latches/bolts released electronically; doors on escape routes; electrically powered locks; dual certificated fire/security doorsets and suitable fire doors.

22: Fire and Rescue Service Key Systems
Restricted key systems for access control box (ACB) or access doors, lifts, smoke ventilation, other types of emergency override mechanism), must be agreed with the appropriate fire and rescue service.

23: Emergency Door Release Devices
Describes how such devices can be abused. Guidance covers vandal-resistant stainless steel self-resetting emergency exit systems; break-glass emergency door release; restricted access to emergency egress route and requirement of Part Q of the Building Regulations (England and Wales).

24: Timber, PVC-U, Aluminium or Metal Windows With Single-Point Locking
SBD certification of the property indicates enhanced security windows. Further guidance relates to: good repair; Easily Accessible windows; beading; glazing types; Anti-Shatter Security Film and secondary glazing.

25: Timber, PVC-U, Aluminium or Metal Windows With Multi-Point Locking
SBD certification of the property indicates enhanced security windows. Additional advice relates to: DOCO confirmation; window replacement; multi-point locking; beading; glazing types and standards; security film; fixed/retractable grilles; secondary glazing and the definition of 'Easily Accessible'.

26: Emergency Egress Windows
All Easily Accessible windows should incorporate key-lockable hardware, unless designated as emergency egress routes within the Building Regulations. Further guidance relates: to designated fire escape routes; non-key locking hardware; Laminated Glass and non-Easily Accessible windows.

27: Boundary Security
As a rule, front boundaries should be open to view, hiding places eliminated and with good Natural Surveillance. Additional guidance covers: height, fences, hedges, shrubs and tree canopies.

28: Side and/or Rear Boundaries
These should be a minimum of 1.8 m overall and capable of raking/stepping to maintain the height over different terrain. Further advice covers: wooden trellis topping; gates and key-locking.

29: Apartment/Flat Boundaries
Provides guidance in respect of communal front doors in respect of apartments/flats (the lobby area treated as a semi-public space) and an access control system that meets the standard quoted.

30: Roof Construction
With roofs vulnerable to criminal intrusion and damage through vandalism, careful consideration must be given to their construction. The relevant standards are set out.

31: Roof Lights
Sets out the certification requirements for roof lights in both low and higher crime, risk applications.

32: Loft Hatches
Details security issues relating to loft hatches located in communal areas; necessity for locking to prevent access via the loft space and secured with products certificated to a standard listed.

33: Intruder and Hold-Up Alarm Systems
Extensive guidance covers Intruder and Hold-up Alarm Systems (I&HAS) to protect against burglary, breeching perimeters and trespass. Further advice covers Crime Prevention measures and standards including: NPCC Security Systems Policy; false alarm prevention; UKAS certification; Information Commissioner's Office (ICO) advice; compliance with BS EN 50,131 and the SBD Alarm Standard.

34: Video Surveillance Systems
CCTV can be cost-effective in preventing Crime and ASB. Advice covers: Video Surveillance Systems; Operational Requirement (OR); Data Protection Act (2018); image storage; ICO guidance for domestic CCTV systems; confidentiality; industry standard; access to images; management and maintenance.

35: Security Lighting
Good Security Lighting should be energy efficient and minimise light pollution. All external doors and the first 10 metres of footpath from the main entrance door, should be well lit for facial recognition. Further advice covers: 'dusk to dawn Lighting', LED Lighting and other energy effective light sources.

36: Secured by Design Sanctuary Scheme
Provides a brief overview of the SBD Sanctuary Scheme, which since its creation in 1998 has successfully helped thousands of victims of domestic abuse remain safely in their homes and maintain their social roots. Contact: specialisthousing.deliveryteam@homesengland.gov.uk.

5.11 SBD—Construction Site Security Guide

1. Introduction

Construction sites often suffer from crimes such as theft, criminal damage, arson and ASB, which can impact on completion dates and development costs. This guide is risk commensurate with advice designed to secure the site throughout development, it applies to all construction sites and all their staff.

2. Secured by Design

This document describes Secured by Design (SBD), and how it encourages and guides specifiers, designers and developers of building and refurbishment work, to adopt Crime Prevention measures and use security products tested and third-party certificated by a UKAS body to Police-Preferred Standards of security. SBD is operated by Police Crime Prevention Initiatives, owned by the UK police service and supported by the Home Office. Further information can be obtained from: https://www.securedbydesign.com

3. Perimeter Security

Construction sites should be contained within secure perimeters with access-controlled entrances. It details the Construction (Design and Management) Regulations 2015 (CDM 2015). Further guidance relates to: perimeter temporary boarded fence ('hoarding'); Crime Risk; safe anti-climb measures; consultation with the local authority planning department; avoiding recessed areas and climbing aids; signage (including infographic warning signs); maintenance programmes to remove graffiti; hate crime-related graffiti, advantages and disadvantages of viewing windows in the hoarding; temporary rigid mesh fencing panels with anti-tamper couplers and temporary fencing support feet.

4. Site Access

Site entrances should be kept to a minimum so as to control entry. Further advice covers: separate entrances for vehicles and pedestrians; clear demarcation of roadway and footpath; safety barriers; Health and Safety legislation to prevent casual entry; recording everyone on-site; turnstiles; identity badges or hard hat stickers; staff vigilance and a challenge culture; visitor escort; staffed gatehouse; well-defined roles/responsibilities for security staff; site operating hours; security awareness within induction training; permanent security presence on larger sites and a CCTV system with audio capability.

5. Site Offices

Site offices tend to be temporary structures or portable cabins that are difficult to secure. Guidance includes: inadvisability of storing valuable items or equipment therein; any Intruder Alarm system should comply with National Police Chiefs Council's Security Systems Policy; alarm-linked use of covert traceable liquids (taggants); security fogging devices; good quality external doors and locks; designated fire escape doors; windows, glazing and security film; shutters and grilles; key security, storage, audit and reporting; cybercrime and further advice regarding the procurement of security products tested and certificated to the Police-Preferred Standards, detailed at: https://www.securedbydesign.com

6. Lighting

Describes how a good Lighting scheme will deter intruders, make the area feel safer at night and assist security patrols. Additional advice relates to: illuminating the entire external perimeter; energy efficient, photo-electric cell (dusk until dawn) Lighting; how it should complement any CCTV system and that exposed cables should be enclosed in tamperproof conduit.

7. CCTV

CCTV with appropriate signage can deter intruders. It references the Surveillance Camera Commissioner's Guidelines and Code of Practice; how CCTV images should be of recognition quality and the Home Office Operational Requirements Manual for CCTV and web addresses. Further advice relates to coverage of vulnerable areas; 31-day recording and storage capability; availability of a member of staff who knows how to operate the system; vandal-resistant housings; portable CCTV systems and Automatic Number Plate Reader (ANPR) enhancement to the CCTV system.

8. Vehicle Protection

Advises against: leaving a vehicle with the engine running; or valuables, tools and equipment left on display. Further guidance covers: safe parking of vehicles; catalytic converters; folding or removable locking posts; temporary CCTV cameras; Internet of Things (IoT) devices; steering wheel locks/immobilisers; recording vehicle identification numbers (VINs), chassis and engine numbers; locking systems; faraday bags; window marking or etching; Property Marking of tools; database-stored photographs to help identify and restore stolen property and after-market vehicle tracking devices.

9. Plant, Machinery and Tools
Details: safe storage of vehicles, plant, machinery and tools; ignition key/immobiliser security; recording and registration of VIN, chassis, engine and serial numbers; tamper-proof plates, labels, hidden transponders/electronic tags, covert traceable liquids; and as a visual deterrent (Nagin, Cullen and Lero Jonson, 2019) visibly marking items to deter theft and disposal. Also recommended: smart tool technology; tool safes/high security storage sheds; high value moveable machinery in a dedicated 3 m high, secure compound; high value fixed machinery; fencing; anti-climb toppings; gates; locking procedures; checking vehicles as they leave; fuel tank alarms and padlocks and chains on gates, tested to one of the Police Preferred Specifications listed.

10. Urban Explorers
Describes an organised group who trespass into high-rise buildings, sewers, tunnels, cranes and building sites to take photos/videos posted on social media. Such trespass is highly dangerous and security personnel should be vigilant and encouraged to challenge and report suspicious behaviour.

11. Fire Prevention
Details: the necessity to keep emergency egress points unobstructed; use of a premises information box (PIB) meeting the standards quoted; flammable gas tanks and the Health and Safety Executive's Fire Safety in Construction guidance document (HSG168) https://www.hse.gov.uk/pubns/priced/hsg168.pdf

12. Additional Advice
Covers on-site security personnel https://www.cpni.gov.uk/optimising-people-security; near completion security; encouraging staff/contractors to report crime and suspicious behaviour; Crimestoppers, https://www.crimestoppers-uk.org and joining the local Business Crime Reduction Partnership (BCRP).

13. Stay Safe Advice
Action Counters Terrorism (ACT) encourages communities to help the police tackle terrorism and save lives by reporting suspicious behaviour and activity. For further details access the websites https://www.counterterrorism.police.uk/what-you-can-do. The advice is to 'Run, Hide and Tell', guidance that can be found at: https://www.counterterrorism.police.uk/what-you-can-do

5.12 Secured Environments—Risk Management Award

'Secured Environments' is a police certification scheme. It is awarded to organisations that are able to show that they have adopted six key principles for protecting themselves against crime.

Upon registering organisations are guided through the process—we offer help and advice throughout—once you have met the six key principles, we can award you with the police approved 3-year accreditation of 'Secured Environments'.

The aim is to help organisations to think about their risks and how to make the best use of measures they already have. It is the type of scheme every organisation can and should sign up to.

Secured Environments
in association with
Secured by Design
Risk Management Award

Secured Environments
Information Pack

www.securedenvironments.com

I think the security industry is still not highly regulated, only recently have they put in measures like performance standards. Secured Environments is [illegible] real heavyweight in terms of a security world, others have not got the academic rigour and background.

Head of Estates, University of Bristol

Secured Environments—Risk Management Award

Principle 1

Principle 1—Commitment
A Secured Environments organisation and its leadership are committed to a secure environment.

Working in insecure environments has serious consequences for employees and the organisation itself. It can be the cause of low staff morale, and high staff absence and turnover. For staff, it can result in physical injury or psychological distress. This can represent a real financial cost to the organisation, as well as affect the confidence of a business, its profitability and even its viability. Costs can also arise from expensive insurance premiums and compensation payments. These are in addition to costs that can arise directly from crimes that are preventable, such as burglary or vandalism or anti-social behaviour. Therefore, it is essential for organisations to strive to create a secure working environment.

This principle asks you to demonstrate how committed the organisation is to creating a secure environment. To achieve this, you need to be able to show that high level commitment exists. Managers need to show that they have taken steps to help to create a safe and secure working environment and staff need to believe that the organisation is committed to creating a secure environment.

Principle 2—Understand

A Secured Environments organisation understands the security risks it faces.

In order to ensure that an organisation is secure, the risks it is exposed to need to be understood. To demonstrate this, senior manager's need to be able to describe the risks posed to the organisation and explain how the risks and incidents are recorded, monitored and analysed. Senior managers should also be able to explain how the information collected is used to create a secure working environment.

Managers need to be able to explain the risks they face, and how incidents are recorded and monitored.

Employees should be able to explain what they need to do to report an incident and provide examples of when they have done this.

Principle 3—Respond

A Secured Environments organisation develops an effective response to reduce the risks to the organisation.

Crime prevention measures are only effective if they are targeted correctly and are based on evidence; here organisations need to demonstrate that they have done this. There should be crime prevention and security measures in place that are appropriate to the organisation based on an analysis of the risks. The designated person should be able to explain why the measures have been chosen and how they address the problems experienced.

Employees should be able to explain their role in creating and maintaining a safe and secure environment. All staff should also be able to provide examples of ways in which the organisation has improved in its response to and management of security.

Principle 4—Management

In a Secured Environments organisation security is managed.

In order to create a secure environment, the organisation needs to have a plan in place to achieve its aims and objectives. All security measures should be proactively managed, this means that security measures should be planned.

Security measures can be expensive, and it is important to track these costs to enable effective budgeting. You need to show that you know what resources (e.g. people, time, money and technology) are used within the organisation to create a secure environment and that this is an efficient use of available resources.

Principle 5—Implement

A Secured Environments organisation effectively implements security.

Many crime prevention measures fail due to poor implementation, you need to demonstrate that you have appropriate processes and procedures in place to ensure that crime prevention initiatives are implemented correctly and provide examples of this. New security measures need to be introduced using project management principles.

The nominated lead from the senior management team needs to have a good understanding of the measures that are being taken to address the organisation's crime and disorder problems. The organisation should also be working with other relevant agencies, for example, the police, neighbours, etc. to improve security and reduce crime and anti-social behaviour.

The role of the employee is fundamental to ensuring a secure environment. All employees need to be able to explain their role in achieving and maintaining a secure environment.

Principle 6—Evaluate

In a Secured Environments organisation, all crime prevention initiatives are monitored, evaluated and fed back to the organisation's leadership.

It is essential that crime prevention initiatives and security measures are assessed to determine whether they are effective. This may occur in a number of ways, for example, through testing existing security measures and processes, auditing of systems and evaluation. Your organisation may use one or all of these methods depending on the size of the company and its security risks. But whatever method you use you need to be able to explain how security measures and crime prevention methods are monitored and evaluated and how the findings are fed back into the strategy. You need to be able to explain how a crime prevention initiative or security measure has been improved based on the findings from an evaluation.

Register

Registering is achieved by emailing securedenvironments@perpetuityresearch.com or by phoning the dedicated team on 01892 538690.

Applicants have 12 months from the date registration in which to apply for a Secured Environments audit, if you do not apply within this time you will have to pay to re-register.

Achieving Secured Environments status does not guarantee you will not be a victim of crime or disorder, nobody can guarantee that. The risk of crime changes over time and therefore responses need to be flexible and updated—following the Secured Environments principles will help with that.

Being accredited with Secured Environments does not exclude you from your legal obligations regarding safety and security at work, where a variety of legal requirements mean that you need relevant policies and procedures in place; the accreditation does not replace those requirements.

The Audit

Once registered to become a Secured Environment, a username and password will provide access to the secure area of the website. This contains:

- Full details of the six principles and the criteria required to receive accreditation
- Case study examples of organisations that have already been through the process
- Websites providing a range of useful information on how to respond to crime and disorder and security issues
- Reasonable support via email and telephone from the team of dedicated advisors to answer any queries as well as offer support to help meet the criteria
- Full details of the audit process and requirements
- An example of the kind of feedback to expect from an audit.

Before the Audit

Documentation required:

- Records of incidents e.g. trespass, bullying, criminal damage etc. or security reports
- Security surveys undertaken
- Other potential security records e.g. insurance claims, intruder alarm activations, etc

- Agendas and minutes from meetings that relate to safety and security on the premises
- Security policies, or guidelines already in place (e.g. anti-bullying policy).

Depending on the size of the organisation an online staff survey may also be undertaken.

The assessor will ask for a full staff list from which they will select a cross section to interview (either one to one or in focus groups) on the day of the audit. This will include:

- The nominated person on the senior management team with responsibility for security
- The designated person with responsibility for security
- Security staff
- Staff from across the organisation.

The Day of the Audit

The assessor will visit the site and interview selected staff in order to see how the principles are met. Further guidance relates to staff availability on the day of the audit and the avoidance of extra costs by having to complete the audit at a later date.

The time taken to carry out the audit will depend on the number of people in the organisation, whilst causing the least disruption.

After the Audit

Following the audit a feedback report will be completed. This will set out performance against each of the set criteria and identify areas for improvement. A successful audit will result in a certificate being awarded; use the Secured Environments logo; and capacity to purchase a plaque for the building.

Case Studies

Four case studies are detailed:

A London Landmark
A Shopping Centre
A University
A Hospital

Requirements: Crime and Disorder Statement

All organisations that register to become a Secured Environment will need to complete a Crime and Disorder Statement. Its purpose is to demonstrate an understanding of crime in the local area and how this impacts on site security. The standard format is described below. Alternative formats may be considered for multi-site organisations and for those where the local Police Designing Out Crime Officer (DOCO) is unable to assist.

The statement will include information on the local area, the site itself and situational security recommendations from a local police expert. The organisation will then have to create an action plan to respond to the recommendations.

Once registered for the accreditation, organisations will be given contact details for their local Police DOCO who will be able to assist the organisation to complete its Crime and Disorder Statement. Depending on the contribution of the DOCO, a site survey may be conducted and/or the DOCO may read/amend/contribute to the statement drafted by the organisation. The Statement consists of four parts:

Local Context

Provides relevant information relating to the surrounding area, premises, estates, offenders, and victims and will paint a picture to inform crime risk for a Secured Environments applicant. This will be completed using publicly accessible data sources or making a request to a police force under the Freedom of Information Act.

Site

This section will be an appraisal of reported, recorded crime and disorder relating directly to the site including statistical data. This will be completed using company records or publicly accessible data sources, or making a request to a police force under the Freedom of Information Act.

Situational Security Recommendations

This section will be completed by the DOCO (or the applicant based on DOCO feedback) and will provide the organisation with appropriate, realistic and cost commensurate situational security recommendations informed by the local context, site appraisal and site survey. It is important to note it is recognised that management of security and the presence of and adherence to appropriate processes and procedures can reduce the need for overt physical security measures in some cases. It will be for the applicant to justify how it mitigates against the DOCO recommended physical security standards, in the action plan that is developed.

Action Plan

In response to the recommendations, organisations will be required to create an action plan—to include people, process and situational improvement. It should include a timeline for action, and current progress against each action e.g. 'completed' or 'working towards'.

Testimonials, FAQs and Member Login

Three other tabs provide links to Testimonials, FAQs and Member Login.

5.13 SBD—Youth Shelters and Sports Systems

Youth Shelters and Sports Systems—A good practice guide

The Secured by Design website reproduces the Development Guide for Youth Shelters and Sports Systems (2002) produced by Roger Hampshire and Mark Wilkinson of and for Thames Valley Police.

The Problems With Youths

In essence, this good practice guide is designed to target young people in the all-important secondary school age group of 12 and over, but before adulthood—important because they have outgrown children's play facilities yet are often woefully under-provided in terms of cost-free venues where they can meet and socialise. Instead, they tend to congregate on park benches, within bus shelters, outside shops and in a host of other locations where their number and supposed behaviour is perceived as intimidatory by adults and younger children. Signs demanding 'No Ball Games' are sometimes installed and calls to the police multiply in number, with the youths in question being moved on—only to return a few minutes later. Moreover, young people resent being seen as the problem, especially because they often consider themselves as the victims.

Nowhere to Go

Young people often gather in the play area in which they grew up, except they are now prone to pay little respect for the equipment which a distinct minority all too frequently vandalise. Simultaneously, a dearth of recreational facilities for adolescents compounds the problem. When asked, young people have proved to be 'surprisingly modest' in their requests: just somewhere dry where they can 'hang out' (meet) their friends. Consequently, the overall desire should be for them to socialise and thereby become integrated into their communities.

The Solution

This guide observes how adults tend to assume that young people require youth clubs or other structured (and necessarily expensive) activities. Yet the responses provided by adolescents contradict such assumptions, simply requesting:

- 'Somewhere to go that is safe, where we won't get hassled'
- 'A place of our own away from adults'
- 'Somewhere to kick a ball about, try out skating or cycling'
- 'Just a place to meet friends'.

What Is a Youth Shelter?

A Youth Shelter is a structure designed to meet the social needs of young people—a sheltered place to sit and talk.

Sports equipment may be included, together with a litter bin to keep the area tidy. A roof provides shelter from the weather, but most Youth Shelters are open on all sides in order to deliver all round visibility (Natural Surveillance) and safety. The guide comments that the most successful shelters are linked to a sporting facility, with the physical and social activities complementing each other.

There is a near universal consensus that the correct location for both Youth Shelters and sports areas has proved to be the most difficult issue. Whilst local residents fear the noise and behaviour they may generate, local authorities and young people are concerned about safety issues. Therefore, the often-difficult aim should be to maximise opportunities for Natural Surveillance in pursuit of both safety and as a deterrent to vandalism, without excessive annoyance to neighbours.

Possible Benefits of a Youth Shelter Project

For young people:

- A place to go and meet that is non-confrontational
- Facilitates young people's development through learning about independence and socialising
- Develops involvement and pride in their neighbourhood
- Is a constructive rather than a destructive experience
- Reduces boredom when associated with a sports facility
- Removes a possible first step towards crime
- Team games provide a physical outlet for emotions and energy
- Improves the employability of the project team
- All round visibility allows users to feel safe.

For the community:

- Reduced damage, graffiti, crime and fear of crime
- Enhanced community harmony
- Any children's play area remains undamaged and usable
- Reduced maintenance costs.

But there are potential problems if the Youth Shelter is poorly designed or located:

- Gangs might dominate the shelter
- Availability of drugs and under-age drinking and drug taking
- Conflicts between groups about use/abuse
- Litter and maintenance costs—who pays? Some parish councils may be reluctant to commit to such projects
- Noise and annoyance to nearest residents
- Graffiti, the possibility of obscenities

- The next generation will need to be given ownership
- Not easy to link to other education or youth service activities
- Solid sides reduce visibility and may increase the risk of bullying and intimidation.

Incidentally, retired Crime Risk Manager and Architectural Liaison Officer (DOCO) with West Mercia Police, the late Jenny Lawrence MBE, described (personal witness) how when she went on to work with Malvern Hills District Council, Youth Shelters tended to have a limited life expectancy, with the next generation of young people being unwilling to use such a facility provided for their predecessors.

The Thames Valley Police authors also observe that Youth Shelters should not be expected to solve broader issues—such as drug taking. A multi-faceted approach will be needed in such circumstances. And perhaps most importantly, Youth Shelters should not be perceived to encourage problems, as alternative locations to gather can always be found.

Consultation

A constant theme throughout this handbook has been the necessity for consultation with the target audience. Youth Shelters and young people are no different in this respect, although the guide describes how youth involvement is often very difficult at the start. A series of measures are suggested to overcome this and if possible, the following school-based activities:

- Design competition
- English language class to devise a questionnaire and write letters to potential sponsors
- Maths students could evaluate data
- Debating society might help run a 'planning for real' exercise
- Geography students could survey potential sites
- Technical drawing students might produce scale drawings of the best designs
- Work experience placements to departments in the local authority where the Youth Shelters idea requires processing
- Consider involving young people on local authority sub-committees.

Lighting

During the consultation process, 75% of the young people registered concern regarding the safety of the potential location as a meeting place. Lighting is an obvious means of reducing such concerns, although the guide suggests the final decision should rest with the young people who will use the Youth Shelter. A number of compromise solutions are recommended, including the illumination of just one side, the main path to the shelter, time-controlled Lighting and 'borrowed' light from an adjacent street or other facility.

Case Studies

The good practice guide continues with a section devoted to examples from across England and The Netherlands and contains a wealth of evidence regarding the best elements in respect of Youth Shelters in: Banbury, Marcham and Witney in Oxfordshire; Aylesbury and Milton Keynes in Buckinghamshire; Bradford, West Yorkshire; Burnley, Lancashire; Thurrock, Essex and The Netherlands.

Sports Facilities Issues to be Considered

This penultimate section includes reference to 'Play Walls' (especially useful as an alternative to the gable end wall of a house having a ball repeatedly kicked against it). Legislation that may be relevant in respect of sports facilities, includes: the National Planning Policy Framework (NPPF); Crime and Disorder Act 1998; the National Playing Fields Association (NPFA) Six Acre Standard; the Local Government Rating Act 1997 and the British Standard for outdoor recreational multi-sports equipment and Youth Shelters, Product Assessment Specification 30: 1998 (PAS 30).

The guide concludes with a 'Checklist' and 'Evaluation Indicators Menu' and includes potential sources of funding, common questions, companies and suggested reading list.

Finally, as Professor Henry Shaftoe states in the foreword, 'Youth Shelters are not the whole solution—young people may also need more structured environments such as youth clubs, organised activities, drop-in cafes, advice and counselling services—but shelters and sports systems can make up an important part of the youth development jigsaw. Failure to invest in facilities that young people want is a false economy'. Indeed, a repeated theme amongst the good practice examples is how incidents of ASB and calls for police attention reduced dramatically once a Youth Shelter had been installed.

5.14 Security Overlay to the RIBA Plan of Work

In 2010 the Royal Institute of British Architects (RIBA) produced its first guidance document on designing for counter terrorism. This has now evolved and in conjunction with the National Protective Security Authority and Police CPI/Secured by Design, published as the *Security Overlay to the RIBA Plan of Work* (2023) accessible via https://www.gov.uk/government/publications.

Security Overlay to the RIBA Plan of Work

This guidance is contained in eight chapters. The Introduction explains how the document is for everyone involved in safe and secure design, construction and operation of any building, including: Architects, Clients, Engineers, Contractors,

Planning and building control professionals, Specialist security advisers, Technology and data system suppliers, Building operators and managers and Stakeholders with an interest in building and infrastructure security matters.

Chapter 1. is devoted to 'Security Over a Building's Lifecycle'—including Security Risk Assessment, Security Requirements, Security Strategy and Security Plan.

Chapter 2. 'Security Stakeholders and Project Team Managers'—covers the role of *DOCOs*, Professional Security Officers, NaTCSO and Counter Terrorism Security Advisers (CTSAs), the National Protective Security Authority (NPSA) and the National Cyber Security Centre (NCSC).

Chapter 3. 'Understanding the Threats'—is devoted to the threats posed to any Physical or Digital Asset and the Assets contained within it—including people and property.

Chapter 4. 'Security Risk Assessment'—explains how the best way to determine the security risks associated with a project on a specific site, is to conduct such an assessment: RIBA Stage 0—Client Requirements confirmed.

Chapter 5. 'Security Requirements'—sets out how these are the client's requirements and how ideally these will be considered at the outset of the project: RIBA Stage 1—Project Brief.

Chapter 6. 'Security Strategy'—captures the design team's response to the security requirements in relation to the topics of crime, National Security Threats such as terrorism, espionage, hostile foreign states, Cyber Security and Information Management. RIBA Stage 2—Architectural Concept approved by the client and aligned to the Project Brief; RIBA Stage 3—Architectural and Engineering Information Spatially Coordinated; RIBA Stage 4—All design information required to manufacture and construct the project completed.

Chapter 7. 'Security Plan'—should set out how the security measures included in the Security Strategy will be managed and maintained. RIBA Stage 6—Building handed over, aftercare initiated and Building Contract concluded. RIBA Stage 7—Building used, operated and maintained efficiently.

Chapter 8. 'References and Guidance.

5.15 Secure Stations Scheme

1. Background

The Secure Stations Scheme allows railway station operators to demonstrate how they are working with partners to reduce crime and safeguard vulnerable people at stations. Accreditation provides reassurance to both passengers and staff that the station is a safe and secure environment.

This national scheme covers all rail networks policed by British Transport Police (BTP). It establishes standards of good practice and accredits individual stations which have worked with BTP and other local partners to improve safety and security.

It recognises that security and safety can be improved though management practices and emphasising the importance of well-trained frontline staff. Security and safety can also be improved through physical design measures, making improvements where possible and managing remaining risks. The design of a new station or major refurbishment provides an opportunity to incorporate best practice into the station design and advises BTP are consulted on such plans. This will increase the likelihood of the station successfully gaining Secure Station accreditation, once completed.

There are four accreditation criteria:

1. Crime statistics for the 12 months prior to the assessment compared to the annual passenger footfall of that station
2. A survey of users, taking account of passenger perceptions and the operator response
3. The operator's management practices must promote and conduct Crime Reduction and Prevention measures, ensure appropriate responses to incidents, the safeguarding of vulnerable passengers and effective communication with passengers
4. The design of the station must conform to standards which prevent and reduce crime and facilitate the safeguarding of vulnerable people, as assessed by the BTP assessor—a DOCO.

Each criterion is assessed by the BTP DOCO. All four criteria must be passed to be accredited.

When the operator and the DOCO agree that the station may meet the accreditation criteria, the DOCO will formally accredit the station. The Department for Transport (DfT) will confirm the successful accreditation and forward the scheme award, in the form of a certificate signed by the Chief Constable of BTP. The certificate should be displayed at the station and advertised in publicity materials and station signage. The DfT will also notify the station's local MP (Member of Parliament) of the accreditation.

The certificate is valid for two years from the date of issue. It can be withdrawn during this time if there is a clear breach of the standard, or a substantial change takes place at the station (e.g. physical redevelopment). After two years the operator may apply for a new certificate, via the same procedure.

2. Station Management

Safety and security risk measures at stations can be managed through monitoring the type and scale of incidents and adopting a proportionate response with appropriately targeted resources. Management of risk should be integral to the management of the station as a whole and as such, should be considered in every aspect of operation.

The station operator should publish a statement of intent, confirming their commitment to setting, monitoring and publicising standards of personal security that individual passengers can reasonably expect and monitoring passenger satisfaction. This should be exhibited prominently in the station.

Further sub-headings relate to safety; suicide prevention and safeguarding; staff deployment and training; staff security and securing passenger property.

3. Station Design and Facilities

In addition to consulting the DOCO regarding new station design/redesign, plans should also incorporate the 'Design Standards for Accessible Railway Stations'. Sub-sections are devoted to: visibility and layout; passenger information and signage; and surveillance.

4. Recording, Monitoring and Responding to Incidents

Recording and monitoring of all types of incidents is essential in ensuring that resources are directed in the best way possible. Passengers and staff must be encouraged to report all types of incidents in order to build up a full picture of crime at the station. This should include instances of crime (e.g. vandalism); non-criminal incidents, e.g. drunken behaviour, and the presence at the station of vulnerable individuals. Crimes and the presence of vulnerable individuals should be reported immediately to the BTP.

5. Passenger Perceptions

Passenger perceptions of the safety and security of a station are an important part of the accreditation process. It is crucial for the credibility of the scheme that stations given accreditation are perceived by passengers to be secure. In order to facilitate the process of gauging passenger perceptions, the scheme will make use of existing data obtained by the National Rail Passenger Survey (NRPS).

6. End to End Journeys

The journey to or from the stop or station often provokes more anxiety than the time spent waiting. The passenger may feel quite secure whilst on the train and on arrival at the station, but may find the walk home or wait at the bus stop a threatening experience. Operators should work with the wider community to improve perceptions of the station and surrounding area.

7. Working Towards Accreditation Award

Stations that could be considered for the 'Working Towards Accreditation' category are those which fail at the outset because of their high number of crimes, or those which have a low enough crime rate to pass the threshold, but fail full accreditation due to the scoring of the management practices and station design audit.

Once a station has failed, the operator and local BTP DOCO should meet with a view to developing an Action Plan. This Action Plan will form the basis for the 'Working Towards Accreditation' award. The Action Plan should be SMART: Specific, Measurable, Achievable, Realistic and Time limited. All Action Plans will be assured by the British Transport Police Independent Accreditor, in order to ensure that a consistent standard is being applied nationally.

Progress towards achieving full accreditation will be monitored by a Secure Stations Scheme Steering Group. As with full accreditation, the Steering Group may withdraw the Working Towards Accreditation award in the case of a serious breach. Unless there are exceptional extenuating circumstances, an operator failing to deliver the Action Plan for the station within the specified time period, will have the award withdrawn. It will then be for the station operator to reapply to the Steering Group for 'Working Towards Accreditation' status.

8. Tackling incidents at stations

Details measures that have previously assisted in tackling crime and other incidents. Sub-headings cover: key principles; partnership working; data collection, analysis and specific initiatives. Further guidance is provided by BTP.

9. Appeal Process

An appeals process is detailed for when a station operator disagrees with the DOCO as to why a station has been failed.

5.16 Protect UK

NaCTSO, CPNI and partners provide counter terrorism provide terrorism security and emergency preparedness advice to businesses and the public. Multi-layering the different measures will provide the best mitigation from an attack. This interactive website contains links from the below six headings to a series of inciteful and highly informative guidance documents.

Risk
There are different approaches to managing risk that can be taken by following the recognised Risk.

Management Process. This section provides you with advice and guidance on these approaches and how they can be used to make your business more secure.

Security
Once you understand the risks that your business, staff and the public might face, you will need guidance on security measures that you can apply. This section provides information on the physical security measures you can install and the active measures that members of staff can adopt.

Response
Should the worst happen and a terrorist incident take place, you will need to be prepared to respond. This section provides support on developing both generic response plans and more detailed guidance for attack types.

Recovery
The long-term impact of terrorism on businesses can be significant. Without preparation your ability to recover will be impacted. This section will support you with that planning.

Awareness
Increasing awareness reduces the likelihood of a successful terrorist attack and improves your ability to respond appropriately. This section covers both individual awareness and building a security culture within your organisation.

Legislation
Changes in legislation can affect businesses and the approaches they must undertake to increase their security.

The contents of this section link to security related legislation. It will not cover all legislation which may impact on you or your organisation.

The website asks that those with an interest and/or responsibility for these issues sign up online to Protect UK, in order to begin their journey to protecting their community and ensuring preparedness.

5.17 Public Cycle Parking

1. Introduction

This especially detailed guidance document produced in partnership with the Bicycle Association and Sustrans, explains how it aims to identify products and installations that are:

- Easy to use (including consideration of disabled people and inclusion generally)
- Safe (for user and their cycle)
- Secure (enables secure locking)
- Long lasting (corrosion resistant etc.)
- Fully compliant with UK legal requirements.

The guidance is intended to be widely used at railway stations, other public transport interchanges, hospitals, educational facilities and other UK public buildings. Cycle parking on public highways is excluded. It references a number of general guidance on cycle parking documents that apply in England, Wales Scotland and at railway stations.

There are three categories of requirements: Guidance; Absolute Requirements—that must always be followed and Context Specific Requirements—in response to local circumstances.

2. Public Cycle Parking Requirements

Acceptable designs include: Sheffield stands (horizontal 'hoops' and all variants); ground anchors; two-tier racks with lower and upper tiers of cycle parking and lockers. Guidance: Design Criteria from LTN (Local Transport Note) stipulates cycle parking should be: safe, direct, coherent attractive and comfortable. Additional advice covers: location; site characteristics; new/improved cycle parking designs; early stage DOCO consultation; access routes; lighting; controlled access; minimum aisle widths/spacing; ceiling heights; type of stand and security classification; quality of manufacture; cycle locker design; installation; sign-off; provision; management and abandoned bikes.

3. Design Principles

Cycle parking and access should be safe, direct, comfortable, coherent and attractive. Additional advice covers: inclusive and accessible design; early engagement; definition of 'cycle traffic'; approaches to the parking area; early stage stakeholder engagement with statutory consultees and others, e.g. DOCOs, and those falling under the heading of railway projects.

4. Location and Situation

Advises how cycle parking should be close to the desired destination; the location easily seen; identifiable as cycle parking; under passive surveillance; sloping locations avoided; Lighting (BS 5489–1:2020 or brighter) of highway and public realm; secure position and visibility; opaque materials for walls, cycle shelters and compounds; height of vegetation; painting of roofs and walls; finishes and CCTV system recording, storage, recognition quality evidence and cameras within vandal-resistant housing.

5. Layout of Site

The facility must allow easy manoeuvrability for different sizes and shapes of cycles. Cycle parking should be well-signed and easy to find within buildings and externally. In large facilities, colour coding or numbering of cycle parking bays can help users to quickly find their parked cycle. Critical minimum dimensions are set out together with sub-sections devoted to: cycle shapes; cycle envelope; aisle widths; spacing of Sheffield stands; space in front of two-tier stands/around cycle lockers; clear signs and markings; specialist cycle parking and considerations for e-bike parking.

6. Equipment Design and Installation

Details: acceptable designs for open-access cycle parking; minimum service life of ten years; easy to use without substantial lifting; three-point locking and SBD Level 2 testing or Level 1 within a secured area; user information about good quality locks and cycle security; aesthetics, longevity, vandalism and maintenance; procurement; public parking; three-point locking; engagement with DOCOs, TfL cycle security experts (in London) and other similarly qualified LA staff; stainless steel stands; ground fixings; tamper-proof fixings; ground anchors; material shape, diameter, thickness and fixings; fob style entry systems; help points and CCTV; hub membership and management; cleaning; servicing;

promoting a sense of safety and security; removal of abandoned cycles; ease of use; Health and Safety Executive (HSE) advice re maximum loads for lifting; the standard for two-tier racks; planning and building regulations; changes of use; modification of existing buildings; heritage features or Listed Building status; Operational Requirements; counter terrorism considerations; BS EN1090 regulations re structural steelwork; cycle shelters and installation contractors.

7. Security Features and Testing

Layers of security assist Crime Prevention by introducing additional measures that increase the time and effort required to steal a cycle. All public cycle parking must meet SBD Level 2 at the point of secure access (to a shelter, compound or hub) or individual stand/locker where the cycle parking area has open public access. SBD is suggested as a benchmark way to demonstrate cycle parking security and further guidance relates to: SBD accreditation scheme; DOCOs' advice; testing and certification regimes; cycle security for the different categories of rail stations; SBD security levels—testing specifications; three-point locking; awarding an SBD rating; 'SBD Specifications for Security Certification Security Level 1 Products': design of secure access compounds (including cycle stores); doorsets and their locking systems; lockable cycle shelters and lockers with registered keyholders; cycle parking that is fully accessible to the public (including short and long-stay locations); cycle stands and racks; lockers and the testing of alternative designs.

8. Quantity of Cycle Parking

The minimum quantity of cycle parking at different types of location is published in LTN 1–20 and must meet typical current peak demand, accommodate future growth and ensure the facility is accessible to all. A proportion of the cycle parking should be accessible to larger types of cycle with a minimum of 1 accessible space provided and a suggestion that 5% of parking capacity is accessible to all. See Table 8.1 in the guidance: Cycle Parking Quantities for New Development.

Additional guidance relates to cycle parking; assessing demand via the Cycle Rail Toolkit; DfT Cycling to Stations Report; National Travel Survey; Cycle Rail Working Group, 2016; Local Cycling and Walking Infrastructure Plans Guidance; The Propensity to Cycle Tool; Planned Cycle Parking Improvements; Planned Network Improvements; Monitoring Cycle Parking Occupancy and Monitoring Informal Cycle Parking Activity.

Appendices

Includes: Example Procedure for Removal of Abandoned Cycles; Certification of Facility as meeting Cycle Parking Standard; Table B1: Standard for Cycle Parking Locations; Table B2: Tests for Equipment and Table B3: Standard for Cycle Parking Equipment Supply and Installation.

5.18 FIA and National Fire Chiefs' Council Code of Practice

5.18.1 Code of Practice for the Provision of Premises Information Boxes in Residential Buildings

This 40-page long Code of Practice (COP) was produced following publication of new regulations, legislation and the recommendations of the Grenfell Tower Inquiry. It relates to recommendations for the provision of Premises Information Boxes (PIBs) and Emergency Response Packs (ERPs) in high-rise residential buildings.

Key headings are: Scope (including the types of high-rise residential premises to which the COP applies); Location and Security of the PIB; Key and Key-pad Security; Signage; Emergency Response Pack (ERP); Maintenance of the PIB and ERP; Exchange of Information; and Definition of Responsibilities.

For UK-based DOCOs, mere mention of the Grenfell Tower Inquiry (following a tragedy in which 72 people lost their lives) explains the relevance of PIBs.

5.19 Safer Parking 'Park Mark' Award

Background and Introduction

In 2004, Secured by Design in partnership with the British Parking Association, created the 'Park Mark' Safer Parking Award scheme. This replaced the original Secured Car Parks award system that operated for more than a decade from 1992. Secured Car Parks had been designed to improve security standards and thereby reduce crime and the fear of crime (FOC) in all car parks and vehicle retention areas. However, by the turn of the century it was apparent that this initiative had reached a point of stagnation—with few new car parks being added to the total number each year. And (for example), a desire for 2,000 Secured Car Parks car parks to have achieved the award by the year 2000 had not been realised.

The Secured by Design website provides a full description of the Safer Parking Award and how car parks that reach the scheme's standards are bestowed with the 'Park Mark' Safer Parking Award (PMSPA).

In essence, this initiative is primarily concerned with the management of criminal behaviour within a car park environment. Consequently, car park owners and operators are expected to minimise crime by means of an active management strategy. And whilst the Safer Parking Scheme is owned by the police services of the UK, it is managed by the British Parking Association via its Development Managers (DMs).

The purpose of the Safer Parking Scheme (SPS) is to:

- Reduce crime and FOC within parking facilities
- Provide guidance to owners, operators and developers of parking facilities, both new and existing, on how to establish and maintain a safe and secure environment through the introduction of proven management processes, physical measures and site security systems, having considered crime and ASB in the immediate location
- Raise awareness to the general public when parking their car, motorcycle, bicycle etc. that the owner/operator has considered and where appropriate taken action, to reduce crime and FOC in the parking facility that they have chosen to use
- Provide a design framework for architects and developers of new parking facilities.

Responsibilities

Police: each police force will have a principal designated Force SPS trained member of staff, the Accredited Assessor, whose role is to maintain consistency in respect of assessment and awards. Larger forces may have a number of Accredited Assessors, all of whom should be Designing Out Crime Officers (DOCOs). This is an especially important point as the role of the Accredited Assessor is to advise on all issues of risk management and car park security.

Development Managers (DMs): are independent persons appointed by the British Parking Association with responsibility for assisting the police Accredited Assessors, advising on management procedures and developing the scheme locally. The DM is also a trained Accredited Assessor.

The Park Mark Safer Parking Award (PMSPA)

Unlike Secured Car Parks, there is only one type of 'Park Mark' Safer Parking Award (PMSPA). The award is issued after an assessment of the Crime Risk within the parking facility. New-build and major refurbishment of an existing car park necessitate compliance with the full requirements of the Safer Parking Scheme initiative and owners/operators are advised to contact the DOCO or DM at the earliest possible opportunity—no later than the design stage.

Initial Award

The PMSPA is initially granted for one year. It does not guarantee a crime-free parking facility. Where car parks are only open for a minimum of 10 hours, clear Signage about this must be displayed. The guide also provides specific figures on the maximum permitted percentages of occupancy outside these times (dependent upon the total number of parking spaces).

Reassessment

The frequency of reassessment will generally be two years. However, local circumstances may dictate annual or extension up to three years. Where previously set at two or three years, the following circumstances may warrant an earlier reassessment:

- Significant increase in crime
- Change of owner/operator
- Material changes to the parking facilities layout or structure
- Change in Crime and Disorder at immediate location
- Significant change in customer profile
- Need to change the original Safer Parking Scheme time/day/season limitations.

The Application and Site Assessment Process

Assessment guidelines and application forms are available from Safer Parking Scheme managers at the British Parking Association. Applicants should then:

- Contact the DM or police Accredited Assessor
- Complete and submit the relevant paperwork
- The DM will contact the local police Accredited Assessor and arrange an agreed assessment date
- The application decision will be given on the day or within 3 working days of the assessment

- If the parking facility fails, a detailed explanation to be sent by DM to the owner/operator within 20 working days.

Costs

After achieving a PMSPA for the first time, the owner/operator becomes a '*Member*' of the Safer Parking Scheme and is required to pay an annual subscription fee. The guidance document lists the range of benefits Members receive.

Assessment and Arbitration

The guide describes how to be granted a PMSPA, a '*parking facility must achieve an effective combination of active management procedures and commensurate preventive security measures*'. Those able to demonstrate zero or very low crime levels '*...will normally be able to achieve an award unless there is a clear and obvious risk that requires attention*'.

Applications for adjudication should be received within one month of the failed assessment by the Safer Parking Scheme Standards Committee at Secured by Design. The final decision will be binding.

Guidance for Parking Facility Owner/Operators.

The following appendices, detailing the different types of parking facilities and necessary security measures, are available to assist everyone involved in the Safer Parking Scheme process:

- Appendix A—Multi-Storey Parking Facilities
- Appendix B—Surface Parking Facilities (Urban areas)
- Appendix C—Surface Parking Facilities (Rural areas)
- Appendix D—Lift-Operated Parking Facilities
- Appendix E—Rooftop Parking Facilities
- Appendix F—Basement/Underground Parking Facilities.

In 2010 new guidelines were devised and agreed in respect of Assessment Guidelines for Parking Facility Owner Operators, New Build Parking Facilities and Meet and Greet Parking Facilities.

Special Considerations

The guidance describes how dispensations may be granted where owners/ operators are unable to

meet the criteria within the respective appendix. For example: a Public Right of Way or other requirements under statutory legislation; where the parking facility is part of another building; or where landscaping (such as foliage) on adjoining land obscures Natural Surveillance.

Promoting the Scheme

Members of the PMSPA scheme are required to prominently display the Park Mark Safer Parking Award certificate, plaque and logo at their facility. Unfortunately, personal witness by this author when conducting SPA reassessments suggested that many car parks had no such signs on display. Furthermore, a sign denoting the accurate name of the car was often an even greater rarity! These issues were reported back to the Safer Parking Scheme Project Board and considerable improvement has subsequently taken place across the UK.

Key Issue

When the 'Park Mark' Safer Parking Award first replaced Secured Car Parks, a small number of CPDAs/ALOs (as they were then exclusively known in the UK) expressed concern and argued that a poorly designed (in terms of criminal opportunity) car park should not receive an award, purely because it is crime-free. This is a moot point, yet it goes to the heart of one of the core attributes for the Designing Out Crime Officer, namely the ability to compromise and negotiate (see Sect. 1.5.10). Because, whilst in the perfect world of pure science such compromise would be seen as inappropriate, for the DOCO it is an inherent necessity. We practitioners are required to make subjective assessments, taking into account all the available evidence. Ultimately, we can always rely on-site specifics as the reason not to confer any type of award. But this should be used very sparingly if our credibility amongst fellow professionals in the built environment and partner agencies is not to be compromised in the long term.

Furthermore, the Safer Parking Award (like Domestic Violence Sanctuary Rooms—see Sect. 6.1; Secured Environments—see Sect. 5.12. above and, indeed, any project where the DOCO demands a higher degree of security than the minimum standard set out in the relevant SBD Development Guide), is a Threat and Risk Assessment (TARA) exercise. This necessitates the DOCO using all their knowledge, training and accumulated experience to deliver the optimum solution.

Full information regarding the 'Park Mark' Safer Parking Award scheme can be accessed at https://www.britishparking.co.uk/initiatives/ParMark

5.20 SBD—Technical Guides

On the Secured by Design website electronic links will take you to the following technical guides are contained on this section of the website:

- **Flat Entrance Doorsets 2019**
- **SBD Alarm Standard 2.0**
- **GAI Specifiers Guide 2019**
- **PSTI 2022—Implications**

5.21 SBD—Application Forms

Also, on the Secured by Design website, electronic links provide direct connections to the following SBD Application Forms contained on this section of the website:

- SBD Residential Application Form
- SBD Non-Residential Application Form
- SBD Self Build 2019 Application Form
- SBD New Schools 2014 Application Form

5.22 SBD—Appeals

The subject of appeals against the decision of the individual Designing Out Crime Officer (DOCO) in refusing to confer the Secured by Design award, has always been a difficult issue. On one level, the client (usually the architect, construction company or developer) is appealing against the decision NOT to confer the SBD award. Consequently, the decision by that individual DOCO is being called into question and few of us take kindly to such an intervention, which by default may also be perceived to criticise our inherent professionalism. A further complication is added by the fact that some might contend the sovereignty of that particular DOCO and his or her police force are also being challenged. Nevertheless, as Secured by Design increasingly seeks to fine tune and further professionalise its service delivery, it is only right and proper that a robust and effective appeals system exists to deal with such issues.

A dispassionate observer might conclude that the overwhelming majority of these appeals would revolve around the reasoning and subjective application by the individual DOCO, in respect of those very Physical Security 'Target Hardening Measures, Technological Innovations and Environmental Elements of Crime Prevention Through Environmental Design (CPTED— see Chapter 5) as applied to a specific project. Historically, that has rarely been the case.

Unfortunately, the very few appeals ever submitted have usually been successful due to administration deficiencies on the part of the individual DOCO. In particular, the failure to keep accurate records of meetings with architects, developers, planning officers etc. and on occasion the complete absence of correspondence between the parties (see Sect. 1.5.12). It should go without saying that such an abject lack of professionalism is unforgiveable.

Appeals are investigated and the ultimate decision made by Secured by Design staff, many of whom are themselves former DOCOs and incidentally, some of the most experienced practitioners in the UK. Consequently, the DOCO who decides NOT to confer a Secured by Design award needs to ensure they occupy extremely firm ground and possess an audit trail of correspondence, one that provides credibility for their decision-making.

Of course, on the opposite side of the argument there is also a professional requirement to stand by your decision-making and refuse to confer the SBD award, where that which was agreed prior to construction has not been fulfilled. The issue of testing and certification failing to support PAS 24 Doors and Windows referred to in Sect. 5.4 SBD Residential Guide 2025, is a classic case in point. However, even in this instance the individual DOCO should be mindful of how it is desirable to check the relevant certification long before the final site inspection (ideally during one of the pre-construction meetings with the architect, developer, his agent etc.—see Sect. 3.3. Unfortunately, resort to this procedure may not ensure the appropriate doors and windows are installed across the site! Vigilance as ever is the watchword.

Finally, it is logical to surmise that no one wants to willingly embark on the SBD Appeals process. This should always be an option of last resort and only chosen when all other means of communication and compromise have broken down. Consequently, it is incumbent upon the individual DOCO to make every effort and more to 'head off at the pass' (resolve) all such appeals before they are made. And that will naturally require the all too necessary Diplomatic Skills and the Art of Compromise and Negotiation referred to in Sects. 1.5.9 and 1.5.10 respectively.

5.23 SBD Guide for Heat Pumps, Battery Storage Systems & EV Charging Cables

5.23.1 The Secured by Design 10 Point Security Guide

This four-page security guide responds to the attraction by thieves of heat pumps, battery storage systems and EV charging cables. Increasingly common in modern homes, they are stolen for their copper piping or cabling, control systems, etc.

It contains 10 sections providing security advice in respect of: Siting and Positioning—in different types of housing; Perimeter Protection—fencing, gates and defensive planting (see Sects. 4.2.3 and 4.2.4); Physical Protection of Units—including secure fixing and security cages; Property Marking—and in addition associated signage (see Sect. 2.13); Tracking and Monitoring e.g. discreet GPS trackers; Lighting and Surveillance—includes dusk to dawn lighting and VSS (video surveillance system) issues; Installation Timing and Site Maintenance (building sites are especially vulnerable); Battery Storage Units (necessity for secure locations, including within gated boundaries); Smart Connected Heat Pumps the relevant legislation, standard and SBD's Secure Connected Device accreditation scheme; and Electric Vehicle (EV) Charging Cables—high-risk targets for theft requiring tamper-resistant fittings.

In conclusion, heat pumps, battery storage systems, smart connected devices and EV charging cables are valuable assets where the risk increases in the installation phase. SBD's layered approach (siting, visibility, physical protection, property marking, surveillance and robust cyber security) allows developers and homeowners to significantly reduce risk, protect investment and support public trust in sustainable technologies. Further details about SBD approved products meeting Police Preferred Specification are available at: www.securedbydesign.com

5.24 SBD Retailers' Crime Prevention Toolkit

Another recent addition to Secured by Design's canon of design guides is the 32-page SBD Retailers' Crime Prevention Toolkit. The Introduction explains how the guide was designed at the request of the Home Office, and through a collaboration with the National Business Crime Centre (NBCC). It provides useful information to assist keeping the business safe from crime and its impact on staff. A synopsis is set out about the work of Police Crime Prevention Initiatives (Police CPI); the National Police Chiefs' Council (NPCC) Prevention Coordination Committee; and the National Business Crime Centre (NBCC). A definition of business crime is also provided.

The Toolkit contains 15 sections: Physical Security—including definitions, Lighting, Bollards for physical security, Roller shutters and grilles, Doors and windows, Locks, CCTV, Intruder Alarm System, Internal; Further Security Considerations—Cyber Crime and preparing for cyber incidents; Safety and Security—Security Awareness, Suspicious Behaviour, It's OK to say, Delivery and Vehicle Security, and Lone Working; Are You Protected?—THRIVE assessment; Prevention—Strategy; Reporting; Offender Management; Training; Employee Support; Corporate Support; Technology—Facial Recognition; Engagement; Data; Appendix One—Retailers self-assessment check list and Appendix Two—Legislation: Law and Liability, Occupiers Liability Acts 1957 and 1984, Data Protection Act 2018 and GDPR.

5.25 Safer Parks—Improving Access for Women and Girls

Published in 2025 and copyrighted to the Safer Parks Consortium, this most recent guidance document was prepared by: Keep Britain Tidy, Make Space for Girls, the University of Leeds and West Yorkshire Combined Authority. It has been revised and updated with Police Crime Prevention Initiatives and accessible on the SBD website.

The introduction describes why parks are essential: somewhere to be active, play, socialise and relax; their importance for health and wellbeing and how such benefits are not so easily available for women. Indeed, the guide concludes that parks are not currently perceived as safe and welcoming places by women and girls.

Sexual harassment in public space is routine for 71% of women in the UK, rising to 86% of 18–24-year-olds. Girls experience unwanted sexual behaviours in public from a young age—77% before their mid-teens; 50% of girls and young women don't feel safe outside alone; ethnicity, disability, age, gender identity and sexual orientation, increase the harassment they face. Whilst concern about safety is perhaps the biggest single barrier to women and girls' enjoyment of parks and green spaces. The evidence suggests that women and girls are less likely to use parks than men and boys, a situation which has a significant impact on their lives.

This guidance addresses the question of what can be done to change this situation, using research by the University of Leeds in 2022, together with a survey of other documents and good practice. This produced a set of design and management principles for both existing parks and new developments, aimed at parks managers, landscape architects, DOCOs and other public realm design professionals across the UK, but also relevant for local and national politicians, developers, council officers and community groups. The principles cover ten core areas under three themes: Eyes on the Park—reflects that the presence of others makes women and girls feel safer; Awareness—addresses design issues that can help women and girls feel more secure; Inclusion—considers the importance of bringing a diverse cross-section of women and girls into our parks and designing spaces with their input.

The guidance addresses the full range of safety provision, advocating design that considers practical issues such as sightlines and escape, and also subtler themes including belonging and familiarity—creating parks which welcome women and girls and ensuring that changes do not result in a bland, sterile or overly-defensive environment. Parks should be beautiful, nature-rich spaces, where improved feelings of safety lead to a positive self-fulfilling cycle of greater use and more reassurance.

The document forms supplementary guidance to the Green Flag Award scheme, which sets the benchmark standard for management of parks and green spaces around the world. It is also part of Secured by Design guidance and also supports Natural England's Green Infrastructure Framework, the Equality Act 2010 and the Public Sector Equality Duty. On an international level, it will help UK parks achieve the UN's Sustainable Development Goals, in its target to provide universal access to safe, inclusive and accessible, green and public spaces.

Key sections include:

Principles—Eyes on the Park, Busyness and Activation, Staffing and Authority Figures

Awareness—Visibility and Openness, Escape, Lighting, Wayfinding and Layout

Inclusion—Belonging and Familiarity, Image, Access and Location Co-production and Engagement

How to implement this guidance

Case Studies—The guide includes 10 inspirational good practice and innovative case studies from Britain and abroad which demonstrate how the principles can be applied: Vienna: Gender Mainstreaming in Parks; Umeå (Sweden) Inclusive City for Women and Girls; Her Barking—Low Cost Interventions for Change; Wakefield—Bringing Women into the Park; Southwark—Lighting for the Community; Toronto (Canada)—Women Managing the Park; Bradford—Active Spaces for Teenage Girls; York—Community Engagement; London—Queen Elizabeth Olympic Park—Information for Safety; and Sweden—Inclusive Skateparks.

The guide concludes with Measuring Success; Research Summary; Bibliography; and Acknowledgements. See also Sect. 6.12 Parks and Open Spaces.

References

British Standards Institution (BSI). (1986). *BS 8220: Part 1: 1986 British Standard Guide for Security of Buildings Against Crime. Part 1. Dwellings*. BSI.

Brown, J. (1999). *An Evaluation of the Secured by Design Initiative in Gwent, South Wales* (Unpublished MSc dissertation). Scarman Centre for the Study of Public Order. Leicester: University of Leicester.

Jones, A., Valero-Silva, N., & Lucas, D. (2016). *The Effects of 'Secure Warm Modern' Homes in Nottingham: Decent Homes Impact Study*. Nottingham City Homes.

Nagin, D.S. Cullen, F.T. and Lero Jonson, C. (eds.) (2019). *Deterrence, Choice, and Crime, 23, Contemporary Perspectives*. Routledge.

Newman, O. (1972). *Defensible Space: Crime Prevention Through Urban Design*. Macmillan.

Pascoe, T. (1999). *Evaluation of Secured by Design in Public Sector Housing—Final Report*. Building Research Establishment.

Secured by Design. (1989). *Secured by Design—The Concept*, South East Region Senior Crime Prevention Officers Conference. Rea Publicity.

Sidebottom, A., Tilley, N., & Eck, J. (2012). Towards Checklists to Reduce Common Sources of Problem-Solving Failure. In *Policing: A Journal of Policy and Practice, 6*(2), 194–209.

Stokes, M. (2022). *An Investigation of the Sustainability of Crime Prevention in the Built Environment: Impact and Implementation Factors* (Published Doctoral Thesis). University of Huddersfield.

Teedon, P., Reid, T., Griffiths, P., Lindsay, K., Glen, S., McFayden, A., & Cruz, P. (2009). *Secured by Design Impact Evaluation Final Report*. Glasgow Caledonian University.

Other Design Guidelines 6

6.1 Domestic Violence Sanctuary Rooms

In recent years across the United Kingdom (and undoubtedly beyond), tangible efforts have been taken to utilise Crime Prevention Through Environmental Design (CPTED) interventions—including Physical Security 'Target Hardening' Measures (see Sects. 4.2.1–4.2.4), Technological Innovations (Sects. 4.3.1–4.3.3) and Environmental Elements (Sects. 4.4.1–4.4.6) as part of the toolbox of control interventions designed to combat, or at least reduce the threat, risk, harm and/or vulnerability from domestic violence.

The original Sanctuary Scheme was an award-winning innovation developed in 1998 by PCs Mark Dowse, Peter Waine, Geoff Bigby and Stephen Sullivan of London's Metropolitan Police. Subsequently, that concept has been further developed and in the West Midlands Police area it became known as Domestic Violence Sanctuary Rooms.

As one such example, following a successful project in the Kings Heath district, in 2008, the 'Birmingham Sanctuary Scheme' went live across the whole of the city. This is a multi-agency partnership that aims to provide a range of security measures designed to assist victims of domestic violence to **Remain Safely in Their Existing Home**. The two major aims of the project are to:

- Increase safety and security for victims of domestic violence
- Reduce the necessity for victims of domestic violence to move home.

This is achieved through a combination of security measures, including the creation of a Sanctuary Room (often known in the wider world as a 'panic room'). The key partners in this arrangement have been Birmingham City Council

M. Stokes, *Designing Out Crime*, Crime Prevention and Security Management,
https://doi.org/10.1007/978-3-031-99840-9_6

Housing Department, Birmingham and Solihull Women's Aid (BSWAID), West Midlands Fire Service and West Midlands Police.

There is a list of eligibility criteria, chief amongst which is that the perpetrator of the violence must NOT reside with the victim. Once these criteria are satisfied, the following security measures can be considered:

- New door locks
- Door viewer ('spy-hole')
- Door chain or opening limiter
- Fire safety equipment
- Provision of a safe sanctuary room
- Provision of a mobile phone.

Following the referral of a victim to the scheme, the local Designing Out Crime Officer (DOCO) will undertake a survey of the premises to determine what measures are necessary. Her or his report is then passed to the Housing Department Sanctuary Support Officer who will in turn instruct a specialist contractor to complete the work. Both the fire service and the police will 'flag' the address on their computerised 'command and control' systems to ensure an immediate response and to gain access, if necessary. And in the interim period between referral and implementation (prescribed as a maximum of 6 weeks), the victim may require temporary accommodation.

Three different levels of safety and security have been designated: Minor Sanctuary, Sanctuary, and Sanctuary Plus. Each level necessitates increasing the minimum-security standard criteria.

Minor Sanctuary

- Main Entrance Doors reinforced with Birmingham Security Bars or London Security Bars, as appropriate
- Provision of a Fire Safety Pack that includes a fire-resistant letterplate bag—into which items pushed through the letterplate (including incendiary items) will be enclosed; and two battery-operated smoke detectors
- Sanctuary Room window fitted with 6.8 mm Laminated Glass
- Window locks fitted where there are no other means of securing windows and a danger of entry exists
- Replacement of exterior door locks, but only where there is a danger that the ex-partner has a set of keys to the property.

Sanctuary

- The door and its frame providing access to the Sanctuary Room should be of composite construction to PAS 24 standard (see Sect. 4.2.1, Doors and Door Security and Sect. 5.4 Secured by Design - Residential Guide 2025), fitted with a slam action locking system. If the room has partition or otherwise insecure walls, consideration should be given to reinforcing those walls
- Where required, a mobile phone to summon the police will be provided in the Sanctuary Room, should the perpetrator attempt to gain access to the property once the occupants have sought shelter therein.

Sanctuary Plus

- In very exceptional circumstances (for example, where an appropriate Sanctuary Room cannot be identified in the dwelling, or existing external doors cannot be adequately reinforced), the renewal of the existing external doors can be considered
- In exceptional circumstances, where the external entrance doors cannot be adequately reinforced and the victim is temporarily residing elsewhere, temporary doors can be installed in order to ensure the property is secure. It is always preferable to consider temporary accommodation such as in a refuge, whilst the work is undertaken
- In exceptional circumstances, installing Laminated Glass in windows elsewhere in the property may be considered
- Window locks will only be considered in rooms elsewhere in the property where these have not previously been installed.

Exceptional Additional Recommendations

This covers recommendations not contained in the above specification—and their justification, e.g. Intruder Alarms.

Whilst the solutions envisaged largely fall under the headings of Physical Security 'Target Hardening' Measures and Technological Innovations, it is intended that the individual DOCO will also use their skills in interpreting the Environmental Elements of CPTED when making recommendations. For example, when assessing the external Perimeter Security (fencing) and the degree of Defensible Space and Natural Surveillance (see Sects. 4.4.6, 4.4.1 and 4.4.2, respectively).

Conclusion

Considered holistically, the above measures deliver a system whereby the aggrieved party can continue residing at their home address, as opposed to being re-housed in a different location. To date, the system of referral and implementation of the above measures has worked extremely well. Moreover, further success can be identified in that the Birmingham team of DOCOs and colleagues from partner agencies received a 2010 Secured by Design Innovation Award and grant for £10,000. Part of this sum was used to host a good practice sharing event held at the Tally Ho Conference Centre in 2011, thereby spreading the positive message to a far greater audience and in so doing, satisfy Dr Ivan Hill and Professor Ken Pease's (2000, p. 140) '*the benign dynamic of reduction leading to further reduction.*

The Sanctuary Scheme Installation Manual

In 2018, Secured by Design published The Sanctuary Scheme Installation Manual: A Place of Safety for Those at Risk, authored by the original exponent Mark Dowse. This sets out the minimum technical specifications for the installation of a sanctuary safe room, together with associated security and fire safety measures.

The manual provides a comprehensive overview of the issues relating to the installation of such rooms, including detailed design advice in respect of the security elements for all types of doors, windows and loft hatches. Specific sections are devoted to:

Section 1—General Security Measures
Section 2—Sanctuary Safe Room
Section 3—Fire Safety
Section 4—Sanctuary Scheme Schedule of Works
See also: https://www.securedbydesign.com/initiatives/sanctuary-scheme

KEY ISSUE

It is imperative that only a fully trained and experienced Designing Out Crime Officer (DOCO) conducts the survey of the dwelling and room that will perform the function of a Domestic Violence Sanctuary Room. This is because such a survey is in effect a Threat and Risk Assessment (TARA—see Sect. 1.5.15) and will dictate the level of security required, taking into account the Commensurate Risk involved. Attempts to reduce this to a simple 'tick-the-box' exercise limited to Physical Security 'Target Hardening' Measures conducted by unqualified personnel, are inherently hazardous and need to be firmly resisted. The primary justification for this line of thinking rests with the threat to life or serious injury that might result. However, there is also the potential for civil litigation that would inevitably follow, if an unqualified person had conducted that same threat and risk assessment.

6.2 'Leaking' culs-de-sac

This chapter should be read in conjunction with Sects. 4.4.6, Permeability; 6.3, Alleygating; 6.4, Closure of Rights of Way on grounds of Crime and Public Space Protection Orders; and 2.10, Space Syntax.

A classic 'Leaking' cul-de-sac providing a footpath link to another road

Background and History

As a simple definition, a 'Leaking' cul-de-sac is a road providing no through route for motor vehicles, yet at least one footpath (sometimes combined with a cycleway and/ or bridleway) links it to another such cul-de-sac or road, public open space, canal towpath, riverbank, etc.

From the perspective of Designing Out Crime, the Permeability afforded by such a design feature causes immediate concern, and with good reason. Designing Out Crime Officers from across the United Kingdom can testify to the higher instances of residential burglary, other forms of Property Crime and ASB that take place in the vicinity of a *'Leaking' cul-de-sac*. Whereas, in a 'True' cul-de-sac with no further pedestrian connection, lower instances of crime are almost always recorded. Consequently, the DOCO most often perceives *the* 'Leaking' cul-de-sac as merely facilitating an Attack and Escape Route for Burglars, other criminals and those intent on ASB/Disorder.

Unfortunately, this is not a message many outside the expertise of Designing Out Crime are willing to hear (including nearby residents—until they become the victims of crime), the issue dovetailing with that of Permeability and the New Urbanist/Urban Villagers' desire for an all but totally open-plan streetscape.

It might be thought that following the development in the UK of so many crime infested 'sink estates' during the 1950s, 60 s and 70 s, where all too obviously their rampant Permeability designed-in Crime Opportunity (see Sect. 2.3), that mistake would not be repeated—and certainly not so soon. But walk through the majority of developments constructed in the UK and wider world over the past 30 years, and this same design feature can be repeatedly identified. Moreover, the rationale and excuse for their inclusion is now most usually based on grounds of Sustainable Transport Links—persuading people to walk or bicycle.

Historically, the simple *'Leaking' Cul-de-Sac* might be viewed as an extension of the original rural common law public footpath, one that enabled even the poorest members of society an easy route to their place of work and to church. Furthermore, this practice was incorporated into the housing design of the Georgian and Victorian periods and can be identified across the UK in the housing layouts of the 1920s, 30 s and 50 s, especially in relation to private (for sale) new build housing schemes. From the 1920s, their inclusion was based on a convenient access to nearby facilities (such as schools, churches, shops, the pub and bus stops), rather than simply enabling workers and worshippers an easy route. Furthermore, during those decades the link between *Permeability* and crime was unknown—an excuse that cannot be used in the twenty-first century.

The Evidence from Crime Science

Barry Poyner was one of the first academics to consider the criminogenic capacity of *'Leaking' Culs-de-Sac* in facilitating *Crime* (Poyner, 1983, 2006). Manifestly in in-depth and highly scientific research arrived with the publication by Professors Shane Johnson and Kate Bowers (2010) of *Permeabilityand Burglary Risk: Are Culs-de-Sac Safer?* The short answer is yes—providing they are not of the 'Leaking' variety. But such a summary does not do justice to this seminal piece of research.

The authors begin with the premise that crime is concentrated in space (Eck et al., 2005) and the role of *Permeability* in facilitating crime. Crime Pattern Theory (Brantingham and Brantingham, 1995, 2008) supports the idea that Permeability and Crime Risk are linked. And Armitage (2007) found that homes located on main roads were at greater risk from residential burglary, whilst those in culs-de-sac were at least risk. However, where the cul-de-sac included a 'leaky' footpath, the risk of crime was unsurprisingly much greater.

Compare this to the concept and methodology known as Space Syntax (see Sect. 2.10). Research conducted by Hillier and Shu (2000) and Shu and Huang (2003) concluded the risk of residential burglary is twice as high in culs-de-sac than in permeable through roads. Despite this assertion, culs-de-sac were thought to be '…very safe places' if linear and part of a larger system of linear roads. And, '*…when combined together to form hierarchical systems of interconnected Culs-de-Sac, they can become extremely vulnerable*' (Hillier, 2004, p. 39).

To test their three hypotheses, Johnson and Bowers (2010) used study data from one police district in the Merseyside police area. This consisted of some 56.8 square km, 118,161 homes and 10,760 street segments. Of these street segments, 9914 contained dwellings, thereby producing 11.94 homes per segment. Police recorded residential burglary data was used for the period April 1998 to March 2002 (four years).

As most of the culs-de-sac in the study area were connected to on average three others and to ensure misidentification as through roads did not take place, each was identified manually—a highly scientific, labour-intensive approach.

Two types of cul-de-sac were identified:

1. Linear culs-de-sac—off through roads with linear geometry
2. Sinuous or isolated culs-de-sac—non-linear with little visibility down the cul-de-sac.

The aggregate results, in terms of residential burglaries per 1000 homes per annum, disclosed that the lowest risk of burglary was in culs-de-sac and private roads. Indeed, the rate of burglary on major roads was three times that of private roads and more than twice that in sinuous culs-de-sac. Furthermore, the risk is higher in linear culs-de-sac than in sinuous or hierarchical types.

Indeed, the Johnson and Bowers research suggests that linking a street section to another major road increases the risk of residential burglary by 8%, and to three other such roads by 26%. Whilst both forms of linear and sinuous cul-de-sac would seem '*...to be associated with a statistically significant reduction in the expected count of burglary*'. The findings demonstrate that connectedness is linked to increased burglary risk—in line with Bevis and Nutter (1977) and Beavon et al. (1994). However, from the data available it appears that some of the early *Space Syntax* research did not differentiate between 'True' and '*Leaking*' *Culs-de-Sac*.

The authors convincingly assert that for future policy decision-making and as *Permeability* may lead to increased *Crime Risk* of residential burglary, it should be limited to that necessary to facilitate local journeys and *sustainable* transportation. Moreover, 'True' culs-de-cac are arguably a beneficial design feature and should be encouraged.

As is repeatedly the case, the Johnson and Bowers research begs that many more questions be answered—especially in relation to *Permeability*. Other research gems (in terms of *Crime Science*) include: how housing density appears to have had a positive effect; burglary is higher in areas with higher levels of ethnic heterogeneity; and a small positive link between levels of unemployment and burglar association. But in summary, the authors powerfully conclude how their results support the hypothesis that increased *Permeability* is inherently associated with *Crime Risk*.

Conclusion

Within the world of Designing Out Crime (CPTED in practice), *'Leaking' Culs-de-Sac* are universally regarded as a design feature that invariably leads to more crime, FOC and ASB. Whereas by contrast, 'True' culs-de-cac have the opposite effect. Consequently, designs that include those of the '*Leaking*' variety should be challenged and alternatives sought. If that fails, measures incorporated to mitigate their *crime* inducing effect (such as Alleygating—see Sect. 6.3) should be considered. Where the aim is to permanently close an existing'*Leaking' Cul-de-Sac* that is also a Public Right of Way, providing evidence of high levels of *Property Crime* can be demonstrated, a *Special Extinguishment Order* under Sections 118B and 119B, *Highways Act 1980* can be utilised. However, *Public Space Protection Orders* (see Sect. 6.4) may be an easier and more practical option.

Some words of caution. Even when it can be proven there is a causal link between a spate of Property Crime and a *'Leaking' Cul-de-Sac* providing the necessary *Attack and Escape Routes* for offenders, much repeated personal experience demonstrates that nearby residents are most often hostile to losing the convenience that the connection affords. The skill therefore with new housing projects lies in producing a street layout design that provides maximum convenience, yet does not compromise safety and security by including *'Leaking'* connection elements.

6.3 Alleygating

Over the past quarter century, of all the initiatives developed and implemented by a Designing Out Crime Officer (DOCO), Alleygating has delivered some of the greatest Crime Prevention benefits—especially in reducing Residential Burglary. Moreover, Alleygating is a magnificent advert of Situational Crime Prevention to police managers (and many other interested parties), demonstrating how such measures can deliver genuine and immediate burglary reduction. Indeed, this is especially true where funding for such projects is available and when combined with the willingness of 'particularly motivated individuals' (see Read and Tilley, 2000; Bullock, Erol and Tilley,, 2006, p. 15) within the relevant community and local government departments.

Alleygating was devised by the then Metropolitan Police CPDA, Sergeant Calvin Beckford (1994). Based in the London Borough of Camden, Calvin's 'manor' exhibited some acute problems with Residential Burglary. Many of these burglaries were executed via the networks of footpaths or alleys that linked the rear gardens or yards of these properties. It should also be noted that across the UK, there are different names for such footpath links including entries, ginnels, backways, snicketts, passages, paths and walks. Moreover, in London many had their key-lockable metal gates removed at the outbreak of the Second World War (supposedly to produce munitions) and were never replaced.

The Alleygater's guide to gating alleys

Even in the early 1990s, Home Office research had shown that two thirds of all residential burglaries were executed from the rear or sides of such properties. Consequently, it could be surmised that a network of footpaths between the rear gardens was facilitating crime—acting as classic Attack and Escape Routes for those with criminal intent. Calvin's solution was to re-close these footpaths or alleys, by means of lockable gates at the point where they join a road—and thus the description Alleygating was born. In the majority of instances, this was merely reinstating a gate that had previously existed—in effect recreating *Defensible Space* (see Sect. 4.4.1) and denying *Permeability* (see Sect. 4.4.6). Residents whose dwellings could be accessed from the footpaths were given keys to the lockable gates, as indeed were local government officials and anyone with a legitimate reason to access the alleys. However, key management was strict and refused to those who simply desired to use the footpath as a shortcut.

The results were spectacular, with a 90% sustainable reduction in residential burglary often realised. Other benefits included reductions in fly-tipping, dog fouling, littering and ASB. See especially Bowers et al. (2004). Despite a small number of complaints, the Alleygating initiative won the support of residents, local government officials, elected members and of course the police. Indeed, the scheme has been replicated across the United Kingdom where this peculiar form of rear access housing design persists in its application through to the present day.

The quality of the Alleygating product will reflect the Commensurate Risk (see Sect. 1.5.15) judged at a given location (evidenced by crime trends and intensity). Nevertheless, it was realised that a minimum standard would be necessary in all environments. For this purpose, guidelines set the minimum height of the gates and any adjacent fencing, the quality and dimensions of the steel required and that of the locking mechanism.

The preferred specification for such gates is as follows:

- The gate should be inward opening
- Made of minimum 3 mm thick galvanised/painted steel
- Outer box frame gate surround (40 × 40 mm) with (ideally) 150 mm long blunted spikes on the top
- Total gate height—minimum 2 metres (6.5 feet).
- The gate should be of open view construction in order to permit Natural Surveillance. (see Sect. 4.4.2)
- Horizontal bars and hinges designed so as not to be used as footholds.
- The locking mechanism should be one of slam action/automatic dead-locking mortise latch (not self-closing)
- The gap beneath the gate should be narrow enough to prevent even a small person from squeezing through.

Planning Permission is not normally required, providing the gate does not exceed 2 metres in height and is not immediately adjacent to the highway. Different rules apply to Listed Buildings and in Conservation Areas. However, the permission of those using the gate(s) (neighbours whose gardens/premises back onto the alley) will be required. Where the alley is an established Public Right of Way, see Sect. 6.4. See also Adamson (2005).

Calvin's The Alleygater Guide (1994) is accessible at: https://www.thecrimepreventionwebsite.com/storage/library/the_alleygaters_guide_to_gating_alleys_bumblebee.pdf) provides step-by-step instructions to create an Alleygating scheme. It is also a classic example of Problem-solving (see Sect. 2.14,

Problem-solving, Problem-Oriented Policing and Partnerships) involving residents, landlords, police, fire service, local authority staff and elected members. Most importantly, two decades later, the majority of these schemes remain in situ and thereby demonstrate long-term sustainability across a wide spectrum of benefits, such as Crime Prevention, financial cost savings, residents' perceptions of safety, etc. See Clarke (2004) for a worldwide review of such closures. See also Sidebottom et al. (2017).

6.4 Closure of Rights of Way on Grounds of Crime and Public Space Protection Orders

In the United Kingdom, Designing Out Crime Officers (DOCOs) were cheered by the prospect of new legislation that would allow Public Rights of Way to be closed on grounds of Crime—such routes having repeatedly proven to act as *Attack and Escape Routes* for those with criminal intent. However, when the Countryside and Rights of Way Act, 2000 came into effect, the reality was somewhat different to what had been promised. The original bill had been dramatically amended when it passed through the House of Lords and consequently, its final effect was simply to provide amendments as Sections 118B and 119B of the Highways Act, 1980.

During the intervening years, this legislation has proved to be extremely difficult to utilise at the practical level. This is largely because it requires evidence of a proven track record of Property Crime within the immediate vicinity of the location it is desired to close. Anti-Social Behaviour (ASB) is not mentioned and the objection of just one person to the closure will result in a full-blown public hearing. Not surprisingly, local authorities (local government) are unwilling to fund the extremely high costs involved with the latter.

A far better solution was delivered by the Clean Neighbourhoods and Environment Act 2005, where Section 6.2 permitted 'Gating Problem Alleys'. However, in 2014, Gating Orders were themselves replaced by Public Space Protection Orders (PSPOs). Created by the Anti-social Behaviour, Crime and Policing Act 2014, PSPOs enable local authorities, the police and other bodies to deal with a particular problem or nuisance (ASB) within a specific area. Suggested examples include public drunkenness, aggressive begging, irresponsible dog ownership and general ASB.

PSPOs were designed around the concept of ensuring that the law-abiding majority could enjoy public places, free from ASB. For this purpose, PSPOs are described as providing maximum flexibility and permit local agencies to work together to develop reasonable, proportionate and necessary responses to deal with ASB.

A PSPO is made by the relevant authority when satisfied (on reasonable grounds) that behaviour in a public place is having a detrimental effect on the quality of life of those living in the community. The behaviour must be persistent or continuing, and it must be unreasonable.

The local authority can make a PSPO for any public space within its jurisdiction, but only after it has consulted with the police and the community representatives it believes appropriate. Once enacted, a PSPO remains in force for a maximum of three years and where believed necessary, will need to be renewed at the end of such a period. Failure to comply with a PSPO is a criminal offence.

Before making the PSPO, the local authority must publish the draft order. An 'interested person' (someone who lives or works in, or regularly visits the restricted area) can challenge the validity of a PSPO in the High Court on either of two grounds: the local authority did not have the powers to make the order, or include prohibitions and requirements; or one of the requirements has not been complied with.

In the same way that the local Highway Authority manages Closure of Rights of Way on Grounds of Crime and the former Gating Orders, they now have responsibility for PSPOs. In this context, PSPOs should only be used where it can be demonstrated that persistent ASB is expressly facilitated by the use of a particular right of way. Such PSPOs will be especially important in enabling the closure of back or side alleys that can be shown to be the source of ASB. Furthermore, rather than the total closure delivered by a Gating Order, central government contends that PSPOs offer instead the potential of restricting specific activities.

When deciding whether to restrict entire access by way of a PSPO, the local authority should consider whether residents and members of the public who use the relevant highway will be inappropriately inconvenienced by its closure, gating and if alternative access routes exist. Nevertheless, this should not prevent the gating of highways where activities are so dangerous such gating is in the best interest of all concerned.

The health implications of the PSPO should also be considered, as gating could potentially encourage the use of cars if the alternatives are too long or lack pedestrianised sections. The closure of a route might even deter people from making trips on foot completely. This should be balanced against the health implications facing pedestrians from the ongoing crime or ASB in the alleyway or other route.

Before using a PSPO to gate a Public Right of Way, local authorities will need to both inform and consider representations from all interested parties. They are obliged to provide a full justification to anyone who objects to a proposed PSPO. Suggested interested parties include:

- All occupiers of premises adjacent to or adjoining the relevant highway
- Any other local authority including community, parish and town councils
- Fire authorities
- NHS Trusts
- Any Local Access Forum through whose area the relevant highway passes
- Statutory undertakers
- Water services providers
- Communications providers.

Further consultation is encouraged with such organisations as the: Auto Cycle Union

- British Horse Society
- Byways and Bridleways Trust
- Cyclists Touring Club
- Open Spaces Society
- Ramblers
- Chiltern Society
- Peak and Northern Society.

Sadly, unlike Gating Orders, this consultation process alone has proved to be highly bureaucratic. Consequently, experience suggests that as Gating Orders expired at the end of their own maximum time period, they were replaced by PSPOs—even when a persistent problem with ASB at the location in question was known to exist.

The Anti-social Behaviour, Crime and Policing Act 2014 also provided new legislation in the form of Community Protection Notices and Dispersal Powers. However, and as any search of the Internet will disclose, it is the PSPOs that have proved to be the most controversial. Indeed, a small number of local authorities have used PSPOs to ban activities such as rough sleeping, begging, skateboarding, ball games, busking and under-16 s gathering in groups of three or more.

6.5 Home Zones and Shared Space

Introduction

Home Zones are the UK version of what the Dutch describe as Woonerf (Woonerven in the plural), which literally translates as 'recreation area' or 'living street'. Developed in the Netherlands during the 1970s, many countries have since utilised the core concepts to create their own versions of Home Zones—on occasion failing to realise that the Dutch have themselves evolved the original design, which some now regard as a little 'quaint', a point returned to later in this section.

Woonerf (Home Zone) in Groningen, The Netherlands

These design features aim to 'reclaim local streets' from the increasing domination by motor vehicles. They should balance the competing needs of everyone who uses the street: children playing, pedestrians of all age groups, cyclists and motor vehicles. Consequently, the over-arching remit of the Home Zone is to create/recreate usable public open space by removing the threats from speeding traffic—at least that of the motorised variety.

Implementing a Home Zone

The Home Zone is achieved either at the concept stage of a proposed development, or (as is usually the case) by altering the existing physical layout of the streets or roads in the designated area, through introducing some or all of the following measures:

- Narrowing the street/road/carriageway

- Incorporating traffic humps, platforms and chicanes
- Reducing sight lines for motorists
- Introducing raised flower beds
- Planting trees
- Installing benches and other forms of public seating
- Reducing the speed limit in order to force motorists to drive more considerately.

This last point is especially problematic as 10 miles per hour is often demanded, although in the UK the minimum legally enforceable speed limit is 20 mph (32 kph)—albeit increasingly the norm in residential areas across the UK.

Past experience suggests introducing these measures is the easy part—especially in a new-build project. However, the built environment is littered with examples of attempts to create a Home Zone, witness closed-off roads that were later reinforced with bollards; and raised flower beds concreted over when the plants were stolen, vandalised, or died through neglect.

Indeed, where the Home Zone is introduced into an existing streetscape and to be especially effective, evidence demonstrates it is imperative extensive consultation with the whole community takes place before any work begins—and both during and after the construction process (see Sect. 1.5.11. Truly Effective Partnership Working; and Sect. 2.14, Problem-solving, Problem-Oriented Policing and Partnerships).

Problems

Home Zones have been criticised for a number of reasons: delayed response times for emergency service vehicles; forcing residents to park their cars some distance away from their homes (of special concern to the DOCO as there is often an inextricable link between such distance and Vehicle Crime); the potential for causing tailbacks of vehicles driving through the Home Zone; and encouraging children to play in the street when these are inherently dangerous places.

The latter is open to considerable debate, especially in early twenty-first century Britain. At one extreme, everyone appreciates that children would not play on the motorway/freeway/autobahn/ autoroute, etc. But in a Home Zone environment, immediately in front of their house and with the constant Natural Surveillance (see Sect. 4.4.2) of parents and neighbours? A cul-de-sac is potentially a 'safer' environment than a play area located several hundred metres or more away from home—especially in the context of enduring concerns about children at risk from strangers ('Stranger Danger').

However, in the Dutch suburb of Hesselterbrink, outside Emmen, some residents described their Woonerf as a traffic engineering measure that 'entailed precious little more than signs and uniform standards.' That said, this opinion appears to have no corroboration and objective criticism of the Home Zone concept is especially rare.

Home Zones in the United Kingdom

Throughout the twentieth century in the UK and abroad, the pattern of street layout was reinforced by increasing numbers of motor vehicles: the road proper (for vehicles) and the pavement (for pedestrians) with a raised kerb separating the two. As young children, we are taught that the pavement is a safe environment and crossing the road involves inherent dangers. Who can forget the 'Tufty Club', 'Green Cross Man' and the army of 'Lollipop' Ladies and Men (school crossing patrols). Moreover, this demarcation became an instructive part of the training of Guide Dogs for the Blind. And as the number of motor vehicles increased, the domination of roads by these machines was reinforced and demarcated as such.

The Children's Play Council (1997) and Transport 2000 (1999) campaigned to create Home Zones, in both new and existing developments. Indeed, the latter defined the key features of a Home Zone as:

- Change in priority: motor vehicles have to give way to pedestrians and cyclists, with drivers responsible for any injuries they cause
- Emphasising the change of status through signing, traffic calming, seating, other street furniture and planting (soft landscaping)
- Residents should be in favour of the Home Zone and must be involved in the design stages.

The Children's Play Council (1997) criteria involved:

- Resident support
- Streets less than 500 metres (1600 feet) long
- Fewer than 100/200 cars per hour at peak times.

Central Government, by way of the Traffic Advisory Unit at the then Department for Transport, Local Government and the Regions (DTLR) authorised nine pilot schemes and pledged to fund more. DTLR Traffic Advisory Leaflet 10/01 Home Zones—Planning and Design (2001), set out the government's position on Home Zones. Department for Transport, Traffic Advisory Leaflet 8/02 Home

Zones—Public Participation, detailed the necessity for public participation. See also Biddulph (2001) re Home Zone planning and design.

Evolution of the Home Zone concept in the Netherlands

Having invented the original concept of the Home Zone, it is important to note that this design idea has not remained static in the Netherlands during the decades since the 1970s.

In the Amsterdam suburb of Ijburg and on territory largely reclaimed from surrounding marshland, the Woonerven have been designed within the confines of the low rise apartment blocks. Motor vehicles are segregated from the Woonerf and often secured in separate car parks accessed from the street by way of electronically operated gates (activated by a remote transmission device kept inside the car).

However, in all other respects the Woonerven possess total Permeability to pedestrian and indeed bicycle traffic, with multiple unrestricted access and entry points—or Attack and Escape Routes. As previously indicated, Permeability is the fault line regarding the relationship between the Designing Out Crime and Planning professions. And whilst such Permeability may cause few problems in the Netherlands, that country's adoption and development of Secured by Design (see Chapter 5) may suggest otherwise.

Interestingly, in another Woonerf, residents appear to have deliberately allowed the trees and bushes to grow around their properties. Presumably, this was to give themselves some privacy, yet it inadvertently provides cover for criminal activity—blocking Natural Surveillance when the dwelling is unoccupied.

Of even greater importance, there was much debate amongst Dutch professionals as to whether such areas of Ijburg could be described as Woonerven—the definition having either changed over time or as some remarked, the concept having expired. However, the below examples from Delft and Groningen indicate that half a century later, Woonerven can be highly effective.

Advice from the Designing Out Crime Officer

As may have been gleaned from this chapter, the creation of a Home Zone is fraught with dangers—especially in the UK where the debate between the advocates of car ownership and usage, versus the rights of children at play, pedestrians, cyclists, etc., has become highly charged. Consequently, in the event that the DOCO is asked to comment on such a proposal to create a Home Zone, she or he is advised to restrict their involvement to the following subject areas:

- Where a Home Zone is proposed in an existing public space, ensure repeated Community Consultation takes place at all stages of the project's development—including during and after the construction phase
- Provide Crime Pattern Analysis/Crime Profile/Crime Impact Statement (see Sect. 2.3) in respect of the area covered by the proposed Home Zone
- Examine the Home Zone proposal to assess what impact it may have in respect of crime, Fear of Crime (FOC) and Anti-Social Behaviour (ASB)
- Inspect the plans for the Home Zone in minute detail. Be especially wary (see Sect. 3.1, Reading Plans) of benches and other seating arrangements that have not been directly requested by the residents, shopkeepers, etc. Have they been included purely as an aesthetic design feature, when in reality they may act as a *Crime Attractor* and/or *Crime Generator*, or encourage FOC, ASB and worse. Will they be within the control/ management of any particular premises?
- Will raised planting areas provide unintended seating and therefore unexpected congregation points (where, for example, alcohol may be consumed?)
- What maintenance schedule has been put in place for the proposed Home Zone? The UK is already littered with often highly valid attempts to 'spruce up' the built environment, yet because no programme of maintenance was included these improvements quickly degenerated, on occasion making the situation worse than it was beforehand
- If at all possible, any dedicated children's play space should be contained within a fenced-off area with maximum opportunities for Natural Surveillance from and into the surrounding environment
- Is any prohibition on the consumption of alcohol proposed?

Conclusion

Home Zones have the capacity to improve community cohesion and deliver a positive impact on the built environment, providing they are designed, installed and maintained correctly. This is especially true where the Home Zone is created as part of a new development, although even in this instance, it presumes community ownership and participation by the residents post-construction, if improper use and ASB are not to become established.

Furthermore, and to counter the opinion sometimes heard that Home Zones have become an historic anomaly of the turn of the twenty-first century, Home Zones feature in the original National Planning Policy Framework (2012) under Clause 35 'Promoting Sustainable Transport'.

Creating Home Zones within existing communities is far more problematic and financially expensive (as with invariably all retro-fitting solutions), due to the existing layout of buildings, gardens, underground services, etc. Experience suggests this will be a long, difficult and expensive process requiring full participation of residents, road users, elected representatives, local authority (local government) departments, emergency services and utility service providers. Nevertheless, where they are executed with full and ongoing Community Consultation, experience in the Netherlands demonstrates they can be highly effective and deliver long-term sustainability.

However, and as with the majority of Home Zone schemes visited by this author, there appear to be scarce resources (revenue funding) to maintain these areas to the standard originally achieved. This is undoubtedly due to the grey area between different local authority departments which Home Zones occupy. Moreover, it also asks the question whether, once Home Zones decay and are not scrupulously maintained, perhaps they become an additional example of Kelling and Wilson's (1982) Broken Windows theory?

In conclusion, Home Zones are more than just a series of simple traffic calming measures, or road restrictions designed to remove on-street parking. They should create a 'living street/recreation area' that is safe for everyone—children playing, simple leisure facilities, pedestrians, cyclists and motor vehicles. However, changing an existing street into little more than a play area courtesy of traffic calming and speed restrictions, is both confusing for everyone and can lead to difficulties and conflict in that environment. And for a Home Zone to be truly successful and sustainable, it requires the full efforts of the Problem-solving, Partnership Approach (see Sect. 2.14, Problem-Solving, Problem-Oriented Policing and Partnerships).

The Netherlands Examples

As the inventors of the 'Woonerf' concept during the 1970s, more than half a century later, any study of Home Zones might want to examine their long-term sustainability in the country of origin. Rather than being specified as a Woonerf, a blue and white sign (containing images of an adult, child with a football, car and house) is located at its entrance and exit. The same sky blue and white colours (but not the images) have been used in some UK Home Zones.

Delft

In Delft, a city famous for its porcelain and the birthplace of the artist Johannes Vermeer ('Girl with a Pearl Earing'), there are Woonerven within a few minutes' walk of the main train station. Narrow streets, e.g. Westerstraat, Tuinstraat,

Graswinckelstraat, Carthuyzerstraat—containing mainly terraced housing of different designs, have been made narrower by the provision of dedicated parking bays—some with electric charging points. And there is an abundance of Symbolic (Psychological) Barriers: plentiful use of different types and colour of brick paviours and traffic calming measures—primarily raised platforms and some chicanes. There are numerous play areas for children and communal seating areas. Community recycling bins mean an absence of multiple bins for each dwelling, thereby avoiding the spectacle that now blights many streets in the UK. Personal witness suggests there are no *Visual Environmental Cues* (*VECs—see* Sect. 2.8) to suggest these are anything other than a wonderfully safe and secure environment in which to reside and socialise.

Woonerf (Home Zone) in Delft, The Netherlands

Groningen

A university city at the heart of a vast agricultural community in the north of the Netherlands, Groningen's Woonerven date from the same 1970s. In terms of aesthetics, those to the north of the city centre in Grote Rozenstraat, Grote Krusstraat and Grote Applestraat are the best, with trees, shrubs, plants and flowers cascading into the streets. As in Delft, there is no apparent hostility to the motor car and many are parked in these same streets. But there does appear to exist an equality of ownership of the space, with pedestrians and cyclists able to move around without fear from speeding cars. It is surely no coincidence that Groningen University's Faculty of Behavioural and Social Sciences is located at the heart of this area, in Grote Rozenstraat.

United Kingdom Examples

Llywynpia Terraces, The Rhondda, South Wales

In the Rhondda, South Wales, the Llywynpia Terraces Home Zones were created in 2000. Bridgestock (2000) observes how in this former Welsh mining community, Front Doors open onto the street. Following demand from residents and in partnership with the Community Design Service, five streets were redeveloped.

Llywynpia Terraces Home Zone, The Rhondda, South Wales

Because the streets were originally only 2 metres wide, where possible residents were persuaded to surrender a small portion of land that each owned on the opposite side of the street containing a brick-built coal store and outdoor lavatory. This increased the street width to some 5 metres (16 feet), the whole surface of which was paved. New street lights, planting and cast-iron bollards were also introduced, with bespoke designs for each street created in consultation with their residents. The improved streetscape created a sense of place, although within a few years, some trees had died due to waterlogging.

The speed limit (now reduced to 20 mph) together with traffic calming measures, force drivers to give way to pedestrians—most especially children. Designated parking spaces were created to break up sight lines and further reduce speed. In one street, parking problems caused by people 'visiting' the nearby courthouse were resolved when it closed in 2011.

Meanwhile, service vehicles can just about negotiate the streets—the bin lorry with 'an inch to spare on either side'. As a pilot scheme, some funding was obtained from the Welsh Office – now the Senedd (Welsh Parliament) and other partners, with £1–1.5 million necessary to transform all five streets.

Frequent visits by this author reveal that not all the terraces received the same extent of Home Zone makeover. Local residents confirm Inverleith Terrace is the best, with extensive car parking provided and the semi-mature trees firmly established. However, even here the bollards required a coat of paint and where accidentally rammed, repositioned with tarmac clumsily applied instead of the original brick paviors (a repeated failing observed with Home Zones in the UK). Similarly, much of the reflective paint used to mark the traffic calming humps has flaked off during the intervening years. Moreover, the mock Victorian lamp-posts (perhaps the one Environmental Cue that indicates some degree of Community Consultation) provide a 'twee' addition to many of the Home Zone schemes visited in England and Wales.

Similarly, Signage and Notices (see Sect. 2.13)—in both English and Welsh are displayed. 'No Motor Vehicles Except for Access' signs remain at many entrances to the terraces. Other signs indicate some terraces are an 'Alcohol Free Zone' with a £500 fine for transgression.

Great George Street, St. Judes, Bristol

Great George Street Home Zone, St. Judes, Bristol, was created in 1990 when Bristol City Council Housing Department embarked on a general environmental and structural improvement of the flats. Phase two involved redefining the public space between the high-density blocks. Overall, it was judged to be in keeping with the Home Zone concept. They aimed to create a 'small social town square', to include improved parking arrangements, increased communal space, planting, seating and play equipment. An open-air meeting took place between designers and residents, with a model of the proposals used to solicit opinions.

Great George Street Home Zone, Bristol, England

Once implemented, vehicle flow through the area slowed causing cars to give way to pedestrians. Average speed is now around 10mph due to these traffic calming measures and others such as chicanes. The former allotment site suffering from neglect and lack of use, became a paved square with undefined road space and cars restricted by bollards and low walls.

The main problem during implementation was caused by underground services (water, electricity, gas and telecommunications) with providers determined to retain the original road surface for access to their conduits. Moreover, double yellow lines (total prohibition on parking) were painted along either side of the newly created Home Zone street, thereby reinstating the road space. This measure seriously compromised the status of the Home Zone. Furthermore, the issue of parking demonstrates a repeated problem: parking on the double yellow lines or in unmarked places, simply to be closer to their intended destination/place of residence.

Repeated visits by this author confirm the small size of Great George Street Home Zone. Vegetation is now fully grown and many of the bollards repainted silver with two yellow bands along their width. However, some remain black in colour and a few completely removed—evidence tarmac clumsily used to fill in the gaps. Meanwhile, the aforementioned double yellow lines have also been repainted. Prolific Signage and Notices (see Sect. 2.13) indicates this Home Zone is a 'No Street Drinking Zone'; CCTV used for Crime Prevention and Community Safety; the location of the Ropewalk Community Room; and how at Elbridge House, only Permit Parking is allowed.

Worthington Street Home Zone, Leicester

Worthington Street was reported to be the first fully implemented Home Zone in the UK created during the mid-1990s. Located in the inner-city district of Highfields, Leicester, the area contains a diverse mixture of ethnic groups. It's 80 terraced houses date from the mid-Victorian era with no front gardens, driveways or rear access parking. Consequently, owners have to park their cars in the street.

Worthington Street residential precinct (formerly a Home Zone), Leicester, England

A key motivation for creating this Home Zone, was closure to traffic causing a migration of vehicles through Worthington Street. The Home Zone was a unique compromise—in that it would not close off the street, but provide space where the residents could relax and socialise. Considerable consultation with the residents is reported to have been conducted.

Unfortunately, a subsequent survey of residents by Leicester City Council received a poor and largely negative response (Bridgestock, 2000). The majority asked for more car parking, suggesting the brick-faced raised planters and flower beds should be removed to provide additional spaces for vehicles to park.

A succession of visits by this author indicates the Worthington Street Home Zone suffers ongoing aesthetic problems. The cast iron bollards have been painted bright blue and white and the tired-looking signage now calls it a 'Residential Precinct.' Whilst the raised planters and flower beds that contained nothing but weeds and rubbish, have been replaced with parking bays. But road repairs have been executed with tarmac, rather than the brick paviors that were an intrinsic part of the original Home Zone. Similarly, the hanging baskets described as a key element of the original scheme have never been witnessed.

Three decades after the Home Zone was created, perhaps the attitudes of the residents have evolved. A change in the demographic make-up of the area may explain the increase in car ownership. But even with the raised planters removed, vacant parking spaces rarely exist. And were residents able to park their cars in neighbouring streets, a DOC perspective would suggest inherent dangers, e.g. vehicles being targeted by criminals when they are not parked in close proximity to the owners to whom they belong.

Magor, Gwent, South Wales

Amongst the Home Zones visited in Great Britain during the past two decades, that at Magor in Gwent, South Wales has been consistently maintained in pristine condition. Created in 2001/2002, the Magor 'Zone' (or 'Parth' in Welsh) encompasses the village centre. Traffic calming measures are employed, including in places the same height for both road and pavement/sidewalk (no kerb distinction) to create what is known as Shared Space. In Magor, this is historic and determined by its narrow village streets. But in other Shared Space environments the road and pavement are delineated by changes in road surface colour and texture. Signage in Welsh and English is plentiful. However, there have also been problems. By following Home Zone principles, Magor refrained from restricting vehicle parking in the village centre. However, with its churches, school, shops, pubs and post office, perhaps it was inevitable there would be demand to park as close as possible to these facilities. The result was double parking and on occasion, obstruction. Parked cars also caused damage to the war memorial/cenotaph at the heart of the village. Consequently, in 2015, double yellow lines (total restriction on parking) were painted around the war memorial and its status as a Home Zone removed.

Magor Parth (formerly a Home Zone) in Gwent, South Wales

At first consideration, this might be seen as a cause for regret. The Signage now simply refers to it being a ‘Parth’ or ‘Zone’. However, reported restrictions on the ability of the police to enforce road traffic law, combined with Community Consultation, led to this change. The latter should be applauded (see also Second Generation CPTED and Sect. 4.1 to of this handbook). Furthermore, it may also indicate that the Home Zone concept has itself evolved into a form more appropriate to the local community.

More worrying are concerns about the lack of height distinction between pavement and road (also referred to in Sect. 6.16, Pedestrianisation of Town and City Centres). In such centres, this merging of heights (a constituent element in creating Shared Space) also causes issues for the training of Guide Dogs for the Blind. Most importantly, repeated personal witness indicates how the absence of a defined kerb in busy town centres, means that pedestrians are liable to walk into the path of motor vehicles—including buses!

6.6 Bus Shelters

At first sight, the capacity of bus shelters to act as generators of crime, fear of crime (FOC) and anti-social (ASB) behaviour might seem a particularly British problem. Indeed, personal witness elsewhere in the world suggests that few other countries appear to suffer from this issue and rarely to the same extent—although the below photo taken by this author is from Amsterdam in The Netherlands.

'Burglar-friendly' toughened safety glass breaks easily and leaves no sharp edges. often used and vandalised in bus shelters

Evidence of such problems post-event can be seen in the form of graffiti (see also Sect. 8.2.1), broken glass especially if Toughened Safety Glass has been used to cover advertising panels) or where GRP (glass reinforced plastic) or polycarbonate are utilised as alternatives to glass (often in response to earlier outbreaks of criminal damage), broken fixtures, plus evidence of the bus shelter being used as a toilet—one of the few examples of combined Visual and Olfactory Environmental Cues (see Sect. 2.8 on this same issue).

It is generally acknowledged in the United Kingdom that such attacks on bus shelters began in the early 1970s. This almost certainly originated as ASB when the bus shelter performed the role of congregation point (Youth Shelter? see Sect. 5.13) especially during the hours of darkness and on occasion evolved into criminal conduct. Over the intervening decades, various strategies have been adopted to combat such vandalism, including the removal of the shelter and even the bus route (where it might be perceived, ASB and crime at the shelter has migrated onto the buses—often because that highly effective Capable Guardian, the 'bus conductor', has disappeared during the past half century). However, both these 'solutions' penalise the law-abiding, overwhelming majority of members of the community at the behest of a criminal minority—neither a morally equitable or sustainable solution.

More logical and effective solutions have concentrated on the fabric of the shelter itself, (its 'micro-environment'—the Situational Crime Prevention/ CPTED/DOC approach). Three decades ago that success appeared to have been achieved by replacing regularly smashed glass panels with ones made of polycarbonate/glass reinforced plastic (GRP). However, offenders soon learnt that these would burn, especially if a solvent-based liquid or gel was first poured over the panel. Consequently, polycarbonate/GRP fell out of favour and was replaced with Toughened Safety Glass. But once again, vandals soon realised that when broken such glass would smash into thousands of small pieces. And more recently, many forms of glass in the public environment have been subjected to graffiti attack by means of a coin being scraped across the surface. Personal witness suggests that one of the worst examples in the world of such coin on glass criminal damage, was on the carriages of the S + U-Bahn systems in Berlin, Germany.

In the great scheme of crime concerns, piles of broken glass, coin-scraped windows and burnt polycarbonate/ GRP panels are rarely evidence of serious offences. But, as Visual Environmental Cues (see Sect. 2.8) they carry a powerful message to ordinary citizens using that bus shelter during daylight hours: ASB and worse occur at such locations (and on the buses?) at certain times of the day—when it is perceived that most criminal behaviour takes place. All told, such criminal damage directly fuels the FOC.

During the early years of the twenty-first century, innovative solutions have removed any form of glazing from the bus shelter. However, the use of steel mesh screens and similar features have tended to produce shelters that are rarely aesthetically pleasing if not implicitly fortress-like and ugly—not a trait the Designing Out Crime Officer wants to support unless willing to fall into the trap laid down by our critics.

More recently, some bus shelters have gone high tech using Lighting, electronic bus timetable display units, loudspeakers relaying music and even TV screens. Many would counsel that this is a recipe for criminal disaster, but to date such 'Super Bus Shelters' appear to have fared remarkably well and not been subject to the vandalism that so many doomsayers predicted. More research and evidence please.

Because problems that relate to bus shelters are often location specific, another solution may be to change the actual position of the bus shelter—to move it elsewhere. Siting it outside a police station (as at Wolverhampton Central in the city of that name) where Capable Guardians lie in wait on a 24/7 basis, may be effective in eliminating vandalism and other forms of crime and ASB, although once again personal witness suggests otherwise. Moreover, not all Capable Guardians want the bus shelter outside their premises and in the instance of (for example), a retirement home for the elderly, there may be a strong case to have the bus shelter moved—especially if it is acting as a generator of ASB (noise in particular).

However, there is a wide body of practical experience amongst DOCOs pointing to the bureaucratic hurdles that have to be jumped to relocate a bus shelter as little as 100 metres. And of course, if a particular shelter is recognised locally as being a generator of ASB and worse, no one will want to have it moved to a position in front of their home/premises!

Finally, all these solutions point to the efficacy of problem-solving, Problem Orientated Policing and Partnerships (see Sect. 2.14), in an attempt to eliminate/drastically reduce the level of ASB and crime that can be generated by a bus shelter (or any other public facility). In this respect, working with 'particularly motivated individuals'(see Read and Tilley, 2000; Bullock, Erol and Tilley, 2006 p. 15), within partner agencies and elected officials, can be highly effective and deliver positive and sustainable results. There is then the potential for the success of such Partnership Working to become ingrained and replicated in respect of new and different challenges within the built environment.

6.7 Place of Worship Safety and Security

By their very nature, the majority of places of worship have existed in situ for many decades, if not centuries past. Consequently, the capacity to Design Out Crime (DOC), Fear of Crime and ASB at the pre-planning stage will be a rare commodity. However, it is an unfortunate feature of modern British society (and beyond) that the premises of all faiths have been targeted by those intent on burglary, theft, criminal damage and worse. The example of the Parish Church of St. Barnabas, Erdington in the Diocese of Birmingham is a classic example in this last regard. Subject to an arson attack which caused considerable fire damage in 2007, the rebuilding/refurbishment of the church was only finally completed in 2013. Considerable DOC advice in pursuit of a Secured by Design award, was provided in respect of that project.

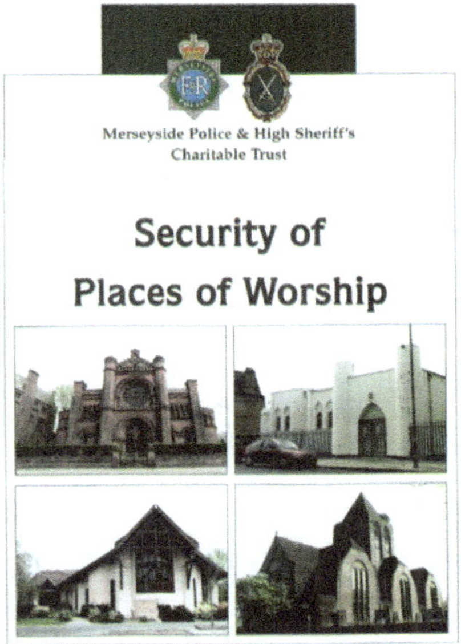

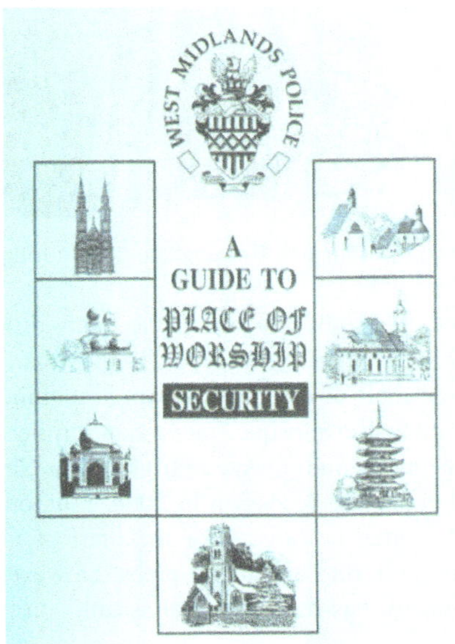

Merseyside Police and West Midlands Police Places of Worship security guides

In *A Little History of the English Country Church* (2007), Sir Roy Strong (former Director of the Victoria and Albert Museum in London) powerfully contends that as long as a single worshiper is willing to say a prayer within the walls of a church, the building is not a museum. This is surely true of any place of worship and acknowledged by those delivering safety and security guidance. As it is often the DOCO who is tasked to provide Crime Prevention advice at these important centres of worship and Community Interaction, current best practice in thwarting criminal attention follows in this chapter. NaCTSO (2009) Counter Terrorism Protective Security Advice for Places of Worship and Crowded Places (2020) have now been withdrawn. See instead https://www.protect.police.uk for potential funding opportunities, DOCOs in England and Wales should access the annual Protective Security Schemes for Places of Worship at https://www.gov.uk/guidance/places-of-worship-security-schemes.

Parish church of St. Barnabas, Erdington, Birmingham, rebuilt following an arson attack

Over recent decades, places of worship of the many different faiths have suffered attack, involving a variety of crime types. In the West Midlands Police area, this has ranged from the theft of ceremonial swords within a Sikh temple, a safe being broken into at a Muslim mosque, racist hate crime and criminal damage attacks at a Jewish synagogue and repeated burglaries at Christian churches. And even before the world economic recession in 2008, religious premises belonging to all the faiths had become the targets for the theft of metal—usually copper and lead used in roofs, guttering and down-pipes. However, across the UK and beyond the greatest concern has been physical assaults on the religious clergy of all faiths.

A Property Crime Wave took place at the close of the 1980s when professional/career/travelling criminals (see Sect. 2.3) were stealing (sometimes to order) specific articles from premises belonging to the Church of England. Such items had often been in the possession of the church in question for several hundred years. Analysis by the Ecclesiastical Insurance Group (the biggest insurer of religious premises within the UK) identified the eventual disposal route for much of the property stolen: Elizabethan silver candlesticks and salvers traded via the Brussels antiques market in Belgium; Restoration period furniture and mediaeval chests were popular in Germany and the Low Countries; brass eagle lecterns had become highly desirable in the homes of the rich in South America; christening fonts used as garden ornaments; and centuries old stained glass had been located in one Tokyo restaurant in Japan!

That Crime-Wave and publication of Lost Treasures of Britain—Five Centuries of Creation and Destruction (1990) again by Sir Roy Strong, were key inspirations for this author in deciding to enter the disciplines of Crime Prevention and Designing Out Crime.

As early as 1993, West Midlands Police organised a Crime Prevention conference and produced its first multi-faith guide (written by this author), both dedicated to Place of Worship Security. Drawing on the earlier guides produced by John Savage and Sergeant Alan Walker, it concentrated on the Physical Security 'Target Hardening' Measures and Technological Innovations of CPTED (see Sects. 4.2.1–4.3.3). However, even at that time, the relevance of the Environmental Elements (see Sects. 4.4.1–4.4.6) was recognised and included within the manual. In particular, the concept of securing the area around the building itself and in effect, deploying the psychological imperative of attempting to persuade potential thieves not to approach the place of worship.

Repeated updates of that guide developed this theme of Environmental Elements and incorporated the classic Layered Security 'Onion Peeling' approach (see Sect. 2.3) and Problem-Solving, Problem Oriented Policing and Partnerships (Sect. 2.14). Indeed, it now includes a chapter devoted to the *Personal Safety* for Priests, Gurus, Imams and Rabbis—a sad necessity in the early twenty-first century. Similarly, whilst there is no desire to turn a place of worship into a fortress, far fewer churches, mosques, synagogues and temples now leave their doors unlocked on a 24/7 basis. This is regarded by many as a regrettable development, but it has delivered immediate benefits in terms of enhanced safety and security.

A 'Crime Prevention Policy' for the Place of Worship

It is strongly advised that each church, gurdwara, mosque, synagogue or temple appoints a sub-committee to establish a Crime Prevention Policy (sometimes referred to as a Risk Strategy or Threat and Risk

Assessment—TARA, see Sect. 1.5.15), thereby ensuring a systematic approach. The sub-committee can create a list of the potential risks to those premises and the people who use them, calling on advice from the local DOCO, or Crime Prevention/Reduction Officer and/or Police Neighbourhood Team officers, wherever possible. Once the Crime Prevention Policy has been created, one person should have responsibility for co-ordinating implementation of the decisions made including:

- Reviewing security precautions and liaising with the DOCO, CPO/CRO, or Neighbourhood Team Officer
- Liaising with other members of the sub-committee
- Conducting the 'start of day' and 'end of day' security checks at the place of worship to ensure

    i. No combustible materials are easily available
    ii. Flammable liquids are removed from the place of worship
    iii. Valuables are safely locked away
    iv. No person is left within the place of worship
    v. Alarm systems are switched off or on—as appropriate
    vi. External security lights are in full working order

- Reporting relevant matters to the local police
- Liaising with the insurance company.

The Perceived Problem
The risks from crime, FOC and ASB will vary, depending upon a range of factors:

- The DOCO, Crime Prevention/Reduction/local police neighbourhood team officer or police community support officer (PCSO) can provide details of local crime trends, or via the https://www.police.uk website
- Specific crimes are more prevalent at certain times, such as vandalism during school holidays
- Items within places of worship may have intrinsic marketable value. For example, even badly damaged wooden chests may be highly prized as antiques
- The ease with which property may be stolen: altar crosses and ceremonial swords are easier to steal than church fonts made of stone

- The vulnerability of security devices. Older safes may be easily forced open, and locks are only as effective as the doors and frames in which they are fitted (see Sect. 4.2.1, Doors, and Door Security)
- Even opportunist thieves will search for keys in nooks and crannies, the very places where they are invariably hidden in places of worship. They will also quickly force locks and climb over internal partitions
- The vulnerability of items to arson and criminal damage. Vandals are more likely to throw stones if there is a supply of 'ammunition'. Whilst arson will be more prevalent if matches (for candles) or fuel (for lawnmowers) are readily available.

Burglary, theft, criminal damage and arson are the most common offences committed against places of worship. And in the wake of the worldwide price hike and demand for metal, copper sheeting and lead became targets for thieves. Similarly, silver, brassware, furniture and electrical items (such as heaters and vacuum cleaners) remain popular with criminals. In churchyards, gravestones and ornaments are often subject to criminal damage, or stolen to become features in landscaped gardens.

Exterior Issues: Perimeter Walls, Hedges, Fencing, Gates and Grounds
A well-cared-for appearance helps to create a feeling of community pride and well-being, as well as helping to deter opportunist theft (see also Sect. 2.3, Crime Opportunity including the Broken Windows theory of Kelling and Wilson (1982); and Sect. 2.8, Environmental Cues). A defined perimeter encourages people to use designated entrances and footpaths, thereby drawing attention (assisting Natural Surveillance, see Sect. 4.4.2) to anyone in the grounds. Similarly, preventing access by vehicles will obstruct criminals' intent on removing large objects like fonts, safes, furniture and copper or lead from roofs.

Stone or brick walls should be well-maintained and graffiti (see Sect. 8.2.1) removed immediately. Unrepaired damage suggests nobody cares and encourages offenders to commit crime. Gates should be properly maintained and ideally, self-closing. It is important that they cannot be removed from their hinges, a feature easily achieved by burring over the top of at least one hinge-pin, or welding a piece of metal to it. Double gates allowing access by vehicles need to be locked when not in use. Locks should be oiled and maintained in good order, with padlocks close-shackled to prevent attack by bolt-croppers. Chains should be constructed of high-tensile steel.

Considerable importance should be vested in the general appearance of the grounds surrounding the place of worship and at the very least, footpaths properly maintained and gravestones clearly visible. Rubbish tips should be out of sight and regularly cleared. To help prevent the theft of gravestones and memorial items, relatives should be encouraged to have each item engraved with the name of the deceased. Vandalism may be deterred by regular visits. Unfortunately, both evidence and experience across the UK demonstrate that acts of vandalism have a history of being committed by children and young people (although by no means exclusively). Consequently, evenings and school holidays are times that require extra vigilance.

Trees, especially those with a low canopy like conifers and Yew trees, obstruct Natural Surveillance (see Sect. 4.4.2) and provide cover for criminals including vandals. Where appropriate, lower branches should be cut from the main trunk to a height of 1.8 metres (six feet) above the ground.

Wooden sheds should not be used to store valuable items such as power tools, heaters and lawnmowers (or their fuel). All outbuildings should be kept locked and windows obscured (net curtain is ideal) to conceal the contents. Spades and ladders can be utilised by criminals and should be stored away—preferably locked to secure fixtures with close-shackled padlocks and heavy-duty chains, or raw-bolted to a solid wall.

Physical Security Issues: Roof, Fall-Pipes, External Doors, Windows, Security Lighting

At places of worship, access to the roof is often relatively easy due to large cast-iron fall-pipes and purpose-built, fixed iron ladders.

Use of Vandal-Resistant Materials Including Anti-Climb Paint

In order to prevent criminal access, anti-climb paint can be used on fall-pipes, above 2.4 metres (8 feet) from the ground. This paint does not dry and is very slippery. It is good practice to advertise, courtesy of written notices (see Sect. 2.13, The Value of Signage and Notices), the fact that such paint has been used. Roof materials can also be painted with bituminous or similar coatings to deter theft. The first rung of fixed ladders should be no lower than 1.8 metres (6 feet) above the ground, and the section beyond this protected by a hinged metal cover.

A common crime associated with churches across the UK is the theft of lead flashing from roofs. Unless regular checks are made of the roof, such a theft will only be discovered when it rains! If replacement or repair is necessary, it is advisable to use materials that have no resale value yet are similarly efficient, if not better. However, this may not be permitted if it is a Listed Building on grounds of its architectural merit, or is subject to regulation by place of worship authorities.

Low-level roofs may also cause problems. Roller devices or plastic 'cacti' can be fitted to discourage the criminal or vandal and thereby prevent access and theft. These must be fitted well out of reach to prevent accidental injury. Such security solutions are rarely aesthetically pleasing and as a Visual Environmental Cue (see Sect. 2.8) they may also raise the fear of crime. Similar solutions can be utilised for parapets where again, anti-climb paint can be used.

Access to cellars should be restricted to one entrance, ideally from inside the place of worship. All other entrances should be permanently sealed by being bricked up. If this is not possible, flaps and doors should be secured internally by means of bars and padlocks. Any retained external entrance to a cellar should be given special attention and secured with a doorset that meets the requirements of SBD—Non-Residential Guide 2025, providing regulations permit.

Doors

Boiler room doors may need strengthening, especially where they also provide access to the main building. Moreover, considerable damage can be caused to the structure of the building if the central heating system is wrecked by vandals during the winter months—or stolen for the resale scrap value of the copper boiler and piping.

External doors are the natural entrances to the place of worship and are consequently the preferred means of attack by criminals. The value of a door does not stop at preventing unlawful entry. Large items stolen from places of worship are almost always removed via a door. It is thus equally important to ensure that when locked, such doors cannot be opened from the inside (unless a Fire Door). Five lever mortise deadlocks to British Standard 3621:2017 are the easiest way to secure existing entrance/exterior doors and should form the minimum-security standard, with a lock fitted one third from the top of the door and a second one third from the bottom (to spread the load when the door comes under attack).

All doors should be of substantial construction with strong hinges and effective frames. External doors need to be at least 44 mm (1.75 inches) thick, with a reminder that the strength of a door is only as good as its frame and hinges (thus the value of the aforementioned pre-constructed doorsets to the standards and certification detailed in SBD Non-Residential Guide 2025 (or the most recent version thereof).

With conventional external doors, a slight movement in its structure or in the hinge fittings can make the door highly insecure, and it should be repaired immediately. Again, good maintenance improves physical security and can act as a psychological deterrent—see Sect. 2.12, Symbolic (Psychological) Barriers.

The lock case of large double doors should be at least 150 mm in length. Old locks are usually inadequate, having simple mechanisms that will not deter even an opportunist criminal. If ancient locks are to be retained and the fabric of the door preserved, additional mortise deadlocks should be fitted (or if a Listed Building alternative security arrangements made).

Windows

Windows are the second most favoured means of access for criminals. Lead framed windows are removed easily, noiselessly and leave no sharp edges. They are also prone to accidental damage (stones thrown up by lawnmowers, etc.) and by vandals. It should also be remembered that many stained glass windows possess substantial antique value and as a result, have increasingly become the specific target for thieves. Consequently, the extent to which windows will need protecting will depend upon:

- The financial and aesthetic value of the windows
- The likelihood of damage—including proximity to any public roads and the current frequency of damage
- The level of Natural Surveillance from nearby residents
- The value of the property within the place of worship. The majority of burglaries are committed to steal specific items of value the criminals already know are on the premises. It naturally follows the fewer the items of value, the less likely the chance of burglary.

Protection against criminal damage and burglary can be achieved by fitting welded-mesh or expanded metal grilles on iron frames, attached to the window and bolted securely, or built into the masonry. Another alternative is to use additional internal barriers of *Laminated Glass* or polycarbonate (GRP—glass reinforced plastic) set in similar well-secured frames. These are more aesthetically pleasing than grilles, but are also expensive and may not offer the same level of protection/deterrence. Where fitted (most usually in toilet or utility rooms), louvre style windows should be replaced ASAP, being inherently insecure with the panes of glass easily and silently removed to provide easy access.

Wherever possible, replacement windows should be double-glazed, constructed and certificated to one of the standards detailed in the SBD Non-Residential Guide 2025 (or the most recent version thereof).

Security Lighting

External Security Lighting (see Sect. 4.3.1) can be an excellent deterrent to criminal attack, providing there is at least some realistic hope the criminals will be seen (see Sect. 4.4.2, Natural Surveillance). No criminal desires to be bathed in light if he fears it might lead to his arrest and therefore, the more appropriate the Security Lighting the greater its deterrent value.

Historically, high pressure sodium units have often been used. However, in recent years, LED technology has rapidly taken their place being more economically efficient in the long-term. Electronic timers require adjustment throughout the year, due to daylight changes and in the event of power failures. Alternatively, photo-electric switches will illuminate such Security Lighting automatically and ensure operation, regardless of changing lighting-up times (daylight saving).

Ground-based security lights are also regularly installed. However, these are vulnerable to theft and criminal damage and therefore, should be firmly fixed and protected by metal grilles. Cabling should also be protected when it can be easily reached. However, the use of metal conduit to protect such cabling is no longer advised, due to vulnerability to theft. Plastic conduit is ideal and should also be used for the aforementioned grilles and wiring for any intruder alarm system (see Sect. 4.3.2).

Passive Infra-Red (PIR) detectors can be used to signal any person approaching the place of worship. They are cheap to purchase, install and energy efficient. However, they are also prone to activation by small animals and swaying branches—and thus have the capacity to raise the FOC.

It should also be remembered that at night and once inside a place of worship, criminals may have ample time in which to execute their crimes. Furthermore, many churches are located some distance away from occupied premises and have thick, soundproof walls, enabling the criminal to work without the need for silence. Therefore, any security within the building should be of the standard necessary to protect the property stored. Security levels need to be realistic and commensurate (appropriate to the perceived threat and risk—see Sect. 1.5.15) to the value of the property being protected.

Internal Security—Safes, Offertory Boxes and Parish Records

Items of exceptional value (such as medieval chalices of silver or gold, illuminated Bibles and ancient parish records) should never be kept inside an unoccupied place of worship, unless considerable resources are spent on additional security and alarms. Nor should they be kept at the home of a senior member of the clergy, unless appropriate security arrangements exist.

These items can be stored in local/national museums, Diocesan Records Offices or bank vaults. Copies can be made for daily use, and the original items only displayed on special occasions with the necessary security regime. However, if despite this advice the intention is to keep such items within a place of worship, as a minimum they should be stored in a modern safe, set and securely fixed into either a substantial wall or preferably, the floor.

Safes vary in price, according to their security grade and insurance rating. Those that are designed to protect documents against fire are usually opened fairly easily by the criminal. An effective safe, firmly fixed in the wall or floor, will be secured by means of either a combination lock, or key and combination lock set. A freestanding safe should be bolted to the floor and if possible, also to the walls. If it weighs over one tonne, this may not be necessary, but in no circumstances should the safe be located in a public area of the place of worship—where it can be seen by visitors.

Offertory and other collecting boxes provide easy pickings for opportunist thieves. However, such crimes can be easily prevented if the following measures are adopted:

- Wooden/free-standing offertory boxes should never be used
- A metal wall box with a flush-fitting heavy metal front should be bolted and grouted into the masonry
- The box should be emptied on a regular basis, allowing only a small amount of cash to accumulate
- Affix a sign stating that the box is emptied regularly
- Never leave the key to such a box inside the place of worship: criminals instinctively know where to search.

Storage of Valuable Items

Parish Registers and Records should not be stored within the place of worship. Moreover, in the instance of the Church of England, the Parochial Registers and Records Measure passed by General Synod in (1978) specifies that all documents over 100 years old should be deposited in the Diocesan Records Office.

Electronic Alarm Systems

When considering the installation of an Intruder Alarm system (see Sect. 4.3.2), it should be remembered that such a device will provide a warning at the earliest possible moment, of entry or attempted entry by a criminal into the place of worship. Furthermore:

- The visual presence of an Intruder Alarm system will often deter criminals, especially opportunists
- The alarm reduces the time to commit a crime
- Physical security alone can be insufficient and too expensive to adequately protect a building
- A good quality alarm system can be cost-effective.

The Intruder Alarm system can be activated by making/breaking electronic contacts on doors and windows, or by inexpensive movement detectors (PIRs). Their installation may also mean qualification for a reduction in insurance premiums.

Two categories of Intruder Alarm system are most often used in places of worship: Remote Signalling Alarms, linked to an Alarm Receiving Centre, where the necessary corroborative activation will summon the police; and 'bells only' systems requiring someone to hear the alarm and telephone the police with additional corroborative information, to ensure police attendance. Advice should be sought from the local police, before installing an alarm and financial quotations obtained from at least three alarm companies. In the instance of Church of England buildings, it may also be necessary to apply to the diocese for a 'faculty'.

CCTV

Closed Circuit Television (CCTV—see Sect. 4.3.3) is increasingly used to enhance security arrangements in places of worship, especially larger venues with visitors constantly passing through. If installed unsympathetically, such systems are rarely aesthetically pleasing—if not ugly and intrusive! Once again, the advice of a police DOCO should be sought.

Inner Secure Areas

Ordinarily, such areas should be limited to those parts of the place of worship that can be viewed from the main body of the premises. All other areas should be kept locked and used to store items of medium value such as kettles, heaters and vacuum cleaners. If substantial doors with locks cannot be fitted, it is better to leave the door unlocked and ensure no item of value is stored inside. A notice marked 'Private' will deter the majority of casual visitors.

Action to Be Taken Following a Burglary or Theft

Confrontation of the Criminal

Under NO circumstances should criminals who are disturbed be confronted! Assault may follow and an extensive knowledge of the law is necessary in order to avoid the possibility of unlawfully detaining a person and the resulting civil litigation and financial penalties.

Action When a Criminal is Disturbed

- Telephone 999 immediately
- A good description of the offender(s) will be of immense benefit to the police, as would that of any vehicle used, its registration number and the direction in which the vehicle and/or person(s) travelled.

Initial Action on Discovering a Crime

- On discovering a burglary (break-in) or other crime has taken place, telephone 101. Nothing else should be done until the arrival of the police or other advice is given. Preservation of the scene will assist the Forensic Scene Investigator/Scenes of Crime Officer/Crime Scene Investigator and photographer
- No area or item should be cleaned or touched, and it is unnecessary to establish what has been stolen until a police officer or PCSO arrives and states it is safe to enter, or other police advice is given.

Information the Police Will Want to Know

- The police will try to establish the exact time and date of the offence. Therefore, they will want to speak to the last person who left the premises intact and the person who first discovered the theft or burglary
- The police will also be interested in the method of entry and the manner in which the property was removed or damaged. In addition, they will require full descriptions of all property stolen, individual values, photographs if at all possible and details of any Property Marking.

Property Marking and Photographing Valuable Items

It is generally accepted that Property Marking can be an effective method of making items identifiable, both to the owner and the police. Moreover, Property

Marking is most often a quick, do-it-yourself task costing very little. In the UK this is usually accomplished by using the postcode (zip-code) of the premises, together (in this instance) with the name of the place of worship. This is done to ensure that stolen property, once recovered and identified, can be returned to the owner.

Property Marking can also have a potential deterrent effect, and it is useful to place small notices in the place of worship, to the effect that all the property is marked (see again Sect. 2.13, The Value of Signage and Notices). Methods of Property Marking include:

Proprietary Marking Systems

Systems like 'Micro-Dot', 'Red-Web' and 'Smartwater' can be employed to mark valuable items of property. Indeed, in the UK the Ecclesiastical Insurance Group uses 'Smartwater' to mark and identify property within 25,000 churches and other heritage buildings that it insures.

Ultraviolet Marking Pens

These are inexpensive pens, obtained from stationers, and the mark is invisible unless viewed under ultraviolet light. The pen can be used on any porous material, such as wooden furniture, vestments, fabrics, books, certain plastics and on unglazed or semi-glazed pottery. In sunlight, such marks will fade after about 18 months when it is necessary to remark the item.

Engraving

This can be achieved by using a diamond-tipped pen, other sharp pointed tools, or an electric engraver to place marks on any metal, plastic or glassware items. Stencils can be used to permit relatively small lettering.

Ceramic Markers

Titanium pens are effective on ceramic surfaces and their markings impossible to remove without damaging the glaze.

Permanent Ink Marking

Utilises a solvent-based ink advertised as 'permanent' and is an effective means of Property Marking items where a fairly large, visible mark is unimportant.

Marking Dies

Suitable for heavy items such as brass eagle lecterns, garden tools, metal flower pedestals and cast iron urns.

Branding
A soldering iron or similar device can be used to apply an identifying mark to wooden or plastic items.

Photographs
A permanent photographic record of valuable items will enable an extensive description to be circulated in trade publications and also assist in later identification. Photographs should be taken against a neutral background with a ruler/tape measure alongside to indicate actual size. Ideally, a colour chart should be included.

Each photograph should contain only one item, yet include hallmarks and any other identifying feature if possible. Flashlight should only be used by a proficient photographer. When items are stolen, such photos are often circulated on websites, e.g. Antiques Trade Gazette, Art Loss Review, Art Theft, Interpol, Salvo, The Art Loss Register and The Art Claim Database—ones specialising in the recovery of stolen antiques and works of art.

Good Housekeeping Measures

Key Security
It is imperative that key security is prioritised at any place of worship. Keys should never be left in locks, even when the building is occupied. There should be as few a number of keys as possible, kept in the possession of named, responsible people. Keys should never be left outside the place of worship, under a mat or in any other supposed hiding place: thieves will always find them!

Professional Descriptions
These are of immense value in respect of furniture and ornamental items, especially as circulation to other police forces relies upon accurate descriptions.

Inventories
An inventory should be kept of all the items in the place of worship. One person should have responsibility for this task, recording accurately any tears, scratches, splits, dents or imperfections as they occur. Two such books of records should be kept, updated regularly and stored in safe locations—never at the place of worship.

Substitute Items

Metal, wooden, or plastic replicas of valuable items are relatively inexpensive and highly realistic. However, it is advisable to indicate to visitors that these are copies (by means of small notices), so as to 'warn off' the criminal and help prevent these same replicas from being stolen!

Fire Regulations

It is important that access to fire exit doors and the free passage to them is not restricted, when parts of the building are locked and other sections open. Further information can be obtained from the local Fire Service.

Reporting to the Police

Small fires (arson attacks) or instances of minor damage should always be reported to the police, as they may be an indication of far worse to come (Repeat Victimisation) and/or indicate a pattern in the neighbourhood area.

Arson Prevention

Arson (deliberate fire starting) is potentially the greatest threat to the very existence of the place of worship. According to the UK's former Arson Prevention Bureau, the majority of fires in churches are begun deliberately. Indeed, the Parish Church of St Barnabas, Erdington in the Diocese of Birmingham and rebuilt following an arson attack, is a timely reminder of the true cost to the community of such crimes. See also the award-winning Keep Your Business In Business (2009) produced by the former West Midlands Arson Task Force.

Community Responsibility

Whilst no longer categorised as the crime of 'Sacrilege' under English law, crimes committed against places of worship strike at the heart of the community. Protecting the building and its grounds can be enhanced under an official 'Churchwatch' (or 'Gurdwarawatch,' 'Mosquewatch,' 'Synagoguewatch,' or 'Templewatch') scheme, or by incorporation within a Neighbourhood Watch area. Once again, the local police should be the first point of contact.

Whether to Lock or Unlock the Place of Worship

The decision on whether to keep the place of worship locked, or allow unrestricted access by visitors, is often a difficult one to make. Indeed, as Sir Simon Jenkins (1999, p. xxxii) describes, accessibility is 'the most vexing topic amongst church enthusiasts.'

The following considerations should be intrinsic to such decision-making:

- The place of worship should always be locked at night
- Isolated places of worship are more vulnerable than those with Natural Surveillance (see Sect. 4.4.2) from nearby premises
- Unlocked places of worship are more vulnerable to theft and criminal damage than those that are locked
- When places of worship are left unlocked, there is a reduced risk of damage if an official is present at all times
- Thefts are less likely if valuables are removed and secure inner areas maintained
- However, income from offertory and other collecting boxes may be substantially reduced in locked places of worship.

The following options are therefore available:

- To leave the place of worship open and unattended at all hours (strongly inadvisable)
- To lock the place of worship at night and leave unattended during the day (inadvisable from the perspectives of safety and security)
- To lock at night and provide passing attention during the day (the hostage to good fortune option)
- To lock at night and provide an attendant (perhaps on a rota system), at all times whilst open
- To lock at night and encourage ways to make the place of worship subject to regular use
- To keep locked at all times with notices displayed indicating from where the key may be obtained
- To keep locked at all times when not in use (the safest option).

Obviously, the final decision will be made by each place of worship, but it should be remembered that High-Crime Areas throughout the country are characterised by low organisational participation amongst residents.

Special Risks
The place of worship may incur special risks at certain times of the year. For example, during religious festivals when precious objects may be on display in the building or grounds. In such circumstances, constant public surveillance by officials of the place of worship is necessary.

Personal Safety

It is a sad reflection of our times that it is necessary to include a sub-chapter dedicated to the Personal Safety of both clergy and worshipers. Most importantly, it is fundamental to understand that the chances of being physically attacked are in reality (and contrary to commonly held public and private perceptions relating to the FOC), extremely low. That risk can be reduced still further if the following simple precautions are appreciated and applied on a routine basis:

- During the hours of darkness or when travelling via or to locations that are believed to contain special risk, always try to be accompanied by another person—especially when on foot
- Try to appear self-confident, assured and travelling with a purpose
- If on foot, walk facing the traffic, in the centre of the pavement and avoid isolated footpaths and short-cuts.
- Instead of walking alone, try to arrange a lift with someone you know
- Use taxis and private hire vehicles from companies you know and trust
- On public transport, sit close to the driver and avoid empty carriages on trains and upper decks on buses
- If driving alone, plan your journey in advance. Ensure you have enough fuel, carry a mobile phone, place bags and valuables out of view and keep doors locked
- If you carry a personal attack alarm, use it if the occasion demands or in its absence, shout at your attacker. It just might frighten them away!

For specific advice relating to Personal Safety, the Suzy Lamplugh Trust (044 207 091 0014, https://www.suzylamplugh.org); Paladin National Stalking Advocacy Service https://www.paladinservice.co.uk); and the National Stalking Helpline (0808 802 0300, https://www.stalkinghelpline.org) are all especially helpful.

Callers at the Door

Some of the most serious assaults against members of the clergy have occurred as a result of the assailant calling at their home. Ideally, callers should be restricted to a specific 'vestry hour' and visits on a 24/7 basis declined/prevented. But in all such circumstances, the following advice is recommended:

- Display a notice by the relevant door detailing the times of the day when callers will be seen
- Install an external light and use a door viewer or side window to identify callers, before opening the door

- Fit a 'panic attack' button adjacent to the door used by callers, linked to the Intruder Alarm system, or stand-alone audible alarm
- Never invite an unexpected caller into the home. Arrange to see them at a later date/more convenient time of the day
- Keep close at hand a list of other agencies that may be able to assist the caller, such as Alcoholics Anonymous, Citizens Advice Bureau, Social Services, etc.
- Operate a 'no money' policy to requests from callers. If you feel obliged to give anything, provide food
- If despite the above advice you decide to invite a caller into your home, set limits as to the duration of such a meeting and ideally conduct it in an office
- Never leave the caller alone and have a pre-arranged code system to summon other members of the household and/or call the police.

Despite the necessity to include the above advice, fear of crime remains a far greater reality than that of crime itself. Nevertheless, being forewarned is forearmed and relatively minor planning can reduce that minute risk even further.

Conclusion
Whilst the instances to Design Out Crime from new building place of worship projects will be few and far between, unfortunately, there is an ongoing demand for safety and security advice from and across the whole spectrum of faith groups and religious premises. Hopefully, the information contained in this chapter will be of some benefit to those called upon to protect places of worship of any faith, their parishioners and clergy.

6.8 Rural Crime

As a practical definition, Rural Crime will embrace the full spectrum of Violent Crime and Property Crime that take place in rural areas. However, it will also include crime types that are specific (at least in part) to the rural environment, e.g. theft of agricultural machinery and plant, wildlife crime, illegal hunting, equine crime, poisoning of wildlife, criminal damage and collection of birds' eggs and plant species, 'nighthawks' (unlawful searching/acquisition of historical objects), criminal use of firearms and explosives, criminal trespass, dumping of rubbish, etc. Like Environmental Crime (see Sect. 8.2) the police may not have responsibility for prosecution. See Harris (2020) Rural Crime: Theory, Community and the 'rural idyl'.

Rural Burglary and Theft

In the UK, a number of forces have established (sometimes at the behest of their elected Police and Crime Commissioner) dedicated Rural Crime Teams. Indeed, in rural communities, there is often a perception (however misplaced) that policing and its resources are concentrated in urban areas and consequently, Rural Crime of all forms goes unchallenged when these take place in farming communities and beyond. The role of these Rural Crime Officers (RCOs—be they police or police staff) is to provide practical reassurance to rural communities. This can be delivered in a number of ways including: physical police presence; disseminating practical Crime Prevention advice that is bespoke to the needs of a specific community and crime types; acting as the 'eyes and ears' of those communities in gathering intelligence; and attending the scene in the aftermath of robbery, burglary, theft, criminal damage, assault, etc. In addition, the National Police Chiefs' Council (NPCC) launched its Wildlife and Rural Crime Strategy 2022–2025.

Theft of Agricultural Machinery and Plant

Tractors, trailers and a diverse range of agricultural machinery and plant are regularly stolen from farms (and construction sites). Thanks to the efforts of 'Operation Opal', NaVCIS (the police National Vehicle Crime Intelligence Service) and manufacturer cooperation, improvements in ignition system technology have improved dramatically in recent years. However, trailer and quad-bike theft are common in certain parts of the country and disposal routes for high value items repeatedly involve export outside the UK.

Wildlife Crime

Dedicated police personnel tasked with the role of Wildlife Crime Officer (WCO) have existed in the vast majority of UK police forces for more than a decade. They will often specialise in particular crime types (as described under the following headings) with priorities based on local incidence and management direction.

Poaching
Whilst specific legislation still exists in respect of Poaching, it is the more serious offences committed by Organised Crime Groups (OCGs) that now solicit police attention. For example: cattle and sheep rustling with the animals often butchered in the field; and explosives used in rivers and lakes to stun fish before wholesale theft.

Hare Coursing, Badger Baiting, Poisoning Wildlife, Illegal Forms of Hunting
The most recognisable forms of Wildlife Crime include: wildlife trafficking, badger baiting and forcing them to fight with dogs; the poisoning of wildlife (especially birds of prey); hare coursing and other forms of hunting with dogs—unless permitted by law. In recent decades, policing such hunting practices has become politically charged.

Equine Crime
Includes theft of horses and the full spectrum of offences committed against them—including maiming, cruelty and acts of dangerous driving by motorists when horses are being ridden on roads and bridleways.

Collecting of Birds' Eggs and Plant Species
Specific legislation relates to these offences and WCOs will often refer to the RSPB (Royal Society for the Protection of Birds) and the Environment Agency for professional guidance and prosecution.

Nighthawks
Treasure seekers who trespass onto land without the permission of the landowner are often known as 'nighthawks'. Usually armed with metal detectors, their actions in searching for objects will be unlawful if they do not have the landowner's permission. Moreover, under the Treasure Act (1996) gold and silver items that they find belong to the Crown and will need to be declared. Naturally, not all items are so declared and are sold to unscrupulous dealers.

Role of the Designing Out Crime Officer
Unless simultaneously tasked as a Wildlife Crime Officer (WCO), the DOCO's role in relation to Rural Crime will vary from force to force. For example, in some locations, DOCOs are required to visit the scene of every rural (if not all)

Residential Burglaries and on occasion Non- Residential Burglaries—depending upon the Modus Operandi, type of premises and nature of the crime.

In other instances, the DOCO's involvement to assist in preventing any of the crime types detailed above is likely to be made on a case-by-case basis—often at management direction. This is where the full spectrum of the DOCO's skillset can deliver benefits—either immediately or in the medium to long-term, e.g. through use of the SBD award system in new or refurbished dwellings, non-residential developments, etc.

Moreover, the DOCO can assist WCOs and Rural Crime Teams in Property Marking initiatives (including forensic- based systems), other Crime Prevention campaigns and 'high visibility' operations where advice and materials are distributed—for example during the National Rural Crime Week, at livestock markets, farming community events, farmers' shows, garden centres and in establishing various Rural Watch schemes.

In rural communities, personal witness and experience demonstrate how such encounters will immediately increase public confidence. And for the DOCO, such visits provide a 'foot in the door' opportunity to liaise with victims of crime when they are most willing to incorporate the Crime Prevention solutions suggested. For example, farmers and how they can make their premises more secure in terms of the Physical Security 'Target Hardening' Measures and Technological Innovations of CPTED. And via the Environmental Elements, e.g. designing out Attack and Escape Routes (including the capacity to divert public footpaths, where appropriate).

Similarly important is the intelligence that can be gleaned from these visits—much of it relevant to the DOC function, although it can also inform the overall police knowledge-base of crime occurrence in specific rural communities. The former will add to the DOCO's skill-set of which measures will and will not work at the practical level in a rural environment, at least according to the victims of crime and local population. This in itself has the capacity to reassure rural communities. And it also possesses the potential of applying the DOC philosophy to Rural Crime, especially in respect of the ability to remove or at least minimise Crime Opportunity (see Sect. 2.3).

6.9 Doctors' Surgeries, Pharmacies and Health Centres

Doctors' surgeries, pharmacies and health centres require specific design considerations, different to those of all other conventional non-residential premises. And not simply because controlled drugs are both stored and often available at such premises. These neighbourhood facilities are, without exception, a microcosm of local life and Community Interaction. And for the most part, such elements deliver highly positive benefits. However, certain functions involve the visit and treatment of individuals whose medical condition has the capacity to cause Anti-Social Behaviour (ASB) if not Violence—especially if they are refused access to the drugs to which they believe they are entitled. For example, intravenous drug abusers.

As with all such recommendations originating from the premise of safety and security, there is no desire to create a fortress at these venues. However, the safety of staff, patients and all other visitors is of primary importance. And by extension, the security of controlled drugs at such locations is imperative.

It is difficult to produce a single strand of recommendations applicable to all types and sizes of doctors' surgeries, pharmacies and health centres. Nevertheless, the recommendations that follow have been collated from past and current experience at such venues. See also Sect. 5.5, SBD—Non-Residential Guide 2025; Sect. 5.8, SBD—Hospitals 2005; and Sect. 9.4, The Village Surgery Cheswick Green.

Such premises will often take the form of stand-alone buildings—that is to say, there is no physical perimeter beyond that of the building itself. This in itself reduces the capacity to incorporate the Environmental Elements of CPTED (see Sects. 4.4.1–4.4.6) and therefore places greater emphasis on the Physical Security 'Target Hardening' Measures and Technological Innovations (see Sects. 4.2.1–4.3.3). In such circumstances, considerable effort must be made to enhance the physical security of the building envelope, as per the recommendations contained in the aforementioned chapters. However, recommendations specific to doctors' surgeries, pharmacies and health centres (and it is believed the minimum necessary) are included in the next section.

Advice from the Designing Out Crime Officer

None of the recommendations made should cause the doctors' surgery or health centre to become a fortress.

- SBD—Non-Residential Guide 2025 should be used as baseline guidance for the premises
- Advice should be commensurate to the perceived Threat and Risk Assessment (TARA—see Sect. 1.5.15)
- Patients and all visitors should normally enter and exit by way of the same combined entrance
- Where provided, a separate staff entrance should have access control and be located in an area inaccessible to members of the public
- Furniture and other fittings within the reception area should be screwed to the floor or walls, to prevent them being used as weapons or to cause damage
- The reception counter should be both high and wide enough to prevent staff from coming into physical contact with visitors and benefit from both Natural Surveillance (and where necessary) CCTV coverage (see Sect. 4.3.3)
- The floor behind the reception counter may be raised
- Any screen should use glazing that complies with BS 5544 and BS 5357
- A personal attack alarm control should always be fitted behind the counter
- Ideally, access to toilets for the use of visitors should be electronically controlled from the reception desk and not located in or near pharmacy areas. Similar electronic control may be advisable to the area containing the doctors' consulting rooms, nurse rooms, etc.
- Controlled drugs should be secured in a BS 2881:1989 storage cabinet
- Regular drug audit trails should take place
- Windows to pharmacies, record rooms, etc. should use either barred protection or steel shutters certificated to LPS 1175 grade SR 2 or 3
- Glazing shall be no less than the minimum set out in SBD – Non-Residential Guide 2025 and compatible with the necessary requirements of Lighting, surveillance and visibility at that location.

6.10 Shopfront Security

Perhaps the greatest negative aspect of Physical Security 'Target Hardening' Measures (see Sects. 4.2.1–4.2.4) is their capacity to be aesthetically ugly. Moreover, this can have consequent effects in terms of raising the Fear of Crime (FOC). Over recent decades, the classic example of this combination of ugliness and

FOC in the streetscape has been the growth in the number of shops installing solid roller shutters outside the front of their premises.

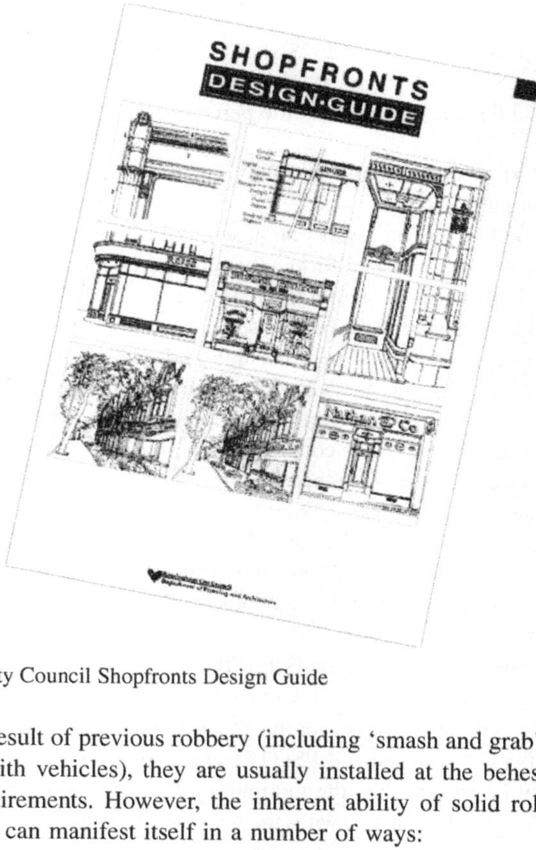

Birmingham City Council Shopfronts Design Guide

Often the result of previous robbery (including 'smash and grab', or 'ram raid' style attack with vehicles), they are usually installed at the behest of insurance company requirements. However, the inherent ability of solid roller shutters to raise the FOC can manifest itself in a number of ways:

- As a surface on which paint can be sprayed or other forms of graffiti (see Sect. 8.2.1) applied
- Kicked or otherwise hit and struck, whereby the roller shutter structure has the ability to create loud noise, itself a form of Anti-Social Behaviour (ASB)
- As a barrier to Natural Surveillance (see Sect. 4.4.2). When the shop is closed and the solid roller shutter down, there is no visibility from either the street into the premises or vice-versa. This in itself has the capacity to raise the FOC

for anyone walking past the shop, especially during the hours of darkness. Moreover, when criminals have broken into the premises from the rear and are now operating inside, their activities cannot be seen by persons walking or driving past—or the fear thereof by those contemplating such attack.

The occasional installation of a solid roller shutter can be aesthetically ugly, especially if later covered in graffiti or otherwise damaged. However, when a whole row of shops is fitted with these measures, it has the capacity to produce the very opposite of a warm and welcoming environment. Moreover, whilst the necessity for roller shutters to protect the contents of particular stores (such as jewellers) may be justified, in the UK, there has been a tendency that once one store has installed this security device, others with little if any desirable content soon follow. Desirable in this context means: concealable, removable, available, valuable, enjoyable, disposable (CRAVED – Clarke, 1999). Yet florists and gift shops have been amongst those eager to install solid roller shutters.

Shopfront Security Design Guides

Realising the necessity to produce guidance in respect of roller shutters and other forms of shop front security, in the UK, a number of local authority (local government) planning departments have produced specific design guidance in this area of business. Indeed, two such authorities, Birmingham and Solihull, sought advice from their Designing Out Crime Officer (Architectural Liaison Officer at that time) prior to publication.

The Birmingham City Council Shopfronts Design Guide (1996) was born from the success of the Moseley Shop Front Design Guide (1992). Moseley is a district of Birmingham with at its heart an architecturally distinct Moseley Village Victorian shopping area.

After consultation with the DOCO, the following passages were included within the Birmingham design guide:

> West Midlands Police deplore the use of solid roller shutters in an ordinary retail environment. Quite apart from the important issues of aesthetics and appropriateness to the risk involved, solid shutters are a target for graffiti, noise pollution when hit or otherwise abused by anti-social elements, and they create an environment which the majority of ordinary, law-abiding people perceive as oppressive and unsafe when the shops are closed. This is partly because Natural Surveillance is impaired when no light from inside the shop can fall on the pavement outside. Moreover, if burglars do still manage to break in, the shutters act as a cover for their activities – an important point to be stressed when advising storekeepers of their inadvisability...
>
> the police prefer the open style shutters/grilles to be used.

> West Midlands Police concur with the guide's preference for Laminated Glass... We never approve of the use of ordinary Toughened Safety Glass. This material is a burglar's best friend because it breaks easily... and it leaves no sharp edges.
>
> Laminated Glass also has a value in making ram-raiding much more difficult to successfully expedite, although because it demonstrates no visibly obvious difference from ordinary glass, its preventive value is somewhat limited. Once again we agree with the guide that stall-risers can have a considerable deterrent value in preventing ram-raiding. Ground to ceiling windows using Toughened Safety Glass are most vulnerable to this form of attack.

Stall-risers refers to the structure (often brick and on occasion reinforced with steel joists) between the ground and where the window starts. Personal witness and recommendation demonstrate how the use of a low-level brick wall as the stall-riser is often sufficient to deter a ram-raid style attack (where necessary, reinforced with steel supports). Whereas, ground to ceiling windows appear to encourage such Modus Operandi, especially with newsagents and off-licences—cigarettes and alcohol often being the intended targets within the 'Target Enclosure' (Ekblom, 2011a).

The Birmingham guide covers a number of subject areas:

- The need for Planning Permission
- The importance of Good Shopfront Design
- Details of Good Shopfront Design
- Shop Signs and Advertisements
- Security Shutters
- Roller Shutter Policy
- Ram-Raiding
- Other Shopfront Elements.

Advice from the Designing Out Crime Officer

Advice supplied by the DOCO and incorporated into such design guides should include the following:

- External security shutters and grilles require Planning Permission. Internal security shutters do not require permission, unless it is a Listed Building
- There will normally be a general presumption against the use of security shutters on Listed Buildings and in Conservation Areas

- Where security shutters are granted Planning Permission, they should not be of the solid, bare or painted metal style—that are repeatedly defaced by graffiti and other vandalism. Instead, they should be 'open' so as to allow light to pass through from both directions and Natural Surveillance from the street into the premises and in reverse. This preferred type includes those of aluminium rib and transparent polycarbonate strip construction
- A preference for fully open style of roller shutter, rather than perforated mesh type construction, which allows little light to pass through and few if any surveillance opportunities—from either direction. Moreover, during daylight hours mesh style roller shutters have the same appearance as solid ones. And they too can be sprayed with paint or attract other forms of graffiti and kicked, or otherwise struck to create noise
- In circumstances where there is a genuine high risk of ram-raiding, the stall-riser can be reinforced with internal steel posts and vertical steel mullions within the glazing. Use of street furniture such as concrete planters, seating (heavy and anchored to the ground) or trees, can also be considered.

In summary, Designing Out Crime influenced Shopfront Security is a highly practical example of how the role of the DOCO is to avoid a Fortress Mentality, and in so doing assist in fostering the concept of 'Elegant Security'—as advanced by Farrell and Tilley (2022). It also demonstrates that relatively minor amendments to the built environment can reduce crime, FOC and ASB.

Furthermore, by making recommendations to install aesthetically pleasing, open style security shutters style roller shutters that allow light and vision from both sides, this once again illustrates how the function of Designing Out Crime is essentially one of compromise (see Sect. 1.5.10). This has the capacity to solicit increased contact and co-operation from colleagues across the built environment, who might previously have feared our advice would include extreme forms of Physical Security 'Target Hardening' Measures and Technological Innovations (see Sects. 4.2.1–4.2.4).

Open style roller shutters allow light and vision into and out of the store when it is closed

6.11 ATMS (Automated Teller/Cashpoint Machines)

Special thanks are given to retired Metropolitan Police 'Flying Squad' Detective, Alan Townsend MBE, for his considerable and much-repeated assistance in the compilation of this chapter. And more recently, Graham Mott, Strategy Director at Link, the European ATM Security Team (EAST) and Chairman of EGAP.

ATM moved a few metres from this corner position where legitimate users felt vulnerable, to a conventional side of building location

Perhaps it was to be expected that the advent of ATMs (Automatic Teller Machines) or Cashpoint Machines as they are often known, combined with their now wholesale installation around the globe, would produce new Crime Opportunities. Indeed, a powerful case can be made that following the invention of the ATM half a century ago, these machines have attracted some of the most determined and innovative forms of Modus Operandi—especially since the beginning of this century. The lure of the bank notes they contain, combined with the number of ATMs, have undoubtedly been powerful motivations in this regard.

The Crime Opportunity (see Sect. 2.3) provided by ATMs can be deduced from their number: in 2011, there were 63,136 ATMs in the UK and 2,177,239 worldwide (Mott & Townsend, 2011). The UK total peaked at 67,000 in 2015 (3.1 million worldwide), and by 2023, this had fallen to 49,421 (House of Commons Library, 2023). And in 2025, 46,712 (link Network, 2025). It should also be acknowledged that the ATM Industry has been involved in an ongoing process of security innovation as a means of defeating criminal attack.

ATM crime

From the perspective of Designing Out Crime (DOC), ATM crime can be divided into four distinct categories:

1. **Targeting the Cardholder**
   - Distraction attacks—criminals distract the customer at the ATM, to steal the card or cash and obtain the PIN, usually by 'shoulder-surfing'
   - Card-trapping—where a device is placed inside the ATM card slot which in turn seizes the customer's card when it is inserted. The PIN is usually obtained by 'shoulder surfing' or a small hidden camera
   - Skimming—a device is attached to the slot, but on this occasion, it copies the card details and a miniature pinhole camera captures the PIN. A counterfeit card is then produced by the criminal(s)
   - Cash-trapping—the criminal blocks the cash dispense mechanism and then extracts the banknotes after the cardholder has left the vicinity
   - The cardholder is the victim of the crime of robbery once the cash has been delivered by the ATM, either in front of the machine, or nearby.
2. **ATM Ram Raids**

The removal of the ATM and its contents from where it is installed, usually using a motor vehicle, agricultural or construction equipment. A ram raid involves an attempt to 'rip-out' the ATM from its position and remove it from the premises, with the intention of breaking into the machine at the location to which it has been moved and at a later time, in order to steal its cash.

Freestanding ATMs (often located within retail stores) are more vulnerable to 'rip-out', with chains and straps being used to lasso the machine and drag it out by means of a 4 × 4 or pick-up truck.

The ATMs targeted are often in remote locations where criminals have calculated police response times to be far longer than might otherwise be the case.

Preventive strategies include: anchoring techniques (using plinths, chains, additional bolt anchors and PAS 68/IWA 14 bollards); tracking devices; banknote staining and degradation systems.

3. **ATM Burglary**

The ATM is broken into in-situ to steal its cash. The attack may be a traditional 'safe breaking' method (using a sledge hammer, angle-grinder or oxyacetylene equipment), or an explosive attack using gas or solid explosives. These last two forms of Modus Operandi (MO) did not appear in the UK until 2012. Moreover, undetonated explosive material poses a significant risk to those who attend the scenes of such crimes.

The overwhelming majority of such attacks will be at night and possibly before a weekend, if the ATM is believed to have been recently replenished. Sites with low levels of protection (such as convenience stores) are often targeted. Organised crime groups (OCGs) involved in this form of crime have been prepared to confront the police and will persist with a successful MO, travelling significant distances where necessary.

The provision of additional Physical Security 'Target Hardening' (see Measures Sects. 4.2.1–4.2.4), alarms and fogging systems cover a range of preventive mechanisms, with the overall aim of delaying the attack and thereby encouraging the criminals to flee in fear of the imminent arrival of the police.

4. **Robbery (ATM Replenishment Attack)**

This involves criminals attacking C/ViT (Cash/Valuables in Transit) crews, the merchant or directly employed staff in the process of replenishing or dealing with cassettes inside the ATM.

Such attacks will normally take place when staff are transporting money through insecure areas, or when the ATM safe is open to receive or remove cassettes. Staff will usually be placed under considerable duress, physical threat or actual violence during such an attack.

Preventive measures to thwart such attacks include CCTV, live threat assessment and improved intelligence. Physical protection involves fogging devices, DNA taggants, banknote degradation and changes in ATM design to prevent more than one cassette being removed when the safe door is open. See also Deakin et al. (2003) re Modus Operandi commonly used by offenders at ATMs when undertaking robberies.

ATM Security Working Group

In the UK, the ATM Working Group was established in 2006 and three years later evolved to become the ATM Security Working Group (ATMSWG). Membership of this body is made up of professionals from the banking and security industries, together with representatives from the police. It has a mandate to reduce crime and enhance the protective measures afforded to ATMs.

A number of strategies have been utilised to reduce theft and robbery at ATMs (Weight, 2009). However, a health warning needs to be given at this point.

Key Issue

Crime Prevention strategies in respect of ATMs must not prevent one form of attack at the machine, whilst simultaneously facilitating others. For example, installing the ATM up a flight of steps, down a narrow passageway, or in a colonnaded area with limited surveillance opportunities, may serve to protect the machine from being 'ripped out' by means of a motor vehicle. But where it also serves to lure legitimate customers into a location where they are themselves vulnerable to attack, then the strategy can be seen to have failed. The skill is in devising interventions that satisfy the demands of all interested parties in Designing Out Crime—including those of Personal Safety for both staff and customers.

Symbolic (Psychological) Barriers

ATMs have provided perhaps the best evidenced example of how Symbolic (Psychological) Barriers (see Sect. 2.12), can have a deterrent effect on crime. More specifically, a yellow (or sometimes red) hatched area is painted on the ground immediately in front of the ATM. Usually measuring some 1 metre by 2 metres (39 inches by 6.5 feet), the aim is to create a Privacy Zone.

Tony Holt and Jon Spencer (2005) report that in their Manchester pilot study area, such Privacy Zones (or Personal Defensible Space) were not only effective in keeping a queue of people some distance away from the person using the ATM, but also led to a 34% average reduction in street robbery and theft from the person. Furthermore, the Privacy Zones were rarely breached—even at night when there were large groups of people, some of whom appeared to be intoxicated. Consequently, it seems that the defining of such Personal Defensible Space acts as a formidable psychological barrier. See Sect. 2.12 for further details regarding this research.

Utilising the Planning Process

One of the first cases where crime and fear of crime (FOC) were accepted as material considerations in the Planning Process, involved an attempt by a bank to install an ATM outside their branch in the Handsworth district of Birmingham (see Sect. 7.5, Planning Appeal Judgements). The local Designing Out Crime Officer (DOCO—Architectural Liaison Officer as the role was known at that time) PC (now Sergeant) Naheed Mushtaq, opposed the application, submitting grounds for refusal based on the then Department of the Environment (DofE) Circular 5/94 Planning Out Crime. She described how the proposed location, being at the junction of a number of streets, provided potential Attack and Escape Routes, together with its proximity to a bus shelter where robbers might lie in wait. The Planning Inspector who visited the scene agreed with those sentiments, and the application to site the ATM was refused.

Relatively few Planning Applications to install ATMs have subsequently been opposed by way of the Planning Process. Furthermore, liaison between the ATM provider and the local DOCO before the machine is installed should avoid the necessity for recourse to the Planning Process. However, in the absence of such liaison, DOCOs have opposed the installation of an ATM on grounds of crime and listed the following key points:

- The ATM would encourage criminal activity
- No Crime Prevention measures had been incorporated into the proposal
- That consequently, the installation ('proposed development') would be contrary to government guidance.

Nevertheless, in recent years, a new practice has been observed, with ATMs being installed prior to Planning Permission being granted. In such circumstances, applications for Planning Permission are made retrospectively. Where this takes place, there is the potential for an ATM to be installed in an unsuitable location, where those withdrawing cash might be placed in considerable danger. This is reported to have happened on at least one occasion, resulting in robbery and grievous bodily harm being sustained by the person withdrawing cash. Indeed, at an ATM without Planning Permission, the machine was broken into and in excess of £10,000 stolen.

Whereas liaison with the DOCO (possessing detailed Crime Pattern Analysis about the area's Crime Risk) would have counselled against its specific location and thereby prevented a serious crime from taking place.

Conclusion

In the final analysis, the use of the Planning System to object to the installation of an ATM (either in the future or retrospectively) should be a course of last resort. Indeed, such recourse to the Planning System goes against the spirit of Problem-Solving, Problem-Oriented Policing and Partnerships (POP—see Sect. 2.14) where the aim is to accumulate positive working relationships across the community of partners, including the ATM Industry.

To repeat, only a minority of ATM Planning Applications have been opposed by way of the Planning System and then on specific grounds that if installed they will act as a generator of crime, FOC and ASB. Most importantly, it is the responsibility of the individual DOCO to ensure that all other options are first exhausted.

The European ATM Security Team (EAST) have published a range of General Advice documents, necessarily 'Restricted' to those in the industry and law enforcement.

Potential privacy zone mats, designs against ATM crime Professor Lorraine Gamman and Dr. Adam Thorpe © design against crime research centre central Saint Martins, University of the Arts London

6.12 Parks and Open Spaces

Background and History

The Permeability debate (see Sect. 4.4.6) is especially relevant in respect of Parks and Open Spaces. In Victorian Britain and much of the world, public parks (particularly those in urban areas) were perceived as potential venues for illicit behaviour. Consequently, many urban parks were fenced, gated and locked between sunset and sunrise, to ensure that only 'respectable' practices took place therein.

In support of such respectability, the Park Keeper ('Parkie' in the language of British children) was a highly visible Capable Guardian (Cohen & Felson, 1979). His role was to eliminate lascivious behaviour, eject the homeless, other 'undesirables' and ensure the park was secured from entry at night. Furthermore, where a law enforcement presence was deemed necessary, specific police units were created—for example, the Royal Parks Police in London (now part of the Metropolitan Police).

In the UK, this system of control began its demise at the outbreak of the Second World War (WWII). Lord Beaverbrook, newspaper magnate and Minister of Information in Winston Churchill's wartime government, used his newspapers to Name and Shame (see Sect. 1.5.13) those who were slow to donate their metal fencing and gates (often surrounding parks and public buildings) for the war effort. However, because the bulk of this metal consisted of cast iron and of no value to any wartime industry, much of it was later dumped at sea!

Following WWII, a period of financial hardship meant that outside the capital, metal railings and gates were never replaced. Proof of this situation can be seen outside some of the oldest police stations in England—such as at Acocks Green in Birmingham, where the sawn-off supports embedded in the stonework remain visible. Financial cutbacks also saw the demise of the 'parkie'. Whilst cultural changes witnessed the demand for more open environments, fewer controls and the 'right to roam'. Unfortunately, the absence of a secure perimeter and/or the void of Guardianship (Reynald, 2009) is often filled by those intent on ASB and crime—for example muggings (street robbery), sexual offences and 'pick-pocketing' (especially in busy parks).

In their investigation into the 249 'neighbourhood parks' in the city of Philadelphia, Pennsylvania, Groff and McCord (2011) conclude the following: parks are often large public areas with little formal or informal control; include Activity Generators and by default Crime Generators (Brantingham and Brantingham, 1995); recreation centres (presence of Place Managers—Eck and Weisburd, 1995), fields and courts for organised activities cause less Disorder; and are susceptible to both positive and negative influences from the surrounding area. Mixed land use in the area beyond the park reduces crime—and therefore planners should consider placing new parks in such areas.

Displaying signage relating to opening hours and rules removes excuses (see Sect. 2.13). And whilst rarely identified, park adoption and Neighbourhood Watch demonstrate visible signs of community involvement, with the capacity that individuals might intervene (Cohen & Felson, 1979; Clarke, 1983). Violence and disorder were concentrated in a specific subset of parks (Hilborn, 2009); crime was more clustered in the areas around parks; a positive association between parks and crime, especially neighbourhood parks; legitimate users prefer a mixture of activities over single activity parks (Project for Public Spaces, 2005): those surrounded by exclusively residential areas experience higher violence;

larger parks have more users and less crime; police focus on Hot-Spots and POP strategies, this is likely to produce reductions in park-related crime (Weisburd et al., 2010). More recently, attention has turned to making parks safer for use by women and girls: https://www.greenflagaward.org/media/2443/saferparks

However, Groff and McCord (2011) explain how their conclusions are not necessarily generalisable to other types of park/parks elsewhere—thereby echoing Jane Jacobs (1961). Nevertheless, many of their conclusions will be supported by DOCOs and other observers.

Role of the Designing Out Crime Officer

A return to the days of parks with secured perimeters and night-time curfews (enforced by means of locking the entrance gates) is unrealistic—although many parks around the world continue to enforce this practice. For example, the majority of London's Royal Parks, The Royal Botanic Gardens, Birmingham Botanic Gardens and the York Museum Gardens. And Newman (1973, p. 114) argued that to facilitate Natural Surveillance '… long thin parks' were 'preferable to square ones'. The DOCO is advised that the following measures have delivered some degree of success in similar locations:

- Perimeter Security (see Sect. 4.4.6)
- Defensive Planting (Sect. 4.2.4)
- Lighting (Sect. 4.3.1). Where the park is normally closed during the hours of darkness, there is no need to illuminate unless for night-time staff and special events. However, where impossible to physically secure, Lighting may attract criminal elements and encourage innocent persons to venture into an environment where they may be subject to attack.
- CCTV (Sect. 4.3.3). May be appropriate for some Parks and Open Spaces. Re-deployable cameras can be especially effective in deterring crime/ASB
- Closure of Rights of Way on Grounds of Crime and Public Space Protection Orders (Sect. 6.4)
- Public Conveniences (Toilets—Sect. 6.13)
- Graffiti, Fly-Posting, Fly-Tipping, Abandoned Vehicles and Excessive Noise (Sect. 8.2.5)
- Key security, locking procedures, lone working (Sect. 8.3, Good Housekeeping Measures)
- Symbolic (Psychological) Barriers (Sect. 2.12)
- The value of clearly visible Signage and Notices (Sect. 2.13).

See also the POP Center response guides, Dealing With Crime and Disorder in Urban Parks, Hilborn (2009); Frace (2020); Sect. 9.8, Washington Square Park, New York; Make Space for Girls Resources library; https://www.queenelizabetholympicpark.co.uk/about-us/how-we-work/handbook-creating-places-work-women-and-girls; and Secured by Design Safer Parks: Improving access for women and girls (2025) for more detailed advice on park safety.

6.13 Public Conveniences (Toilets/Bathrooms)

In the British (or at least English) psyche, the very notion of design guidance in respect of public conveniences or bathrooms, toilets, washrooms, or whatever epithet you choose to use, is liable to attract considerable mirth. However, this did not deter Colin Cockfield, the now retired Force Architectural Liaison Officer with Staffordshire Police from producing the first such Design and Good Practice Guides (1998, 2003).

Cast iron men's pissoir manufactured in Glasgow, Scotland and now relocated to Bewdley Station, Worcestershire on the Severn Valley Railway

Colin correctly championed the Partnership Working approach (with key providers such as Local Government, the National Parks and various other public agencies) in order to remove problems occurring at existing public conveniences and to ensure they did not appear at new ones—to be 'Designed Out' at the pre-planning stage. Certainly, in the UK and similarly identifiable around the world, problems associated with old, dilapidated and rarely used toilet blocks appear to have intensified over recent decades. Graffiti (see Sect. 8.2.1), vandalism, sexual activity and drug taking/dealing are some of the activities reported to take place at such locations. Moreover, evidence of the occurrence of crime and ASB is often apparent long after the event (see Sect. 2.8, Environmental Cues), thereby having the capacity to raise the fear of crime (FOC).

In an attempt to facilitate legitimate use, prevent such criminal activity and publicise worst and best practice, Colin produced the Public Conveniences Problem Reduction Guide (1998, 2003). This has been subsequently revised and republished as the Publicly Available Toilets: Problem Reduction Guide (2010) by Gerry Brophy of Hertfordshire Constabulary Crime Prevention Design Service, in partnership with the British Toilet Association. No single social agency can deal with the problems generated at such locations. And indeed, Colin and now Gerry have advocated the classic Problem-solving, Problem-Oriented Policing and Partnerships (see Sect. 2.14) multi-agency approaches.

Removable plastic urinals for men in Amsterdam, The Netherlands

The recommendations for best practice regarding new build and refurbishment of existing toilet facilities (together with dealing with problem sites) are covered under the following subject areas in these design guides.

Background

Consultation must take place with the public, most especially the immediate neighbours, local businesses, the local authority (local government), police and any other body that may wish to contribute

- Facilities should be provided where there is a confirmed need.

Location

- The location chosen should be suitable for the intended use
- Orientation—a well-frequented location reduces Crime Opportunity, e.g. vandalism (criminal damage)
- Natural Surveillance (see Sect. 4.4.2)—the entrance should face the street, footpath, car park, etc., and not be hidden from view at the rear
- Artificial Surveillance (Technological Surveillance, e.g. CCTV, see Sect. 4.3.3). Toilet facilities located in an area already under CCTV surveillance (e.g. a car park) will deter vandalism and other crime
- Shrubbery (soft landscaping, including genuinely low growing shrubs—see Sect. 4.2.4, Defensive Planting) should be used in order to define pedestrian movement and enhance the site, without providing unnecessary cover that may raise the fear of crime
- Perimeter Security – fencing, railings and gates. Avoid features that could act as climbing aids
- Pathways—should be visually open, direct and well used
- Parking—locating toilet facilities in a car park both enhances Natural Surveillance and extends usage to a wider range of people
- Cycles, buggies, prams and dogs—recommends provision of secure anchorage outside toilet facilities for the security of these items
- Accessibility—ideally at ground floor level, the design should facilitate the use by those with disabilities (Equality Act 2010 compliant) and for maintenance purposes
- External Lighting (see Sect. 4.3.1)—should not conflict with other Crime Prevention measures (e.g. CCTV). Install appropriate anti-vandalism luminaires
- Street furniture—secured seating a short distance away from the toilet building may be a useful addition, but requires careful consideration
- Location—siting toilet facilities close to other buildings may enhance surveillance opportunities, but only when those buildings are occupied

- Signage and Notices – clear, up-to-date, appropriate (including opening hours) and where well-maintained, can increase usage and therefore reduce Crime Opportunities (see Sects. 2.3 and 2.13).

Building exterior

- External walls—smooth surfaces encourage graffiti and should be avoided. Use special non-stick paints and coatings to deter taggers
- Roofing—pitched roofs are recommended to deter climbing
- Guttering and downpipes—should be designed so as not to act as climbing aids
- Windows—use Laminated Glass and install as high as possible to make access difficult. Roof lights and sun tubes are good alternatives
- Ventilation—correct positioning of the entrance/exit should permit the ideal level of natural ventilation and reduce both condensation and smell
- External Lighting—provides comfort for the user and is a proven deterrent to Crime. Use flush-fitting security luminaires
- Servicing and storage—incorporate separate access (from the exterior) to the storeroom where cleaning materials are kept
- Entrance/exit—access doors should maximise the capacity for Natural Surveillance and reduce anonymity, by facing any passing pedestrian and/or vehicular traffic
- Single use cubicle facilities (accessed directly from the exterior of the toilet block) help deter internal loitering and their design also enables disabled access.

Building Interior

- Internal CCTV—should only be sited to view washing, drying or cubicle access areas and must NOT provide any viewing of urinal areas or the insides of cubicles
- Personal surveillance—attendants deter vandalism, allow prompt reporting of incidents and help maintain standards of cleanliness
- Ceiling, floor and walls—should be painted off-white to maximise light reflection. Non-porous, stipple-effect surfaces help deter graffiti attacks. Floors should be slip-resistant

- Internal Lighting—provide low energy, flush-fitting, vandal-resistant luminaires. Where possible, natural light via the roof space should be utilised to reduce risk of damage and the capacity for voyeurism
- Interiors—use vandal-resistant furniture/accessories
- Cubicle Walls—should be clad in stainless steel or similar to be drill-resistant and to prevent voyeurism (by holes being made in the walls)
- Cubicle Doors—again *Equality Act 2010* compliant with capacity to be opened from the outside in an emergency
- Toilet pan—stainless steel toilet pans are vandal-resistant, but less aesthetically pleasing. Solid surface fixtures may be preferable
- Urinals—privacy screens are recommended to prevent voyeurism
- Flush systems—concealed cisterns with electronic non-touch operation, deter vandalism and are more hygienic
- Toilet tissue dispensers—should be key-lockable to deter vandalism
- Sanpro bins—consider ducted disposal chutes or wall-mounted bins for feminine hygiene products
- Hand-washing/drying—solid surface or stainless steel basins are a deterrent to vandals, as are automatic taps and built-in hand-washers.

The Hertfordshire Constabulary document provides additional guidance in respect of a range of other issues including Accessible Toilets, Changing Places, Family Cubicles, Automatic Public Conveniences, Street Urinals, Refurbishment of Publicly Available Toilets and Income Generation. It also includes a Self-Assessment Checklist.

Men's outdoor urinal in Bruges, Belgium

General Advice

- Baby changing facilities should be installed in both female and male toilets to promote family use and deter sexual activity
- A policy of regular maintenance and inspection linked to instant repair, should be adopted to avoid degeneration
- Charging for the facilities (payment on entry) provides revenue for maintenance, whilst discouraging loitering and internal damage

- Changes should be looked upon as an opportunity to promote the local area. Many people take a positive or negative view of a neighbourhood from the toilet facilities provided
- Well-cleaned and maintained facilities deter misuse and encourage legitimate use (see the Broken Windows theory of Kelling and Wilson, 1982). Human behaviour is affected by all our senses. If a public convenience, looks, feels and smells unpleasant then it will tend to be avoided by the law-abiding majority, yet simultaneously attract those who are visiting for non-legitimate purposes
- Pre-constructed, easily maintained single cubicle facilities can be purchased outright and permit revenue generation by charging and advertising, or leased from the service provider
- Many of the problems associated with toilets can be dealt with by good design and solving one problem can often eliminate another.

In the United Kingdom, a great number of public conveniences have been closed in recent years, whether on grounds of crime, ASB or because local authorities (local government) have reduced finances and contend they can no longer afford to maintain (or repair) such facilities. This has had subsequent effects in two specific areas of concern. First, the general public (especially the elderly and those suffering from particular medical conditions) bemoan the lack of access to public conveniences in town and city centres. Second, late at night as the pubs and clubs close, young adults who have vacated such premises have nowhere to 'relieve' themselves. The net result is all too easily imagined!

Historic men's urinal in Amsterdam, The Netherlands

For once this is not a singularly British problem, although the efforts of (amongst many others) the French (with their 'pissoirs') and the Dutch and Belgians (with their hydraulic 'Urilifts' or removable plastic urinals—craned into place on a Friday afternoons and removed on Sunday mornings) demonstrate how the problem can be reduced. However, at 'popular' locations in Amsterdam, personal witness testifies that these same removable plastic urinals will often overflow by Saturday night, which somewhat defeats the public health remit of such facilities. And they also have a tendency to stink even before they are first used on a Friday evening.

Urinoir hydraulic toilet emerges from the ground at night in Amsterdam, the Netherlands (and now also in the UK)

Until 2018, all of these urinals catered solely for men. But as of 2018 the hydraulic 'Urinoir' has appeared in Amsterdam with facilities for both 'Dames' and 'Herren.'

In addition, North Yorkshire Police, under the heading Crime Prevention Measures and Advice relating to Misuse of Drugs in Public Toilet Areas, particularly within Retail and Fast-Food Business Premises, issued a leaflet (undated, but published during the early years of the twenty-first century) that lists the following recommendations:

- Conduct regular patrol/checks of the toilets by designated members of staff (in pairs where there is a perceived problem). Blood splashes on walls and floors may indicate misuse of drugs is taking place (see especially Sect. 2.8, Environmental Cues)
- Locate staff at or near the toilet entrance
- Secure cistern lids to prevent syringes/needles ('sharps') being discarded or stored in same
- Ensure ceiling tiles are secured to prevent disposal/ storage of drug syringes/ needles
- As vinegar is often used in the injecting process, remove same (including sachets) to an area under staff Surveillance
- All staff should be fully conversant with safety measures in relation to the safe handling of rubbish (including syringes)

- If injured by such a syringe/needle, try to make the wound bleed outwards, run under a cold tap and seek medical advice immediately
- If drug abuse on premises is suspected, inform the police
- Change Lighting in toilets to blue or blue ultra-violet to deter intravenous injection (but see below paragraph)
- Fit digital or electronic access control locks to toilet doors with only the staff being able to authorise their opening on customer request
- Fit coin-operated locks to toilet doors
- Install CCTV cameras at the entrance to toilet facilities
- If drug paraphernalia is found, 'Do Not Touch', unless trained to do so.

Incidentally and following on from one of the previous bullet points, this author has witnessed at toilet blocks around the world the use of ultra-violet light with the aim of preventing intravenous drug injection. In theory, such light makes it difficult for the addict to locate a vein and inject himself, although Colin Cockfield rightly described how such '*...use could actually create more problems*' (2003, p. 25). For example, graffiti drawn with certain pens or sprays can look spectacular under UV light. And such Lighting is unsettling for those with autism and makes it awkward for the partially-sighted to negotiate the environment. Whilst mixing white and blue light negates any positive effects. Moreover, the ingenuity of the criminal/addictive mind knows few bounds and drug treatment professionals report that marking the required vein with a pen in advance of entering a room bathed in UV light, negates any positive benefits.

Similarly, the provision of a 'sharps bin' (to receive discarded syringes resulting from intravenous drug injection) is equally contentious. Put simply, will its presence encourage such activity and other criminal conduct?

Conclusion

Public conveniences are an important part of the streetscape. They should not be designed or adapted to appear as fortresses and thus increase the FOC. In the UK, a recent trend has been for male and female toilets to be combined facilities—albeit without urinals! However, following complaints in certain locations (such as department stores) some toilets have reverted to single gender allocation. And a recent (2025) decision by the UK Supreme Court may exacerbate this situation. Furthermore, as the above guidelines suggest, careful consideration of their design can make such toilet facilities less attractive to improper use in the form of crime and ASB, whilst simultaneously more fitting for legitimate visitors wishing to use them.

6.14 Pedestrian Subways

History and Background

There can be few better examples in the United Kingdom built environment of 'the road to hell being paved with good intentions' (or perhaps more appropriately the subway to hell) than the rise and demise of the pedestrian subway. A product of the 1960s design boom and accompanying prioritisation of the free-flow of vehicular traffic, they were at least (in part) constructed with the best of intentions—namely pedestrian safety.

Excellently designed and well-lit pedestrian subway at Grant Park, Chicago, USA

From the outset and despite TV public information films advising their inherent safety, a great number rapidly fell into disrepair. Poorly drained pedestrian subways repeatedly flooded—always a major concern in the UK. But it was their ability to raise the fear of crime (FOC) that caused many pedestrians to avoid using them. Human beings are hardly a species of troglodytes and the thought of descending into a dark and dank environment (especially at night) or when alone, caused many people to avoid their use. Nor is this a particularly British problem, as personal witness suggests a worldwide disdain to use pedestrian subways.

This situation is often reinforced by: poor Lighting (see Sect. 4.3.1); meagre Natural Surveillance (Sect. 4.4.2); a gathering point for Anti-Social Behaviour; role as an ad hoc toilet; and providing all too many walls for graffiti 'artists' to pen their monikers (see Sect. 8.2.1). Moreover, pedestrian subways were almost always utilitarian affairs with little if any artistry invested in their creation. For example, as you descended into the subway there was no guarantee of a clear line of sight through to the other side. And those that acted as a multiple connection point beneath a crossroads junction, were often poorly designed in this last respect.

All told, far too many pedestrian subways acted/act as FOC Generators and are consequently avoided (especially at night) by the pedestrians they were designed to assist. This can be identified internationally with pedestrians who prefer to take their chances with motor vehicles and cross at ground level.

Furthermore, this is not a gender specific issue—as one senior Crime Prevention Officer discovered to his cost when he asserted such a viewpoint on national television (in the early 1990s) and received the all too expected criticism from a litany of commentators! Once again, personal witness (and conduct) indicates men are as likely to avoid subways at night or when few other people are using them. Moreover, following a homophobic Hate Crime attack in a subway outside a gay pub in Birmingham city centre, England, news quickly spread within the city's Gay community to stop using these supposedly safe 'design' features of the built environment.

Crime in Pedestrian Subways

This author's own investigation for West Midlands Police conducted in 2000 disclosed that during the previous three years, very little reported crime had taken place within the Birmingham city centre network of pedestrian subways. In addition to the attack on the gay victim referred to in the previous paragraph, one of the few reported crimes involved an off-duty police officer who was robbed at night in another subway. However, perhaps the low incidence of crime is caused by a majority of people instinctively avoiding such environments (see especially Sect. 6.16, Pedestrianisation of Town and City Centres). And this author can personally testify how when off duty and in pursuit of a robber armed with a knife into a network of pedestrian subways, he realised the hero he would NOT be if found in a pool of blood the following morning!

The Removal Solution

Thankfully, this scenario is now largely impossible, because (in parallel with many other local authorities) during the first decade of the twenty-first century, Birmingham City Council embarked on a policy of restricting vehicle access to the city centre, removing pedestrian subways and embracing the 'desire lines' for those on foot who wish to cross the roads safely, courtesy of electronically controlled crossings. This reversed a policy initiated during the city's major regeneration of the 1960s—although as the former capital of the nation's motor vehicle industry, the primacy of the motor car can be identified in Birmingham's road design plans as far back as the 1920s. In this context, the last four decades of the twentieth century can now be viewed as an aberration in the city's history (having been recorded in the Domesday Book of 1086 as a small village worth only 20 shillings (£1!) Furthermore, the removal of these subways coincided with a policy of gradually excluding all unnecessary motor vehicle traffic from the city centre.

Beginning with the removal of the unattractive pedestrian subway interchange beneath Smallbrook Queensway, this multi-million pound venture (the local 1960s architecture was structured on these different levels) has seen pedestrian crossings share the lowered carriageway. Vehicle traffic has undoubtedly been slowed, but in return FOC has been all but eliminated and pedestrians are now happy to cross into the Chinese Quarter and Theatre District that lie beyond the city's former inner ring-road.

During 2002, the pedestrian subways around Colmore Circus (including those immediately in front of West Midlands Police Headquarters at Lloyd House) were replaced with both traffic lights and electronically operated pedestrian crossings. And with the opening of the rebuilt Bull Ring Shopping Centre, a network of inhospitable underground walkways has been removed.

The elimination of the pedestrian subway system at St Chads Circus in 2007, was an especially difficult task as the area also contained the extractor fan plant room for the underground road tunnels that form part of the Queensway inner city ring-road). However, despite the artistry of the 'Kennedy Mosaic Mural' (now resited in Digbeth), this strange multi-level subway system was notorious for drug addiction (used syringes were a common sight), few people used it and even fewer mourned its passing. And between 2015 and 2021, redevelopment of Paradise Circus necessitated the closure of its grim and foreboding pedestrian subway complex.

A further motivation can be identified in the desire to break the 'noose' created at the heart of the city in the 1960s, with the construction of the Queensway inner ring-road. This same issue can be seen across the UK in town and city centres that were 'modernised' during this era. Coventry City Council identified the same problems, in that pedestrians are unwilling to use the subways (and footbridges) to access the areas beyond the ring road—especially at night. Consequently, during 2015 access between Coventry Railway Station and the city centre was redesigned to eliminate pedestrian subways.

Counsel from the DOCO

For the Designing Out Crime Officer asked to advise in such circumstances, from the perspectives of crime, FOC and ASB, the removal of pedestrian subways provides a great opportunity to demonstrate their skills. Moreover, removal greatly improves the aesthetics of the town or city centre. The sole issue remains the protection of pedestrians and cyclists from motor vehicles. However, a key element within the rationale of reintegrating pedestrians with vehicular traffic, is that the crossings will reduce speeds and produce a safer environment (see also Sect. 6.5, Home Zones and Shared Space).

Historically, pedestrian bridges have been used as an alternative to subways. However, such bridges possess their own issues—like those engaging in throwing objects onto the roofs of vehicles passing below and as potential locations for suicide (see Stokes, 2025). Furthermore, they too must take account of the needs of the less able-bodied.

Where pedestrian subways cannot be removed or are the only option in a new environment, the following advice may be given:

- Pedestrian subways should be the option of final resort
- They should be short, straight and designed with wide entrances in order to maximise Natural Surveillance
- A person on foot approaching a pedestrian subway should be able to see through to the other side and beyond, with no potential hiding places
- Entrances to the subway should be chamfered, in order to reduce areas of concealment and maximise Natural Surveillance. Radius (convex) entrance/exit walls can reduce the length of the subway and the capacity for inappropriate loitering
- Multiple access and exit points should be avoided
- If intended for use at night, the pedestrian subway should be well lit with white light (see Sect. 4.3.1)

  If not to be used at night, the pedestrian subway should be gated at both ends and locked when dark
- Walls should be coated with white (to maximise light) non-porous and stippled surfaces on which it is difficult to apply/easy to remove graffiti
- A regular maintenance policy should ensure graffiti and other criminal damage are removed immediately
- A reassessment should be undertaken annually to establish whether the pedestrian subway remains necessary in that built environment.

6.15 Retail Parks, Business Parks and Industrial Estates

The original Secured by Design—Commercial Development Guide (1992) included Retail Parks, Business Parks and Industrial Estates within its target audience (together with office buildings, hotels, public buildings, and car parking). And SBD – Non-Residential 2025 reiterates that mandate. Both versions authoritatively assert that these particular features of the built environment demand their own specific Designing Out Crime requirements.

Such non-residential developments are especially vulnerable, largely because they are rarely occupied on the same 24/7 basis expected at residential properties. In the instance of industrial buildings/estates and business parks, that absence of occupation may exist for several days (e.g. at weekends and during public holidays). Consequently, security measures must take account of the inherent vulnerability produced by there being no Capable Guardian (Cohen and Felson, 1979) site at such times—potentially over three or more days.

Each development should have a 'defined perimeter', most usually consisting of walls, fencing and/or hedging in order to produce Defensible Space (see Sect. 4.4.1). In addition, Symbolic (Psychological) Barriers (see Sect. 2.12) may also be used. Landscaping should be complementary to other security features and must not obscure Lighting (see Sect. 4.3.1), CCTV coverage (Sect. 4.3.3) and Natural Surveillance (Sect. 4.4.2).

For the Physical Security 'Target Hardening' Measures and Technological Innovations of the individual buildings, see Sects. 4.2.1, Doors and Door Security; 4.2.2, Windows and Glazing; 4.2.4, Defensive Planting; 4.3.2, Intruder Alarms; and 4.3.3, CCTV (Closed Circuit Television).

SBD—Non-Residential Guide 2025 (see Sect. 5.5) should be the first reference point for the Designing Out Crime Officer. And the following recommendations have been found to be especially effective across a range of different sites.

Key Recommendations from the Designing Out Crime Officer

- Wherever possible, a secure perimeter should be maintained (and regularly inspected) including the use of gates (pedestrian and vehicular) for when the site is unoccupied. The height and structure of the secure perimeter (with no footholds or other climbing aids) should be based on an assessment of the local Crime Risk
- With industrial estates, electronically-operated vehicle gates are advised at the main (ideally sole) entrance/ exit to the site (either stand-alone or in conjunction with a staffed gatehouse) AND also at the entrance to each of the

industrial units. Manually-operated gates are rarely used on the necessary routine basis—except in the aftermath of a criminal attack, although even then such a practice is difficult to maintain

- Where there is no gated entrance to the site (as with the majority of retail parks), it is strongly advised that manually operated gates are erected at either the entrance or key locations, to prevent intrusion by 'boy racers' and similar unwanted visitors at night
- Where Public Rights of Way compromise the security of such a development and there is a proven link with the commission of crime, consideration should be given to closing or diverting such rights of way by means of the relevant legal powers (see Sect. 6.4, Closure of Rights of Way on grounds of Crime and Public Space Protection Orders)
- As with rear gardens to domestic dwellings, security can be enhanced if the general layout incorporates back-to-back service yards and loading areas
- Car parking should be overlooked from the occupied premises and ideally (for staff) be secured behind gates. Parking bays should be laid out so that they are perpendicular to the building—allowing Surveillance along both sides of parked vehicles from inside the premises. See also Sect. 5.19, 'Park Mark' Safer Parking Scheme
- Lighting is a proven deterrent to crime and fear of crime. See Sect. 4.3.1 for detailed information in this regard and practical recommendations
- Rear access to individual premises should be tightly controlled. All visitors should be directed to the main entrance, located at the front of the building
- Unless visitors are expected to wait at the front entrance, a well lit and welcoming reception area (even if unstaffed) should be secured, to prevent unaccompanied access to the rest of the premises
- Intruder Alarms and CCTV systems (see Sects. 4.3.2 and 4.3.3) are often mandatory requirements of the insurance company. As a consequence, those intent on crime or ASB feel vulnerable, whilst the law abiding majority feel safe and secure
- Larger sites may demand their own security/CCTV control room. Ideally, this should be located close to the main/sole entrance (to reinforce the message of safety and security)—but mindful of the need to protect both security staff and the recorded images produced by the CCTV system (see Sect. 4.3.3)
- Accounts offices and cash handling facilities should be located well inside the premises (to benefit from the Layered Security approach) with the necessary Commensurate Risk (see Sect. 1.5.15) assessment made and security measures incorporated. Similar advice applies to high value IT equipment—especially those items that are easily portable

- Liaison with the local DOCO may solicit the opinion that the provision of a police office in a high-profile location would be welcomed by police senior management
- Good Housekeeping Measures (see Sect. 8.3). This includes key management, visitor escort and maintaining an Operations Manual. It should also be common practice that areas not normally open to the public are secured. Gates to service yards and loading areas should be closed and locked except when deliveries are being made/collected.

6.16 Pedestrianisation of Town and City Centres

From a Designing Out Crime perspective, the increasing tendency for pedestrianisation (denying access to motor vehicles) in town and city centres across the UK and beyond, appears to deliver solely positive outcomes. With cars, vans and lorries banished from such streets, surely a far safer environment in which to shop, work and socialise will result? And a continental style 'café culture' can be fostered, with public seating arranged as per the 'elegant architects' models, complete with trees and happy people' (Pease and Farrell, 2009, p. 18) that rarely feature a private motor vehicle and abound with images of people sat at small tables sipping their lattes and cappuccinos.

Furthermore, by using strategically placed bollards and concrete planters, even ram-raiders can be thwarted. If only, for as ever the best laid plans of urban planners… Consequently, in the unlikely event (these are most often prestige projects where crime, fear of crime (FOC) and anti-social behaviour (ASB) are rarely considered in advance) of the Designing Out Crime Officer (DOCO) being asked to provide advice regarding the pedestrianisation of such streets, the following counsel may be helpful.

Background
First, a concise history. Roberts (1981) comments that pedestrianised streets have a venerable history in Britain, beginning in the thirteenth century and emblematic of the post 1945 new towns. Indeed, in the 1960s, the solution to urban problems of traffic congestion and social disintegration was largely seen in terms of grand designs (McKay & Cox, 1979). For example, the 1963 'People and Cities' conference, jointly staged by the Town Planning Institute and British Road Federation, heralded the coming motor age with enthusiasm. Meanwhile, the interests of pedestrians and cyclists were ignored as can be witnessed in the 'race-track' road

design of many British city centres (downtown areas), epitomised until recently in Birmingham (England) city centre.

Beginning in the 1960s, the supremacy of motor vehicles in England's 'second city' meant that pedestrians were forced to endure a near subterranean lifestyle in a city centre infused with pedestrian subways (see Sect. 6.14). By the late 1970s, these had come to be identified as what might be called Crime Generators (Brantingham and Brantingham, 1993) especially at night. However, towards the end of their existence hardly a soul would use them during the hours of darkness and as a result their role was reduced to simply generating FOC.

Barrier controlled pedestrianised High Street, Solihull, West Midlands, England

Teeman (1999) reports how London Street, Norwich city centre was the first street in Britain to be pedestrianised in 1967. Whilst Stenhouse (1977) observes how by the late 1970s, most towns had some traffic-free pedestrian precinct and that congestion, noise and parking problems had begun to make other patterns of shopping more attractive. Roberts (1981) records that there were 1304 known pedestrianised streets, arcades and shopping centres in Britain. Yet the true heritage of modern-day pedestrianisation can be identified in mainland Europe.

Tanghe, (1984) detail how the two kilometres long Strøget in Copenhagen city centre was the first street (in reality, a series of linked streets) in Europe to be pedestrianised in 1962. It witnessed a 35% increase in pedestrian use and in 1973 was extended to include Grey Friars Square. Most importantly, the authors convincingly argue that the pedestrian must be considered the equal of the motorist and that the city can be liberated—not by eliminating motor vehicles altogether, but by humanising vehicular traffic and subordinating it (see Sect. 6.5, Home Zones and Shared Space).

Often influenced by the Danish experience, some degree of pedestrianisation followed in Frankfurt's Romer Square, in Munich, Vienna, Grenoble, Milan, Rotterdam's Lijnbaan, and in Cologne, Verona and Kyoto (Lottman, 1976)—all during the 1970s and 1980s. Stockholm's 1.5 kilometres long Drottninggatan was also pedestrianised in the 1980s and in Helsinki, following such success in Mikankatu, another city centre street, Aleksanterinatu, was pedestrianised (to include tram access) during the summer of 2002 (personal witness). The latter, including the same tram access, was replicated in Birmingham's Corporation Street and Stephenson Street in 2016. Interestingly, in Tokyo, pedestrianisation was semi-adopted for certain streets on Sundays only (Lottmann, 1976), a theme which has some parallels within the British context.

If the UK was slow to mimic their European counterparts regarding pedestrianisation, there was no such lethargy in the creation of enclosed shopping centres (malls). Ignoring indoor markets, the first such enclosed shopping centre was created in Germany in 1926, the USA in 1959 and Britain in 1967. Thereafter, an explosion in their number took place across Great Britain. In Nottingham, the Victoria and Broadmarsh shopping centres opened in 1972 and 1975 respectively, the latter cutting across the former street layout as part of a wholesale redesign of the city centre south-side. As in Birmingham, pedestrians were forced underground as new routes were created for motor vehicles. Indeed, one such new route, Maid Marian Way, was voted the second worst street in Britain by listeners to BBC Radio 4's Today programme in 2002. However, the shortlist of favourite streets was all pedestrianised—and chosen by the poll organisers, the now defunct Campaign for Architecture in the Built Environment (CABE).

Creeping/exponential pedestrianisation (Meades, 1999) achieved traction in Britain during the early 1990s and three decades later is the rule rather than the exception. From Fort William in north-west Scotland, to Penzance in Cornwall, Folkestone in Kent to Rhyl in North Wales, local authorities have increasingly banished motor vehicles from the heart of their towns and cities. Furthermore, those with the necessary finances have invested in lavish pedestrianisation schemes, utilising multi-coloured brick paviors, semi-grown trees, raised flower beds and public seating. But to reiterate, the one thing universally missing from this built environment decision-making appears to have been any thought as to its effect on crime, FOC or ASB. In essence, evidence from Environmental Criminology and Crime Science was ignored and advice of a DOCO rarely sought.

The motivation for such Pedestrianisation is usually perceived as traffic congestion, combined with a desire to create a more user-friendly shopping and leisure environment. However, contemporary accounts state such congestion was common in Victorian times—albeit by horse drawn means. Photographs of many of the world's cities at the end of the nineteenth century often bear witness to congestion. Yet at that time there was no movement towards pedestrianisation. Indeed, one of the most congested areas of any city, the West End of London, was the first to install parking meters in 1926, yet to this day many streets are open to vehicular traffic. That said, this part of London is very much the exception to the national rule and congestion charging to restrict vehicular traffic was introduced in 2003, whilst cities like Birmingham and Bristol have since introduced emission charging.

Despite the lack of a definitive causal link between road congestion and pedestrianisation, from the perspective of improving road safety such a policy has numerous supporters. Except, few such schemes ban all motor vehicles completely—and not just in Britain. Indeed, the pedestrianised streets of most European countries permit some degree of motor vehicle access. Thus, pedestrians still have to keep a wary eye for buses, trams, taxis, delivery wagons, vehicles belonging to the disabled, bicycles, emergency service vehicles and law breakers. Indeed, the ingenuity of criminals to access pedestrianised areas (often believed to be protected by the aforementioned bollards and raised planters) in order to execute ram-raid style attacks, is highly inventive—evidence repeated use of fork-lift trucks or bulldozers to remove ATMs (Automatic Teller or Cashpoint Machines, see Sect. 6.11) and lorries to smash through roller- shuttered jeweller's shop fronts.

Another motivation for pedestrianisation can be identified as a response to the demands of the retail community. In the 1980s and 90s, storeholders were concerned about the growth of out-of-town shopping malls, whose free parking and lack of traffic congestion were perceived as attracting shoppers away from traditional town and city centre shops. Small retailers were often opposed to pedestrianisation, arguing that unless motor vehicles could park on the road and close to their shops, trade would undoubtedly suffer. Nor was this an idle fear.

Following the pedestrianisation of Solihull's (England West Midlands) High Street in 1994, within 12 months both the butcher's and greengrocer's shops had closed. The fourth and most powerful motivation appears to be aesthetics and prestige. Once a number of town and city centres had embarked on a policy of pedestrianisation, it became very much the fashionable thing to imitate.

Consultancies arose producing glossy brochures of what could be achieved in a particular environment and central government was only too prepared to finance such studies, although not their implementation. Birmingham City Council's Gun Quarter Study (1993) is a classic example, although its recommendations have never been realised. The national chain stores, banks and building societies seem to have supported such measures. Whilst the concerns of small retailers were ignored. And for local authorities, there was the belief they were producing a safer built environment. Moreover, distinct parallels can be made with town and city centre CCTV schemes, the installation of whose cameras was often near simultaneous with that of pedestrianisation (see Sect. 4.3.3).

Yet in some town centres, like Ludlow in Shropshire; Ledbury, Herefordshire; Devizes, Wiltshire; and Keswick in England's Lake District, unwilling to embark on such wholesale pedestrianisation projects, the centuries old practice has continued. Namely, closing off streets to vehicles on market days, during fetes, fairs and 'when the circus comes to town'.

The Negative Side of Pedestrianisation

Apart from the aesthetics issues caused by altering a town or city centre layout (the street plan of which may have existed for 700 years), the hours of darkness point to the major problem with pedestrianisation: perceptions of crime, FOC and ASB. Indeed, a 'Jeckyl and Hyde' environment may be created: during daylight hours a bustling shopping environment; but at night a landscape devoid of virtually any human activity—vehicle or pedestrian (although in mid-winter some pedestrianised environments on weekday afternoons '*possess a windswept derelict look*' (Teeman, 1999, p. 6) exacerbated by maintenance issues.

There is an increasing amount of anecdotal evidence (although sadly little as yet emanating from Crime Science or Environmental Criminology), that pedestrians (especially men and women over 35 years of age) perceive that during the hours of darkness, there is safety in numbers in town and city centres. Remove the motor vehicles, and part of that overall number is also removed, causing at least some of those on foot not to venture into parts of the pedestrianised environment, which in turn further reduces the number of street users. Cumulatively, this will cause a reduction in both Movement (Permeability, see Sect. 4.4.6) and thereby Natural Surveillance (see Sect. 4.4.2). And whilst during daylight hours, pedestrianisation will usually lead to an increase in the number of people on foot in that environment, once the shops close and night descends, the reverse is often true. This in turn helps to explain the Technological Surveillance provided by CCTV—filling the void caused by reduced surveillance opportunities, now motor vehicles are banned and fewer pedestrians enter that environment. Thus, it could be argued pedestrianisation can (and often does) lead to more Crime, FOC and ASB (forms of Disorder in particular).

The greatest irony is that a common argument in favour of pedestrianisation is the creation of a safe environment during the hours of darkness, one that will encourage the return to cafes, wine bars, pubs and clubs (for leisure purposes) of the over 35 s. In recent decades, this group has either abandoned (once the shops had closed), many town and city centres in Britain, or traditionally never ventured there. Plus, in the wake of the transfer of control of alcohol licensing powers in 2003, there is currently much debate within the local authorities (local government) about the creation of that aforementioned 'European style outdoor café culture'—a description which may cause consternation in what is often cold, wet Britain! One author observes how 'the lack of an holistic, all embracing urban policy is, of course, not surprising' (Meades, 1999, p. 6).

As an additional observation, pedestrianisation creates considerably more space in which crime and ASB can take place. There is a limit to the number of fights and 'squaring up' that can occur on a ribbon of narrow pavement, once the pubs have closed and ejected their patrons. Remove motor vehicles (moving and parked) from this environment and the spatial capacity for ASB and worse multiplies considerably. Personal witness repeatedly confirms the accuracy of this assertion.

Strøget, Copenhagen, Denmark: Europe's first pedestrianised street.

Analysis of police Command and Control ('ISIS') incident logs for Solihull's newly pedestrianised High Street disclosed an average 12% annual increase in disorder over the 5 years 1998–2002. Moreover, as access to police vehicles is often denied by such gated pedestrianisation schemes and officers on foot patrol are usually an unviable logistical resource for police managers, instances of ASB are prone to go unchallenged. Worse still, many of these incidents are captured and recorded on the very CCTV cameras that accompanied pedestrianisation and whose original rationale was to prevent such behaviour from taking place. When CCTV recorded pictures are shown on national television (often following criminal conviction), this can reinforce the perception that town and city centres are unsafe places, once the shops close and darkness falls. And indeed, this use of CCTV is a repeated theme amongst town and city centre managers.

One final issue concerns the increasing number of Licensed Premises (serving alcoholic drinks like beer, cider, wine and spirits) that have opened in town and city centres over the past 30 years. The total number of pubs in the UK has plummeted, in line with the demise of heavy industry, economic downturn and a change in drinking habits and culture. Yet, many town and city centres have witnessed a doubling or trebling in the number of venues serving alcoholic drink. The overwhelming majority of these target young drinkers (the 18–30 year olds) and include half price or similar 'Happy Hours' (banned in Scotland since 2009) and other drink promotions. And despite changes in the law in 2003 governing licensing hours and transferring powers from the police to local authorities, there continues to be a tendency for pubs and clubs to decant their customers at near identical closing times into those often-pedestrianised streets. The net result of 'well-oiled' people, buying and eating food from burger wagons and similar whilst waiting for taxis or other forms of transport, can be easily imagined. And in a pedestrianised environment, the potential for crime and ASB is magnified. Despite the presence of street pastors, street wardens and CCTV cameras, these issues have the capacity to spill out into the surrounding streets.

Advice from the DOCO

As previously described, recent history suggests that the DOCO is rarely called upon by a local authority or other interested party to provide advice regarding the effects on crime, FOC and ASB accompanying any proposed pedestrianisation scheme, or changes to an existing project. However, if their advice is sought, the DOCO is advised to restrict their guidance to the following subject areas:

- Existing Crime Pattern Analysis for the area

- The extent of Permeability (Movement) throughout the proposed pedestrianised environment
- What public seating is envisaged, under whose control and can it be predicted to act as a generator of ASB or worse?
- What level of Street Lighting is proposed within the pedestrianised environment? (see Sect. 4.3.1)
- Will the whole of the pedestrianised area be subject to CCTV coverage, so that a 'suspect' can be tracked from camera to camera (Continuity of Evidence) and public seating or other potential congregation points monitored? (see Sect. 4.3.3)
- What plans are there for nearby and easily accessible car parks, taxi ranks, bus stops and pickup areas from Licensed Premises in the vicinity?
- Will a Taxi Marshalling Scheme operate on busy nights before and after pubs and night clubs close?
- To what extent will emergency service vehicles be restricted from the pedestrianised streets? Barriers with manually operated locks will by default bar such vehicles from the streets in all but emergency situations and may delay rescue/transfer to hospital.
- Does the increasing adoption of Shared Space (especially in the form of kerbless environments) compromise pedestrian safety?

Ludlow in Shropshire, England where the absence of pedestrianisation helps to ensure a vibrant town centre

Conclusion

The pedestrianisation of the UK's town and city centres has accelerated in recent decades and is now very much the norm. Rarely was any thought given as to how this might generate Crime Opportunities, FOC or ASB. And whilst the motivations for the creation of Shared Space were well founded, the reintroduction of trams into the built environment (resulting in road traffic collisions by motorists unfamiliar with such vehicles), points to the challenges of changes made to the street scene and how these can impact upon the safety of users—especially pedestrians.

As a result, the primary skill for the DOCO is to ensure that decisions regarding the extension of such pedestrianisation schemes (including minor changes, e.g. the introduction of new public seating) recognise that these can have negative consequences. And to counsel either against this happening on a full 24/7 basis, or design in measures that will reduce the capacity for crime, FOC and ASB. In particular, the general obstruction of access to emergency service vehicles should be strongly challenged. See also the following Sect. 6.17, Counter Terrorism Design.

6.17 Counter Terrorism Design

A pertinent development emanating from *Designing Out Crime* has been the realisation that it is possible to extend its philosophy and methodology to embrace the concept of *Counter Terrorism Design(CTD).* Whilst this very title expresses the intended aim, in reality, it is the ability to mitigate the effects of terrorist attack, especially those caused by improvised explosive devices (IEDs) that has delivered profound, positive and tangible results. For all too obvious reasons, this chapter will only describe the methodology that is at present easily accessible within the public domain, there being no desire to compromise the recommendations and responsibilities of those charged with such tasks.

Protecting the Houses of Parliament, London

The United Kingdom's experience with large vehicle borne IEDs began during 'The Troubles' in Northern Ireland, which took place over the final three decades of the twentieth century. As part of these same tragic events, the true extent of explosive effect and potential for its reduction was demonstrated by repeated bombings in Northern Ireland (Ulster) and in cities across Great Britain. However, it was the large vehicle borne IEDs at St Mary Axe (1992) and Bishopsgate (1993)—both in the financial heart of the City of London. And at South Quay (Canary Wharf, London) and Manchester city centre (both in 1996), that demonstrated their massive effect. Of all the other vehicle borne IEDs that have detonated with such catastrophic consequences elsewhere in the world, the one causing particular concern (in terms of its explosive effect and the intense structural damage that led to so many deaths and serious injuries), was at the Alfred P. Murrah Federal Building in Oklahoma City, USA in 1998.

Two specific parallels exist between Designing Out Crime and Counter Terrorism Design. First, there is the desire for early stage engagement. In both instances, once the design of a building has been agreed and past the planning consent stage, the capacity to design out the Crime Opportunities—see Sect. 2.3 (including Terrorism), has been lost. Retrospective bolt-on features (such as Anti-Shatter Film adhered to glazing), can be applied following construction.

However, invariably they will be:

- Far more expensive both financially and in terms of carbon cost (from a long-term perspective)
- Require replacement after a set period of time (every 10–15 years in respect of Anti-Shatter Film)
- Compromise the aesthetics of the original design
- Rarely afford the same degree of protection as would have been the case had they been an integral part of the original design concept
- Furthermore, it has been estimated how CTD measures factored in at the design stage will add between 20 and 25% to the original project costs
- However, compare this to the statistic that following the Manchester city centre bombing in 1996, in excess of 80% of the businesses affected did not recover and thereby survive!

The second parallel concerns the desire to keep the criminal/terrorist away from the ultimate target—most often the building in question. It has only been appreciated in recent decades that Physical Security 'Target Hardening' Measures and Technological Innovations (see Sects. 4.2.1–4.3.3) are merely the final line of defence in preventing many forms of crime, most especially Property Crime. The real skill lies in persuading the criminal not to approach the intended target – thus the value of the Environmental Elements of CPTED—see Sects. 4.4.1–4.4.6 and also Davey et al., (2005); Clare and Morgan (2009); and Ekblom (2012). This has always been appreciated in Counter Terrorism Design, where it is known as the Stand-Off Area.

In the instance of Counter Terrorism Design, four issues can be identified as highly desirable elements to be designed-in at the concept stage:

Design Stage CTD Measures

1. **Stand-Off Area**

It is both logical and indeed commonly understood, that the further away from the target the IED actually detonates, the less impact it will have—in terms of explosive effect and by extension, mortal injuries. Consequently, the design of a building and the grounds in which it stands are imperative in preventing any form of IED from reaching the intended target. In this regard and to prevent vehicle borne IEDs from approaching the building, the inclusion of the following design features has been shown to be highly effective in reinforcing that same Stand-Off Area:

- Barriers
- Bollards (including rising bollards and those tested to stop such vehicles)
- Jersey (concrete) barriers
- Berms (raised earth mounds)
- Moats/ditches
- Chicanes.

2. Structure

Experience gleaned following previous bombings relates directly to the building's structure, facades and glazing. For example, it is now known that framed buildings properly 'tied' (as at Bishopgate in 1992) are far more resilient to the effects of bomb blast than those that have no frame (as at St Mary Axe in 1993). And where a vehicle borne IED is placed in close proximity to a beam of manifest structural importance (as in the instance of the Alfred P. Murrah Federal Building), the consequences can be catastrophic in terms of collapse, injuries and fatalities.

Other design features include the elimination of:

- Overhanging upper floors and basements that extend away from the main building
- Stone cladding
- All but essential glazing and then, only where truly necessary
- Car parks within or adjacent to the building/key part of the site being protected.

3. Glazing

Over recent decades, perhaps the greatest learning curve in relation to the blast effects caused following an explosion has been in the subject area of glazing. In a blast situation, conventional float or annealed glass has the capacity to cause traumatic injuries. Toughened Safety Glass ('burglar friendly glass'—as it breaks silently and leaves no sharp edges) is little better (see also Sect. 4.2.2, Windows and Glazing). Nor is Georgian wired glass (most often used in Fire Doors) a viable alternative to use, if the aim is to reduce the effects of explosive blast.

Meanwhile, opinions differ in respect of polycarbonate and glass reinforced plastic (GRP) 'glazing'—salthough recent innovations suggest these are now vastly improved products when compared to their original forms.

Indeed, to have no glazing (in effect, no windows) whatsoever is the greatest safety option—although this is hardly a practical or aesthetically desirable choice in the majority of circumstances.

Laminated Glass has a proven history of saving lives when vehicle borne IEDs explode. It works by having a PVB (Polyvinyl Butyral) or similar interlayer, a form of plastic, within the glass (minimum thickness 6.8 mm). In the event of an explosion, the PVB interlayer holds the glass together and prevents shards from flying in all directions, most especially into the building itself (see also vii, Anti-Shatter Film, below). Its ability to strengthen the glass in this way also explains why Laminated Glass is the preferred means of preventing Property Crime like Non-Residential Burglary and Residential Burglary.

4. **Bomb Shelter Areas**

The efficacy of internal evacuation within the building intended as the target (or adjacent buildings) was demonstrated in the instance of the taxi borne IED that exploded outside the BBC Television Centre, London in 2000. Where internal evacuation can be effected to properly designed Bomb Shelter Areas, this is a safer option than disgorging people onto the surrounding streets where they can be at far greater risk—especially from flying glass.

However, not all buildings are suitable for the inclusion of Bomb Shelter Areas—which require a windowless room and sound structure, see https://www.protectuk.police.

Retrospective Enhancement CTD Measures

Unfortunately, Counter Terrorism Design is usually delivered in the form of retrospective Physical Security 'Target Hardening' Measures, often incorporated long after the building was originally built. In the instance of historic buildings, this is forgivable, in that terrorist attacks were never envisaged. However, where the building has been recently constructed, if nothing else such retrospective enhancement will prove to be highly expensive in terms of financial investment.

A. **Post-construction Stand-Off Area**

The creation of a Stand-Off Area once a building has been built can be extremely difficult, especially in city centres where land is at a premium and/or it is a 'stand-alone' building with public (or worse still) vehicular access up to the shell of the premises.

Wherever possible, the measures recommended at 1. above should be incorporated, although on occasion a full Threat and Risk Assessment (TARA—see Sect. 1.5.15, Commensurate Risk) might suggest the only practical measure that can be taken is to move to a safer location. Hopefully, if the latter option is chosen, CTD measures will be 'designed in' at the concept stage.

B. **Structural Strengthening**

This can be achieved by replacing cladding and ensuring that the frame of the building is properly tied.

C. **Anti-Shatter Film**

In the 1970s and 80 s when the danger of flying glass shards from the aftermath of an explosion was first truly appreciated, 'anti-bomb-blast' net curtains were the chosen solution. However, by the late 1980s and in the absence of Laminated Glass, such curtains had been replaced by the far more efficient and adhesive, Anti-Shatter Film.

Such Anti-Shatter Films are adhered to the glass on the (most often) inside face of the building. They may be 'daylight films' (affixed to only the exposed glass); or 'anchored film' (adhered to the whole glazing surface including that which lay beneath the beading, or inside the frame). Due to the degradation effects of ultra-violet light, all such films need to be replaced approximately every 10–15 years, an expensive proposition and thereby reinforcing the objective of designing in Laminated Glass at the concept stage of a building's creation.

D. **Hostile Vehicle Mitigation**

This includes rising and static bollards/barriers, tied barriers, jersey barriers, Hesco bastions, gates, fencing, chicanes, ditches, moats, berms, planters, bicycle parking, seating and litter bins. Further details can be found at: www.protectuk.police.uk, www.mfdgroup.co.uk, www.trl.co.uk, www.marshalls.co.uk and other websites.

Many of these physical security devices (bollards in particular) have been tested in accordance with the impact requirements of PAS 68 (and since 2013 IWA 14—although the product range is not as extensive within the latter). Furthermore, design-led security solutions continue to evolve at considerable speed in this most sensitive of arenas, in order to provide both more cost-effective and aesthetically pleasing (far less ugly and intrusive) CPTED Physical Security 'Target Hardening' Measures (see Sects. 4.2.1–4.2.4) and Technological Innovations (Sects. 4.3.1–4.3.3).

E. **CTD Good Housekeeping Measures**

Counter Terrorism Design has also spawned a series of recommendations that might be described as falling under the heading of Good Housekeeping Measures (see Sect. 8.3). For example, the routine wearing at upper chest level (by all staff) of identity passes; visitor registration and escort; and changing staff culture so as to 'challenge' anyone on site/in the premises who is not recognised, or not wearing an identity pass. This heading might also include the Operational Requirement (OR) and Code of Practice (COP) in respect of the premises/ facility's CCTV system (see Sect. 4.3.3). Where such measures are detailed within a single document or contained within a folder, this is often referred to as an Operations Manual.

Protect UK

See Sect. 5.16, for further information regarding the Protect UK website.

This in turn provides links to the following headings:

Guidance

- Risk
- Security
- Response
- Recovery
- Awareness
- Legislation.

Threat and Risk

- Security risk management
- Threat analysis.

News and Views

- News
- Forum
- Webinars.

Learning

- e-learning
- Digital toolkits.

About Protect UK

The Secured by Design (SBD) website provides links to a number of guidance documents.

Publicly Accessible Locations (PALs) Guidance

Produced by NaCTSO, PALs guidance covers numerous subject areas including: attack methodology; bomb threats; good housekeeping; chemical/biological/radiological (CBR) attacks; Marauding Terrorist Attacks (MTA) Run Hide Tell; vehicle as a weapon; physical security; search planning; hostile vehicle mitigation; personnel and people security; cyber security; etc.

Protecting Crowded Places: Design andTechnical Issues

This guidance was first published in 2012 and subsequently updated. Access to it is now restricted.

Security Overlay to the RIBA Plan of Works

See Sect. 5.14, for detailed information regarding the *Security Overlay to the RIBA Plan of Works* (2023). In 2010, the Royal Institute of British Architects (RIBA) produced its first guidance document on designing for counter terrorism. This has now evolved and in conjunction with the National Protective Security Authority and Police CPI/Secured by Design, published as the aforementioned 2023 document. It is accessible in full at: https://www.gov.uk/government/publications.

References

Adamson, S. (2005). Burglary Reduction in Action: The Hartlepool Experience. *Crime Prevention and Community Safety: An International Journal, 7*(2).

Armitage, R. (2007). Sustainability Versus Safety: Confusion, Conflict and Contradiction in Designing Out Crime. In G. Farrell, K. Bowers, S. Johnson, & M. Townsley (Eds.), *Imagination for Crime Prevention: Essays in Honour of Ken Pease. Crime Prevention Studies* (Vol. 21, pp. 81–110). Willan.

Beavon, D. J. K., et al. (1994). The Influence of Street Networks on the Patterning of Property Offences. In R. V. Clarke (Ed.), *Crime Prevention Studies* (Vol. 2, pp. 115–148). Criminal Justice Press.

Beckford, C. (1994). *The Alleygater's Guide to Gating Alleys.* Metropolitan Police Service 2 Area CPDA Group.

Bevis, C., & Nutter, J. B. (1977). *Changing Street Layouts to Reduce Residential Burglary.* Paper Presented at the Annual Meeting of the American Society of Criminology, Atlanta, Georgia.

Biddulph, M. (2001). *Home Zones: A planning and design handbook.* Policy Press.

Birmingham City Council. (1993). *The Gun Quarter: Planning and Urban Design Framework.* Llewellyn-Davies.

Birmingham City Council, Department of Planning and Architecture. (1996). *Shopfronts Design Guide.* Birmingham City Council.

Bowers, K., Johnson, S., & Hirschfield, A. (2004). Closing Off Opportunities for Crime: An Evaluation of Alley-Gating. *European Journal on Criminal Policy and Research, 10*, 285–308.

Brantingham, P., & Brantingham, P. (1993). Nodes, Paths and Edges: Considerations on the Complexity of Crime and the Physical Environment. *Journal of Environmental Psychology, 13*, 3–28.

Brantingham, P., & Brantingham, P. (1995). Criminality of Place: Crime Generators and Crime Attractors. *European Journal on Criminal Policy and Research 3*, 3, special issue on Crime Environments and Situational Prevention, 5–21.

Brantingham, P., & Brantingham, P. (2008). Offender Mobility and Crime Pattern Formation from First Principles. In L. Liu & J. Eck (Eds.),*Artificial Crime Analysis Systems: Using Computer Simulations and Geographical Information Systems* (pp. 193–208). Idea Press.

Bridgestock, M. (2000). *Homezones—Good Practice in Britain and Henbury, Bristol Feasibility Study.* University of the West of England and Sustrans.

Bullock, K., Erol, R., & Tilley, N. (2006). *Problem-Oriented Policing and Partnerships: Implementing an Evidence-Based Approach to Crime Reduction.* Willan.

Children's Play Council. (1997). *Home Zones: Reclaiming Residential Streets.* National Children's Council.

Church of England General Synod. (1978). *Parochial Registers and Records Measure.* Church of England General Synod.

Clare, J., & Morgan, F. (2009). In J. D. Freilich, & G. R Newman, *Reducing Terrorism Through Situational Crime Prevention in Crime Prevention Studies* (Vol. 25). Lynne Rienner.

Clarke, R. V. (1983). Situation Crime Prevention: Its Theoretical Basis and Practical Scope. *Crime and Justice, 4*, 225–256.

Clarke, R. V. (1999). *Hot Products: Understanding, Anticipating and Reducing Demand for Stolen Goods* (Police Research Series 112). Home Office.

Clarke, R. V. (2004). *Closing Streets to Reduce Crime: Should You Go Down This Road? Problem-Oriented Guides for Police Response Guides Series No. 2.* US Department of Justice, Office of Community Oriented Policing Services.

Cockfield, C. (1998, revised 2003). *Public Conveniences Problem Reduction Guide.* Staffordshire Police.

Cohen, L. E., & Felson, M. (1979). Social Change and Crime Rate Trends: A Routine Activity Approach. *American Sociological Review, 44*, 488–608.

Davey, C.L., Hodge, M., & Wootton, A. B. (2005). Designing Against Terrorism. *Engineering Designer, 30*(1) (January/February), 10–13.

Deakin, J., Medina-Ariza, J. J., Spencer, J., & Ainsworth, P. B. (2003). *Both Sides of the Coin.* Greater Manchester Police.

Department for Communities and Local Government (DCLG). (2007). *Manual for Streets.* DCLG.

Department for Transport, Local Government and the Regions (DTLR). (2001). *Home Zones—Planning and Design, Traffic Advisory Leaflet 10/01.* DTLR.

Eck, J., & Weisburd, D. L. (1995). *Crime Places in Crime Theory. Crime and Place: Crime Prevention Studies* (Vol. 4, pp. 1–33). Criminal Justice Press.

Eck, J., Chainey, S., Cameron, J. G., Leitner, M., & Wilson, R. E. (2005). *Mapping Crime: Understanding Hot-Spots.* National Institute of Justice.

Ekblom, P. (2011a). *Crime Prevention, Security and Community Safety Using the 5Is Framework.* Palgrave Macmillan.

Ekblom, P. (2012). *Design Against Crime: Crime Proofing Everyday Products.* Lynne-Reiner.

Farrell, G., & Tilley, N. (2022). Elegant Security: Concept, Evidence and Implications. *European Journal of Criminology, 19*(5), 932–953.

Frace, C. (2020). *Importance of Safety in Public Parks for Crime Prevention.* California Polytechnic State University.

Groff, E., & McCord, E. S. (2011). The Role of Neighborhood Parks as Crime Generators. *Security Journal* 1–24.

Harris, B. A. (2020). *Rural Crime Prevention: Theory, Community and the 'rural idyl'* Routledge.

Hertfordshire Constabulary Crime Prevention Design Service and the British Toilet Association. (2010). *Publicly Available Toilets: Problem Reduction Guide.* Welwyn Garden City: Hertfordshire Constabulary.

Hilborn, J. (2009). *Dealing With Crime and Disorder in Urban Parks. Problem-Oriented Guides for Police Response Guides Series No. 9.* US Department of Justice, Office of Community Oriented Policing Services.

Hill, I., & Pease, K. (2000). The 'Wicked' Issues: Displacement and Sustainability. In S. Ballantyne, et al. (Eds.), *Secure Foundations: Key Issues in Crime Prevention, Crime Reduction and Community Safety.* Institute for Public Policy Research.

Hillier, B. (2004). Can Streets be Made Safer? *Urban Design International, 9*, 31–45.

Hillier, B., & Shu, S. (2000). Crime and Urban Layout: The Need for Evidence. In S. Ballintyne, K. Pease, & V. McLaren (Eds.), *Secure Foundations: Key Issues in Crime Prevention, Crime Reduction and Community Safety*. Institute of Public Policy Research.

Holt, T., & Spencer, J. (2005). *A Little Yellow Box: The Targeting of Automatic Teller Machines as a Strategy in Reducing Street Robbery*. Perpetuity Press.

House of Commons Library (2023). *Statistics on access to cash, bank branches and ATMs* Lorna Booth for House of Commons Library. Accessible at: https://researchbriefings.files.parliament.uk/documents/CBP-8570/CBP-8570.pdf

Jacobs, J. (1961). *The Death and Life of Great American Cities*. Random House.

Jenkins, S. (1999). *England's Thousand Best Churches*. Penguin Group.

Johnson, S. D., & Bowers, K. J. (2010). Permeability and Burglary Risk: Are Culs-de-Sac Safer? *Journal of Quantitative Criminology, 26*(1), 89–111.

Kelling, G., & Wilson, J. (1982). Broken Windows: The Police and Neighbourhood Safety. *The Atlantic Monthly* (pp. 29–38). Washington DC.

Lottman, H. (1976). *How Cities are Saved*. Universe.

McKay, D., & Cox, A. (1979). *The Politics of Urban Change*. Croom Helm.

Meades, J. (1999). We're on the Road to Nowhere. Article in *The Times, Weekend* (pp.10–11), dated Saturday September 11, 1999. News International.

Moseley Society and Birmingham City Council. (1992). *Moseley Shopfront Design Guide*. The Moseley Society.

Mott, G., & Townsend, A. (2011). *About the ATM Industry*. ATM Security Working Group (ATMSWG).

NaCTSO. (2009). *Counter Terrorism Protective Security Advice for Places of Worship*. NaCTSO.

NaCTSO (2020). *Protecting Crowded Places from Terrorism*. NaCTSO.

Newman, O. (1973). *Defensible Space: People and Design in the Violent City*. Architectural Press.

Pease, K., & Farrell, G. (2009). *The Carbon Cost of Crime and Its Implications*. Secured by Design.

Poyner, B. (1983). *Design Against Crime: Beyond Defensible Space*. Butterworth.

Poyner, B. (2006). *Crime Free Housing in the 21st Century*. Jill Dando Institute of Crime Science.

Project for Public Spaces. (2005). *How to Turn a Place Around: A Handbook for Creating Successful Places*. Project for Public Spaces.

Read, T., & Tilley, N. (2000). *Not Rocket Science? Problem-solving and Crime Reduction* (Crime Reduction Research Series 6). Home Office.

Reynald, D. M. (2009). Guardianship in Action: Developing a New Tool for Measurement. *Crime Prevention and Community Safety: An International Journal, 11*(1), 1–20.

Roberts, J. (1981). *Pedestrian Precincts in Britain*. Transport and Environmental Studies.

Shu, S., & Huang, J. (2003). Spatial Configuration and Vulnerability of Residential Burglary: A Case Study of a City in Taiwan. In *Proceedings of the 4th International Space Syntax Symposium*. London.

Sidebottom, A., Tompson, L., Thornton, A., Bullock, K., Tilley, N., Bowers, K. & Johnson, S. (2017). Gating Alleys to Reduce Crime: A Meta-Analysis and Realist Synthesis. *Justice Quarterly, 2018*, 35(1), 55–86.

Stenhouse, D. (1977). *Understanding Towns*. Canlat.

Stokes, M. (2025). *A Tale of 3 Bridges: Barriers to Suicide Prevention*. Presentation given at the conference of the International CPTED Association in Palm Springs CA on 11 May 2025.

Strong, R. (1990). *The Lost Treasures of Britain*. Guild Publishing.

Strong, R. (2007). *A Little History of the English Country Church*. Vintage Books.

Tanghe, J., Vlaeminck, S., & Berghoef, J. (1984). *Living Cities: A Case for Urbanism and Guidelines for Re-urbanism*. Pergamon.

Teeman, T. (1999). Traffic-Free Zone. Article in *The Times, Weekend* (p. 11), dated Saturday September 11, 1999. News International.

Transport 2000/Streets for People/Children's Play Council. (1999). *At Home in My Street: Exploring Home Zones in the Netherlands and Germany*. Transport 2000 and the Children's Play Council.

Weight, A. (2009). *Best Practice for Physical ATM Security*. ATM Security Working Group (ATMSWG).

Weisburd, D., Telep, C., Hinkle, J. C., & Eck, J. E. (2010). Is Problem-Oriented Policing Effective in Reducing Crime and Disorder? Findings from a Campbell. *Criminology and Public Policy, 9*(1), 139–172.

West Midlands Arson Task Force. (2009). *Keeping Your Business in Business*. West Midlands Fire Service and West Midlands Police.

Planning Interventions 7

Preface
Unlike Fire Safety (in the United Kingdom and most other countries), there is no legal sanction in respect of a failure to incorporate Physical Security 'Target Hardening' Measures, Technological Innovations or Environmental Elements of Crime Prevention Through Environmental Design (CPTED—see Sects. 4.1.–4.4.6) within the built environment. According to reports at that time, creating such legal sanction for Crime Prevention measures was very briefly discussed in Cabinet (UK government executive) one Thursday morning in the early 1990s. However, as the then government was in the process of dismantling 'red tape' perceived as a hindrance to the business community, the prospect of statutory Crime Prevention measures was rejected.

All of which puts the Designing Out Crime Officer (DOCO and indeed CPTED Practitioner) in an apparently powerless position. Yes, there are the Diplomatic Skills, The Art of Compromise and Negotiation, Truly Effective Partnership Working and Naming and Praising—skills discussed in Sect. 1.5. But what if these fail to deliver a satisfactory result? One solution is to utilise the Planning System, which does have legal sanction—highly powerful legal sanction in the UK that ultimately can impose heavy fines and commit people to prison, if they fail to comply with a Judge's ruling.

In 1994, the former Department for the Environment issued DofE Circular 5/94 Planning Out Crime and over subsequent years a growing number of decisions by the Planning Inspectorate (a body whose powers are now determined by the Ministry of Housing, Communities and Local Government) have asserted that

M. Stokes, *Designing Out Crime*, Crime Prevention and Security Management,
https://doi.org/10.1007/978-3-031-99840-9_7

crime, fear of crime (FOC) and anti-social behaviour (ASB) are 'material planning considerations'. Some of these cases are discussed in Sect. 7.5, Planning Appeal Judgments.

7.1 National Planning Policy Framework

To provide an important historical background, prior to 2012, this chapter would have been devoted to Planning Policy Guidance (PPGs) and Planning Policy Statements (PPSs), as produced by the central government Department for Communities and Local Government (DCLG) and its predecessors, the Department of the Environment and Office of the Deputy Prime Minister. Indeed, out of more than two dozen PPGs and PPSs, those of particular relevance to the UK Designing Out Crime Officer (DOCO) were:

- PPS1: Delivering Sustainable Development
- PPS3: Housing
- PPS4: Planning for Sustainable Economic Growth
- PPS7: Planning for the Historic Environment
- PPS12: Local Spatial Planning
- PPG13: Transport
- PPG17: Planning for Open Space, Sport & Recreation

However, the publication in 2012 of the *National Planning Policy Framework(NPPF)* swept away all the PPSs and PPGs, reducing 1300 pages of policy to the 65-page *NPPF* and 27 page *Technical Guidance* documents.

Indeed, it was claimed that the *National Planning Policy Framework '…is a key part of Government reforms to make the Planning Systemless complex…'* See also: https://www.gov.uk/guidance/national-planning-policy-framework

As with the previous PPGs and PPSs, the *National Planning Policy Framework* sets out central government guidance in respect of planning issues that relate to the environment. Consequently, in certain instances, the *NPPF* impacts directly on the advice that can be given by the DOCO, especially in respect of the Environmental Elements of Crime Prevention Through Environmental Design (CPTED—see Sect. 4.4.1–4.4.6). In this context, the following sections of the updated *NPPF* (2024) may be of particular relevance to the DOCO and should be referenced where appropriate.

7.1.1 Achieving Sustainable Development

Despite references under this sustainability heading of '*...a presumption in favour of sustainabledevelopment,*' '*...a sufficient number and range of homes...*' and '*by fostering well-designed, beautiful and safe places...*' there is little mention of crime, fear of crime (FOC) or anti-social behaviour (ASB – disorder) within the *NPPF*. However, this is not uncommon in planning legislation (see Sect. 2.11, The Carbon Cost of Crime).

Paragraph 8
Under other headings, the following elements may have some relevance to the individual DOCO.

7.1.2 Promoting Healthy and Safe Communities

Paragraph 96:
Provides more scope for the DOCO and describes how:

> *Planning policies and decisions should aim to achieve healthy, inclusive and safe places which:*
>
> *(a) promote social interaction, including opportunities for meetings between people who might not otherwise come into contact with each other—for example through mixed-use developments, strong neighbourhood centres, street layouts that allow for easy pedestrian and cycle connections within and between neighbourhoods and active street frontages;*
>
> *(b) are safe and accessible, so that crime and disorder and the fear of crimedo not undermine the quality of life or community cohesion—for example through the use of well-designed, clear and legible pedestrian and cycle route, and high-quality public space, which encourage the active and continual use of public areas...*

7.1.3 Open Space and Recreation

Paragraph 105

Planning policies and decisions should protect and enhance public rights of way and access, including taking opportunities to provide better facilities for users. For example, by adding links to existing rights of way networks including National Trails'.

There is a great danger that this last recommendation could facilitate *Attack and Escape Routes* for criminals. *See* Sect. 6.4, Closure of Rights of Way on grounds of Crime and Public Space Protection Orders.

7.1.4 Promoting Sustainable Transport

Paragraph 109

Transport issues should be considered from the earliest stages of plan-making and development proposals, using a vision-led approach to identify transport solutions that deliver well-designed, sustainable and popular places. This should involve:

(b) ensuring patterns of movement, streets, parking and other transport considerations are integral to the design of schemes, and contribute to making high quality places;

Paragraph 111

(d) provide for attractive and well-designed walking and cycling networks with supporting facilities such as secure cycle parking (drawing on Local Cycling and Walking Infrastructure Plans);

Paragraph 112

If setting local parking standards for residential and non-residential development, policies should take into account:

(a) the accessibility of the development;
(b) the type, mix and use of development;
(c) the availability of and opportunities for public transport;
(d) local car ownership levels; and
(e) the need to ensure an adequate provision of spaces for charging plug-in and other ultra-low emission vehicles.

Paragraph 113

In town centres, local authorities should seek to improve the quality of parking so that it is convenient, safe and secure, alongside measures to promote accessibility for pedestrians and cyclists.

Paragraph 117

(C) create places that are safe, secure and attractive—which minimise the scope for conflicts between

pedestrians, cyclists and vehicles, avoid unnecessary street clutter, and respond to local character and design standards.

7.1.5 Achieving Well-Designed Places

Paragraph 131

The creation of high quality, beautiful and sustainablebuildings and places is fundamental to what the planning and development process should achieve. Good design is a key aspect of sustainabledevelopment, creates better places in which to live and work and helps make development acceptable to communities. Being clear about design expectations and how these will be tested, is essential for achieving this. So too is effective engagement between applicants, communities, local planning authorities and other interests throughout the process.

Paragraph 135

(f) create places that are safe, inclusive, and accessible and which promote health and wellbeing, with a high standard of amenity for existing and future users and where crime and disorder. And the fear of crime, do not undermine the quality of life or community cohesion and resilience.

7.1.6 Conclusion

Historically, all such planning reform has taken some considerable time to become established and applied routinely. As an example, *Department of the Environment Circular 5/94, Planning Out Crime*, was being quoted a decade after it had been repealed and replaced by the now similarly defunct PPS1. However, the *National Planning Policy Framework* has suffered few if any such errors—probably because its introduction marked what can now be interpreted as a revolutionary point in the evolution of planning legislation.

Furthermore, in 2012, Lord Taylor of Goss Moor and chair of the National Housing Federation, was appointed by the DCLG to review planning guidance in support of the *NPPF*. The 'Taylor Review' was published in 2013 and recommended the wholesale cancellation of some 103 existing policy documents, including *Safer Places: the Planning System and Crime Prevention* (2004) and the subject of the next chapter; much of the material in a further 28; and retention of a further 106 more. In addition, Lord Taylor recommended six areas for the preparation of new guidance, three of which will be of interest to the DOCO: 'Environmental Quality' (including noise and light pollution); 'Neighbourhood Planning'; and the 'Duty to Co-operate' (see Sect. 2.14, Problem-solving, Problem-Oriented Policing and Partnerships). The *NPPF* was updated in 2018, 2019, 2021, 2023 and 2024.

Despite the recent revolution in the manner in which planning legislation is now applied, there remains a key role for the DOCO in passing comment on those planning applications where there is the capacity to reduce Crime Opportunity, FOC and ASB. Consequently, it is essential that the DOCO establishes and maintains an effective working relationship with their local planning authority(ies). This will enable the latter to alert the DOCO about relevant applications where police comment is especially valued. And in return, the DOCO can provide evidence derived from crime figures, Crime pattern Analysis, Crime Science and their consummate skill-set, experience and training.

7.2 Safer Places: The Planning System and Crime Prevention

7.2.1 History

Safer Places: the Planning System and Crime Prevention (2004), was officially repealed in 2014. However, a much-reduced version of Safer Places has been expected for more than a decade—as per the recommendations of the Taylor Review (2013). Consequently, the key elements of this important and groundbreaking document (certainly from the perspective of Designing Out Crime—DOC) are reproduced in this chapter.

Safer places: the planning system and crime prevention

As previously described, *Department of the Environment (DofE) Circular 5/94, Planning Out Crime* provided the earliest all but statutory indication by central government that the Planning System had a part to play in DOC, FOC and ASB. As is often the case with statutory instruments, it took several years before it began to be routinely applied by key decision-makers—and ironically is still referred to many years after it was repealed, together with the now similarly defunct Planning Policy Statement 1 (PPS1, 2005—see Sect. 7.1) and the companion to *Safer Places*.

7.2.2 Foreword

Safer Places (the abbreviated title will be used throughout this chapter) is a key example of 'joined-up government' in that it was produced by both the then Office of the Deputy Prime Minister (ODPM—which had assumed responsibility for environmental issues) and the Home Office. From the outset, it acknowledges that *'Safety and security are essential to successful, sustainable communities'*. Moreover, it accepts that *'for far too long, too little attention has been paid by planners and designers to Crime issues'*. Most importantly, *'By highlighting the need to consider Crime Prevention as part of the design process, this guide aims to deliver safer places.' (2004, p.5).*

Other nuggets of gold contained in just the ministerial Foreword include:

> *It need not cost more either, and proper investment in the design of a development brings numerous social and economic benefits over its lifetime. And,*
>
> *This guide challenges developers, designers and all those who influence the design and layout of developments, to think in a holistic manner about each development. A key principle is that there is no universal solution to every problem. Each location is unique, and so what works in one place may not work in another. It is therefore important that the many professional disciplines work closely together and when they do, that they pay close attention to the principles and practical details in this guide and apply these carefully to meet the needs of the local area.*

Finally, and perhaps of greatest importance to the DOCO, the Foreword states:

> *'This guide is not a manual, nor is it intended to be a substitute for using experts in Crime Prevention, specialist Urban Designers and other skilled professionals. It is about encouraging greater attention to the principles of Crime Prevention and to the attributes of Safer Places.'*

7.2.3 Section 1

The first section of *Safer Places* describes how sustainable communities are ones that succeed economically, socially, environmentally and respect the needs of future generations:

> *They are well-designed places where people feel safe and secure; where Crime and Disorder, or the Fear of Crime, doesn't undermine quality of life or community cohesion. (2004, p.7).*

Section 1 goes on to affirm that Designing Out Crime and designing in Community Safety should be central to the planning and delivery of new development. It reiterates how Sect. 17 of the Crime and Disorder Act 1998 (and still law) demands that local authorities exercise their functions and do all they reasonably can to prevent crime and ASB.

Perhaps of greatest importance, *Safer Places* repeats the established wisdom that makes sense financially and how once a development has been completed, the main opportunity to incorporate Crime Prevention measures will have been lost. As all DOCOs are aware, retro-fitting security solutions to correct or manage a badly designed development is invariably far more expensive financially.

Safer Places is not solely about Crime Prevention, but rather the promotion of safe, sustainable and attractive environments. It is concerned with good planning, in creating a strong and positive sense of community identity and its particular role in tackling crime and the FOC. Crucially, it aspires to an evidence-based approach yet concedes the evidence is not always 'clear-cut'. Indeed, there are tensions between the primacy of Crime Prevention and other concerns of sustainable development—although the experienced DOCO would contend a compromise can almost always be agreed.

Planning can contribute directly to Crime Prevention through the following 'tools':

- Development plan policies
- Supplementary Planning Guidance (see Sect. 7.4)
- Pre-application discussions and negotiations
- Development control decision-making
- Planning conditions and planning obligations
- Planning can also influence crime prevention by working in partnership with the police and agencies such as Community Safety Partnerships.

This section also includes the magnificent statement, (and one that should be repeated by DOCOs to managers at every opportunity):

> *For many, there can be no better place to start than their local Police Architectural Liaison Officer or Crime PreventionDesign Adviser (2004, p.9. and now meaning DOCO).*

Section 1 concludes with page 10 devoted to the *Principles of Crime Prevention* and (Ekblom, 2001—see Sect. 2.6).

7.2.4 Section 2: The Attributes of Safer Places

Safer Places set out seven attributes of sustainable communities, all highly relevant to Designing Out Crime.

Access and Movement

Defined as *'Places with well-defined routes, spaces and entrances that provide for convenient movementwithout compromising security (2004, p.16)'.*

- Any access or service routes must not generate vulnerability to the rear of any premises (such as rear access footpaths)
- Routes must not provide potential offenders with unnoticed access to potential targets or multiple escape routes. All routes must be well-defined so as not to undermine private or Defensible Space (see Sect. 4.4.1)
- All proposed routes must be safe to use at any time of the day or night
- Routes for different users should be integrated and not segregated from each other
- All users must be able to understand which routes they use
- It must also be easy to understand how to travel through an area
- Routes must be designed to maximise the scope for Natural Surveillance (see Sect. 4.4.2)
- Routes should be straight and wide without blind spots where potential offenders could be hidden
- Routes must not become a focus for ASB
- Where Street Lighting (see Sect. 4.3.1) is to be provided, this must be to the locally adopted standard
- The number and nature of all connections must be considered

- All routes proposed must be necessary and lead to places where people want to go.

Structure

Defined as *'Places that are laid out so that Crime is discouraged and different uses do not cause conflict'* (2004, p.20):

- The types of building must be selected and designed with security in mind
- All 'uses' in an area must be compatible with each other, and potential conflicts must be identified and resolved within the design
- Vehicle parking should be within the curtilage of the building it serves and be over-looked. Where this is not possible, car parking provision must be located very close to the building(s) they serve and again be well over-looked
- The layout of the development must be appropriate for the identified Crime Risk
- All public open space must be clearly defined and serve a purpose. It must also support an appropriate level of legitimate activity
- Any open space that has been associated with ASB must be redesigned to eliminate its continued future use as a venue for behaviour of this kind
- Consideration should be given to the redevelopment, removal or re-use of buildings and spaces that historically have proven vulnerable to crime
- Buildings should be oriented to maximise Natural Surveillance.

Surveillance

Defined as *'Places where all publicly accessible spaces are over-looked' (2004, p.24):*

- Active Rooms with an increased number of windows should overlook the street and other public spaces
- Well lit spaces are crucial to reducing the FOC and should be sensitive to the needs of residents and users
- Lighting should be used to ensure good Natural Surveillance is available during the hours of darkness
- Surveillance should not be relied upon as the sole strategy for tackling crime and ASB
- Offenders should not be able to travel to and from locations without fear of being seen
- Criminal activity should attract the attention of onlookers
- Buildings and spaces should be designed to allow for all round surveillance

- Open and bright spaces reduce the number of potential hiding spaces and reduce the FOC
- Where in-curtilage vehicle parking is not provided, parking should be located in areas where surveillance is plentiful and provision is close to owners' homes
- Blank walls and recesses should be avoided
- CCTV (see Sect. 4.3.3) may be introduced as a means of improving surveillance, but not as an alternative to getting the design right in the first place
- An unmonitored CCTV system delivers no response to activities and is of little value.

Ownership

Defined as *'Places that promote a sense of ownership, respect, territorial responsibility and community' (2004, p.30):*

- All space (i.e. public, semi-public, semi-private or private) must be clearly defined and adequately protected in terms of its use and ownership
- It must be clear where one type of space ends and another begins
- The creation of Defensible Space empowers residents to take control of areas close to their home and is an important principle that should be used extensively throughout any development
- ASB can flourish in communal space if people do not identify the area as belonging to them and natural self-policing does not occur
- Private space should not be easily accessible to people who have no right to be there
- Low fencing, hedges, and bushes can be used as demarcation of space where ultimate security is not an aim
- High fencing that actively impedes access is most appropriate in places that are vulnerable to crime, such as the rear of dwellings
- Sensitive placement and appropriate selection of physical barriers such as gates, fences, walls and hedges, creates safe places that are also attractive
- Landscaping can be used to make places safer as well as more attractive, provided it does not restrict Natural Surveillance. To ensure that this principle is adhered to, continued maintenance of the overall height of bushes and the canopies of trees is necessary
- Most people respond to changes in paving, surface texture, materials, colour, landscaping and signage. These measures help to differentiate between public and private space for the benefit of all users.

Physical Protection

Defined as *'Places that include necessary and well-designed security features' (2004, p.34)*:

- Effective security creates a delay to the intruder, and therefore, the robustness of doors and windows should not be overlooked as a means of defeating the criminal
- The target hardening principles and standards utilised by Secured by Design (see Sect. 5.1) should be used to inform the applicant of the nature of security hardware currently available
- Designing-in effective physical security from the outset is cheaper, simpler and more successful than retrospectively attempting to install measures later
- Local Development Control Officers may be asked to impose formal planning conditions requiring the use of security enhanced doors and windows, such as those referred to in Secured by Design. You should always consider these products as an extremely effective, image enhancing and cost-effective way of creating physical security
- In multi-occupancy buildings, a combined audio-visual access control system should be installed
- In multi-occupancy buildings, access control should be used on each level to further restrict movement into private areas
- Retrospective measures such as barbed wire and broken glass embedded in the top of a wall, can be avoided by the adoption of DOC principles during the design stage of a new development.

Activity

Defined as '*Places where the level of human activity is appropriate to the location and creates a reduced risk of crime and a sense of safety at all times*' (2004, p.36):

- The public realm should be designed to be enjoyed by different cultural or age groups at the same time. This can be done by providing a range of complementary activities and designing the environment to minimise conflict
- The creation of places that become devoid of activity at certain times of the day or night, whilst remaining accessible to offenders, should be avoided (see Sect. 6.16, Pedestrianisation of Town and City Centres)
- As many law-abiding people as possible should be attracted to make use of space in the public realm
- Decisions about the appropriate level and type of activity must be made with the local context in mind

- Criminals should not be able to go about their business unnoticed
- Certain types of places, such as public squares and town centres, thrive on attracting a large number of people. The key is to create a high-quality environment and alternative opportunities and activities for those who might otherwise become involved in crime or disorder (ASB)
- A town centre residential population brings activity, Surveillance and ownership and should be encouraged. However, bars and clubs are best located away from these areas
- Local shopping areas should be designed to provide a safe environment whilst limiting the scope for ASB.

Management and Maintenance

Defined as, '*Places that are designed with management in mind, to discourage Crime in the present and the future' (2004, p.40)*:

- Anyone with 'ownership' of space, whether public, semi-public, communal or private must be aware of their personal maintenance responsibilities
- Places must be tidy, attractive and well cared for, as this indicates crime and disorder will not be tolerated and generates a feeling of safety in the user
- All signs of any previous Disorder or of neglect, such as broken windows, litter, abandoned vehicles or graffiti, must be removed promptly
- Whilst getting the physical design and build of areas right is paramount, management and maintenance will remain important factors in the creation of safer places
- Responsibilities of care should be agreed at the earliest opportunity during the development process and should be documented, to prevent future disagreements between the relevant parties
- When an organised human presence, such as security staff, vehicle parking attendants, concierges or gardeners is part of the overall management strategy, care should be taken to ensure their long-term sustainability
- The provision of secure storage for maintenance equipment is required, as they are both a target for theft and a tool for crime.

7.2.5 Section 3: The Planning and Crime Prevention Toolkit

This section describes how the Toolkit is used to incorporate *Crime Prevention* and promote Community Safety.

KEY ISSUE

Section 3 also references Sect. 17 of the Crime and Disorder Act 1998 (still valid despite the repeal of Safer Places) with the duty placed on each local authority to:

> *without prejudice to any other obligation imposed upon it... to exercise its functions with due regard to the likely effect of the exercise of those functions on, and the need to do all that it reasonably can to prevent, crime and disorder in its area.*

Planners are advised to take account of the local geography and its Crime Risks when considering changes to the built environment, employing Crime Pattern Analysis and Crime Risk Assessment where appropriate. The input of a DOCO (although in Safer Places the description Architectural Liaison Officer is used) should be sought when considered appropriate—and as early as possible. Development Control is especially relevant in utilising planning policies to prevent crime.

Consultation with the DOCO should be sought by architects, developers, etc., at the pre-application (planning) stage. Thereafter, Sect. 3 contends that Crime Prevention can be a material consideration (see Sect. 7.5, Planning Appeal Judgements) in the determination of Planning Applications. Indeed, it goes on to describe how:

> *Local planning authorities should consider whether proposed development could be amended or planning conditions imposed that would contribute to the prevention of Crime and Disorder. Where proposed development would undermine Crime Preventionor the promotion of Community Safetyand the conditions are relevant to land use planning, the application could be refused Planning Permissionwhere refusal is consistent with the development plan (2004, p.49).*

Crime prevention considerations should be 'given weight' where the potential *Crime Risk* is demonstrated, or expected to be high and/or the consequences for the community, vulnerable groups and individuals, serious.

Finally, consideration is given to *Sect. 106 Agreements* (under the *Town and Country Planning Act 1990*), which can be used to incorporate planning obligations, whereby the developer has to finance the provision of (for example) *CCTV*, Neighbourhood Wardens, late night bus services or even police facilities.

Annex 1: The Case Studies

This section of *Safer Places* forms the bulk of the original document and almost certainly will not be contained within any new advisory guidance.

Annex 2: Reconciling evidence of what works, knowledge of Crime Reduction and Community Safetyprinciples and values

This part of *Safer Places* owes its provenance to a speech given by Professor Paul Ekblom to the Designing Out Crime conference in London on 11 April 2003. It covers a wide range of important issues, including the move to evidence-based policies and practices and the relative absence of research regarding 'land use and layout' (the CPTED arena). He provides a synopsis of the 'Target Enclosure and Permeability' debate—including *'Leaking'* culs-de-sac (see Sect. 6.2) and Space Syntax (see Sect. 2.10). The highly context-sensitive nature of many Crime Reduction initiatives is explored, together with the value of location-specific knowledge. Professor Ekblom also powerfully observes how,

> *'... it is interesting Secured by Designhas increasingly moved towards a flexible, adaptive stance rather than one of rigid prescription' (Ekblom, 2003b).*

In addition, this overview covers: conflicting predictions—such as Jane Jacobs' (1961) *Eyes Upon the Street* surveillance processes; the need for creative solutions, interventions based on synergistic combinations (situational and offender-oriented); Crime Impact Analysis (see Ekblom, 2001); virtual reality techniques to identify crime locations in advance; the *'Conjunction of Criminal Opportunity'* framework (Ekblom, 2001—see Sect. 2.6); and in conclusion,

> *'...the kind of expertise to be developed to get the best out of environmentally-based crime reduction as a whole is less one of the technician applying a limited set of diagnostic tools and more one of the consultant with all principles at his or her fingertips, ready to configure and adjust on the fly, in specific circumstances' (Ekblom, 2003b).*

Interestingly, those of us who worked on the production of Safer Places can confirm how the fault-line between the Planning and Designing Out Crime communities was (all too predictably) drawn on the basis of Permeability (see Sect. 4.4.6), an issue that rings true to this day.

The *Safer Places* document can be found via the gov.UK website.

7.3 Design and Access Statements

Ted Kitchen, the former Professor of Planning and Urban Regeneration at Sheffield Hallam University, states in the forward to *Design and Access Statements: How to Use Them to Prevent Crime* (2010) that if at all considered in the design process, Crime Prevention is too often merely an afterthought. Consequently, developers and their agents are unwilling to utilise the opportunity to Design Out Crime (DOC) at the very concept stage when it can have the most *sustainable* (and cost-effective) result.

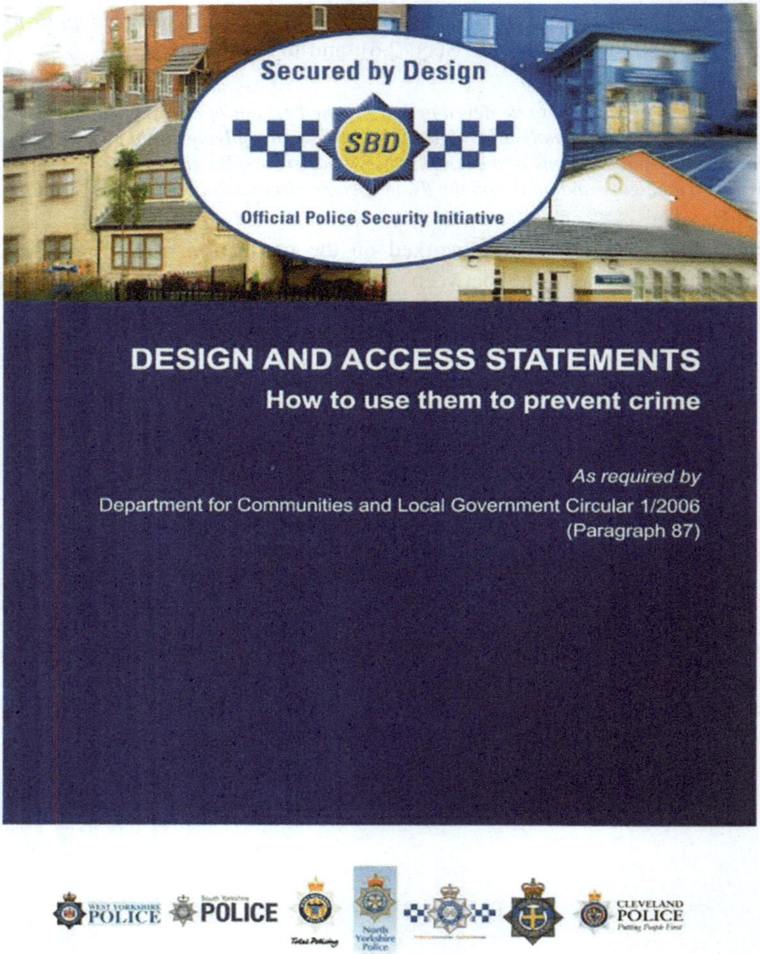

Design And Access Statements: How to use them to prevent crime

The professor powerfully contends that Design and Access Statements provide the means by which the development community has to confront the issue of crime, long before construction begins. Moreover, by considering the forms of crime likely to be attracted to the proposed development, strategies can be devised to prevent their incidence, rather than having to invest in massively expensive, bolt-on (retrospective) Physical Security 'Target Hardening' Measures and Technological Innovations (see Sects. 4.2.1–4.3.3). And most importantly from the perspective of the Designing Out Crime Officer (DOCO), Professor Kitchen observes that such strategies can be determined from crime statistics and police advice.

7.3.1 Background

A Design and Access Statement can be described as the written report that should ideally accompany and support all relevant outline and full planning applications (unless they relate to a material change of use).

It should seek to explain and justify the proposal and demonstrate how opportunities for Crime and Disorder (ASB) have been designed out of the proposed development. Failure to comply with this requirement may cause the planning authority to refuse the application or seek further information—both of which will cause delays. Design and Access Statements that satisfy the planning authority will describe the Crime and Disorder problems in the area and explain precisely how these are being addressed by the developer.

7.3.2 Guidance

Design and Access Statements for outline and detailed applications should demonstrate how Crime Prevention measures have been considered in the design of the proposed development and how the design reflects the attributes of safe and sustainable places.

Thus, the Design and Access Statement has an important role in validating the planning application and its future determination. And most importantly from the perspective of DOC, it must include reference to Crime Prevention.

7.3.3 CABE Guidance

The former Campaign for Architecture in the Built Environment (CABE) produced the first guidance document, CABE: Design and Access Statements: How to write, read and use them (2007). Access: Page 8, Paragraph 1 states:

'Statements should demonstrate how development can create accessible and safe environments, including addressing Crime and Disorder and Fear of Crime. These may be particularly relevant to address under layout and landscaping themes. Early consultation with the police will help identify key issues in your local area, and measures to help address these.'

7.3.4 In Practice

For the DOCO, the submission of a Design and Access Statement provides the opportunity to challenge if not rebut the claims of the developer in question at the planning stage. Moreover, a Crime Prevention aware developer will have consulted the relevant DOCO prior to submitting the statement. Once again, this process demonstrates how over time the influence of the DOCO in preventing/reducing crime, FOC and ASB are (some would argue were) being enhanced.

Finally, and for the benefit of DOCOs, Thames Valley Police (2008) produced a guide to *Design andAccess Statements*. This was followed by the West Yorkshire Police (2009) booklet *Design and Access Statements: How to use them to prevent Crime*. This last document was subsequently reproduced by Secured by Design (2010) and contains the Foreword written by Professor Ted Kitchen with which this section began.

Also in 2010, Thames Valley Police published The Compendium of Crime Prevention and Reduction in the Planning System. This is a boxed set of documents devoted to definitions, websites, acknowledgements, primary planning policy, planning companion guides, supporting documents and general advice. Some of this content has since been replaced by central government. Nevertheless, it provides a fascinating testimony as to where the DOCO is positioned in terms of delivering an individual service to the highest standard of professional etiquette.

See also Sect. 3.3, Site Surveys and Environmental Visual Audits (EVAs).

7.4 Supplementary Planning Guidance

The creation of *Supplementary Planning Guidance(SPG)* in respect of *Community Safety* (the heading under which *crime, Fear of Crime(FOC)* and *Anti-Social Behaviour (ASB)* often fall in the planning vocabulary), has a noble if it would seem short-lived history that took place during the early years of the twenty-first century.

Design for Community Safety: Supplementary Planning Guidance from Dudley Metropolitan Borough Council

As an example, Dudley Metropolitan Borough Council (2002) *Supplementary Planning Guidance* entitled *Design for Community Safety* is a collaborative publication with West Midlands Police. Its purpose is to provide a corporate approach to improving *Community Safety*, acknowledging the contribution made by *Secured by Design* (see Chapter 5). To this end, its key objectives are to:

- Provide planning guidance that enables security issues to be considered at all stages of the design process
- Assist developers to adopt designs for new developments that take the security of people and property into account
- Establish a framework of principles to assist those responsible for planning and design to make considerations about safety and security
- Provide a wide and varied pattern of land use to create environments that are lively and well used to help deter criminal activity
- Minimise the incompatibility of land uses that are instrumental to higher *crime* rates
- Promote and encourage good quality design in all development.

The Dudley SPG recognises that tackling crime and the FOC is one of the highest concerns for local people. Over 64 pages, it details both good and bad practice, using a wealth of illustrations of both effects. Moreover, the document repeats the powerful mantra contained within Sect. 17 of the Crime and Disorder Act, 1998:

> *'Without prejudice to any other obligation imposed on it, it shall be the duty of each authority to which this section applies to exercise its various functions with due regard to the likely effect of the exercise of those functions on, and the need to do all that it reasonably can to prevent Crime and Disorder in its area.'*

Furthermore, this SPG convincingly asserts that good quality design can contribute to reducing *crime*, despite the other social influences. It also cites a list of local and national documentation that supports this stance.

The Performance Criteria describe how previous planning standards have constrained design creativity, and lists those key areas where good design principles can impact on safety and security.

- Context: people feel safer when there is activity; when they can be seen by residents, drivers and other users; accessible local features encourage greater use and build a greater sense of community; developments that respond to natural features often create innovative designs that are more robust; and a variety of uses can encourage greater activity that enhances Natural Surveillance (see Sect. 4.4.2)

- Layout: a better connected network of streets and public spaces contributes to Personal Safety and the security of property; streets should be highly connected, well overlooked and busy; pedestrians and cyclists should be put before motorised transport; existing, well used pedestrian routes need to be preserved and designed into layouts at an early stage; minimising long detours on foot; public footpaths complemented by parallel cycleways; and routes positioned away from the rear access to buildings, as this provides *Attack and Escape Routes* for criminals
- Arrangements between Buildings, Streets and Gardens: the necessity for clearly defined public, private and communal space definition; creating a sense of ownership and to personalise spaces, whilst projecting the image of a well-kept and loved environment; good quality, attractive boundary treatment and gated entrances should give the impression of a safe and more private situation, causing criminals to feel more vulnerable
- Natural Surveillance: residential developments should acquire Secured by Design awards to maintain high standards of physical security; dwellings should have at least one Active Room (not a bedroom), fronting the street at ground floor level; development should not turn its back onto the street (or canals and watercourses); no blank gable facades on corner sites; and the amount of Natural Surveillance from the street should not be compromised
- Landscape Design: poor landscape design can compromise safety and security; needs to be formal in urban areas whilst naturalistic in more rural environments; design out overgrown shrubs and other thick barriers that are in close proximity to public footpaths; reduce hiding and entrapment spots; provide clear sight lines; plant thorny or spiny shrub species in front of vulnerable boundaries and buildings; design communal facilities so that nearby dwellings provide supervision and a safe route for users to come and go; play areas should not be located so as to increase noise pollution to nearby properties, but sufficiently close enough to ensure observation
- Revitalising Towns through Mixed Use: locate mixed use development close to public transport and shops, reducing the need to travel; provide a mix of uses for people with different incomes, family size, ages, gender and mobility; walkable perimeter of 5–10 minutes; minimum 25% residential component; private rear parking and gardens should have controlled access
- Shop frontages: external roller shutters lead to monotonous, unattractive frontages which reinforce the FOC and attract criminal activity; all shutters, (internal or external), should be as transparent as possible to maximise window shopping potential

- Street Furniture: the public realm is frequently cluttered with poorly sited signage, street furniture, service and phone boxes where the car-oriented environment has become dominant; necessity for better co-ordinated approach between stakeholders such as highways engineers; higher quality design is encouraged, with more innovative and contemporary designs that do not always rely on the heritage image; high quality products are visually attractive and longer lasting
- Lighting: in most cases, Lighting (see Sect. 4.3.1) helps people feel more secure and reduces the FOC; it also increases chances of detection; good Lighting design plays an important part in creating a more comfortable and attractive environment that reduces the FOC; low level Lighting is not suitable in remote locations, because they do not illuminate the face of a potential attacker, therefore reducing the power of identification
- CCTV: has a role in the dissuasion and monitoring of Crime in town centres, car parks and sensitive areas where crime rates are high; the installation of such equipment should not compromise the visual amenity of a place; with Listed Buildings, CCTV installation requires planning consent; CCTV camera position should not be detrimental to the integrity of a building
- Car Parking: multi-storey car parks need to be more attractive so as to contribute positively to the urban environment and be sympathetic to the scale and character of nearby buildings; large surface car parks should be subdivided into smaller areas; landscape planting to define spaces, but must not obscure views, cars or create potential hiding spaces or litter traps; residential parking should be within close proximity of dwellings; residents to have unrestricted views over their vehicles; alternative locations for garages need to be an acceptable distance from a dwelling
- Maintenance of the Environment: people's perceptions are affected by the appearance of places, hence a tidy, well-maintained, urban environment is essential to sustain confidence and help to control vandalism, crime and FOC; use high quality, longer lasting materials; products should be vandal-resistant; ensure the correct use and choice of planting that takes into consideration growth rates, heights and spread to avoid creating potential hiding spaces.

7.4.1 Words of Warning

Almost two decades after this author's endorsement of the Dudley Supplementary Planning Guidance, from a DOC perspective and with the benefit of hindsight, some of the advice espoused in the document is open to debate. For example, the confidence placed in Natural Surveillance (see Sect. 4.4.2) and *'busy highly connected routes'*. Unsurprisingly, these issues all relate to that continuing fault-line between the Designing Out Crime and Planning professions – Permeability (see Sect. 4.4.6). Indeed, after publication one fellow DOCO argued that too much had been given away by this author's endorsement. And ill-informed academic criticism of the SPG has also been discovered.

Nevertheless, this SPG is a classic example of Problem-Solving, Problem-Oriented Policing and Partnerships (POP) working (see Sect. 2.14), often necessitating compromise by all the parties involved. In this instance, such compromise led to an increase in Partnership Working in pursuit of the Secured by Design award.

7.4.2 Conclusion

More than a decade after Safer Places (2004) and PPS1 (2005) were withdrawn (and the power to create Supplementary Planning Guidance similarly abolished), the necessity for SPGs in respect of Community Safety is believed to have all but disappeared—at least by central government (although anecdotal evidence suggests they are still being used). However, they testify both to the continuing efficacy of the Corporate Approach to Crime Prevention and Community Safety and how within the Planning community, there are many professionals who actively agree with the sentiments of Designing Out Crime, CPTED and Situational Crime Prevention.

7.5 Planning Appeal Judgements

In Great Britain (and further afield), a powerful element was added to the toolbox of the Designing Out Crime Officer (DOCO) once crime, fear of crime (FOC) and anti-social behaviour (ASB) were interpreted as 'material planning considerations'—attracting the potential legal sanction/threat of sanction that contravening the Planning Process provides. To date, considerable success has been achieved in this respect, but there have also been disappointments—as the following Planning Appeal Judgements testify.

Three separate searches conducted by Compass Research (a sister company of Planning Resource and part of the Haymarket group), produced more than 2,100 Planning Appeal Judgements in which during the past quarter century in England, Scotland and Wales crime, FOC or ASB have been material considerations.

Reading through these judgements provides a snapshot of the planning concerns at large in Great Britain at the beginning of the twenty-first century. Indeed, there are repeated references to ATMs, fencing, roller shutters, gated communities, Gypsy and Traveller sites, licensed premises (serving alcohol) Metal Theft and (most recently) Street Hubs. Furthermore, one of the primary conclusions reached is that in the vast majority of instances the Planning Inspectors have proved to be (however unwittingly) expert *Street Level Crime Scientists* (see Sect. 1.5.3) And overwhelmingly they display enviable wisdom and judgement in often trying and difficult circumstances. All references in this chapter to 'Inspector' mean a Planning Inspector.

7.5.1 Evidence to be Supplied by the Police

Most pertinently, the following recommendations are drawn in respect of the necessary information that needs to be supplied by the police (often the DOCO) where it is decided to pass comment in respect of a planning application, or retrospective planning application to which the police wish to object (often in partnership with the local authority/local government):

- Crime, ASB, road traffic and incident data should be supplied in detail. Expressions such as High-Crime Area are too nebulous and require evidence
- The data supplied must be relevant to the location of the proposed development, e.g. occurring in that street and/or immediate vicinity
- The data must be relevant to the proposed function at that location, e.g. in the instance of an ATM (Automatic Teller/Cashpoint Machine—see Sect. 6.11) incidence of robbery and theft from the person would be an important consideration
- When analysing these Planning Appeal Judgements a repeated theme is the failure to provide relevant crime data—and on occasion any crime data! Furthermore, often it is easy to conclude that had such data been provided, a different judgement might have been reached.

7.5.2 ATMs (Automatic Teller/Cashpoint Machines)

In 1999, Planning Permission was refused in respect of an exterior ATM at a bank in the Birmingham suburb of Handsworth. On appeal, the Inspector decided there was no harm to highway safety, no evidence of an increase in crime and the appeal was allowed. However, in 2000, a second ATM proposed for a bank only metres away from the first, produced a different conclusion. A Planning Inspector visited the location and agreed with local police Architectural Liaison Officer (ALO—now DOCO), Constable (now Sergeant) Naheed Mushtaq, that the proximity of the ATM to surrounding streets (potential Attack and Escape Routes) and bus stops (where potential robbers might lie in wait) plus the high incidence of robbery in the area (especially at night), were relevant considerations and Planning Permission was refused.

This case provided a watershed in opposing or placing conditions on the installation of ATMs at inappropriate locations. Unfortunately, for more than a decade some ATM installers have been requesting Planning Permission retrospectively (after the machines have been installed). And there is further evidence that elements such as CCTV and Natural Surveillance (see Sects. 4.3.3 and 4.4.2) and provision of security bollards are being used to obtain approval. For example, during 2011 there were two instances where applications to install ATMs were upheld by separate Inspectors. Each decided there was no harm to crime or disorder, in view of the accompanying security measures. However, at a post office in Islington, London an Inspector dismissed the appeal to install an ATM, concluding there was an 'unacceptable community safety risk'. That decision was influenced by limited Natural Surveillance, potential hiding spaces, nearby ATMs and police adviced the area was a Crime Hot-Spot.

In 2012, retrospective application for an ATM on a supermarket within a Dudley Conservation Area was refused. The Inspector described the side street location as unacceptable and was influenced by the police providing evidence of offending behaviour. Similarly, the retrospective application for an ATM at a London post office was dismissed because it was adjacent to a busy bus bay where those using the ATM might park causing disruption, danger and inconvenience to bus passengers. However, in another 2012 London example, the appeal was allowed because ATMs were already installed in nearby frontages and whilst the police claimed a high level of crime locally, evidence to this effect was not produced.

In Greenwich in 2017, an appeal relating to an ATM in a High Street shop was dismissed, where the police DOCO stated good Natural Surveillance was vitally important—evidence reinforced by the ATM Security Working Group Best Practice document (see Sect. 6.11). An adjacent bus shelter severely inhibited opportunities for Natural Surveillance. Police opinion that the ATM would be a Crime Generator was accepted. Whereas at Lambeth, London, both council and police concern (a DOCO's report) were dismissed regarding the location of an

ATM on the chamfered corner of a building with restricted visibility for users of the machine.

In 2019, the appeal to install an ATM in a Listed Building, adjacent to a pub in a Bristol Conservation Area was dismissed. The ATM and signage being visually prominent, adding clutter and harm the Listed Building and area. Crime Risk was alleged, but no evidence of crime or FOC was provided. Similarly, at Haringey, London, an appeal regarding an ATM outside a shopping centre supermarket was allowed. Both the police and council had raised FOC concerns, but the Inspector decided additional safety, anti-theft measures and Natural Surveillance from a nearby bus shelter mitigated these risks. Whereas a few miles distant in a Bloomsbury Conservation Area, the appeal regarding a street corner ATM was dismissed. Crime was discussed, with weight given to a formal objection by the Metropolitan Police, highlighting how the area suffered from a high number of rough sleepers and users of existing ATMs had been subjected to aggressive and intimidating begging. A further concern was the potential for distraction thefts in an area containing many hotels, tourist attractions and student accommodation with ATMs used for foreign exchange.

In Camden, London in 2021, a retrospective application to retain an ATM installed 2 years before in a mid-terrace, roadside ground-floor shop, with flats above in a very busy urban area was dismissed. Crime and Community Safety were considered, with inadequate evidence provided about the lighting of the site and footpath. Physical barriers obscured views of the ATM from the road and footpath; little active use nearby; and Natural Surveillance insufficient to deter loitering—a police concern. It was considered the ATM could have an adverse effect on Crime and Community Safety.

In addition, over the past decade, a similar diversity of judgements can be identified in respect of former telephone kiosks converted into ATMs and more recently, Street Hubs. In Northampton in 2013, the application to replace a telephone kiosk with a public telephone/ cash dispenser located in a Northampton Conservation Area was refused. This was based on its utilitarian appearance, highway safety and 'realistic and pertinent strong police reservations relating to loutish, criminal and ASB'—citing Paragraph 35 of the NPPF which required development to have a safe layout minimising conflicts with traffic.

Refusal was also decided in respect of a public telephone/cash dispenser to be located opposite Bromley Police Station, due to its effect on the security and safety of the general public and cash in transit staff. However, in 2023, a BT Street Hub incorporating 75-inch LCD advert screens in Birmingham City Centre was acceptable in a lively 'mixed townscape setting of predominantly modern commercial, retail buildings' and a *'barely perceptible change'*. Nor was any substantive evidence provided that Street Hubs were correlated with increased incidence of adverse behaviour.

7.5.3 Nightclubs

The turn of the millennium also witnessed the instance of a proposed nightclub in Brighton city centre. In 2000, both the local authority and police (via the highly experienced ALO, Detective Sergeant Peter Hardy) opposed this planning application. Their grounds were based on the existence of an inordinate number of Licensed Premises ('saturation') already serving alcohol in the city centre and an increase in the size of the nightclub. Despite its Conservation Area location, the Planning Inspector concluded there was no evidence of harm to its character or appearance, no loss of residential amenity and police and local authority opposition on grounds of Crime and Disorder were unwarranted. Most pertinently, such opposition was only submitted when the maximum capacity was increased to 2,400, suggesting the case might have been stronger had it been opposed when Planning Permission was originally sought.

More recent examples have proved to be far more positive and acknowledged the '*...unacceptable risk of increased Anti-Social Behaviour and Crime and that this would be harmful to the environment and character of the town centre.*' This verdict was given in 2010 by an Inspector re the conversion of a bingo hall into a nightclub with the capacity of 1,000 patrons. And in 2011, a retrospective application by an unauthorised nightclub *'containing several hundred customers'*, was refused. Staying open as late as 4am harmed the sleep and amenity of nearby residents and was unacceptable. Whilst the police objection detailed incidents of personal injury and *'Drug Crime'* taking place after 1am.

However, in 2013, the conversion of a restaurant and bar into a nightclub in the city of Wakefield was allowed. Instances of 'Disorder and ASB' at existing clubs and bars in the area were alleged, but the Inspector was unconvinced that these were directly attributable to nightclubs. Whereas, in the coastal town of Harwich, a similar nightclub conversion was judged unduly harmful due to late night opening and noise. Its 3 am closure at weekends, external smoking area and objections by local residents were instrumental in dismissing this appeal. Likewise, in another 2013 case at Darlington, County Durham, the appeal to convert a two-storey Listed Building into a nightclub and public house was dismissed, largely on grounds of aesthetics. Interestingly, the Inspector also commented that issues of Crime and Disorder were more appropriately regulated by the licensing regime.

In 2018, an appeal relating to a nightclub in a Manchester residential area was dismissed. The Enforcement Notice requiring its cessation was upheld, for causing harm to the living conditions of nearby residents by way of noise, disturbance and public safety. Nightclub use took place every night between 2300 and 0300 hours, the Inspector noting persistent and consistent complaints causing a noise abatement notice to be issued in 2017.

In 2022 at Acton, London, an appeal relating to an unauthorised Shisha café/ nightclub was dismissed. The claim it had been a *'community centre with ancillary shisha bar from a vacant property'* since 2019, was rejected on the basis of photographic and publicity material evidence. The Planning Inspector found that the unauthorised change of use had resulted in unacceptable harm in generating noise, smell and disturbance complaints. There was also compelling third-party evidence of violence, unsocial behaviour and operation during the COVID lockdown. Also in 2022, in Enfield, London, a proposed restaurant and bar on the first floor of a storage warehouse within a strategic industrial estate was denied permission by an Inspector who found no justification for main town centre use in this location.

More recently, in 2023, an appeal was allowed regarding an east London nightclub was upheld. Plans to redevelop a brownfield site with a mix of uses including flats above a nightclub, were approved by a Planning Inspector, who was satisfied closed windows and mechanical ventilation would ensure suitable living conditions for occupiers and avoid complaints. The site was currently occupied by a self-storage facility, film studio and well-established 1600 persons capacity nightclub. Examining the technical evidence, the inspector concluded there was no conflict with development plan policies.

7.5.4 Pubs and Bars

The introduction of the Licensing Act 2003 transferred powers relating to the sale of alcohol from the police to local government. This has caused occasional disagreement between the new Licensing Authorities and the Planning Inspectorate, as can be identified in some of the following cases.

In 2006, a change of hours application to extend permitted opening until 3.30 am in respect of a 'function room' at a public house in London was refused by the Planning Inspector, despite approval by the local Licensing Authority. It was argued that the Licensing Act 2003 provided a more up to date means of dealing with late night opening hours. However, in dismissing the appeal, the Inspector decided the Planning System required a broader approach and extending the licensing hours would increase the level of disturbance experienced by local residents – some of whom had already complained about Excessive Noise (see Sect. 8.2.5) and smoke from a barbecue at the pub.

At Bristol in 2008, an extension to midnight for a bar and restaurant was allowed. The local authority and residents (including those in the flats above) were concerned that the extended hours would lead to noise, disturbance and ASB. But the Inspector noted there had been '*...no complaints from the police in terms of criminal activity or from the council's Pollution Control Team in respect of noise.*'

In 2013, conversion of the first floor into a function room at a cafe bar in a residential part of Liverpool was refused. The Inspector observed the '*significant number of recorded incidents of affray, assault, drugs, discharge of firearms and evening disturbance in the area*' and concluded the scheme '*would not create a safe environment... Crime and Disorder and the fear of crime would undermine the quality of life of local residents and would not result in community cohesion.*'

Also in 2013, at a café bar in Cardiff, an application to open until 3 am was rejected due to the adverse impact on public safety. The premises lay within an area dominated by 'vertical' (customers stand rather than sit) drinking establishments very popular with late night revellers. The council with police support argued that the saturation of licensed premises in the area had led to an increasing number of incidents of violence and disorder. The Inspector noted the strong objection from the police highlighting the widerange of incidents caused by excessive consumption of alcohol. It created many problems for policing with the road having the highest number of recorded assaults and public order offences of any street in the centre. The clear link between violent offences and consumption of alcohol was overwhelming and allowing opening until 3 am was inappropriate.

At Birkenhead, Merseyside in 2018, a Planning Inspector allowed the conversion of a ground floor nightclub/ bar into a lap-dancing club. Located in a row of mainly bars and nightclubs, with no retail uses nearby a reason for refusal was it '*does not promote a positive image of the area.*' However, there was no such requirement in the development plan or otherwise. As such, a lap-dancing club was compatible with existing night-time uses and make a positive contribution to the vibrant night-time economy. The Inspector awarded costs against the local authority who had rejected advice from its officers.

7.5.5 Gated Communities

One of the most controversial design features to have recently appeared in the UK has been that of the *'gated community'*. The controversy most often involves a particular type of residential design, one where electronically-operated pedestrian and/or vehicle gates control access into what is in effect a closed environment—thereby preventing entry by ordinary members of the public.

In 2008, in Manchester, the Planning Inspector decided that a *'general presumption against Gated Communities was outweighed by security needs'*. This involved 10 houses and 18 apartments (nine within a Grade II Listed former nursing home). Reference was made to Safer Places (2004—since repealed) and the Inspector accepted: the site had previously been a Hot-Spot for crime and ASB; that a number of residents had been influenced in deciding to purchase by the gated entrance; and that an access road and footpath passed close to the rear of the houses. Consequently, whilst the council's desire to encourage connectivity and social cohesion were laudable aims, in these particular circumstances, he found that consequences would be entirely unacceptable. Also, in 2008, and despite no evidence of Crime Risk, an Inspector allowed the erection of 'low and lightweight gates' at the entrance to a cul-de-sac development of 11 houses, off a busy urban highway in the county of Surrey. He decided: their erection would not result in social division or exclusion; the gates would not be obtrusive or oppressive; were consistent with the existing street scene; and would supply a secure area in which to play for children living on the estate.

However, in 2011, at a development of 167 dwellings adjacent to the River Thames in Twickenham, the Planning Inspector refused to vary a planning obligation that required the vehicle and pedestrian gates be kept permanently open. The Inspector agreed with the council that keeping the gates closed but unlocked conveyed a powerful message: that the general public was at best unwelcome and at worst, excluded. The council's policies sought to ensure all developments were connected by footpaths and cycleways to adjoining areas and streets. Ensuring the gates were in the open position at all times improved Permeability (see Sect. 4.4.6) and a lack of segregation.

In 2013, in south-west London, the appeal to erect a gate to a private alleyway at the rear of urban terraced housing was dismissed. The Inspector described the design and appearance as *'stark and oppressive'* and *'would also impinge on views out of an adjacent property'*. In 2020, near Lichfield in Staffordshire, an Inspector dismissed an appeal relating to security gates and CCTV. Accepting the gates were intended to restrict the use of private rights of access into and out of the estate. However, the council's planning policies supported the creation of sustainable communities, social cohesion and inclusion. The gates would be seen from the highway and whilst the pedestrian would remain unlocked, this would not be readily apparent. It would increase the feeling of segregation and exclusivity, and insufficient evidence had been provided to demonstrate that it would alleviate the fear of crime (FOC) or burglary on the estate.

During 2023, a trio of appeals were dismissed by different Planning Inspectors. At Croydon, London, a retrospective application for gates at the only access to a completed 9 dwelling development, all served by a private drive in a residential suburb, was dismissed. Here, no concerns were found regarding their character and appearance of the area, nor were they harmful to social cohesion. However, whilst they were only closed at night, the Inspector found the restricted visibility would cause harm to highway safety. In Eastbourne, Sussex, an Inspector dismissed the erection of automatic entrance gates. Designed to be permeable (allowing views through), they would nevertheless form a substantial physical barrier across the vehicular entrance to a 50 dwellings scheme. It was believed the proposal would harm social cohesion, by physically segregating the development from nearby residential streets of culs-de-sac connected by pedestrian paths. Plus, the proposal was exclusionary to persons not authorised to access. The proposed gates would harmonise with area character and appearance, as the permeable design would allow views through. It was also noted that there was no evidence of Crime Risk; and an absence of explicit support from the DOCO.

In Chiswick, London, an appeal was dismissed relating to electronically operated gates at the entrance to a residential cul-de-sac off a busy road. It was shown that only a car (rather than a larger vehicle) could turn back out of the cul-de-sac when the gates were closed. This could result in dangerous movements by large vehicles. The gates would restrict pedestrian access to a square containing Grade 2 Listed buildings; that reducing Permeability would not be inclusive design and conflict with the Equality Act, 2010; and there was no evidence of the claimed threat from crime.

7.5.6 Fencing and Gates

In 2000, an Inspector supported an appeal to move a 1.8 metres high fence 2 metres closer to a public footpath which ran through a housing estate in Whitminster, Gloucestershire. Whilst it was agreed that this would lead to a reduction of some visibility, the Inspector concluded it would not create hiding areas which might otherwise increase the Crime Risk.

A decade later in 2011, an Enforcement Notice to take down galvanised steel palisade fencing at a tyre dealer's premises in north London was upheld, despite the appellant's claim it would breach her Human Rights. Its height, length and prominent siting gave the impression of the premises being barricaded and it looked wholly inappropriate within a shopping area on a main road. There had been burglaries, thefts and physical violence against employees and the Planning Inspector accepted provisions of the Human Rights Act 1998 were material considerations. But he concluded that rights under this legislation could be adequately safeguarded by means other than palisade fencing.

In 2016, at Tower Hamlets, London, a Planning Inspector supported the erection of a 2.4 metres high gate across an alleyway linking two residential areas (an Alleygating scheme). However, the fencing created a significant sense of enclosure when walking along the route, eroding a proportion of Natural Surveillance at ground floor level, reinforced by overgrowing shrubs. The Council did not consider levels of ASB and crime were high enough to outweigh its objectives in ensuring permeable and accessible movement. However, the Inspector noted a significant and sustained history of cases indicating a persistent level of crime with supporting police evidence. It was decided that a secured gate was necessary—left open in daylight hours despite this being a technical breach of policy.

In 2020, in Camden, London, the appeal was dismissed for new metal gates and fencing to the existing front boundary low wall in a Conservation Area. The appellant suggested recorded crime in Camden was about 40% higher than the London average, but the Inspector held the use of such a headline figure obscured as much as it illuminated, given the size of the borough and the diverse character of its different areas, as well as the all-encompassing nature of the Recorded Crime label.

Most recently, in 2024, at a Hampshire engineering company, an appeal was dismissed where it was argued that the prevalence of Rural Crime (see Sect. 6. 8) justified a new home in the countryside. The security needs of a rural business to combat serial theft convinced the Planning Inspector of an essential need for a mobile home in the Hampshire countryside. The appellant lived beyond a quick response time to prevent burglary and the Inspector was satisfied of an essential need for a rural worker to live permanently at the appeal site. Nevertheless, the appeal was rejected due to unmitigated impacts on a protected habitat.

7.5.7 Roller Shutters

The negative effects of solid roller shutters on the streetscape are discussed in Sect. 6.10, Shopfront Security. Unsurprisingly, Planning Appeal Judgements reflect this tension between aesthetics and often incompatible security measures.

In 2007, a Planning Inspector allowed the retention of two existing roller shutters and a new shopfront located within a Conservation Area in Lewisham, London. Despite the Inspector concluding the design did not comply fully with the council's policies (a ban on perforated roller grilles in Conservation Areas), there was an *'acceptable exceptional use of material'*.

In 2011, an appeal was dismissed in respect of an unauthorised solid roller shutter installed at the front of a hot food takeaway in the City of Stoke-on-Trent. Declaring it *'prominent and unsightly'*, the Planning Inspector also made reference to the now defunct Safer Places guidance document (see Sect. 7.2) and how such features may actually increase the perception regarding the likelihood of crime—a constant concern from the philosophy of Designing Out Crime.

In 2019 at Wisbech, Cambridgeshire, an Inspector dismissed the retrospective application for roller shutters on the front of a fishing tackle shop in a town centre Conservation Area. Solid and silver in colour and housed in brown protruding boxes, he agreed they created a *'dead frontage'* and focus for ASB such as fly-posting or graffiti. Nor was there a compelling reason why solid shutters would be the only security feature, undermining the appellant's case of addressing crime and fear of crime. The Local Plan set out how the use of external roller shutters should be avoided. Whilst the crime statistics provided were Reported Crimes taken from a police website and *'did not stand up to scrutiny'*.

In 2022 at Keighley, West Yorkshire, the Inspector refused to sanction external roller shutters on a town centre Victorian shopfront, finding no evidence they would reduce crime and ASB as a public benefit in the heritage balance. She noted police advice that roller shutters can raise the fear of crime, cause people to avoid an area, and encourage ASB. And how alternative means suggested by the council in supplementary guidance to shopkeepers on securing their premises, would be less harmful to the character and appearance of the building and surrounding Conservation Area.

7.5.8 Anti-Scaling Cacti, Rotary Spikes and Razor Tape

In 2000, at a vehicle repair garage on an industrial estate in Leicester, the Planning Inspector supported a retrospective application to install wall-mounted anti-scaling cacti, judging there was no harm to the appearance or character of the area and quashing the Enforcement Notice issued by the council.

Compare this to another 2000 case where razor tape was installed without Planning Permission, above a wall (more than 2 metres high) at a public house in a Warwickshire Conservation Area. The appeal was dismissed, the Inspector deciding the need for increased security and deterrence by visual means, was not an overriding consideration, especially as more aesthetically pleasing means of protection were available.

In 2007, an Enforcement Notice relating to rotary spikes on gates to a service yard (together with CCTV cameras on a 4-metres high pole) in Surrey was upheld by a Planning Inspector, on the basis they harmed *'visual amenity'*. Whereas in 2010, retention was allowed of 400 mm rotary spinners fixed to the top of a fence surrounding the service yard of a supermarket in Weston-Super-Mare, North Somerset. In this instance, the Inspector agreed their height and siting deterred crime and ASB; and that there was no harm to visual amenity, due to their relatively small size and no overbearing impact.

More recently, in 2017 at Plaistow, an unauthorised rotating anti-climb system installed on top of the garden wall of a house in a mixed-use area of London, was described as an *'intrusive and hostile addition'* by an Inspector. Usually associated with commercial properties it was unsympathetic, intimidating, and a detraction from the period residential character of the property and the pedestrian environment of the pavement outside. Noting alternative and less intrusive security measures used in the area, the Natural Surveillance available from the house windows, and no evidence to support claims of a higher-than-average crime rate, the Planning Inspector uphold the Enforcement Notice. Nor did the Inspector consider Human Rights to be infringed in terms of the protection of the family home.

In 2020, at a Doctor's Surgery in Ashton Under Lyne, Greater Manchester, an appeal in relation to coiled barbed wire used as a security feature on a doctor's surgery, was judged to be an intimidating feature and the appeal dismissed. Attached around the edge of a single storey flat-roofed extension to the rear, it could be seen from the main entrance to the surgery, to passers-by and from other nearby streets, it was clearly *'intended as an overt security measure but lacked any aesthetic qualities'*. Indeed, the Inspector felt it contrasted starkly with the public realm improvements that had taken place in the immediate surroundings and detracted from the overall appearance of the surgery building and the residential character of the area. Moreover, it would conflict with the adopted Local Plan policy re high quality design and paragraph 127 of the NPPF which aimed to create high quality places where the fear of crime did not undermine the quality of life or community cohesion and resilience.

To summarise, there are far fewer of these cases than might be expected. This suggests that in society at large there is an understanding that: the installation of barbed wire, razor tape, etc. will not be tolerated; Enforcement Notices are readily complied with; and/or that Planning Control turns a blind eye to such installation in less than fully public locations. Personal witness suggests that all three elements are in contention.

7.5.9 Theft of Metal from Church Premises

In 2011, an Inspector allowed an appeal regarding a Grade 1 Listed Anglo Saxon Church located within a central Conservation Area in the City of St Albans. The isolated position of the church and a previous history of theft (including copper and lead from the roof), necessitated its replacement with less valuable alloy-coated stainless steel. The Inspector was influenced by: an 'Ecclesiastical Exemption' for Listed Building Consent; Intruder Alarm and CCTV installations (see Sects. 4.3.2 and 4.3.3 respectively); guidance from English Heritage; and the evolution of the building over the centuries (the copper roof that now required replacing had only been installed in the middle of the last century). See also Sect. 6.7, Place of Worship Safety and Security.

7.5.10 Fear of Crime

Fear of Crime has proved to be an especially difficult material consideration. Mentioned in numerous cases and often as a justification for all manner of inappropriate physical security measures, it repeatedly features in opposition to probation and bail hostels, care and children's homes, rehabilitation centres, Gypsy sites, amusement arcades and casinos. Usually, little if any evidence is provided to justify such hostility and the Inspectors reject these highly subjective viewpoints.

Following on from this last theme, the seminal case dates from 1995 when a Planning Inspector dismissed the appeal by West Midlands Probation Committee for an extension to a bail/probation hostel in a residential part of Walsall, West Midlands. Two years later, the case was appealed to the High Court, where the Inspector's decision was quashed, and it was ruled that fear of crime and ASB must be based on sound reasons in order to carry any weight in the Planning

Process. Smith v. FSS and Another (2005) developed this theme, adding that fear of crime could be a material consideration, but the weight it could be accorded depended on the extent to which the fear could be objectively justified.

In 2010, the appeal to change newsagents into a fish and chip shop in Bury, Lancashire was upheld, the Inspector deciding that council and police concern regarding fear of crime and ASB were *'not founded on clearcut evidence'* and *'the information fell far short of establishing that it would give rise to high levels of anti-social behaviour'*. Whilst in 2011, an Inspector supported the appeal by Surrey Police to construct a custody centre with 24 cells on an industrial estate. She observed there was no evidence from similar facilities to substantiate fear of crime following the release of detainees.

In 2022, an Inspector allowed the appeal in respect of a probation hostel in Windsor, deciding fear of crime was unchanged by a three-bedroom extension that it met a need and did not materially affect safety/perceived safety, of residents. The Inspector found no evidence that the hostel had led to any particular prevalence of ASB, Crime and Disorder in the locality.

Also in 2022, but in Birmingham, a Planning Inspector dismissed an appeal against an Enforcement Notice requiring premises used as a homeless hostel to desist. The hostel provided accommodation for a large number of residents over 18 years of age accommodated as single individuals. There had been a large number of objections and the council's evidence suggested the police had logged a range of incidents, which indicated the hostel gave rise to a greater level of crime and fear of crime than one would expect given the residential nature of the surrounding area. The appellant had submitted a hostel management plan which set out a zero-tolerant approach to noise and general disturbance. However, after 2 years of use, the hostel continued to generate complaints and police intervention.

In 2023 at Mansfield, Nottinghamshire, an Inspector decided a footpath link was necessary for good design and dismissed the appeal. A house-builder had failed to secure permission to block up a footpath access into a recreation ground, approved as part of a development of 32 houses. The appellant claimed the access would result in ASB, crime and disorder or fear of same. The footpath leading to the attractive park was straight, wide and well-lit by lamp posts, overlooked and followed Secured by Design principles. The Inspector concluded that the approved plans would not compromise security and safety of the development, and blocking up the access would conflict with local and national policies seeking the creation of inclusive and accessible well-designed places.

7.5.11 Conclusion

Whilst the above examples provide an overview of the Planning Appeals and Judgements process, a major conclusion which can be drawn from these cases is how each must be judged on their own merits. One final point to highlight. If the person seeking planning approval decides to appeal the decision of the Planning Inspector, considerable costs will be incurred by both parties. To date, this author is not aware of any instances where the local authority or police have decided to mount such an appeal—the legal costs being prohibitive (see especially the Brighton nightclub and Walsall bail/probation hostel examples detailed above). And without that fight the appeal is already lost. If nothing else, this ably demonstrates how use of the Planning System is but one mechanism in the police toolbox of potential interventions and not the magic bullet cure some have imagined.

7.6 Appearing Before Planning Hearings, Public Inquiries, Inquests, etc.

Amongst those Designing Out Crime Officers (DOCOs) who have had the rare ‘pleasure’ of giving evidence at a Planning Hearing, Public Inquiry, Inquest, etc., many describe it as the most intimidating incident of their career. This is because these situations often fall within the civil law process and may include barristers (wearing wigs and gowns) representing each side in the legal wrangle. Consequently, DOCOs (even those with police experience of giving evidence before the criminal courts) are liable to be cross-examined—and this after their qualifications, background and experience have been called into question (sometimes the longest part of the interrogation).

The following advice has been gleaned from that small number of DOCOs (including this author) who have appeared before a public hearing, as detailed above.

7.6.1 Create and Regularly Update a Planning Hearing/ Public Inquiry/Inquest Toolkit

- The toolkit should include your curriculum vitae (CV)

- List your qualifications, training courses attended (in person and online), continuing professional development (including the SBD National Training Event, PCDA RPEL, team meetings, etc.).
- Membership of any professional organisations
- Conference presentations made
- Any personal research published.

7.6.2 Preparation for the Planning Hearing, Public Inquiry, Inquest, etc.

- Establish and rehearse your qualifications in advance. Repeating how you have regularly attended the relevant training courses is insufficient. You will need to support this with evidence of your experience in the field of Designing Out Crime, together with any further qualifications and memberships that are pertinent
- The breadth of your experience in this field should include a preamble that describes your background and qualifications
- Your stance will be predicated on a belief in the validity of Designing Out Crime, Situational Crime Prevention, CPTED measures, Secured by Design and other interventions. Use crime and ASB data, police response figures (where available), any Dispersal Orders, whether contrary to existing policy, location analysis and any evidence regarding the fear of crime. Investigate and rehearse subjective elements such as Defensible Space, Natural Surveillance, Territoriality and Permeability.

7.6.3 Conduct at the Planning Hearing, Public Inquiry, Inquest, etc.

- When appearing before a Planning Hearing, Public Inquiry, Inquest, etc., your CV will be read out during the opening introduction, followed by examination via your counsel, cross-examination by opposing counsel, the chair or coroner's questions and after other witnesses have been called, the decision or finding
- Do not stray (or be led) into subject areas where you have no training, direct responsibility or consummate experience, e.g. Health and Safety
- If asked, state that the recommendations you made were commensurate to the risks perceived at that time

- Your strategy when challenged should be based on preparation, remaining calm under pressure and defending your opinion.

In conclusion, the observation that failing to plan is planning to fail, has a double meaning in the context of preparing to give evidence before such a venue. To reiterate, to date, the appearance of DOCOs before such venues has been a relatively rare occurrence. Nevertheless, it should be perceived and treated as the ultimate test of our professionalism and experience. And remember, you may have no more than six weeks to prepare for this potential ordeal.

References

Campaign for Architecture in the Built Environment (CABE). (2007). *Design and access statements: How to write, read and use them*. CABE.

Department for Communities and Local Government. (2004). *Safer places: The planning system and crime prevention*. DCLG.

Department for Communities and Local Government. (2005). *Planning Policy Statement 1 (PPS1): Delivering sustainable development*. DCLG.

Dudley Metropolitan Borough Council (2002) *Design for community safety: Supplementary planning guidance*. Design and Development Division.

Ekblom, P. (2001). How to think about crime and its reduction—Rigorously, systematically and inclusively. In P. Ekblom, C. Lehman, & K. Pease (eds.), *Reducing offending II: An updated assessment of research evidence on ways of dealing with offending behaviour.* Home Office Research Study. Home Office.

Jacobs, J. (1961). *The death and life of great American cities*. Random House.

Ministry for Housing, Communities and Local Government (2024). *National Planning Policy Framework*. MHCLG.

Secured by Design. (2010). *Design and access statements: How to use them to prevent crime*. Secured by Design.

Taylor, Lord Taylor of Goss Moor (2013). *External review of government planning practice guidance*. Applies to England. Ministry of Housing, Communities and Local Government.

West Yorkshire Police (2009) *Design and Access statements: How to use them to prevent crime*. Leeds, West Yorkshire Police.

Additional Considerations 8

8.1 Legal Liability

8.1.1 Background

In the United Kingdom, the different legal systems of England and Wales, Scotland and Northern Ireland all accept that the occupiers of property have a Common Duty of Care in respect of people who visit those premises—regardless of their intent. In practical terms (and this will come as something of a revelation to people from certain parts of the world), this means that even those intent on crime and anti-social behaviour (ASB) have legal rights in respect of their health and safety—for example, although not exclusively when breaking into (burgling) premises!

Thankfully, it is not within the scope of this handbook to debate the merits of offenders' rights in such circumstances. However, recent examples in respect of *Commercial Cannabis Cultivation(cannabis farms*—see Appendix B in this handbook) point to the necessity for legislation in this arena. In the UK, there have been repeated instances of the criminal operators of such cannabis farms wiring the metal letter box/plate to the electricity mains, and devices designed to inflict grievous bodily harm (GBH) on intruders (be they the criminal competition, utility providers or police officers) are not uncommon—and ongoing.

In England and Wales, the legislation covering this subject area is contained within the Occupiers' Liability Acts 1957 and 1984 (1960 in Scotland; 1957 as amended in 1987 for Northern Ireland). This is of great relevance to the Designing Out Crime Officer (DOCO) because it places a restriction and indeed a necessary responsibility (not to mention a potential Legal Liability) on the advice and recommendations that they can provide. Not that we would expect any DOCO

M. Stokes, *Designing Out Crime*, Crime Prevention and Security Management,
https://doi.org/10.1007/978-3-031-99840-9_8

to be recommending the installation of barbed wire, razor tape, carpet gripper or broken glass embedded into the top of a brick wall.

As detailed elsewhere in this handbook, where enhanced Physical Security 'Target Hardening' Measures and Technological Innovations (see Sects. 4.2.1–4.3.3 are believed necessary, elements which neither contravene the Occupiers' Liability Acts, 1957 and 1984 or produce an overtly fortress-like visual effect, can be recommended. For example (and especially relevant to houses with rear gardens), wooden trellis on top of a fence—designed to provide insufficient strength when used as a climbing aid and if attempted will break, create noise in doing so and hopefully, thereby attract attention. Similarly, and as discussed in Sect. 4.2.4, Defensive Planting, there is a wealth of thorny plants, shrubs and trees that can be used to reinforce a given boundary and provide a natural barrier that might otherwise be proscribed by law.

8.1.2 Occupiers' Liability Act 1957

The original Occupiers' Liability Act 1957 codified elements of the Common Law, classes of visitor to premises and the Common Duty of Care owed to them by the occupier or person in control of the premises. Occupiers are permitted to decide the extent to which visitors can enter premises and the length of time.

However, it is the exceptions to the act in relation to trades people and the necessity to realise that children are less careful than adults, which are especially important. Indeed, case law from over the past century and more can be identified to influence policy to this day. For example, in Glasgow Corporation v Taylor (1922), the corporation was found liable after a 7-year-old child died from eating poisonous berries picked from a bush in the park. Had the child been accompanied by a parent or other guardian, current legal wisdom suggests that verdict might have been different: see Phipps v Rochester Corporation (1955).

The occupier has far less responsibility in respect of a person visiting in pursuit of their professional duties and who is expected '...will appreciate and guard against any special risks...' For example, the case of Roles v Nathan (1963), where two chimney sweeps died from carbon monoxide poisoning after cleaning industrial flues. The Court of Appeal held that the occupier was not liable, as an engineer had warned the sweeps about the danger of carbon monoxide when the boilers were still lit.

8.1.3 Warning Notices

Whilst warning notices are of some relevance in absolving responsibility, once again case law has defined their application. Staples v West Dorset District Council (1995) held that there is no need for a warning sign where danger is all too obvious to the visitor. But Rae v Mars (UK) Ltd (1990) contends that if the danger is extreme or unusual, a warning is insufficient and a barrier or additional notice needs to be displayed.

8.1.4 Vicarious Liability

As a general rule, occupiers cannot be held responsible for the actions of an independent contractor, providing the work has been given to a competent person. However, in Bottomley v Todmorden Cricket Club (2003), the Court of Appeal decided the defendant should have ensured the existence of public liability insurance. Of course, the relevance of vicarious responsibility cannot be underestimated, in that the advice given by the DOCO (at least in the UK) has the potential to similarly implicate their employer—most often named in law as the Chief Constable of their police force.

8.1.5 Landlords' Liability

Under Common Law, landlords were not liable in respect of injuries caused to visitors by their failure to repair property. However, Sect. 4 of the 1957 Act created such a responsibility.

8.1.6 Occupiers' Liability Act 1984

This Act of Parliament was deemed necessary following the case of British Railways Board v Herrington (1972), where the House of Lords (forerunner to the present Supreme Court) ruled that occupiers owed a Duty of Care to trespassers. The new legislation extended the Common Duty of Care to such trespassers, albeit they can only claim for death and personal injury—but not damage to personal property.

8.1.7 Extension of Duty of Care

At Common Law, a trespasser had to prove that he had been intentionally or recklessly injured: Addie v Dumbreck (1929). However, after the aforementioned case of British Railways Board v Herrington, the Law Commission recommended a new Duty of Care be created for 'uninvited visitors'.

For the occupier to be liable in respect of trespassers, he must be aware/ have reasonable grounds to believe the danger exists and reasonably be expected to protect visitors from it. Once again, case law has refined this legislation. In White v the Council of the City and District of St. Albans (1990), the Court of Appeal rejected the claimant's contention that when taking a short-cut across the defendant's land and falling into a trench, the fact that the land was fenced off did not infer an admission of liability.

More recently, in Ratcliffe v McConnell (1999), the plaintiff (who was drunk) jumped into a shallow swimming pool and suffered serious injuries. Warning Signage and Notices (see Sect. 2.13) were displayed and the Court of Appeal decided that he should have known the risk and acted accordingly. That said, warning notices in themselves may not be enough: the sign must be sufficiently clear so that the risk is obvious to the trespasser. Furthermore, Sect. 2 of the Unfair Contract Terms Act 1977 allows landowners to permit access to their land by educational and recreational bodies, without fear of liability.

8.1.8 Advice from the DOCO

Perhaps most importantly, warning signs are usually considered inadequate for children, who may be unable to read or appreciate the danger contained therein. This situation is exacerbated in an area with a high ethnic concentration whose first language may not be English. At the practical level, extreme care should be exercised by the DOCO when recommending the use of the more extreme forms of Physical Security 'Target Hardening' Measures and Technological Innovations, of Crime Prevention Through Environmental Design (CPTED) for fear of the potential for civil litigation.

See also Sect. 3.4 Writing the Report/EVA: the 'Crown Jewels' of its written presentation (regarding the relevance of legal disclaimers) and Sect. 2.13, The Value of Signage and Notices. And one final point, when citing the Occupiers' Liability Acts, 1957 and 1984, ensure the two acts are referenced as the latter merely changed one element in the former and both remain valid and on the statute book.

8.2 Environmental Crime

Environmental Crime is often dismissed as 'low level' crime or anti-social behaviour (ASB), suggesting it is not truly worthy of consideration. However, local authorities (LAs—local government) and other bodies, public and private, collectively spend £ millions clearing up the aftermath of such behaviour. In much of the world, the elements that might be described as *Environmental Crime* have been increasingly criminalised over recent decades—although cultural attitudes vary greatly depending on the individual nation and the often-differing perceptions of their peoples.

From a Designing Out Crime perspective, in the UK prosecution is most often the responsibility of the LAs. However, it should be remembered that because ASB is often the precursor to substantive criminal actions, design that works to eliminate the opportunities (see Sect. 2.3, Crime Opportunity) for such conduct can produce multiple benefits.

8.2.1 Graffiti

Graffiti is one of the most problematic subject areas, if only because different perceptions of the issue often exist within the same community, and indeed across the globe. Even within mainland Western Europe, toleration appears to be far greater than in say the United Kingdom and North America. Indeed, at a pan-European Designing Out Crime conference held in Amsterdam in 2008 and organised by Paul van Soomeren, British members of the audience perceived that city to be awash with graffiti, whereas local residents and those from much further afield had no such concerns—hardly a scientific study, but it does illustrate the issues currently at play.

Indeed, one person's graffiti and its association with criminal damage, is another's 'tagging' or 'street art'—even the descriptions are loaded with meaning. Is there a halfway house of compromise? Reports regarding the success of 'graffiti walls' are at best, mixed. Nevertheless, Theresa Wignal (2008) draws a distinction between graffiti and murals. She describes how murals painted on walls can be traced back to cave paintings such as those at Lascaux in France. Today in locations as diverse as New York, Philadelphia, Toronto, Amsterdam and China, murals are *'adding cultural artefacts to communities'* (Wignal, 2008, p.1). And whilst the distinction between murals and graffiti is not described, Wignal contends that the former can prevent the latter.

In 2004, Brisbane City Council published its *GraffitiPrevention Guidelines.* These identified *Graffiti-Friendly* structures and produced both a *Graffiti Prevention Hierarchy* and *Graffiti Prevention Model.* Separate sections are devoted to Access Control, Canvas Reduction and Easy Maintenance Selections—including protective and 'sacrificial' coatings, and surfaces to avoid. Pearls of wisdom include how graffitists are attracted to blank featureless surfaces where there is little casual surveillance. And that commissioned art should be applied to a minimum height of three metres above the ground (the usual range of graffitists).

In Koln (Cologne), Germany, the KASA (Koln Anti-Sprayer Action) initiative heralded a partnership between the police and the Office for Public Order to combat the spray-painting of graffiti. And similar initiatives are known to exist around the world in cities as far apart as New York and Berlin. Indeed, during the early years of this century personal witness suggested the Berlin metro network had the worst incidence of tagging by scraping coins into carriage window glazing. Whilst in the UK, Network Rail (2009) published a warning notice describing how 'graffiti vandals' were using Hydrofluoric Acid contained in 'bingo dabbers' or large markers to burn their tags into glass. The warning element related to the fact that this type of acid is extremely corrosive and will burn into skin, even after washing. Extreme care was advised accordingly. Meanwhile, in 2016 (and whilst attending an European Union-funded crime conference) the incidence of graffiti in the Greek capital of Athens appeared worse than in any other city visited.

In 2009, the Department of Electronic, Electrical and Computer Engineering at the University of Birmingham, England was creating a database of known graffiti tags based on 'deBonet's statistical image characterisation and superquadric mathematics.' Whilst the ICA's (International CPTED Association) Barry Davidson, had in a previous role in Calgary, Canada maintained a world-wide database of identifiable offenders.

Graffiti and/or street art? Ghent, Belgium

For the Designing Out Crime Officer, the following graffiti-reduction advice is suggested:

- Walls should be coated with white (to maximise light) non-porous and stippled surfaces on which it is difficult to apply and easy to remove graffiti
- There should be a regular maintenance policy to ensure graffiti and other forms of criminal damage are removed immediately (Kelling and Wilson, *Broken Windows* theory, 1982)
- Do not publicly advertise 'anti-graffiti campaigns.' Past experience suggests these can backfire on a spectacular scale and both attract and exacerbate the graffiti problem.

8.2.2 Fly-Posting

Many commentators perceive fly-posting to be a form of graffiti and in the UK offenders can be prosecuted under local byelaws. However, and as with graffiti, different attitudes to fly-posting may be encountered.

For example, in Birmingham, England, David Harding (2004) reported how a lack of communication in one local authority led to the removal of artist Tom Bloor's 'pop art wallpaper' in a pedestrian subway by one council department, when it had in fact been authorised and funded by another as part of the 'Clash' visual arts event. However, the last example is very much the exception and as a general rule, the DOCO should provide the same advice as in the previous sub-section. See also Sect. 6.13, Public Conveniences (Toilets/Bathrooms) regarding the importance of stippled and other wall surfaces in the prevention of graffiti and fly-posting.

Further sub-types of both fly-posting and graffiti can be identified in 'Stencilling' and 'Stickering.' In the former, paint is applied (usually as an aerosol) through a previously made stencil onto a receptive surface. Whilst in the latter, self-adhesive 'stickers' of often professional manufacture are fixed to a sign, door, window, wall, etc.—sometimes in their hundreds.

8.2.3 Fly-Tipping

This refers to the dumping of rubbish (including, tyres, waste soil, building rubble, tarmac and poisonous/contaminated substances) other than at a recognised disposal site. Heavy penalties under both local byelaws and national legislation exist, and it is not unknown for the relevant agencies to use (see Sect. 4.3.3) redeployable CCTV cameras or other covert means to detect and convict offenders.

For DOCOs, the skill is in eliminating or at least reducing the Crime Opportunity (see Sect. 2.3) that might facilitate easy dumping of such materials. For example, isolated parking areas with no Natural Surveillance, especially where these are adjacent to main roads. Indeed, personal witness across the UK suggests the absence of Natural Surveillance (or indeed Formal or Technological Surveillance—of being witnessed during the commission of their crimes) is not surprisingly a key determinant in the decision to offload rubbish at a given point—an intrinsic part of their Crime Script. Thus, their desire for isolated locations. But simultaneously, they seek out a place where they can pull over off the main road and thereby complete the deed in a relatively innocuous manner.

8.2.4 Abandoned Vehicles

Until the close of 2007 and certainly within the United Kingdom, abandoned motor vehicles (usually although not exclusively burnt out cars that had been stolen and then 'torched') were a constant problem across the UK and especially in those areas close to urban centres. However, there followed the world-wide hike in the price of metal and instances of such abandoned motor vehicles plummeted—presumably because either the original owner or subsequent criminals saw the value in trading in the car for cash at a scrap metal dealership or 'canibalising' it for spare parts.

Crime Prevention/Reduction has proved to be an important element in the strategy for dealing with the massive increase in metal theft that has accompanied the rise in values for all such substances—including steel. Moreover, local government now has competent policies in place for dealing with abandoned vehicles. Truly effective (see Sect. 1.5.11) Problem-solving, Problem-Oriented Policing and Partnerships (see Sect. 2.14) like the former West Midlands Arson Task Force were instrumental in driving forward policies with a mandate to secure forensic evidence by ensuring minimal use of water is used to fight fires started in known stolen vehicles.

For the DOCO and in the knowledge that the scourge of abandoned vehicles will undoubtedly return, the ultimate desire is to ensure that all new development (and the refurbishment of existing projects) removes as many opportunities as is possible where such cars, etc. may be dumped and set alight. Rear parking courtyards are perhaps the most obvious example, together with service roads and isolated parking bays. This is a difficult task as the committed offender is invariably highly innovative. However, that is no reason why we should not attempt to thwart his actions by minimising the more obvious and desirable of locations. Indeed, it can be powerfully argued it is a key skill of the DOCO (and one to be communicated across policing and the agencies of law and order) for the necessity to innovate as fast if not faster than the criminal, if Crime Prevention is to be truly successful.

8.2.5 Excessive Noise

A growing problem in the UK and presumably throughout much of the world, complaints of excessive noise most often relate to loud music facilitated by electrical amplification. Thankfully, due to the British weather this only occasionally migrates into the garden and is then associated with outdoor parties and raves, with such loud music continuing into the early hours of the morning. Far more common are the instances of loud music and other noise migrating through the walls of neighbouring properties, a situation exacerbated by increased housing densities and on occasion the absence of noise-reducing materials in walls.

In the UK, the responsibility for the prosecution of offenders in relation to excessive noise has been increasingly accepted by local government and their Environmental Health departments. However, there is also a role for the DOCO—especially when reminding developers of the necessity for sufficient insulation against noise in party walls (walls between dwellings). See especially Sect. 5.4.9 Secured by Design—Residential Guide 2025.

8.3 Good Housekeeping Measures

Cottesbrooke Infant School, (see Sect. 9.3 in The Good Practice Case Examples, of this handbook), details how one of the benefits which emanated from that project was the inclusion of an appendix devoted to Good Housekeeping Measures. This is not an 'Operations Manual' (although it may motivate and/or feed directly into the creation of such a document), but rather a series of recommendations made by the Designing Out Crime Officer (DOCO) for the post-construction security management of the premises in question.

The creation of such a Good Housekeeping Measures appendix requires the DOCO to think beyond construction of the project to those practices that need to be adopted once the building is occupied. In this regard she or he will most often be guided by their skills as both a Crime Prevention/Reduction Officer and as a champion of Crime Prevention Through Environmental Design (CPTED)—in other words, as a practitioner of Designing Out Crime. DOCOs should already possess the holistic skills necessary to formulate the Good Housekeeping Measures, believed to be relevant to a given development project, post-construction. The addition of any rationale derived from Crime Science (see Sect. 2.1–2.15 could be similarly persuasive.

Previous experience suggests the specific issues relating to Good Housekeeping Measures might include:

- Key management and security (see especially Appendix 1 to the Secured by Design—New Schools Development Guide, detailed in Sect. 5.7
- Procedures to be adopted when opening and closing the premises in question
- Visitor identification, registration, escort and safety
- Fire Safety (as it relates to any recommendations made by the DOCO)

- Necessity to manage Perimeter Security—even when the premises are in use
- Intruder Alarm and fire alarm system management (see Sect. 4.3.2)
- CCTV system management, including the necessity to regularly check image quality and maintain recorded images for 31 days (see Sect. 4.3.3)
- Procedures for calling the police following criminal attack, uninvited persons on premises, etc.
- Signage and Notices displayed (see Sect. 2.13)
- Energy efficiency (e.g. Security Lighting—see Sect. 4.3.1 and use of Internal Lighting
- Procedures to be adopted on receipt of suspicious packages
- Promoting a culture of vigilance appropriate to the premises in question, with suspicious behaviour reported to the police

This is not an exhaustive list, and the individual DOCO will no doubt encounter many other subject areas that need to be covered.

Perhaps most importantly, Good Housekeeping Measures both implies and requires a degree of interaction between the DOCO and the eventual managers and occupiers of the building. It places day-to-day responsibility with the latter for the security of the premises and provides a link between the DOCO and the end user, thereby illustrating how safety and security do not finish once construction is complete. Furthermore, whilst potentially relevant to both residential and non-residential premises, a separate appendix relating to Good Housekeeping Measures is more likely to be considered necessary in respect of non-residential premises, with its wide spectrum of potential end uses for the development in question.

8.4 Good Housekeeping Measures - in Practice

Whilst mention of the subject of Good Housekeeping Measures can be made during the opening dialogue with the client, for the DOCO, the most appropriate moment to forward such recommendations to the end user will almost certainly be following completion of the project. That way the correct person responsible for their implementation can be identified and a rapport established—from which a two-way flow of information will hopefully develop, including opinions as to the recommendations originally made by the DOCO. Such feedback might be considered especially valuable, as it is at present a rare commodity.

Finally, an annex devoted to Good Housekeeping Measures is not a mandatory requirement. It is merely a suggestion that the DOCO or CPTED Officer/Adviser might wish to include in respect of a specific development project.

References

Harding, D. (2004). The Graffiti gaffe—*Council given £2m urban art grant rips down work.* In Metro national newspaper, July 29, 2004. London: Metro.

Kelling, G., & Wilson, J. (1982). Broken windows: The police and neighbourhood safety. *The Atlantic Monthly* (pp. 29–38). Washington DC.

Rail, N. (2009). *Graffiti using Hydrofluoric Acid, Infrastructure Group Safety Bulletin.* Network Rail.

Wignal, T. (2008). Murals List Communities—The opposite of Graffiti, in *CPTED perspective, volume 7*—issue 3 August 2008. Alberta: International CPTED Association.

9 The Good Practice Case Examples

9.1 The Four Towers, Duddeston Manor, Nechells, Birmingham

Home Office Minister, Baroness Blatch and Councillor Marje Bridle, Head of Housing at Birmingham City Council inspect newly refurbished The Four Towers in 1993.

M. Stokes, *Designing Out Crime*, Crime Prevention and Security Management,
https://doi.org/10.1007/978-3-031-99840-9_9

Background

The four, 12-storey, residential tower blocks at Duddeston Manor in the inner-city district of Nechells, Birmingham, England were the first such high-rise flats/apartments to be constructed in the city. They were officially opened in 1954 by the then Minister for Housing and Local Government, Harold Macmillan, who 5 years later became Prime Minister.

Their stretched X-shape design was popular with Modernist architects after the Second World War. Whilst the round porthole style windows and white tubular railings on the roofs, mean that Queens, Home, High and South towers (each containing 66 flats) portray a nautical theme. Moreover, because high-rise public housing was a new concept in the 1950s, the design of these flats was revolutionary in many other respects. For example, for almost 40 years, all four blocks were heated courtesy of a boiler room in the basement of Home Tower. And despite having two electrically operated lifts (elevators) and staircases per block, it was believed necessary to incorporate additional internal fire escapes. To this end, an enclosed staircase (five per block) adjacent to each flat's external balcony provides a further escape route to the ground—although there is no evidence to suggest they have ever been used for that purpose. Another feature at The Four Towers and the norm in cities across the UK and beyond, was the total Permeability (see Sect. 4.3.6) that lay beyond the blocks themselves.

Central government funding (in particular the 'progressive height subsidy') meant that over two decades Birmingham City Council built a total of 464 tower blocks, albeit none of them replicating the balcony-linked fire-escape system. However, by the late 1980s, The Four Towers were experiencing high levels of crime and ASB never foreseen by proponents of high-rise living—world-renowned architects like Le Corbusier and Gropius. The flats were difficult to let, many tenants staying less than 12 months; Residential Burglaries were soaring (often targeting vacant flats via the fire-escapes); ASB by non-residents was endemic; and a murder was committed inside one block. Indeed, like many other high-rise inner-city tower blocks across the UK (Nuttgens, 1989) in Europe and beyond (Power, 1997), The Four Towers had taken on a distinctly unsavoury countenance.

Community Consultation and Partnership Approaches in Action

As is repeatedly demonstrated with Community Consultation and Truly Collaborative (and therefore effective) Partnership Working (see Sect. 1.5.11 and 2.14), timing and the willingness of key partners to actively participate, are essential to deliver long term, sustainable development. The creation in the 1980s of the new police role of Architectural Liaison Officer/Crime Prevention Design

Adviser (DOCO) and the SBD award (1989) in the UK, coincided with forward-thinking decision-makers in local government wanting to ensure future development was free of Crime, Fear of Crime and ASB—and especially commonplace on those developments of the 1950s, 1960s and 1970s.

In 1989, Birmingham City Council successfully bid for £4 million of central government Estate Action funding to refurbish The Four Towers. They sought a partnership arrangement with the police to Design Out Crime on a sustainable basis and the new discipline found a classic example to prove its value and confer the first major SBD awards by West Midlands Police. Nevertheless, it is important to note that The Four Towers were not selected for refurbishment and the SBD process due to the Hot-Spots and Hot Wards analysis identified by Bennett and Durie (1999) in their study of Residential Burglary in the City of Cambridge, England. Almost four decades after construction, the primary motivation was the necessity for refurbishment, although Crime, Fear of Crime and ASB were contributory factors.

Ultimately, this was classic Problem-Solving and Problem-Oriented Policing (POP—see Sect. 2.14), with police-provided Crime Pattern Analysis (and DOC advice) to justify expenditure on the CPTED measures. Similarly, acknowledgement should be given to '*particularly motivated individuals*' within The Four Towers Tenants' Association (Santa Wilder and Jean Ray); Birmingham City Council Housing Department (Barry Fulford, Bill Bellman, Nick Berry and Michael O'Connor), Planning and Architecture Department, Landscape Practice Group and West Midlands Police DOCOs (Brian Hewett) in ensuring the project was successful (Read & Tilley, 2000; Bullocket al., 2006; Kirby and Read, 2004).

Designing Out Crime in Practice at The Four Towers

First and foremost, The Four Towers Tenants' Association both initiated and maintained their interest. This is a key message for all development programmes—and especially refurbishments where residents remain in situ long after completion, whereas the project managers, builders and delivery agencies will have moved on. Indeed, it is an essential element of sustainability: active participation by the residents/tenants. For as Morris (1996) recommends, long-term solutions require the support of the community. Similarly, residents should be engaged by local officers, encouraged and supported in adopting the longer-term Problem-solving approach.

In terms of CPTED, there was a heavy concentration on what Designing Out Crime proponents might describe as Physical Security 'Targeting Hardening' Measures and Technological Innovations: high quality 'Mul-T- Secure' entrance doors to each of the individual flats/apartments; railings and gates on all the ground floor balconies to prevent access to the fire escape system by burglars; CCTV cameras in vandal proof housings (including the interior communal areas and lifts); communal entrance doors at ground level, activated by a key-fob operated electronic access control; and a 24/7 staffed concierge presence. Moreover, the historical significance of the tower blocks was not ignored, evidence the insecure steel-framed windows that were replaced with PVCu units. And perhaps most significantly, for the very first time, West Midlands Police knowingly recommended the inclusion of CPTED Environmental Elements.

For example: the creation of Defensible Space (see Sect. 4.4.1) outside the blocks with distinct physical barriers—aesthetically decorative but highly functional brick walls, steel gates and railings helping to define public, semi-public, semi-private and private space; Symbolic (Psychological) Barriers, (see Sect. 2.12)—brick pillar gate-posts, gates (that are never closed), climbable fencing and changes in road surface texture and colour; Natural Surveillance (see Sect. 4.4.2) employing the windows from the flats themselves; Territoriality (see Sect. 2.14); restrictions on Access and Movement (Permeability see Sect. 4.4.6) and all these features enhanced by improvements in the external lighting (see Sect. 4.3.1).

Following refurbishment completion in 1992, West Midlands Police conferred their first major SBD awards. The then Chief Constable, Sir Ron Hadfield, presented an award for each tower and all parties waited to see the results, especially the effects on crime.

Pre- and Post-Completion Crime Analysis

There are inherent dangers in comparing before and after crime figures. From a Crime Science perspective, other influences inevitably exist, e.g. changes in demographics, the economy, employment, residents, local crime trends. Nevertheless, there is no simpler exercise than comparing before and after recorded crime statistics—whilst mindful of the health warning that accompanies these figures and differences within repeated British Crime Surveys and subsequent Crime Surveys for England and Wales.

Evidence of Anticipatory Benefits (Smith et al., 2002) was reported (but not evidenced) when the Residential Burglary figures began to fall many months before contractors arrived on site.

Post-refurbishment and with almost 100% levels of occupancy (the tenants paying a premium on their rents for the security measures), during 1992 and as the refurbishment neared completion the number of burglaries tumbled. In 1992, 19 burglaries were reported; in 1993 none (100% elimination); whilst in 1994 just one attempt. Indeed, over the subsequent quarter century, burglaries have averaged 1.9 per year (0.008 per dwelling P.A) demonstrating an 89.2% decrease compared to the 1992 figures. Indeed, the worst performing Property Crime category has been Vehicle Crime with 10.28 recorded offences per year (0.039 per dwelling P.A). Moreover, there has been a sustainable decrease in all the Property Crime categories of at least 58% making The Four Towers one of the best-evidenced examples of long-term sustainable Crime Prevention—in the world!

When the SBD award was originally launched in 1989, its key intention was to influence new-build, private for sale housing. However, a depressed housing market and the private sector's antipathy (at that time) towards SBD and security measures in general, meant that during those early years it was RSL (residential social landlord) refurbishment projects like The Four Towers that proved the efficacy of both the SBD award and the concepts like CPTED and DOC. A key conclusion can be drawn: historically, local authorities, housing associations and other RSLs have demonstrated an economic interest in ensuring the long-term safety and security of their developments and tenants.

Lessons Learnt from The Four Towers

a. **Entrance Doors to the Individual Flats**

Somewhat uniquely, financial worries were of little concern in respect of security measures during the 1989–1992 refurbishment. Top of the range Mul-T-Secure doors—complete with frame armour, hinge bolts and a four-point bolt-action locking system—were installed at the entrance to each flat in LA ownership. These doors proved to be highly effective, witness how following completion of the 1989–1992 refurbishment no credible burglary was committed by forcing them open. Indeed, the doors proved to be too secure, soliciting complaints from specialist police departments like the former Drug Squad and West Midlands Fire Service. Even their door-opening hydraulic jacks could not overpower the frame armour—an example of over-engineering the Physical Security 'Target Hardening' Measures. This is a serious issue when lives are at risk and points to the primacy of safety over security considerations, reinforced following the 2017 Grenfell Tower tragedy in London. However, such concerns were absent when the project was commissioned. Subsequent experience with multi-storey refurbishment demonstrates the security of these dwelling entrance doors is imperative. Too little, or poor quality 'Target Hardening' and the flats are prone to criminal attack. The insecurity of such doorsets led to the creation of PAS 24 and other standards being developed (see Sect. 4.2.1). Furthermore, in projects where investment has favoured 'Target Hardening' solely the communal entrance doors to the blocks, burglary by neighbouring tenants has been identified—especially where drug addiction and the necessity to 'get a fix' are evidenced or suspected.

The financial costs associated with Secured by Design are another consideration. Professor Rachel Armitage (2013) describes how research details SBD as cost-effective or cost neutral: Armitage (1999); Association of British Insurers (2006); Teedon et al. (2009). Armitage (2005) also calculated the additional cost per dwelling of incorporating SBD measures as £795. Whilst Davis Langdon (2010) estimated that this had reduced to £240 for a ground floor apartment and £70 for one located on an upper floor (see Sect. 2.11).

b. **The CCTV System**

From the early 1990s until the present day, considerable concern relates to the privacy and civil liberties aspects of CCTV (see Sect. 4.3.4). However, limiting camera coverage to ground floor, communal areas and lifts was a mistake. Once past such communal entrance doors ('tail-gating' or threatening legitimate callers were repeated MOs) unwanted visitors could not be tracked through the building. A 2006 upgrade of the CCTV system (from 12 to 35 cameras per tower block, including 31-day, digital recording and motion detection) provided surveillance of each landing area, allowing uninvited guests to be tracked from floor to floor—and addressed by means of two-way audio communication. However, during the most recent 2019–2021 refurbishment, the CCTV system was upgraded yet there remains no external camera coverage. It is therefore interesting to muse how the extension of the CCTV system to include external areas, e.g. car parks, might have delivered added benefits—especially relevant in relation to Vehicle Crime and the now withdrawn 24/7, 365 day concierge system monitoring all four tower blocks and providing a human presence in offices and a control room located on the ground floor of Queens Tower.

c. **CPTED Issues**

Reviews of Secured by Design—Brown (1999); Pascoe (1999); Armitage (1999); Teedon et al. (2009); Armitage and Monchuk (2009); Jones et al. (2016); Stokes (2022) often point to the lack of, or diminution of the Environmental Elements of CPTED in many such projects. It seems that whilst all DOCOs, are comfortable with the 'tick the box' culture that formerly dominated the Physical Security 'Target Hardening' Measures and Technological Innovation elements of SBD, there is sometimes a tendency to ignore the Environmental Elements that require highly subjective interpretation and application. See especially Armitage (2017). Nor is it just DOCOs. Some partners have been hostile to certain concepts—most especially Defensible Space (Newman, 1972, 1973, 1996). Indeed, one of the repeated (and highly inaccurate) criticisms directed against DOC and SBD, is that of a'Fortress Mentality'. Yet as Dr Tim Pascoe (1999) powerfully concludes, the benefits of DOC and SBD could be intensified by making areas more private, giving priority to Defensible Space criteria and other measures that reduce access and the flow rates of Permeability (Movement).

At The Four Towers and in the context of the very early years (1989–1992) of both DOC and the SBD award, a dispassionate observer would conclude that the application of the Environmental Elements of CPTED (see Sects. 4.4.1–4.4.6) is very good. The use of brick walls and ornamental metal fencing and gates has produced both Defensible Space and Territoriality. And Natural Surveillance is afforded from the flats that overlook external areas from up to 12 storeys.

Nevertheless, a snaking public footpath with poor sight lines leading to a bus stop on the main road, bisects Queens and Home Towers. Whilst South Tower is located a quarter of a mile (400 m) away from the other three. Similarly, the pedestrian gates designed to be left permanently open are purely symbolic. The value of such Symbolic (Psychological) Barriers, especially in what was an inner city High-Crime Area, is open to considerable debate (see Sect. 2.12and Shaftoe and James, 2004). However, more than three decades later and following the evolution of DOC and Crime Science over the intervening years, the degree of CPTED measures incorporated by a wealth of partner agencies between 1989 and 1992, can be looked upon as particularly encouraging.

Future Developments

Following government changes to housing benefits, in 2015, the on-site concierge presence at The Four Towers was withdrawn and replaced by a centralised control room serving the city's remaining tower blocks. The electronic, access-controlled communal door entry system, has all but eliminated the potential for the insidious crime of Distraction Burglary—where the elderly or those suffering from physical or mental issues are targeted. However, as Thornton et al. (2003) describe, ethnic minority communities report higher levels of worry about crime.

On a sad note, the formerly active and instrumental Four Towers Tenants' Association ceased to exist in 2006. This was apparently due to tenants/members having passed away and it reflects the changing demographics of inner-city areas like Nechells and across the UK—in terms of age, employment and ethnicity. Nevertheless, it is heartening that with its proven track record of Sustainable

Crime Prevention, The Four Towers at Duddeston Manor continues to provide safe and secure homes for all British citizens—new and old.

Postscript

Since retiring from West Midlands Police and as part of his own continuing professional development, this author completed a PhD research project. This investigated the effectiveness of CPTED (Jeffery, 1971) measures and the SBD award process over the very long-term following the refurbishment of The Four Towers and (the comparison) Severn, Thames and Medway towers. During those refurbishments, CPTED measures were included following advice from police DOCOs. However, changes in the funding structure and SBD's minimum standards meant different grades of CPTED were incorporated. In particular, new and extremely high-quality multi-point locking entrance doors to each flat were installed at The Four Towers. Whereas, at the comparison, a simple additional locking mechanism (a BS 3621 five lever mortise deadlock) was fitted to existing flat entrance doors.

A Case Study Approach was used to test the effectiveness of Situational Crime Prevention (Mayhew et al., 1976), CPTED and SBD at both the case study and comparison sites. And rather than SARA (Eck and Spelman, 1987) this impact study chose 5Is as a means of analysis '*…reflecting the requirement to handle the rich complexity of preventive action*' (Ekblom, 2011, p. 86). The investigation evolved to include four research aims:

1. **Has There Been a Reduction in Recorded Crime at The Four Towers Compared to the Comparison?**

Close to a quarter century of anonymised crime data were supplied by West Midlands Police for both sites that detailed 1,506 Recorded Crimes. It charts how burglary was all but eliminated towards the end of 1992 as the process of installing the entrance doors to each flat and communal entrance doors was completed and the on- site, 24/7 staffed concierge began operating. A sustained reduction in burglary of 89.2% at The Four Towers was maintained over 23 years. Of crucial importance, over the final 18 of these 23 years, there were 50% more police-recorded burglaries at The Four Towers than at the comparison. This is explained by new entrance doors being installed at the comparison, causing burglaries to fall and overtake (no reported burglaries) the reductions at The Four Towers during the last 5 years of analysis.

2. **Has Any Such Net Reduction Been Sustained over a Period of 25 Years?**

The rarity of forced-entry burglary at both The Four Towers and comparison (once the existing insecure doors were replaced) suggests a combination of secure flat entrance doors and fob-reader, electronically operated communal entrance doors delivered much of the Sustainable Crime Reduction. The one crime type that appears impervious to CPTED treatment is that of assaults, which increased during the study period—possibly due to changes in the law, Home Office recording rules and police recording policies and practices. For example, Common Assault becoming a recordable offence.

3. **What Impact Have the Individual Elements of Design Had on Crime in the Study Area?**

Despite the inherent difficulty in attributing impact to specific measures, 'Target Hardening' appears to have had the greatest effect. In particular, the entrance doors to individual flats at both sites. Whilst, enhanced security of the communal entrance doors provided an added layer of security. CPTED elements like the 24/7 staffed concierge system at The Four Towers are more contentious. Crime rates had all but converged before 2015 when both sites were linked to an off-site 24/7 control room, suggesting the concierge was having no additional Crime Reduction effect. Similarly, the CCTV cameras were never extended to external areas and tenants at both sites dismissed their prevention and detection capacity. Gill et al. (2005) report the preventive role of CCTV is not well-evidenced. Least effective are the CPTED Environmental Elements in the grounds and car parks. There is no quantitative or qualitative evidence to support the creation of Defensible Space, Territoriality, Natural Surveillance (Newman, 1972, 1973) or Symbolic Barriers (Shaftoe & James, 2004; Sect. 2.7). And one retired DOCO described symbolic barriers as 'the weakest tool in the tool-chest' (Stokes, 2022, p. 364).

Sadly, CPTED measures cannot prevent crime committed within the home by members of that household. Nevertheless, the entrance doors to each flat at both The Four Towers and comparison now meet the requirements of Domestic Violence Sanctuary Rooms (see Sect. 6.1). At The Four Towers, a highly effective security measure in reducing Residential Burglary, relates to the balcony-to-balcony linked fire-escapes. This is an important aspect of design trade-offs/conflicts (Fire safety/security—Ekblom, 2011). During the 2019–2021 refurbishment, all the balconies were fully enclosed with PAS 24:2016 windows to form an integral part of the building—together with secure, outward opening Fire Doors at ground floor level.

4. **Which Specific Crime Prevention Interventions Have Impacted on Reductions in Crime and How Effective Were They?**

With two layers of CPTED (flat and communal entrance doors) protecting the Target Enclosure (each flat) and using PAS 24 as the minimum-security standard for doors and windows, Residential Burglary can be reduced to a minimum. There is also the Grenfell Tower legacy and potential conflict between Fire Safety over Security. At the CSS, some elements of 'Target Hardening' put in place between 1989 and 1992 (for example frame armour and hinge bolts on the doors), can be perceived (in retrospect) as compromising safety. However, the original multiple, fire-escape system within each block (included during the original construction between 1951 and 1953), now appears especially forward-thinking and a lesson to future generations who wish to build high.

Contribution to Knowledge
This author's research investigation produced a number of key findings including:

1. **Potential Conflict Between Safety and Security**

Careful/creative design (rather than compromise) is necessary to ensure there is no conflict between safety and security. If in doubt, safety takes primary importance.

2. **Knowledge of Crime in High-Rise Tower Blocks**

Case comparison analysis of Recorded Crime data, tenants' quantitative and both tenants and professionals' qualitative data provides detailed knowledge about crime in high-rise tower blocks.

3. **Analysis Over a 25-Year Timeframe**

This was an especially unique investigation, examining crime, fear of crime and ASB over a quarter-century timeframe (including issues of Durability and Sustainability) and the problems encountered.

4. **Effectiveness of CPTED and SBD**

Over the very long-term, CPTED measures and SBD delivery at The Four Towers produced substantial and sustainable reductions in crime—most especially Property Crime and Residential Burglary.

5. **Security of Entrance Doors to Each Flat and Communal Entrance Doors**

Ground floor communal entrance doors and those to each individual flat must be of sufficient quality (and Durability) to deliver Sustainable Crime Reduction over the very long-term.

6. **Significance to Senior LA and Police Managers and to DOCOs, Planners and Housing Officers**

The benefits of Partnership Working, in enabling senior LA and police managers to assess where resources should be concentrated and what works in practice, are profound.

7. **Desirability of Consulting the Tenants/Residents**

Prior existence of The Four Towers Tenants' Association points to the inherent desirability of such a consultative group. Such consultation led to immediate and Sustainable results at The Four Towers.

8. **Effectiveness in Reducing Residential Burglary**

Amongst the six Crime categories examined, Residential Burglary in high-rise blocks was the most susceptible to CPTED and SBD Interventions delivering a Sustainable Reduction of 89.2%.

9. **The Once in a 30-Year Opportunity to Get Things Right**

Failing to get it right at the comparison meant the entrance doors to each flat had to be replaced within a decade. Whereas at The Four Towers durability was maintained for close to three decades.

10. **5Is as a Research Tool**

The 5Is (Ekblom, 2002) performed extremely well as a methodological means of research, compared to a more simplistic form of analysis. This reduced the capacity to miss data or nuances of seminal importance. Ultimately, this investigation adds to the canon of existing research regarding the effectiveness of CPTED and SBD.

11. **Victim's Perspective**

The plight of victims living in blighted inner-city environments explains why tenants of high-rise tower blocks, deserve to be heard and both their safety and security concerns addressed.

Conclusion

Within the United Kingdom, The Four Towers was one of the very first projects to be taken through the SBD award process. Recorded Crime began to fall once refurbishment was announced and accelerated after security measures were introduced. Most commendably, a reduction of 89.2% in Residential Burglary has been sustained over close to 25 years after refurbishment was completed in 1992, thereby providing considerable justification for the claim that it is one of the best instances of long-term, evidenced Sustainable Crime Prevention in the world. It represents both a classic example of Sustainable Crime Prevention achieved through the implementation of CPTED measures delivered via SBD and simultaneously, an effective Problem-solving and Problem-Oriented Policing/Partnerships initiative. All parties were committed to its success, including Birmingham City Council Housing Department and the then West Midlands Police DOCOs. And most importantly, the tenants themselves were a constituent part of the design and refurbishment process.

The Four Towers following their most recent refurbishment 2019-2021

9.2 Brindleyplace: City Centre Commercial, Retail and Leisure Complex

Background

There is a greater length of canals within the city of Birmingham, England than there is in Venice, Italy. However, this is largely due to the difference in geographical size of the two cities and you will not find the good citizens of 'Brum' relying on vaporettos or gondolas for their public transport needs!

Brindleyplace, in Birmingham City Centre, is a prestige commercial, retail and leisure development on a site formerly occupied by decaying warehouse buildings. It is bordered by the interchange of two such canals; the acoustically magnificent Symphony Hall (one of the best such venues in the world and home of the world-renowned City of Birmingham Symphony Orchestra); the International Convention Centre (ICC); Utilita Arena Birmingham; and the bars and nightclubs of Broad Street (also a Business Improvement District). In 1998, a G8 Summit was held at the ICC and the then United States President, Bill Clinton, was photographed drinking a pint of beer on the balcony of the Malthouse public house, overlooking Brindleyplace and the Birmingham Canal.

The phased construction of Brindleyplace began in 1992. From the outset, it was built to achieve a series of Secured by Design (SBD) awards, each conferred as that constituent part was completed. Indeed, nine SBD awards (including one for the National Sea-life Centre) plus a 'Park Mark' Safer Parking award (see Sect. 5.1.s) have been conferred at this project, with the final one presented in 2009 following completion of number 11 Brindleyplace.

It is testimony to the developers, Argent and those advising on security issues (in the first instance Security Consultant and retired police chief superintendent Derek Williams), that they embraced SBD from the very outset and in doing so, eliminated the maximum number of Crime Opportunities, fear of crime (FOC) and ASB. Indeed, Argent has subsequently redeveloped the nearby Paradise Circus area.

Three decades later, the waterfront development of Brindleyplace with its public squares and tree-lined streets continues to attract the wide spectrum of patrons for whom it was originally designed. This bears witness to the safe and secure environment that has been created and maintained, an issue that should not be underestimated in view of the aforementioned Night Time Economy within Brindleyplace and the immediately adjacent Broad Street area.

At the concept stage, it was envisaged Brindleyplace would appeal to an 'upmarket clientele'—in terms of its offices and leisure facilities that would be let. Consequently, safety and security were of primary importance. However, the proximity of Brindleyplace to one of the most economically-deprived council wards in the United Kingdom, combined with a desire by both elected members and planning professionals for an open-plan, permeable environment (see Sect. 4.4.6, Permeability) provided unprecedented challenges from the perspectives of CPTED, Designing Out Crime and Secured by Design.

It was relatively easy to introduce the Physical Security 'Target Hardening' Measures and Technological Innovations (see Sects. 4.2.1–4.3.3) to the building elements at Brindleyplace. However, the desired Permeability of the site layout demanded special attention in terms of the Environmental Elements of Crime Prevention Through Environmental Design (CPTED—see Sect. 4.4.1–4.4.6). Enhanced CCTV, including the development's own security control room and 24/7 manned guarding) were volunteered by Argent and professionally maintained once installed.

Fundamentally, the greatest challenge at Brindleyplace was the longevity of its construction period. Whilst building work began in 1992, from the very beginning, it was understood that the whole project would take the best part of two decades to complete. It was therefore highly unlikely that any single Designing Out Crime Officer (DOCO)—originally described as Architectural Liaison Officer when the project commenced and Crime Prevention Design Adviser (CPDA) by the time of completion in 2009, would oversee the development through its life cycle—as was the case. Nevertheless, this longevity supported the claim that Brindleyplace was a major catalyst in the regeneration of Birmingham City Centre.

Partnership Approach at Brindleyplace

Truly Effective Partnership Working (see Sects. 1.5.11 and 2.14, Problem-Solving, Problem-Oriented Policing and Partnerships) involving 'particularly motivated individuals' (see Read & Tilley 2000; Bullocket al. 2006; Kirby and Reed, 2004), were key elements in producing a safe and secure environment at Brindleyplace.

In this context, the original developers and subsequent owners have now supported Secured by Design at Brindleyplace for more than three decades. Apart from the aforementioned Derek Williams, other key players in this project (for example, Birmingham City Council Planning and Architecture Department), demonstrate once again how such Partnership Working predated by several years the statutory requirement to work in this way, created by Parliament under Section 17 of the Crime and Disorder Act 1998.

Secured by Design at Brindleyplace

Brindleyplace could in no measure be described as a fortress. Indeed, it is very much the hub of a massive turn of the century redevelopment. As a result, Brindleyplace has an open-plan, permeable layout for pedestrians, if not motor vehicles. Nevertheless, exemplary quality Physical Security 'Target Hardening' Measures and Technological Innovations were recommended and are much in evidence in those buildings where no general public access is required. These take the form of electromagnetic door locking systems with proximity access control. Laminated Glass was used extensively throughout Brindleyplace, whilst recessed doorways and other potential hiding places were removed from the designs at the pre-planning stage. Nor was there any ill-advised planting of soon to be overgrown shrubs, or trees that would restrict the gaze by CCTV cameras (see also Sect. 4.2.4, Defensive Planting; 4.3.3, CCTV; and 4.4.2, Natural Surveillance).

The potential to create Defensible Space (see Newman, 1972, 1973, 1996 and Sect. 4.4.1) outside the buildings themselves on such a development, is almost impossible. However, Natural Surveillance was maximised, together with some of the highest levels of ambient artificial Lighting (see Welsh & Farrington, 2008; and Sect. 4.3.1). The use of Symbolic (Psychological) Barriers (changes in road surface colour and texture) is abundant (see Sect. 2.12). Territoriality (Sect. 4.4.3) is also maximised by maintaining a pristine and ENCAMS (Environmental Campaigns) award-winning environment (see Kelling and Wilson, 1982 Broken Windows theory).

Most importantly, it was always known that such a permeable layout would require additional security measures. Even before the first of five Home Office CCTV Challenge competitions, the developers installed a state of the art CCTV system—repeatedly updated over the past three decades. And Brindleyplace also employs security officers who constantly patrol the development from their on-site control room and have regular and effective liaison with the local police.

Pre and Post-Completion Crime Pattern Analysis

Levels of recorded crime at Brindleyplace have remained especially low. Prior to development there was little cause to visit this run-down area of the city and thus, little crime took place. However, it now possesses the 'tasty pickings' and indeed, up-market clientele that might ordinarily be considered easy prey for both opportunist and more professional/career/travelling Criminals (see Sect. 2.3).

Nevertheless, crime (anecdotal and recorded—across all categories) has remained low at this location, compared to other parts of the citycentre. Undoubtedly, the security patrols, CCTV and high level of ambient artificial Lighting make this an attractive destination for the overwhelmingly law-abiding, majority of the public.

This in itself produces an extraordinarily high level of Movement generation, which in turn adds to Natural Surveillance and however speculatively, acts as a further deterrent to crime and anti-social behaviour (ASB). Indeed, such city centre areas are perhaps the only locations where there is sufficient Movement to guarantee Natural Surveillance (*Eyes Upon the Street*—see Jacobs, 1961) for much of the day and thereby justify high levels of Permeability (Sect. 4.4.6).

Evaluation

Evaluation of a project like Brindleyplace would normally take place in the immediate years after its completion, when it is bathed in the glow of novelty and is still perceived a roaring success. However, over three decades, Brindleyplace demonstrates how Designing Out Crime and Secured by Design can be applied by a succession of DOCOs, providing they maintain some continuity of the original advice given and the client is receptive to their ideas.

That Brindleyplace remains a prestige development is largely due to the truly long-term, sustainability elements of DOC delivered through the highly cost-effective SBD award process. The original developer, Argent and its subsequent owners, Birmingham City Council Planning and Architecture Department plus a succession of West Midlands Police DOCOs, forged a Partnership Approach, the results of which are there for everyone to witness and experience.

Indeed, the then West Midlands Police Assistant Chief Constable, Anil Patani, observed when conferring the final *SBD* award *in* 2005, '*Brindleyplaceepitomises the way in which police, business and local authorities can create an environment where people can be safe, feel safe and therefore feel good about being in Birmingham.*'

Lessons Learned at Brindleyplace

a. **Permeability**

Whilst the issue of near total Permeability has been laid to rest at Brindleyplace, there are fears that it may return to haunt the development at a future date. A state-of-the-art CCTV system and security patrols are excellent features, but there is always the worry that at some point they may fall into disuse or be withdrawn on grounds of cost—being weaker elements on the Intervention Sustainability Spectrum (see Sect. 2.15).

Nevertheless, two of the primary skills of the DOCO are those of diplomacy and the ability to compromise (see Sects. 1.5.9 and 1.5.10). Even *Safer Places: the Planning System and Crime Prevention* (2004—see Sect. 7.2), studiously avoided becoming embroiled in the Permeability debate. And Newman conceded (1996) that Defensible Space was far more essential for residential developments. Moreover, Permeability was never an original concern for the public areas within the Brindleyplace development and the compromise has delivered remarkably low rates of crime over 30 plus years.

b. **Terrorist Attack**

Concern about a terrorist attack at this location was in evidence back in 1992 and subsequent bombings at St Mary Axe, Bishopgate, South Quay (all in London) and Manchester City-Centre ensured this was a foremost consideration when liaising with the developer throughout the course of Brindleyplace construction. However, a post-completion survey by police Counter Terrorism Security Advisers (CTSAs, see Sect. 6.17), identified one potential issue. This was immediately remedied and had only come to light since the beginning of this century and as new terrorist threats, tactics and counter-measures have evolved.

Similarly, concerns had been raised regarding the 'cloister' style design of the external areas of some of the larger office blocks, the columns of which also act as a barrier to Natural Surveillance. However, this was one design point on which both the developer and the architects were not prepared to compromise at any stage during the original consultations.

Consequently, the diplomatic skills of the DOCO (and the realisation that some issues could not be resolved in their entirety), were well demonstrated in this regard. Moreover, it could be argued that were Designing Out Crime to be entirely successful in achieving every modification it desires, buildings would lose their aesthetic qualities—a point at least worthy of debate.

Conclusion

More than three decades have passed since the original meetings to incorporate CPTED, Designing Out Crime and Situational Crime Prevention measures into the long-term construction of Brindleyplace. And almost as long since the first building was completed, occupied and that first Secured by Design award conferred. During and since completion, crime, FOC and instances of ASB have remained commendably low at this location—itself testament to the Partnership Approach, Designing Out Crime and SBD processes in providing long-term, Sustainable Crime Prevention solutions. Perhaps most importantly, whatever time of day or night you visit Brindleyplace it feels a safe and welcoming environment—no mean feat for an inner city (downtown) area at the heart of Birmingham's nightlife culture.

Brindleyplace, Birmingham, England

9.3 Cottesbrooke Infant School, Acocks Green, Birmingham

Introduction

Cottesbrooke Infant School, located in the Birmingham suburb of Acocks Green, was completely rebuilt during 1995/1996. Following liaison between the city's Education Department and West Midlands Police, the project was assessed from the perspective of Designing Out Crime and subsequently processed through to a Secured by Design award (detailed in Chapter 5). Of perhaps far greater importance, this experience provided the catalyst for the city's subsequent Private Finance Initiative (PFI) and Building Schools for the Future programmes, to achieve Secured by Design status. Furthermore, the lessons learnt from Cottesbrooke Infant School have been repeatedly fine-tuned, thereby demonstrating how a safe and secure school does not have to be a fortress, visible or real.

This school was the first educational establishment in the West Midlands Police area to be taken through the Secured by Design (SBD) award process. Indeed, it was one of the earliest such schools in the whole United Kingdom.

Background

In retrospect, the mid-1990s were portentous years in the history of school security and the safety of both children and staff—in both the UK and wider world. In 1993, two year old James Bulger was abducted and murdered on Merseyside; two years later a London head teacher was fatally stabbed outside his school; in 1996, 16 infant children and their teacher were murdered within their school at Dunblane, Scotland; and in the same year 4 adults and 3 children were seriously injured in a machete attack at an infants' school in the City of Wolverhampton. Outside the UK further tragedies took place on school premises and as we are only too aware, many more (often involving firearms) have since occurred around the globe.

Against this background, it was inevitable that school security and the safety of children and teaching staff, rose to prominence. In the West Midlands Police area and before most of these tragic events, liaison had taken place between Chris Gilbert, the Capital Projects Officer with Birmingham City Council Education Department and the then Force Architectural Liaison Officer (DOCO and this author). Very little new school building was taking place at that time anywhere in the UK, but these negotiations led directly to one of the few such projects, Cottesbrooke Infant School, achieving the SBD award.

The school had existed at this location in the largely residential suburb of Acocks Green (a Worcestershire village before its incorporation into the city in 1911) since 1932. But even before the 1990s, this dilapidated collection of prefabricated (and largely wooden) buildings was in a poor state of repair, necessitating a complete rebuild. This was achieved by way of a phased construction and demolition programme, with the new school being fully open in 1996.

The Partnership Approach at Cottesbrooke Infant School

Yet again, Truly Effective Partnership Working (that is to say genuinely collaborative, see Sect. 1.5.11) required the key elements of timing and a willingness by 'particularly motivated individuals' (Read & Tilley, 2000; Bullocket al., 2006, p. 15; Kirby and Read, 2004), to deliver long term, sustainable development and Crime Prevention. Such key individuals were instrumental in the rebuilding of Cottesbrooke Infant School—many from within the local community.

However, further motivations can also be deduced. Quite apart from the high-profile tragedies already referred to, *On the Third Day* (1993) a West Midlands Police (authored by DOCO and now retired Detective Superintendent Phil Ball) and West Midlands Fire Service initiative, had identified how an Arson attack (deliberate fire starting) took place on school remises every three days within the Force area.

Meanwhile, by 1995, Secured by Design had become recognised throughout much of local government as a means to deliver Sustainable Crime Prevention—and thereby reduced costs—both capital and revenue. In Birmingham, repeated positive experience with the ongoing housing refurbishment programme meant the city's Housing Department and elected members (councillors) had become vociferous advocates of Secured by Design. This message spread to other council departments—most notably that of Education. Moreover, by working with the police to Design Out Crime, Cottesbrooke Infant School once again demonstrates how the Partnership Approach was working several years before the Crime and Disorder Act 1998 made this a statutory responsibility.

Cottesbrooke Infant School, Acocks Green, Birmingham, England

Secured by Design in Practice at Cottesbrooke Infant School

Whilst Secured by Design was well established by the mid-1990s, the first specific design guide for schools had yet to publish. Consequently, the original and all-embracing (any construction project that was not a dwelling) SBD Commercial (1992) was used as the basic design model for the new school building, with security measures being adapted and designed-in during the meetings with the Capital Projects Officer.

The potential for the Environmental Elements of Crime Prevention Through Environmental Design (CPTED—see Sects. 4.2.1–4.4.6) at Cottesbrooke Infant School, was somewhat restricted by the limitations of the existing site. An oblong-shaped plot located in a built-up suburb with roads on two sides, limited many of the recommendations to little more than simple Physical Security 'Target Hardening' Measures and Technological Innovations (see Sects. 4.2.1–4.3.3).

Nevertheless, the Environmental Elements of CPTED can be identified in the 1.8 metre (6 feet) high hooped- top metal railing perimeter fencing to create Defensible Space (Newman, 1972, 1973, 1996—Sect. 4.4.1) and allow Natural Surveillance (Sect. 4.4.2) from the adjoining streets and houses that overlook the school. Similarly, advice was given to avoid trees with low canopies and any form of vegetation that would restrict such Natural Surveillance—especially relevant in respect of trespass during out of school hours.

Physical Security 'Target Hardening' Measures requested and incorporated into the new building included: the aforementioned fencing—joined to the building at specific points to prevent circumnavigation/permeability (Sect. 4.4.6) by intruders and create secure areas (private Defensible Space in which the children can play safely); lockable gates; reducing the number of gates and door entrance points; Laminated Glass in windows and doors; a secure reception area with electronic control of the main exterior entrance door; door and window security that complied with the then SBD minimum standards; improvements to the external Security Lighting, including Street Lighting (Welsh & Farrington, 2008 and Sect. 4.3.1); plus the installation of both fire and Intruder Alarm systems (see Sect. 4.3.2) linked to the city council's own central station/control room.

Pre and Post-Completion Crime Analysis

All recorded crime categories were substantially reduced when compared to their 1995 and earlier year comparisons. However, Cottesbrooke Infant School was never a Crime Attractor or Crime Generator (Sect. 2.5) and consequently it would be easy to presume too much from single digit Crime figures—'*insufficient statistical robustness*' being a constant warning from the disciplines of both Crime Science and statistical analysis (see Weisburd et.al., 2020).

Most importantly, there has been no evidence of geographical or any other form of Crime Displacement activity (Sect. 2.1) over the past three decades. Indeed, local recorded Crime figures suggested that if anything, there had been a Diffusion of Benefits to the immediate surrounding area of Acocks Green.

Cottesbrooke Infant School: The Dividend of Compromise

Considerable compromise was demonstrated by the Architectural Liaison Officer (as DOCOs were then known) in the rebuilding of the school. For example, following completion and conferring the SBD award, a car belonging to a member of staff was stolen from the campus. Despite an original recommendation by the DOCO to close off the staff vehicle entrance from the main Yardley Road (and combine it with a service access point in the secondary Cottesbrooke Road), at least one staff member was hostile to this change. Indeed, even following this theft, the entrance remains insecure to this day—itself testament to how Designing Out Crime most often has just a single window of opportunity (within perhaps a 30-year time period) to introduce solutions.

At one level, this is a classic case of convenience (for the staff on this occasion) being prioritised over security. And simultaneously, it demonstrates how just one crime can frustrate an otherwise safe and secure environment. However, a willingness to compromise on this point (and realising that greater politics were taking place behind the scenes) helped to ensure Secured by Design was subsequently adopted for Birmingham City Council's new school building programme. Put simply, once they received confirmation that Designing Out Crime and Secured by Design were not inflexible doctrines and could be adapted to local requirements, a positive working relationship was established.

It could be argued that a more rigid position should have been taken from the outset. But losing the battle over a single vehicle access point was instrumental to ensuring the war was won in the adoption of Secured by Design. See also Sects. 1.5.9, Diplomatic Skills; and 1.5.10, The Art of Compromise and Negotiation. Furthermore, a Machiavellian perspective would contend that the failure of the vehicle access point at Cottesbrooke Infant School has subsequently been repeatedly used by this DOCO (as an example of how crime can take place when their advice is not heeded) in negotiations regarding the design of new school buildings—and many other diverse projects!

Fine-Tuning School Security Measures

Cottesbrooke Infant School provided a base level for security measures at educational establishments within the West Midlands Police area. Until the publication of SBD—Schools in 2004 and when considered appropriate (from the perception of Commensurate Risk, see Sect. 1.5.15) such measures were enhanced and evolved at new school building projects. For example, experience with staff vehicle access points confirmed the need to reduce such entrances to a safe, secure minimum and ensure they are located in a position that affords maximum Natural Surveillance.

Similarly, CCTV systems (Sect. 4.3.3) incorporating colour cameras WITH digital recording became a requirement at all Birmingham's schools. In addition, it is always desirable for reception staff to be able to view visitors to the school, both physically and via any CCTV produced images (a clear line of sight from the reception desk to the school entrance) itself dependent upon the ergonomics of the reception area and desk design. Also, the recommendation that the CCTV recording equipment is kept separate from the monitor(s) in a secure storeroom, was a constant rejoinder after the Cottesbrooke project. And as of 2009, it was agreed to standardise the minimum CCTV system requirements across the whole schools' estate.

Perhaps most importantly, following liaison from West Midlands Fire Service, fire suppressant (sprinkler) systems became a mandatory consideration. Interestingly, it required a three-pronged initiative by the Fire Service, Local Education Authority and Force ALO (DOCO) to convince one company charged with delivering the PFI school building programme of the value of such sprinkler systems. No doubt they were concerned at the cost of such systems. However, at least one Chief Fire Officer has questioned whether it would require a school fire disaster of Aberfan proportions to concentrate minds? This is a reference to the horrific school tragedy of 1966, when a massive spoil heap from the Merthyr Tydfil Colliery slipped and engulfed the Pantglas Junior School in the Welsh village of Aberfan, killing 144 people—including 116 children and 5 teachers.

Rejecting the Fortress Mentality Accusation

As previously described, the accusation of a Fortress Mentality is sometimes levelled at Designing Out Crime—although rarely with any foundation. Indeed and to repeat, the 'fortress' tag can be easily rebuffed by asserting that if after the involvement of the DOCO a particular project becomes a fortress (visible and/or real), then we have failed. Because, if nothing else a visible fortress is likely to raise the Fear of Crime.

In school design, a balance has to be struck between avoiding a fortress appearance, whilst simultaneously ensuring the safety of children, staff and legitimate visitors. There has been considerable debate about the requirements for perimeter protection (primarily fencing type and height) at the front of such school premises, with 'site specifics' being used to judge each case on its individual merits. Cottesbrooke Infant School (and schools subsequently built under the Private Finance Initiative and thereafter the Building Schools for the Future programme), demonstrate the necessity for a secure perimeter. However, they also attest to the undesirability of spiked topped railings and more oppressive security measures—like barbed wire, razor tape or electric fencing. And to date, there has (in the UK) been no necessity to recommend the routine use of electronic security measures like walk-through metal detectors, or hand-held x-ray scanners, as detailed by Green (1999) in her schools security publication for the US National Institute of Justice. However, sadly this situation is now changing in the UK and other countries.

The Value of 'Good Housekeeping Measures'

One of the unique features of the Cottesbrooke Infant School project was how it spawned a new chapter in the recommendations listed by the DOCO: namely, that of Good Housekeeping Measures (Sect. 8.3). This precursor to the creation of an 'Operations Manual' for the project in question post-completion, requires the DOCO to think beyond construction of the project to those practices that need to be adopted once the building is occupied. For example: key security; procedures for opening and closing the school; visitor identification, registration and escort; Signage and Notices (Sect. 2.13); CCTV management (Sect. 4.3.3); Intruder Alarm management (Sect. 4.3.2); energy efficiency (e.g. Security Lighting—Sect. 4.3.1) and fire safety. Much of this advice was subsequently incorporated into the West Midlands Police Knowing Your School (1996) booklet and the good practice manual, Keep Your School In Business (2006) produced by the former West Midlands Arson Task Force, both a beacon of excellence and a manifestly effective partnership arrangement between West Midlands Fire Service and West Midlands Police (see Appendix I).

Most importantly, Good Housekeeping Measures implies and requires a degree of interaction between the DOCO and the eventual managers and occupiers of the building. It places a degree of responsibility on them for the security of the premises and provides a link between the DOCO and the end user, thereby illustrating how safety and security do not finish when the building is handed over and first occupied.

Conclusion

Cottesbrooke Infant School remains a worthy recipient of the Secured by Design award. It represents a classic example of Sustainable Crime Prevention (as delivered by SBD) and a highly effective illustration of Problem-solving, Problem-Oriented Policing and Partnerships (POP—Sect. 2.14).

A considerable degree of compromise was demonstrated by the DOCO. However, this was instrumental in producing the dividend by which the Local Education Authority adopted Secured by Design for the whole of its new-build schools programme. More than 20 schools in Birmingham were subsequently processed through to SBD award status. In addition, Cottesbrooke Infant School initiated the Good Housekeeping Measures (embryonic Operations Manual) table of recommendations and provided a template from which future SBD Schools' security measures could be fine-tuned. And instead of the spiked-top metal palisade fencing, friendlier looking/more aesthetically pleasing hooped-top railings and more recently, weldmesh-style fencing have been employed on such projects, in order to help counter the charge of a Fortress Mentality. Furthermore, many of these elements have been replicated in successive SBD Development Guides, most recently Secured by Design—New Schools 2014 (Sect. 5.7).

9.4 The Village Surgery, Cheswick Green

Background

The Village Surgery, Cheswick Green

An early (1994) recipient of a Secured by Design (SBD) award, The Village Surgery, Cheswick Green presented immediate problems for the Designing Out Crime Officer. Chief amongst these was (and remains) its location as a 'stand-alone building'. This means that there is no perimeter and no barrier to pedestrian access beyond that of the walls to the structure, or to circumnavigation around it. Indeed, even vehicular access to all but the fabric of the building can be achieved on three sides. Whilst the fourth side lies adjacent to a public right of way footpath that had to be maintained.

Back in 1994, Crime Pattern Analysis revealed very little recorded crime within the suburban 'village' (built in the late 1960s/early 1970s) in which the surgery was to be constructed—a situation that remains to this day. However, the area around the adjacent six small shops, pub and village hall had become and continues to act as an ad hoc meeting point for school age teenagers. Furthermore, since the village was constructed there have been occasional instances of petty vandalism at this location, largely although not exclusively by way of graffiti (see Sect. 8.2.1).

Another feature of this development and one consistent with many Designing Out Crime (DOC) projects during the early years of the SBD award, was that the DOCO in question had been contacted well beyond the concept stage and indeed after Planning Permission had been granted. Consequently, the capacity to influence the design was all but limited to traditional Physical Security 'Target Hardening' Measures and Technological Innovations (see Sects. 4.2.1–4.3.4) and with virtually no scope for the Environmental Elements of Crime Prevention Through Environmental Design (CPTED—see Sects. 4.4.1–4.4.6).

Physical Security

The Architect had already chosen wooden framed windows for the building and prior to PAS 24, no form of multi-point locking system could be sourced for windows constructed of such material. Consequently, the method proposed by the DOCO (and used) to secure these windows was to apply additional surface-mounted key-operated window locks—two per opening window unit. Aesthetically, this is somewhat ugly and had any of the opening windows been a Designated Means of Escape, key-operated window locks would have been unlawful as they contravene Fire Regulations. In addition, screw- fixed post-installation window locks do NOT have a track record of success in preventing burglary, possessing the inherent capacity to be easily prised off should the window come under even minor attack/force, from a jemmying type instrument like a screwdriver or chisel.

A far more effective and key recommendation incorporated into The Village Surgery, was the use of 6.4 mm thickness Laminated Glass within the doors and adjacent windows. In the early 1990s this was a relatively new commodity, but has proved its worth over the intervening years. Some vulnerable windows were also later secured with roller shutters and 'Trellidors' (security grilles).

All the exterior doors were fitted with additional five lever mortise deadlocks to BS (British Standard) 3621, with a single main entrance/exit door for the public to use. Security Lighting (Sect. 4.3.1) was installed around the building and genuinely low growing, shrubs (see Sect. 4.2.4, Defensive Planting) advised and planted beneath those windows perceived vulnerable to attack.

The Village Surgery was officially opened in 1994 by the then Mayor of Solihull, Councillor Eric Pemberton and Lady Mayoress Trish Pemberton who cut a ceremonial ribbon in honour of the occasion. Chief Superintendent Chris Gray presented the SBD award to head of practise, Dr Janet Leese.

Lessons Learned

Within weeks of being opened, the covered 'pram store' adjacent to the public entrance at the front of the building became an unofficial gathering point for a small number of young people, a minority of whom embarked on acts of graffiti. In retrospect, the DOCO should have recommended, this feature was removed when it was first identified on the drawings. There was no great desire for it on the part of the client and the DOCO (this author) might have realised how it would act as a magnet to young teenagers, convincing (hopefully, although as previously indicated there is no power to compel) the architect of the inherent dangers for its inclusion in the overall design. In any event, less than a year later, Dr Leese replaced the pram store by extending the waiting room into the space thus vacated.

Nevertheless, over the intervening years, petty acts of vandalism (primarily the aforementioned graffiti), have continued to take place outside the surgery. In addition, a metal safety barrier in front of the main entrance door has been badly bent as a result of young people (of successive generations we can now surmise) sitting on it over the years. Indeed, this is the first Environmental Cue (see Sect. 2.8) that catches the eye of the visitor, with an inherent capacity thereby to raise the fear of crime FOC)—if not acting as an actual generator of ASB.

Two issues (both of which are within the remit of Designing Out Crime) might reduce the opportunity for such petty vandalism at this location. First, the provision of additional youth services at the adjacent village hall (a building that has similarly suffered such damage)—or perhaps a Youth Shelter (Sect. 5.13. SBD—Youth Shelters and Sports Systems). However, during the enduring financial climate, such investment seems unlikely.

Second, during the hours of darkness, the public area around The Village Surgery suffers from poor levels of ambient light. Enhanced Street Lighting in the vicinity of the surgery, shops and village hall would improve this situation. Farrington and Painter (2002) and Welsh and Farrington (2008) detail the efficacy of improved Street Lighting in preventing Crime, ASB and reducing the FOC (see Sect. 4.3.1, Lighting). As a qualification, such enhanced Lighting should be appropriate so as to bathe in light only the public areas around these buildings—and not cause light pollution beyond these grounds or into the night sky. However, at least one local resident has opposed increased Lighting at this location.

Conclusion
The Village Surgery is believed to have been the first doctors' surgery in the United Kingdom to receive a SBD award. Despite the problems detailed in the previous section, close to three decades after it opened the surgery remains the focus of village life and has the aura and appearance of being well looked after and cared for—in no small part due to the passionate efforts of Dr Leese and her husband, who was regularly called upon to act as the post-event Capable Guardian. Indeed, parallels can be drawn with Mr Leese's role in quickly remedying the aforementioned acts of petty vandalism, and the Broken Windows theory of Kelling and Wilson (1982— see Sect. 2.8). However, the surgery also testifies to the problems of a stand-alone building in an area frequented by relatively few people (Sect. 4.4.6, Permeability) and suffering poor levels of Natural Surveillance (Sect. 4.4.2) during the hours of darkness—largely the result of inadequate levels of external Lighting. At the time of going to press, representations have been made that will hopefully remedy this last situation, or at least replace the badly bent safety barrier in front of the entrance.

9.5 Christchurch Bus Interchange, New Zealand

Christchurch Bus Station. Photo courtesy of Sue Ramsay

On 22 February 2011, the city of Christchurch and the surrounding area on South Island, New Zealand, were devastated by an earthquake measuring 6.3 on the Richter scale. In the course of this monumental tragedy, 185 people lost their lives. And in the city centre buildings like both Cathedrals, Canterbury Provincial Council Buildings, the Carlton Hotel, 26 storey Hotel Grand Chancellor and Christchurch Bus Exchange, were ruined beyond repair.

As part of the strategy of redevelopment, Crime Prevention Team Leader, Sue Ramsay, was tasked with providing the CPTED (Crime Prevention Through Environmental Design—see Chapter 4 input for many of the public buildings that now required rebuilding. Working for Christchurch City Council and the Christchurch Earthquake Recovery Authority (CERA), Sue has been actively involved in the rebuilding design from a CPTED perspective, to ensure that the replacement buildings are sustainable in terms of their capacity to minimise instances of crime, fear of crime (FOC) and anti-social behaviour (ASB).

She describes how the Design Out Crime Advisory Service delivers CPTED activity through four headings: the CPTED Framework ensures it is included in planning documents and cross-organisation co-operation; CPTED Assessments respond to local concerns, existing environments and involve stakeholders; Rebuild CPTED reviews plans, produces a technical abstract and close involvement with the building professions; whilst SafeGrowth hands CPTED to the community and incorporates Second Generation CPTED (see CPTED Introduction). When applied to public places and buildings, this demands a focus on the safety of people as well as the security of buildings, the identification of issues and treatments, contextual and both process and method driven.

In Christchurch, the CPTED method employed by Sue Ramsay began with the environmental scan, followed by site assessments, safety audit, safety map, issue and asset analysis—to produce the improvement plan. The '*CPTEDHierarchy*' (in descending order of priority) is design, management and technology. This methodology was developed by Dr Frank Stoks, who kindly allowed Christchurch City Council to utilise same.

Christchurch City Council has integrated CPTED with its partners: the police, Ministry of Justice, Neighbourhood Support, NGOs (non-governmental organisations), Fire Service and Community Patrols. And with other disciplines: urban planning, lighting engineering, master planning, universal design, architecture, Injury Prevention Through Environmental Design (IPTED), urban design and site planning.

In the Concept and Developed Design stage, 30 safety issues were identified across 11 areas of focus. 44 recommendations were made, of which 24 were accepted, 2 partially accepted, 5 not implemented, 1 deemed not applicable and 12 resolved in detailed design. Moreover, 80% of those agreed to be actioned were included in the design. Whilst under the heading of Construction

Issue Design, 30 recommendations were made together with 1 via Streetscape Redesign.

The first of the public 'anchor projects' to reach fruition, Christchurch Bus Interchange was opened in May 2015. Having cost $53 million (£29 million) to rebuild, the interchange is immediately identifiable by its 'heroic' roof design—a curved structure that resembles a wave. Moreover, the influence of CPTED can be identified in a host of different features.

For example, the original deep window recesses have been reduced so that they are now shallow, thereby eliminating '*entrapment zones*'. This has also maximised Natural Surveillance, both internally and externally. Similarly, the veranda is constructed of glass sheets, allowing light to fall on the entire footpath. It is especially important to have optimal pedestrian Lighting in this location, because both Christchurch Bus Interchange and the adjacent late night entertainment precinct have elevated risk profiles and high pedestrian counts.

The 16 bus bays are diagonally positioned and facilitated by extra-wide footpaths. This design feature assists in separating queues from pedestrian movement—a sadly absent feature at many bus and coach stations around the world. The footpaths also benefit from a 'change in materiality' (mixture of paviors and slabs) to subtly define the different functions.

The build-out in front of the entrance is designed to provide more than sufficient gathering space (especially relevant for young people), without impeding access into the building. It is also desirable that the Bus Interchange has nearby taxi facilities, in order to support multi-modal travel. Sue recommended the taxi rank be located at the eastern end of the Interchange, to serve bus passengers during the day and the entertainment precinct at night—safe and readily available transport options being important components in late night environments.

The public toilets (washrooms) have been positioned at either end of the interchange, with designs based on CPTED/IPTED recommendations. Automatic glass entrance doors from the concourse into the foyers of the toilet areas, provide excellent Natural Surveillance. This permits users to 'see ahead of travel' and enables them to exercise a choice not to enter the environment if it appears unsafe—see also Sect. 6.13, Public Conveniences (Toilets/Bathrooms).

Toilet cubicles have their own hand-washing facilities. This design also supports the concept of 'clarity of purpose', because where hand washing facilities are shared, users are unable to judge the purpose of other people in the space. In this layout, there is no valid reason to be here except to pass to or from the cubicles. The design reduces the uncertainty and fear that legitimate users may have and 'forces conspicuous behaviour': a person who is acting in an unexpected way becomes highly obvious to others.

CPTED recommendations also included the provision of various seating options throughout the interchange concourse, to allow people to sit where they feel comfortable and provide oversight. Bench seats are located at the terminus of each of the 16 bus bays, thereby providing Natural Surveillance and assisting with the accessible journey. Users with different physical abilities are able to stage their travel within the interchange, by resting near the bus stop whilst waiting.

A designated cycle storage area has been designed to eliminate Crime Against the Person and Bicycle Theft. The area is well lit and surveilled by monitored CCTV cameras. Automatic opening glass doors are wide enough to permit easy access by cyclists, whilst the prominent corner location maximises Natural Surveillance—further enhanced by the combination of floor to ceiling height glass and architectural fins. The walls are visually permeable and attractive, allowing internal light to contribute to footpath illumination.

Another incorporated recommendation involves 'ambassadors'. These Capable Guardians are trained in conflict resolution and tasked to seek opportunities to offer assistance, rather than their previous passive security presence. Here, the aim is to enhance a non-threatening environment where all users are welcome—helping to ensure acceptable standards of behaviour are maintained.

The pre-earthquake Bus Exchange was identified as a Crime Hot-Spot. However, since the new Christchurch Bus Interchange opened, Crime Pattern Analysis indicates that dramatic reductions have been achieved. Indeed, to date the only observable problem has involved some minor incidents of graffiti within the toilet cubicles.

Somewhat uniquely, this project (and the ongoing redevelopment of Christchurch city centre) shows how the CPTED and Designing Out Crime processes can be applied in the post-apocalyptic world following a natural disaster. However, even in a first world country like New Zealand (one where there is exemplary CPTED experience and input as delivered by Sue Ramsay), the redevelopment process will take many years to complete. In the meantime, during its first year of operation Christchurch Bus Interchange delivered 70% less calls for police attendance than its predecessor (when adjusted for passenger numbers). And Sue has been honoured with a New Zealand Public Transport Innovation Award for her work on the project.

9.6 Burlington Hotel Entrance, Birmingham

Background

Burlington Hotel Entrance, Birmingham, England

The Macdonald Burlington Hotel (previously known as the Midland Hotel) is one of the landmark buildings located at the heart of Birmingham city centre, with a history dating back to Victorian times. It was completely refurbished in the mid-1990s (when the name was also changed) and the developer sought to remove a design problem. Namely, that the entrance lay at the centre of a small network of part covered and part open footpaths and shops known as Burlington Arcade, linking the major thoroughfares of New Street and Stephenson Street. During daylight hours or when the surrounding shops were open, this network suffered few problems, although it appeared scruffy and uncared for. However, once night-time descended it was subject to the classic 'Jeckyl and Hyde' scenario in that it became a no-go zone for savvy (environment aware) pedestrians (see also Sect. 6.16, Pedestrianisation of Town and City Centres) and unofficial public lavatory—see Sect. 6.13, Public Conveniences (Toilets).

Working for the developers, Horton Estates, the Malcolm Payne Design Group met with the local Designing Out Crime Officer (DOCO—Architectural Liaison Officer at that time) and a schedule of proposals was agreed in 1993.

In essence, the number of entrances into the former arcade was reduced from three to two. Members of the public can and still use this conduit to walk between New Street (itself now pedestrianised) and Stephenson Street—until 2012 a major bus and taxi route through the city centre, since replaced by an extension to the city's tram-line extension. Furthermore, a series of additional design features have been incorporated, with the effect of deterring those who have no deliberate reason to use this route, most especially those intent on crime and/or anti-social behaviour.

Designing Out Crime in Practice at the Burlington Hotel

There is a natural camber between the two streets—difficult to appreciate from two-dimensional plans). During the years of the former Burlington Arcade, the link was achieved courtesy of a sloping footpath. Post-refurbishment, it is delivered by way of a staircase or lift (elevator). Other design changes include:

- Before refurbishment, the public areas were paved with tarmac and appeared to be an extension of the public realm beyond. Whereas, they are now enclosed within the premises and covered with a mixture of high quality, predominantly off-white and mixed colour floor tiles that act as a Symbolic (Psychological) Barrier (see Sect. 2.12) and emphasise the semi-private nature of this area
- The shops themselves are now up-market designer stores, scrupulously maintained and thereby acting as a similar Symbolic Barrier to those intent on ASB (the presumption being that such activity is less likely to be tolerated/is more inappropriate in such an environment)
- Number of entrances reduced from three to two
- Most importantly, both entrances (onto New Street and Stephenson Street) can be secured from late at night until early morning (between midnight and 6 am). This is achieved by way of aesthetically designed, 4 metres wide, full height, electronically-operated Laminated Glass bi-fold doors
- A 'Security Office' is situated at the top of the flight of stairs and adjacent to the hotel entrance, with its door marked accordingly in large letters to reinforce this message (see Sect. 2.13, The Value of Signage and Notices) and in full sight of anyone approaching from the main New Street entrance

- CCTV cameras (see Sect. 4.3.3) cabled back to, monitored and digitally recorded in the security office, cover all entrance system public areas.

Collectively, these are some of the 'classiest' Designing Out Crime measures ever witnessed and thereby hangs the crucial point: scrupulously maintained 'classiness' is but one form of demonstrating that people care. It can be understood as a constituent element of how to avoid the consequences implicit to Kelling and Wilson's (1982) Broken Windows theory. And as a result, the psychological imperative of influencing the decision-making process of those intent on crime or/and ASB, to the extent that they will desist from committing their acts. Furthermore, the local DOCO was so impressed that he gladly awarded it a Secured by Design award.

Lessons Learned

In a word, nothing. A very small number of critics from the persuasion of uninhibited Permeability (see Sect. 4.4.6), were understood to be uncomfortable with the fact that restrictions were being placed on public access during the middle of the night. Such viewpoints need to be robustly challenged, as they do not represent the poor night porter or security guard confronted in the wee small hours by some obnoxious and often violent drunk, packs of the same—or worse! Furthermore, personal witness following the summer disturbances of August 2011, demonstrated how shopping malls with a high degree of Permeability were subjected to criminal damage and theft that might otherwise have been easily prevented.

Conclusion

The Burlington Hotel entrance system illustrates how Designing Out Crime and CPTED have practical application in some of the most complex of commercial and retail locations. In addition, three decades later, it now has a track record of success, demonstrating that a compromise can be achieved between Permeability and safety/security, to deliver a result that genuinely designs out Crime Opportunity, ASB and by extension, fear of crime.

9.7 Drytown Water Park, Palmdale, California

Drytown Water Park, Palmdale, USA

Introduction

The City of Palmdale is located some 50 miles north of downtown Los Angeles in California, USA. Until the fall/autumn of 2007, Lisa Scates was the unsworn CPTED (Crime Prevention Through Environmental Design) Adviser employed by the Palmdale Police Department. Since 2008, she has worked as the Statewide RASS (Responsible Alcohol Sales and Service) Co-ordinator with the Montana Department of Revenue based in Helena, Montana).

Background

DryTown Water Park recreational facility lies on the outskirts of the city. It attracts visitors of all ages (although especially children and young people) to its swimming pools, water chutes and other facilities. Indeed, one of its friendly welcome signs as you enter the complex states, 'You will get wet!' Unfortunately, some of its original design elements had the capacity to act collectively as a Crime Generator (Brantingam & Brantingham, 2008) when the water park first opened, largely due to the absence of some crucial Physical Security 'Target Hardening' Measures, Technological Innovations and Environmental Elements of CPTED (see Chapter 4. What DryTown Water Park now demonstrates is how following the involvement of the CPTED Adviser, dramatic reductions in Crime can be achieved through: relatively minor changes ('tweaking') of the physical layout; introducing Physical Security 'Target Hardening' Measures; Signage and Notices (see Sect. 2.13); and new management practices.

The Crime Problem

After opening in 2004, crime at the water park began to steadily increase and by 2006 a distinct pattern had emerged. In particular, the changing rooms were being targeted (something of a worldwide phenomenon as any health club or sports facility user can testify) and management was accurately concerned that this might dissuade legitimate and highly valuable customers from attending the venue. Thefts were also affecting the Palmdale Police Department's crime figures—thus Linda's involvement and largely at the behest of her Police Chief who was, '*...constantly talking about CPTED'*(Scates, 2007).

Survey and Recommendations

Linda was invited to survey the premises, after which she put forward a number of recommendations. Pre-eminent amongst these was the necessity for key-operated locks on the locker doors in the changing rooms—an unbelievable absence in the first instance, as all Designing Out Crime Officers (DOCOs) would surmise. However, a detailed number of additional recommendations (from both the Physical Security 'Target Hardening' Measures and Environmental Elements perspectives of CPTED) were also recommended and put into effect by the DryTown Water Park management. The list included the following:

- Key-operated lockers—rented for $5 a day from the general store, with a $1 refund when the key is returned
- Locating the lockers within full sight of the public areas of the changing rooms to maximise opportunities for Natural Surveillance (see Sect. 4.4.2)
- Altering management practices so that the changing rooms are routinely inspected and Defensible Space (see Sect. 4.4.1) reinforced
- Layout of the commercial facilities within the water park 'tweaked' so as to deny Crime Opportunities (see Sect. 2.3)
- Perimeter fencing secured to prevent casual intrusion onto the site.

Conclusion

The implementation of these security measures led to the all but elimination of criminal activity at the DryTown Water Park—due to Linda's recommendations and highly commendable action on the part of the leisure facility management.

As such it is an excellent good practice case example, in that DryTown Water Park ably demonstrates at the commercial level how retrofitting can deliver immediate benefits in terms of reduced Crime and perhaps most crucially for facility managers, increased levels of customer satisfaction—which should in turn lead to more customers and therefore revenue.

Nevertheless, such retrofitting is expensive—both financially and in terms of the customers who were dissuaded from visiting the facility once they had been subject to a criminal offence (Scates, 2007). No business can afford to be subjected to Repeat Victimisation, because once it reaches epidemic proportions (see Sects. 2.9 and 2.14), that commercial undertaking may well cease to trade. Sceptics beware: as a parallel, following the Manchester, England city centre bombing in 1996 (see Sect. 6.17), it is reported that in excess of 80% of the businesses affected by the explosion never recovered!

9.8 Washington Square Park, New York

Washington Square Park, New York, USA

Historically, parks owe their ancestry to the gardens of antiquity (e.g. the Hanging Gardens of Babylon; the Zen Buddhist gardens of Kyoto); and as depicted in the Bible, the Qur'an, Buddhist, Hindu and Shinto texts, and across Arabia, Persia, India, China and Japan. More recently, parks were the hunting grounds of the kings and princes of mediaeval Europe and laws were enacted to proscribe the activities of ordinary citizens within them. In more recent centuries they became the pleasure grounds of the middle classes/bourgeoisie, before being opened to all 'law-abiding' members of society. Indeed, the Park Keeper ('Parkie') was the epitome of the twentieth century Capable Guardian—at least in the UK (see Sect. 6.12).

The Lonely Planet travel guide to New York City (2016) provides a concise history regarding the trials and successes of Washington Square Park. In particular, its role as a location for public executions, military parade ground and cemetery during the early nineteenth century (20,000 bodies are buried under the park). Now dominated by the 72 feet high Stanford Washington Arch and Central Fountain, repeated visits to Washington Square Park have witnessed the sheer scale of its multi-disciplinary use.

The park is heavily frequented by the good citizens of New York City, many engaged in watching the street theatre on display, or simply basking in the sunshine. Children play safely, many in a specially fenced-off area for the under-8s. Similarly, a separate facility for the walking of dogs (further segregated depending upon the size of the pooch) is utilised by visitors—canine and human. Perhaps most prophetically (and confirmed during all visits), Washington Square Park exudes an atmosphere of contentment as the overwhelming majority of citizens enjoy themselves. The capacity to appreciate such atmospheres (by way of positive and negative Environmental Cues, see Sect. 2.8) is an innate skill for the Designing Out Crime Officer.

As you walk through Washington Square Park, in addition to the professionaly maintained nature of the grounds and flowerbeds (a fairly unique feature in many of the world's parks in recent decades due to labour costs), a key element that attracts attention is the sheer volume of Signage and Notices (see Sect. 2.13). Quite apart from this all too obvious rule setting (rare in Western Europe which in itself points to cultural differences on opposite sides of the Atlantic), what such notices demonstrate is that both the park authorities and users care about this environment. And whilst some of the notices are prone to cause amusement (like an advertisement for a 'Halloween Fancy Dress Parade for Dogs'!) the overall message is clear: people take an interest in the welfare of the park—and such concern is a core issue at the heart of Kelling and Wilson's (1982) *Broken Windows* theory.

Following a decade of improvements, the $30 million renovation was completed in 2014. Improvements include: restoration of the Stanford Washington Arch and Central Fountain; new and expanded planting beds; repaved paths; new benches; Lighting; enhanced playground; a stage; pétanque courts; dog runs; new chess plaza; new park house; toilets ('comfort station'); landscaping; and fencing. Moreover, recent visits confirm that Washington Square Park continues to be one of the best such examples of public park restoration encountered through personal witness, anywhere in the world.

In these early decades of the twenty-first century, Washington Square Park (like most parks) has an open plan, permeable design—meaning no gates or physical restrictions to access at any time of the day or night, (although the Park Rules prohibit access after midnight). Nevertheless, human behaviour, rule setting and New York Police Department (NYPD) ensure ASB and worse are minimised—despite some public order issues in 2021. Indeed, there are no Environmental Cues (see Sect. 2.8) to suggest illicit activities take place at any time. All told, Washington Square Park is the 'paradigm of public space we see today' (Lonely Planet, 2016). It should also be remembered that in the 1950s, Jane Jacobs, author of The Death and Life of the Great American Cities (1961), which contains the contentious *'Eyes Upon the Street'* assertion (see Flint, 2009) was, with others, instrumental in saving the park from being developed into a freeway!

9.9 Piccadilly Gardens, Manchester

Located at the heart of Manchester's city centre, for more than 150 years, Piccadilly Gardens has been occupied by first, the Manchester Royal Infirmary (hospital) and from the twentieth century onwards, its new role as a public transport interchange. In the 1990s, the area became increasingly derelict and was generally perceived to be unsafe—especially during the hours of darkness.

Following the infamous bombing of Manchester city centre in 1996, the subsequent regeneration included the creation of public space. An international competition to redesign Piccadilly Gardens was won by Landscape Architects EDAW, Engineers Arup, Architects Tadao Ando and Chapman Robinson, and Lighting Engineers Peter Fink. Working collaboratively, they aimed to consider both context and space, together with high levels of investment and preparation of the finished product as a constituent part of 'New Manchester Spaces'. The project was completed in time for the 2002 Commonwealth Games.

Piccadilly Gardens, Manchester, England

Background

Why Places Fail?

The former Piccadilly district was previously considered the centre of the city. Indeed, Piccadilly Gardens were created from residual (leftover) space. However, they were isolated by the public transport layout and declined in parallel with adjacent uses and environmental conditions. Such incremental problems included poor accessibility, visibility (Natural Surveillance—see Sect. 4.4.2) and Lighting (see Sect. 4.3.1); restricted use, drugs, alcohol and anti-social behaviour (ASB) problems—all compounded by being badly maintained and managed.

The contention is that such spaces fail for a variety of reasons (evidenced by crime and ASB): their creation from leftover space; no clarity of purpose; being at odds with socio-economic change; fear of such change; and not being designed to reflect their current use. Furthermore, from the perspectives of Designing out Crime and Crime Prevention Through Environmental Design (CPTED), we might contend that where there is no immediate ownership or control of a particular space, there is a greater tendency for it to be occupied by those intent on ASB and crime. In large areas of public open space, these issues are often especially problematic.

The Piccadilly Gardens project aimed to create:

- Vibrant and safe city space
- A catalyst to restore economic vitality

- City space worthy of the city centre
- A place of pride and status.

With these goals in mind, the design and management response concentrated on tackling the issue of residual space, ensuring wider area integration, removing traffic, Permeability (see Sect. 4.4.6), visibility, active frontages', activities, Lighting, delivery mechanisms, the management plan and bid process. Whilst the actual construction process addressed the maintenance of public safety, economic vitality and confidence.

The theme of 'active frontages' involved connecting Piccadilly Gardens to the surrounding streetscape to encourage Natural Surveillance and a 24/7 overview to and from those premises. Clarity of movement was delivered through the perception of safety, quality of experience, ease of movement and good Signage and Notices (see Sect. 2.13). Collectively, these were designed to encourage activity with an over-arching aim that the venue should be a place to gather (for both play and formal events), seasonal colour, restaurants, fountains and whilst tolerant of fun, intolerant of abuse.

Night Time Economy issues were addressed courtesy of high-quality Lighting, small-scale installations and balancing transport with pedestrian traffic to encourage night-time gathering. Whilst management of both the daytime and night-time use is delivered through policing, CCTV (see Sect. 4.3.3), the 'City Rangers' and BID (Business Improvement District) involvement.

Piccadilly Gardens: A Quarter Century Later

The depths of winter might not be the best time to make an initial inspection to Piccadilly Gardens. However, a quarter century after this regeneration project was completed, subsequent visits indicate that for the most part this large area of public open space has passed the tests of long-term success, popularity and sustainability.

In essence, the pedestrian interchange of Piccadilly Gardens exudes a friendly environment and even on cold winter days, people sit relaxing on the public seating around the fountains. Indeed, those same fountains are one of the key visual attractions of this scheme: visit the internet and you will see they are especially well 'used' on hot summer days. Similarly, once darkness descends, the lighting is especially excellent and ensures the perception of a welcoming environment is maintained. Nevertheless, there are some problems.

Much of the grassed areas are often little more than mud during mid-winter visits, largely the result of heavily subscribed public events. In addition, the public seating requires constant maintenance as the wooden slats have a tendency to rot.

Such factors easily fall within the *Broken Windows* theory definition described by Kelling and Wilson (1982) and need to be repaired quickly if the perception of decline is not to gain momentum. And more recently, Piccadilly Gardens has attracted the homeless and those taking synthetic drugs like 'spice', 'mamba' and 'monkey dust.'

Designing Out Crime Officers (DOCOs) might be especially concerned about the bare concrete wall which shields Piccadilly Gardens from the public transport interchange (trams and buses). Such bare surfaces provide a ready background for graffiti—sometimes described as 'street art' (see Sect. 8.2.1). And as the majority of DOCOs would have confidently predicted, there have been repeated problems with this wall acting as a canvas for graffiti fly-posting, stickering, etc. Consequently, this whole structure may be removed or redesigned.

Conclusion

Piccadilly Gardens demonstrates how public space can be redesigned to provide a safe and welcoming environment at all times of the day and night. Recent history suggests the problems at this location relate to maintenance and are largely attributable to the British weather and sheer volume of people using the venue. Indeed, after being out of action for a year, work was completed in 2017 to repair the much admired fountains. However, it is also reported that there was a distinct absence of Community Consultation during the concept stage of this project.

Finally, amongst the Grade II Listed statues at Piccadilly Gardens, stands one of Sir Robert Peel—former Home Secretary and Prime Minister, recognised as the founder (in 1829) of modern policing in the UK. It is interesting to muse what Sir Robert might have thought of policing expanding to incorporate CPTED and Designing Out Crime. Presumably, he would have seen it as satisfying a primary role of the police in Crime Prevention.

Statue of Sir Robert Peel at Piccadilly Gardens, Manchester. Could the founder of modern policing have imagined it might expand to embrace Designing Out Crime?

9.10 Reitland Park Tram Station, Amsterdam

Reitland Park tram station, Amsterdam, The Netherlands

Reitland Park Tram Station was constructed in 2004 and is located 2 miles (3 kilometres) east of Amsterdam City-Centre on the main high speed tram route that runs to the suburb of Ijburg. In the Summer of 2006 Paul van Soomeren of the DSP-groep research company organised a Designing Out Crime seminar held in the Dutch capital and widely attended (at their own expense and in their own time) by Designing Out Crime Officers (DOCOs), CPTED Advisers and academics from across Europe and the USA. Part of that seminar involved a study tour of the Reitland Park Tram Station).

Although occupying a suburban location, the tram station is not immediately overlooked and even at the busiest times of the day there is rarely a constant flow of pedestrian activity. Trams stop to pick-up and set-down passengers in both directions approximately every 10 minutes. However, there is no staffed ticket office at Reitland Park—in effect there exists no Capable Guardian (Cohen & Felson, 1979). Nevertheless, this tram station has been designed to maximise the opportunities for Natural Surveillance (see Sect. 4.4.2) with multiple access points (by stairs, lifts or escalators) for pedestrians at either end of the platforms. It is also a transport interchange, with the slower surface level trams stopping immediately above the tram station.

Whilst the whole station and platform area are below surface level, it is simultaneously open to the elements and to both Natural Surveillance and Technological Surveillance by CCTV cameras (see Sect. 4.3.3) from above, from the trams, platforms and all angles. Even the shelters that provide some degree of protection from the elements for waiting passengers, are constructed primarily of glass thereby providing a minimum barrier to Surveillance. Furthermore, there is a distinct absence of the type of 'street furniture clutter' that could provide cover for anti-social behaviour (ASB) and worse. As such, the capacity for crime or raising the fear of crime (FOC) at such an unstaffed transport hub has been minimised.

Interestingly, Reitland Park Tram Station is situated in an environment with what might be described as total Permeability (see Sect. 4.4.6). Yet, for once this works to the advantage of the overwhelming majority of citizens, who are law-abiding travellers for whom the multiplicity of *Attack and Escape Routes* have an inherent capacity to reduce the FOC. In reverse, the decision-making process of any potential criminal (influenced by mulitple *Visual Environmental Cues—VECs*, see Sect. 2.8) would realise they have no element of surprise; can be seen by their potential targets from afar; and are likely to have their actions noticed and recorded by the CCTV (see Sect. 4.3.3) system. Furthermore, return visits at night revealed how Reitland Park Tram Station is extremely well illuminated during the hours of darkness—now with economic and carbon efficient LED lights.

In conclusion, Reitland Park Tram Station demonstrates how even a relatively isolated public transport interchange can receive the Designing Out Crime treatment and in so doing reduce Crime Opportunity, FOC and ASB.

9.11 Edgbaston Cricket Ground, Birmingham

Edgbaston Cricket Ground, Birmingham, England

The Birmingham base for Warwickshire County Cricket Club and international 'Test' matches played by the England Cricket Team is located in the suburb of Edgbaston. Indeed, Warwickshire County Cricket Club was formed in 1882 and moved to its current location at Edgbaston Cricket Ground 12 years later. In 1902, the ground was given Test status and the first match played against Australia in May of that year (for those with an interest in cricket, that test match was drawn).

For more than a century, Edgbaston Cricket Ground has been repeatedly modified and expanded. The largest ever expansion took place between 2008 and 2011, when total capacity was increased from 21,500 to 25,000 spectators. This has enhanced its reputation as one of the great venues for International Test Cricket in the UK and indeed, throughout the world.

The £32 million stadium complex includes a conference and banqueting suite catering for as many as 700 guests, the luxury '1882' and 'Jaguar' Club suites, management offices and both press and spectator facilities. Lift (elevator) access to all five-tiered levels means it is Equality Act 2010 (formerly the Disability Discrimination Act 1995) compliant in respect of access for the less able-bodied.

Collectively, Edgbaston Cricket Ground is now a genuinely 365-day venue with events (including weddings, for which this venue is especially popular) taking place every day of the year.

The local DOCO, PC Gareth Pemberton, was actively involved in this project from the design concept stage, courtesy of an introduction from his Chief Superintendent. Whilst as an international sporting venue and in accordance with the requirements of Crowded Places (2010) and the then RIBA guidance for counter terrorism (2010) documents, the services of a police Counter Terrorism Security Adviser (CTSA) were also extensively utilised during the course of this project. Their involvement witnessed a planning condition imposed by Birmingham City Council that mandated the development address issues of Hostile Vehicle Mitigation (HVM) and incorporate these into the overall design. In this pursuit, considerable efforts have been made to design in features that prevent vehicle-borne devices, encroachment, penetrative vehicle impact, duress and deception (see Sect. 6.17).

The new stand and its facilities were officially opened by Prince Philip Duke of Edinburgh in 2011 and just in time for the International Test Match between England and India.

Future development of the site includes a planning application to build 225,000 square feet of residential accommodation. It is further envisaged that over the next decade construction will include an 11-storey hotel, commercial office space (150,000 square feet), retail (45,000 square feet), leisure and Night Time Economy units (11,000 square feet). Police DOCO and CTSA involvement will be an intrinsic part of this future development.

Edgbaston Cricket Ground is a pristine example of how both the Designing Out Crime and Counter Terrorism Design processes can be applied to protect an existing internationally renowned sports and leisure facility, as it is further expanded. Moreover, it demonstrates that such application can deliver enhanced safety and security, both in the present and with a watchful eye on future development of this highly prestigious complex.

A decade after the SBD award was conferred, this venue remains a textbook example of the Layered Security ('Onion Peeling') approach. Moreover, for the trained eye of those tasked with replicating such elements of safety and security, a site visit is especially recommended.

9.12 Zero Carbon House, Balsall Heath

The Zero Carbon House, Balsall Heath, England

A 2010 recipient of the Secured by Design award, the Zero Carbon House in the inner-city district of Balsall Heath, Birmingham, represents a cutting edge approach to constructing a home and one which also embraces the Designing Out Crime process. The Zero Carbon House (meaning no fossil fuels were used in its construction) was designed by the architect John Christophers, who himself works for Associated Architects, fulsome supporters of Secured by Design (SBD) for more than three decades. The project had three aims: to work with the existing house; good architectural design; and to produce a carbon neutral family home.

In this instance, there is an even greater achievement, in that the Zero Carbon House was the first extension and modification of an existing property in the United Kingdom to have achieved Level 6 on the DCLG's (Department for Communities and Local Government) Code for Sustainable Homes (2006). Level 6 was the highest standard set and means the house is 'net Zero Carbon'—renewable energy (including heating and electricity) providing all the needs of the home. The then British government had set the target of reducing carbon dioxide emissions by 80%, as of the year 2050 (courtesy of the Climate Change Act 2008). And by 2016, all new homes were to have met Level 6 of the code.

The original two-bedroom Victorian house was built in the 1840s. However, John Christophers and Jo Hindley's purchase of the property was made with the intention of extending it both sideways and vertically, to create two more bedrooms together with a studio loft. This necessitated considerable discussion with the local planning authority, in order to incorporate the nine areas of sustainability detailed by the code, most especially those relating to:

- Carbon dioxide emissions and energy use
- Materials including their 'embedded energy'
- Water use, minimising surface water run-off
- Waste (facilities to make recycling easy)
- Pollution from insulating materials
- Health and well-being (good use of daylight)
- Managing the process of construction
- Ecology and preserving wildlife.

At the Zero Carbon House, a number of these areas of sustainability have been achieved by incorporating the following measures:

Energy

Air-tightness: sealed to a level 28 times better than the existing requirements, but permitting vapour permeability in order to prevent condensation.

Heavy construction: stores heat in order to keep the house warm in winter and cool in the summer.

Passive solar heating: the glazing is manufactured to absorb solar energy during the winter months, whilst the leaves of the existing mature Ash tree, located in the rear garden, provide shade during the summer. In addition, the house design incorporates low-energy/high-efficiency fluorescent lighting with A++ fittings.

Solar hot water panels: using energy from the sun to heat water, stored in a large cylinder and supplemented by a low-powered gas, oil or LPG fired condensing boiler.

Solar roof: photovoltaic (PV) panels convert the sun's energy into electricity with any excess exported back into the National Grid.

Thermal insulation: upgrading the existing (and new) insulation to make it 16 times more efficient.
Ventilation system: recovers 91–93% of the heat from the extracted stale air and uses this to warm fresh air from outside.
Windows: triple glazing provides 14 times the insulation qualities of existing windows.
Wood-burning stove: wood from the garden assists in topping-up heating for the house and hot water, but only in the coldest of months.

Water

Rainwater harvesting: tank-stored rainwater collected from the roof is used to supply the house.
Grey water system: allows bath and shower water to be used for the garden.
Low water usage: WCs, showers and taps designed to minimise the use of water, have been fitted throughout the house.

Materials

Highly ecological building materials: including reclaimed timber, recycled newspaper insulation, reprocessed waste timber insulation boards, lime plaster with recycled crushed glass, lime render, reclaimed door handles, hydraulically-compressed load-bearing earth blocks, rammed clay floor screeds and non-polluting materials.

Secured by Design and the Designing Out Crime Process in Practice
The Code for Sustainable Homes (2006 but currently repealed) also incorporated security requirements—thus the relevance of Secured by Design and the Designing Out Crime processes.

As John Christophers says, '*We were really excited about doing this project as it shows people that it does not have to be a hassle to create something Zero Carbon. By doing this project we are saying 'yes, this can be done'. We realise that some people just want to build houses in the same ways that they've always built them, but there must be a change.*'

Indeed, it was John who made the initial approach to the local Designing Out Crime Officer (known at the time as Architectural Liaison Officer), Sergeant

Dave Cotton. John and Dave subsequently met and the security requirements of SBD were discussed and agreed.

Many of the construction techniques used in creating the Zero Carbon House were of such a revolutionary nature that they required special authentication.

John states, '*The contractors, SpellerMetcalfe Ltd., were superb in delivering a construction that allowed the Building Research Establishment(BRE) to certify that the house had reached the right standard to be called Zero Carbon, as they had to undertake a number of specific assessments.*'

As is most often the case when dealing with an existing site layout, the capacity to influence the Environmental Elements of Crime Prevention Through Environmental Design (CPTED) was extremely limited. Consequently, the DOCO concentrated on the Physical Security 'Target Hardening' Measures and Technological Innovations (see Sects. 4.2.1–4.3.3), whilst mindful of the need to maximise Defensible Space, Natural Surveillance, Territoriality and deny Permeability (Sects. 4.4.1–4.4.6). In pursuit of this goal, the extension to the existing house denies access to the rear garden from the road, which can only be achieved by a covered 'entry' where it joins the next door house with a secure door at the front. The rear garden is itself surrounded by a secure perimeter fence and hedge, minimum height 1.8 metres (6 feet).

Naturally, the most fundamental of the 'Target Hardening' elements relate to the front entrance door and the windows/glazing. The original Front Door (which accessed directly into the house) has been replaced with a full height, triple-glazed window, thereby greatly improving the insulation qualities of the existing property.

The new, PAS 24 doorset (including all security furniture) is located within the extension and forms part of a wooden facade across the width of the ground floor. From a distance this facade looks like it might be the entrance to a garage. Similarly, all the triple-glazed windows in the house are to BS 7950 (the precursor to the now mandatory PAS 24: 2016), the inner most panes of which are of Laminated Glass.

Other security elements incorporated include energy efficient PIR-activated Security Lighting and an Intruder Alarm system (see Sects. 4.3.1 and 4.3.2, respectively).

Whilst this part of Balsall Heath is not a Conservation Area, the extension to the original dwelling planned by the architect to create the 'carbon neutral eco house' is of a post-modern style, deliberately different to that of the Victorian original and indeed, most housing in the area. This accords with architectural and planning good practice in the UK, where a clear distinction is most often demanded between the original design and more recent additions. However, different attitudes to such issues exist elsewhere in the world.

Considerable negotiation and compromise took place between the architect and planning officers, especially in respect of the studio loft. From the front elevation, it almost appears that a second (read 'third' in North America) floor has been added. This is because the roof of the extension had to be deliberately oriented in order to maximise the energy-raising potential of the sun's rays for the solar panels, whilst simultaneously preventing over-shadowing by the new roof onto the adjacent house. Consequently, officers within Birmingham City Council's forward thinking Planning Department demonstrated commendable judgement in approving this revolutionary design, one that delivers the rarely achieved accolade of being both highly different from and simultaneously sympathetic to the dominant architectural design style of the immediate area.

Conclusion

Simultaneously to being at the cutting edge of architectural and environmental design and living, the Zero Carbon House also demonstrates how Secured by Design, Designing Out Crime and CPTED principles and methodology can be applied to both revolutionary design and relatively small projects. Furthermore, it is all too obviously more sustainable to reuse rather than rebuild, in terms of both economic cost and the energy required for new construction projects.

In addition to their path-breaking role in terms of environmental sustainability, the fact that the architect and homeowners, John Christophers and Jo Hindley, actively sought out and incorporated Secured by Design measures, merits special praise whilst simultaneously providing a fillip to the Designing Out Crime community. And perhaps most importantly, it reinforces the growing awareness that Crime, Fear of Crime and Anti-Social Behaviour are themselves intrinsic threats to sustainability requiring their prevention, or at the very least drastic reduction. See also Sect. 2.11, The Carbon Cost of Crime. Further information is available at: https://www.zerocarbonhousebirmingham.org.uk.

9.13 Marie Curie Hospice West Midlands, Solihull

An example of how the Designing Out Crime and Secured by Design award processes can be applied to one of the most sensitive of developments in the built environment, is demonstrated by the Marie Curie Hospice, West Midlands. Located close to Solihull town centre, in 2013 this purpose-built development replaced the original Warren Pearl Hospice which for many decades occupied and had been converted from two former large, detached houses.

Marie Curie Hospice, West Midlands, England

Unlike the original hospice, the new building has been designed for its specific roles. This translates as providing a spacious and comfortable layout, one that caters for the varied needs of patients, their family members, visitors and staff. Consequently, Symbolic (Psychological) Barriers—see Sect. 2.12 (changes in decor, glass, textiles and colours) are used to separate the public and patient areas. And these are reinforced by the design and location of the wide, curving reception desk, sliding glass doors and extensive artwork.

Similarly, all patient bedrooms now have en-suite facilities; specially designed separate bathrooms for the less able-bodied; the capacity for family members to stay overnight; and each room overlooks the grounds. In order to minimise its development footprint in an edge of town, rural and Green Belt location, thoughtful design has utilised space in the roof to locate administration, training and meeting rooms at this first-floor level.

Further evidence of this sympathetic and complementary design can be identified in the external areas, most of which constitute well-crafted gardens with seating, water features, 'man-shed' and bird feeders. Moreover, the site is bordered on two sides by separate arms of the Solihull Bypass, which have no footpaths and thereby permit a secure perimeter to become established from existing and new thorny shrubs (primarily Hawthorn, Craetegus Monogyna—see Sect. 4.2.4, Defensive Planting).

The front perimeter is relatively open plan, albeit with electronically-operated barriers controlling vehicle access to and from the site. Meanwhile CCTV cameras, linked to the reception, provide Technological Surveillance opportunities over the car parks.

The Secured by Design—Hospitals Development Guide was used as the template for the new hospice, from which key security elements were drawn. Taken collectively, this 4-year-long project represents the leading-edge of both hospice evolution and the practical application of SBD. These have been combined to create a warm, welcoming, safe and relaxing environment for everyone within its confines. Indeed, Marie Curie Hospice, West Midlands demonstrates how Designing Out Crime, CPTED and SBD can be applied to the most sensitive of projects to deliver an environment that is both safe and secure.

At the 2013 Secured by Design award ceremony, West Midlands Police area commander, Chief Superintendent Sally Bourner, presented the SBD certificate to hospice manager, Liz Cottier. Liz said, '*Our new hospice has received such wonderful, positive comments from all who have visited since our opening. However, receiving the Secured by Design award will also be so reassuring to all who use our facilities.*' Thankfully, since that award ceremony Marie Curie Hospice, West Midlands has proved to be a safe and crime-free environment.

Sustainability has been a key theme of this handbook. More than a decade after it opened, visual audits indicate the Marie Curie Hospice remains in pristine condition with no DOC issues—a situation confirmed by Dave Rice, Facilities Manager. Indeed, having become established, the planting softens the external appearance and the grounds now open annually under the National Gardens Scheme. Whilst a new 'Fairy Garden' has been created for children dealing with the trauma of losing their family members.

Inside Marie Curie Hospice, West Midlands, England

References

Armitage, R., & Monchuk, L. (2009). *Re-evaluating Secured by Design (SBD) Housing in West Yorkshire*. University of Huddersfield.

Armitage, R. (1999). *An Evaluation of Secured by Design Housing Schemes Throughout the West Yorkshire Area*. The Applied Criminology Group, The University of Huddersfield.

Armitage, R. (2005). *Secured by Design—An Investigation of Its History, Development and Future Role in Crime Prevention* (Unpublished PhD Thesis). University of Huddersfield, Huddersfield.

Armitage, R. (2013). *Crime Prevention Through Housing Design: Policy and Practice*. Palgrave Macmillan.

Armitage, R. (2017). *Burglars' Take on Crime Prevention Through Environmental Design (CPTED): Reconsidering the Relevance from an Offender Perspective*. Macmillan.

Association of British Insurers. (2006). *Securing the Nation*. Association of British Insurers.

Bennett, T., & Durie. (1999). *Preventing Residential Burglary in Cambridge: From Crime Audits to Targeted Strategies* (Police Research Series, Paper 108). Home Office Policing and Reducing Crime Unit.

Brantingham, P., & Brantingham, P. (2008). Offender Mobility and Crime Pattern Formation from First Principles. In L. Liu & J. Eck (Eds.),*Artificial Crime Analysis Systems: Using Computer Simulations and Geographical Information Systems* (pp. 193–208). Idea Press.

Brown, J. (1999). *An Evaluation of the Secured by Design Initiative in Gwent, South Wales* (Unpublished MSc dissertation). Scarman Centre for the Study of Public Order. Leicester: University of Leicester.

Bullock, K., Erol, R., & Tilley, N. (2006). *Problem-Oriented Policing and Partnerships: Implementing an evidence-based Approach to Crime Reduction*. Willan.

Clarke, R. V., & Eck, J. (2003). *Become a Problem-Solving Crime Analyst in 55 Small Steps.* Jill Dando Institute, University College London. Available online at: www.jdi.ucl.ac.uk/publications/manual/crime_manual_content.php

Cohen, L. E., & Felson, M. (1979). Social Change and Crime Rate Trends: A Routine Activity Approach. *American Sociological Review, 44*, 488–608.

Ekblom, P. (2011a). *Crime Prevention, Security and Community Safety Using the 5Is Framework.* Palgrave Macmillan.

Ekblom, P. (2002). Towards a European Knowledge Base' and 'The Five Is: Experimental Framework for a Knowledge Base for Crime Prevention Projects. In *European Crime Prevention Conference 2002* (Vol.1, pp. 62–97). Copenhagen: Danish Crime Prevention Council. https://www.designagainstcrime.com/web/towards_european_k_pease.pdf

Eck, J. E., & Spelman, W. (1987). *Solving Problems: Problem-Oriented Policing in Newport News.* Police Executive Research Forum.

Farrington, D., & Painter, K. (2002, November). *How to Evaluate the Inpact of CCTV on Crime.* Paper given at Conference on Evaluating the Real Evidence on CCTV in Leicester.

Flint, A. (2009). *Wrestling with Moses: How Jane Jacobs Took on New York's Master Builder and Transformed the American City.* Random House.

Gill, M., Spriggs, A., Allen, J., Argomaniz, J., Bryan, J., Hemming, M., Jessiman, P., Kara, D., Kilworth, J., Little, R., Swain, D., & Waples, S. (2005). *The Impact of CCTV: Fourteen Case Studies.* Home Office Online Report. Home Office www.homeoffice.gov.uk/rds/pdf s05/rdsolrl505.pdf

Green, M. (1999). *The Appropriate and Effective Use of Security Technologies in U.S. Schools.* National Institute of Justice.

Jacobs, J. (1961). *The Death and Life of Great American Cities.* Random House.

Jeffery, C. R. (1971). *Crime Prevention Through Environmental Design.* Sage.

Jones, A., Valero-Silva, N., & Lucas, D. (2016). *The Effects of 'Secure Warm Modern' Homes in Nottingham: Decent Homes Impact Study.* Nottingham City Homes.

Kelling, G., & Wilson, J. (1982). Broken Windows: The Police and Neighbourhood Safety. *The Atlantic Monthly* (pp. 29–38). Washington DC.

Kirby, K. & Reed, D. (2004) *Developing a Problem-oriented Organisation.* London: Association of Chief Police Officers.

Lonely Planet. (2016). *New York City.* Lonely Planet.

Mayhew, P., Clarke, R., Sturman, A., & Hough, J. (1976). *Crime as Opportunity* (Home Office Research Study 34). HMSO.

Morris, S. (1996). *Policing Problem Housing Estates* (Police Research Group Crime Detection and Prevention Series Paper 74). Home Office.

Newman, O. (1972). *Defensible Space: Crime Prevention Through Urban Design.* Macmillan.

Newman, O. (1973). *Defensible Space: People and Design in the Violent City.* Architectural Press.

Newman, O. (1996). *Creating Defensible Space.* U.S. Department of Housing and Urban Development.

Nuttgens, P. (1989). *The Home Front.* BBC Books.

Pascoe, T. (1999). *Evaluation of Secured by Design in Public Sector Housing—Final Report.* Building Research Establishment.

Power, A. (1997). *Estates on the Edge: The social Consequences of Mass Housing in Northern Europe.* Macmillan Press Ltd.

Read, T., & Tilley, N. (2000). *Not Rocket Science? Problem-solving and Crime Reduction* (Crime Reduction Research Series 6). Home Office.

Scates, L. (2007). Incorporating CPTED into the Palmdale Planning Process, presentation given to the *Conference of the National Institute of Crime Prevention*, Las Vegas, on 1 August 2007.

Shaftoe, H., & James, S. (2004). *Do Symbolic Barriers Prevent Crime and Offer an Increased Sense of Safety?* Routledge.

Smith, M., Clarke, R. V., & Pease, K. (2002). Anticipatory Benefits in Crime Prevention. *Crime Prevention Studies, 13*, 71–88. Criminal Justice Press.

Stokes, M. (2022) *An Investigation of the Sustainability of Crime Prevention in the Built Environment: Impact and Implementation Factors. Published Doctoral Thesis. University of Huddersfield.*

Teedon, P., Reid, T., Griffiths, P., Lindsay, K., Glen, S., McFayden, A., & Cruz, P. (2009). *Secured by Design Impact Evaluation Final Report.* Glasgow Caledonian University.

Thornton, A., Hatton, C., Malone, C., Fryer, T., Walker, D., Cunningham, J., & Durrani, N. (2003). *Distraction Burglary Amongst Older Adults and Ethnic Minority Communities* (Home Office Research Study 269). Home Office.

UK Government. (1995). *Disability Discrimination Act.* London. Accessible at www.legislation.gov.uk/ukpga/1995/50/contents

Weisburd, D., Britt, C. Wilson, D.B., & Wooditch, A. (2020) *Basic Statistics in Criminology and Criminological Justice.* Springer Nature.'

Welsh, B. P., & Farrington, D. C. (2008). Effects of Improved Street Lighting on Crime. *Campbell Systematic Reviews, 2008*, 13. https://doi.org/10.4073/csr.2008.13

West Midlands Police. (1996). *Knowing Your School: Straightforward Advice and a Simple DIY Survey to Help You Manage the Risk to Your School.* West Midlands Police Reprographics Department.

[illegible] in the [illegible]
[illegible]
[illegible] Sentencing [illegible]
[illegible]
[illegible] (2012) [illegible]
[illegible] Journal [illegible]
[illegible]
[illegible]
[illegible]
[illegible] Home Office [illegible]
[illegible] Woodhouse [illegible]
[illegible]
[illegible]
[illegible]

10 And to Conclude

10.1 The Future of Designing Out Crime

Evidence from Crime Science

Close to half a century has passed since Designing Out Crime was first embraced by the police services of the UK and beyond, academics, other social agencies and consultants. In the first instance, we can thank Jeffery (1971) and Newman (1972) for initiating this process through their development of Crime Prevention Through Environmental Design (CPTED). But the academic credibility for our discipline comes from the work of eminent Crime Scientists like Rachel Armitage, Kate Bowers, Patricia and Paul Brantingham, Ron Clarke, Lawrence Cohen, Derek Cornish, Paul Cozens, John Eck, Paul Ekblom, Graham Farrell, David Farrington, Marcus Felson, Lorraine Gamman, Martin Gill, Ron Glensor, Hermann Goldstein, Shane Johnson, Gloria Laycock, Kate Painter, Tim Pascoe, Ken Pease, Barry Poyner, Henry Shaftoe, Nick Tilley Patricia Mayhew, Danielle Reynald and many more.

Historically, CPTED, Designing Out Crime and Situational Crime Prevention have been predominantly associated with volume property crime—offences usually perceived as most susceptible to these approaches. However, the research findings of numerous Crime Scientists across the globe provide academic credibility to the use of such interventions to prevent more serious forms of Crime—namely, Terrorism, Homicide, Robbery, Kidnapping, Firearms, Sexual offences and Elder Abuse—with their expansion to include all forms of criminal activity. Stressing that link with Crime Science has the potential to open up a whole new era of long-term, genuinely *Sustainable Crime Prevention*, with practical application in the built environment and beyond).

M. Stokes, *Designing Out Crime*, Crime Prevention and Security Management,
https://doi.org/10.1007/978-3-031-99840-9_10

Contraction

Nevertheless, it would be wrong to presume that Designing Out Crime is well embedded within mainstream society or policing—even in the UK with its relatively high number of police DOCOs. From in excess of 500 ALOs/CPDAs in England, Scotland, Wales, Northern Ireland, the Channel Islands and the Isle of Man at the turn of the century, in less than a decade that figure had fallen to 304, many of whom had other police responsibilities (Salford and Huddersfield Universities, 2009). Seven years later the total number had more than halved to 137 (Armitage 2016), although this author's own research calculated the number of practising police DOCOs in 2020 as 199 (Stokes, 2022). Meanwhile, Crime Prevention Officers/Crime Reduction Officers (CPOs/CROs) in which role many DOCOs learnt their craft, are an all but extinct species.

This rapid contraction in service delivery followed the closure (and subsequent demolition) of the Home Office Crime Reduction College at Easingwold in North Yorkshire (one of only two such centres of excellence in the world). However inadvertently, this sent out a message that Crime Prevention and Designing Out Crime were no longer valued and without the free places on the courses the college formerly delivered, police demand fell considerably. Similarly, the repeal of *Safer Places: the Planning System and Crime Prevention* (2004) and *Planning Policy Statement1 (PPS1): Delivering Sustainable Development* (2005), removed recognition for the role. Unsurprisingly, the perception then grew that the lights of Designing Out Crime were being switched off.

Future Promise

Since 2010, the concept of a UK National Police Crime Prevention Service (NPCPS) has become a fading memory (at least for now). Nevertheless, there are many beacons of hope. For example, with SBD having taken responsibility from the College of Policing for the delivery of Crime Reduction and Designing Out Crime training, it is hoped that this might increase the number of fully trained DOCOs. This will be especially necessary if government plans (first announced in 2017 and repeated in 2024) to build 300,000 new homes per year are realised.

In the absence of an NPCPS, a strategic argument could be made to bring Crime Prevention and Designing Out Crime within the CID (Criminal Investigation Department) umbrella (including Public Protection), from where it originally emerged in the 1960s. This would acknowledge the necessity for Prevent Strategies in respect of Serious Crime types, examples of which dominate the Professor Ron Clarke inspired eight Crime Prevention Matrices in the Appendices section of this handbook: *Violent Crime, C/VIT (Cash/Valuables in Transit) Robbery, Car*

Key Burglary, Metal Theft, Garage Forecourt Bilking, Night Time Economy, Serious Acquisitive Crime and Commercial Cannabis Cultivation(Cannabis Farms).

Indeed, where this has taken place it has proved to be far easier to cascade Designing Out Crime from within the CID or Public Protection. Evidence of this newfound value placed in Crime Prevention can be seen in the following statistic: during the first six months of transfer into Force CID, one team of police DOCOs received more requests for advice and strategic thinking in respect of Serious Crime issues, than in the previous decade! (personal witness).

Another line of thought considers the future lies with DOCOs being independent, self-employed consultants. Outside the UK and the English-speaking world, this is often the norm. But as membership of the ICA (International CPTED Association) demonstrates, such individuals are very few in number, thereby implying that worldwide and overwhelmingly the opportunities to Design Out Crime are being lost. Furthermore, as the number of police-employed practitioners contracted in the UK, there is no evidence that this resulted in work for and a growth in the number of private sector consultants offering this service.

The Value of Design

It has been an enduring theme throughout this handbook (and a strong case can be made) for DOCOs to be aware of the aesthetics of their recommendations and avoid the pitfalls leading to, the allegation of creating fortresses and/or a Fortress Mentality. In essence, the pursuit of 'Elegant Security'—see Farrell and Tilley (2020). A knowledge of the history of Architecture, Planning and Design together with their evolution into the modern age might avoid the subjective and disastrous pronouncements made by a tiny number of DOCOs referred to in Sect. 1.5.9, Diplomatic Skills. Self-motivated academic study or at the very least a keen interest in 'Design' and all its manifestations in the built environment, should be recognised as a necessary requirement by the DOCO. Such learning will also enhance the DOCO's expertise when they appear before Planning Hearings, Public Inquiries, Inquests, etc. (see Sect. 7.6).

Meanwhile, in recent years, the term Designing Out Crime has been embraced by the Design community (see Ekblom, 2012). At Central St. Martins, University of the Arts, London, Professor Lorraine Gamman's Design Against Crime Research Centre designed (and redesigned) products to remove any inherent Crime Opportunity. To date, they have scored notable successes with such items as: the 'Chelsea' chair, whose seat is cut in such a way that a bag can be safely hung from the seat; thief-resistant shoulder bags;

ATM (Cashpoint Machines) Privacy Zone art (see Sect. 7.6); and a range of theft-resistant bicycles. See Gamman (2013); Gamman and Thorpe (2012, 2014) and the websites https://www.designagainstcrime.com, https://www.stopthiefchair.com, https://www.bikeoff.org, https://www.inthebag.org.uk, https://www.grippaclip.comand https://www.goneshopping.org.uk

Meanwhile, at the University of Salford, Professors Caroline Davey and Andrew Wootton head the Design Against Crime Solution Centre and have produced: Davey et al. (2012) Engaging Young People in Design Against Crime; Wootton et al. (2009) City Centre Crime: Design Thinking for Safer City Centres; Davey et al. (2005) Designing Against Terrorism.

Secured by Design

On a similarly positive theme, in the UK and now well into its fourth decade, Secured by Design (see Sect. 5.1) continues to be the most successful and *Sustainable Crime Prevention* initiative in the world and requires examination and approval by a police DOCO. Likewise, the Safer Parking 'Park Mark' Award (see Sect. 5.19) and Domestic Violence Sanctuary Rooms (see Sect. 6.1) are in essence Threat and Risk Assessment (TARA—see Sect. 1.5.15) models requiring the skill-set of a fully-trained and experienced DOCO. Consequently, any attempt to reduce these to simple 'tick-the-box' exercises conducted by untrained staff should be robustly challenged if embarrassment, civil litigation and worse are to be avoided.

Suicide Prevention

The efficacy of Designing Out Crime has relevance beyond its traditional applications in crime, fear of crime (FOC) and anti-social behaviour (ASB). As a highly positive extension, CPTED and Designing Out Crime have been discovered to possess a cross-discipline application in respect of Suicide Prevention. Suicide—not a crime in England and Wales since 1961, (in Northern Ireland not since 1966 and never in Scotland) but always a tragedy for the deceased, family, friends and all those who deal with the aftermath.

British Transport Police (BTP) covers the rail networks of England, Scotland and Wales and is the lead police service in dealing with Suicide. Its highly effective Suicide Prevention and Mental Health (SPMH) unit monitors those who contemplate and all too sadly choose the railway system to take their lives. As an indication of success, approximately 2,000 potential Suicides are now identified on the railway system each year and placed on Suicide Prevention Plans (BTP,

2018). Of these, those who return to the network to attempt Suicide has fallen to less than 0.25%—demonstrating minimal displacement, maximum diffusion of benefits and how those who wish to take their own lives are highly location and method-specific.

All Suicide and Attempt Suicide locations are surveyed by a BTP DOCO who submits a series of recommendations to prevent repeat tragedies at that location. In addition to the standard CPTED measures (e.g. fencing, Lighting, CCTV, access restriction, etc.), these will often include the use of Symbolic (Psychological) Barriers (see Sect. 2.12) e.g. Pegamoids/'Witches Hats' and hatched markings on station platforms; holographics; and the removal of memorials. The aim is to disrupt the thought processes of those contemplating Suicide. See also the implied deterrence value of ATM Privacy Zones detailed in Sect. 6.11. And Sect. 4.2.1, Lighting in respect of the ineffectiveness of blue light in Suicide Prevention. Such Suicide Prevention methodology is deliberately not advertised, in order to prevent replication of the Modus Operandi by those contemplating taking their own lives. See Hawton et al. (2001); Smith and Cornish (2012); Pease and Farrell (2014); House of Commons Library (2024).

Use of CPTED and DOC methodology can also be identified in the prevention of Suicide from bridges. First at Clifton Suspension Bridge, Bristol; Erskine Bridge west of Glasgow; Union Bridge, Aberdeen; and most recently (2024) Golden Gate Bridge, San Francisco. At all four locations Physical Security 'Target Hardening' Measures have led to reductions in suicide of at least 50%. Three of these bridges are 'listed' on the basis of aesthetic/historical importance. Whereas the Erskine Bridge (over the River Clyde in Scotland) has no such classification: it has witnessed the all but total elimination of Suicide—and with no evidence of displacement (Stokes, 2025). And at the Golden Gate Bridge whose Suicide Prevention 'The Net' went live in January 2024, Suicides have reduced by 73% and during one 3-month period in 2024 no suicides were reported. The results provide *'early but clear evidence that the safety nets are associated with an immediate and substantial reduction in suicides'* (Washington Post, 2025). Furthermore, at all four locations these 'Elegant Security' measures (as set out by Farrell & Tilley, 2022) can be identified as being instrumental in delivering reductions in Suicide—and by default the original prognosis of Clarke and Mayhew (1988). Consequently, they amount to extremely powerful evidence of the effectiveness of CPTED, Designing Out Crime and Situational Crime Prevention—and with minimal displacement effects. It is also heart-warming to know that whilst the Clarke and Mayhew (1988) research empowered CPTED and the efficacy of denying crime opportunity, in return CPTED measures are now being used to reduce suicide see Clarke and Lester (2013).

Finally, amongst all policing initiatives, the processes of DOC will almost inevitably fall within the designation of Problem-Solving and Problem-Oriented Policing and Partnerships (see Sect. 2.14)—even if perceived as future issues. Consequently, DOCOs should be advocates, ambassadors and champions of Problem-solving, POP and the situational processes whereby crime, FOC and ASB can be prevented on a long-term, sustainable basis.

Erskine Bridge over the River Clyde west of Glasgow, Scotland with its barriers all but totally eliminating suicide at this former hot-spot location

'The Net', constructed of 20 feet (6 metres) wide marine grade stainless steel netting erected along both sides of the 1.5 miles long Golden Gate Bridge, San Francisco, USA, witnessed a 73% reduction in suicide during 2024—its first year of operation

10.2 Suggested Reading

There follows a list of suggested reading for the DOCO. Very much this author's selection and as previously advised, it may be wise to 'dip inside' these works to ascertain the information required, or at least obtain a flavour of the ideas articulated. Some of these theories now form the bedrock of Crime Science. Full publication details can be found in the Bibliography of this handbook.

Rachel Armitage: a decade after her first review of SBD housing projects in West Yorkshire (1999), Professor Armitage (with Dr Leanne Monchuk), revisited the original findings. Their 2009 research confirmed the effectiveness of SBD, and how long-term *sustainable* benefits are even greater than first believed. Amongst her canon of works also read *Sustainability versus Safety: Confusion Conflict and Contradiction in Designing Out Crime, Imagination for Crime Prevention: Essays in Honour of Ken Pease* (2007); *Crime PreventionThrough Housing Design* (2013); *Crime Prevention Through Environmental Design* (2016); and *'Burglars' Take on Crime Prevention Through Environmental Design* (2017).

Ron Clarke: Co-recipient of the 2015 Stockholm Prize in Criminology, Professor Clarke is credited with Professor Derek Cornish (1986) as inventing *Rational Choice Theory* (see Sect. 2.4). Also read: *Situational Crime Prevention: Successful Case Studies* (2010) the truly revolutionary: *The British Gas Suicide Story and its Implications for Prevention* (Mayhew & Clarke, 1988); *Twenty-Five Techniques of Situational Prevention* (Cornish & Clarke, 2003); *Hot Products: Understanding, Anticipating and Reducing Demand for Stolen Goods* (Clarke, 1999); *Become a Problem-solvingCrime Analyst in 55 Small Steps* (Clarke & Eck, 2003); and *Understanding Risky Facilities* (Clarke & Eck, 2007).

Lawrence Cohen and Marcus Felson: *Social Change and Crime Rate Trends: A Routine Activity Approach* (1979). Professors Cohen and Felson created *Routine Activity Theory* and maintain it is fundamental to our understanding and the importance of Crime Opportunity in the criminal's decision-making process.

Paul Cozens: *A Review and Current Status of Crime Prevention Through Environmental Design (CPTED)* (Cozens & Love, 2015) provides definitive examination of the evolution of *CPTED*. Also recommended, *Crime Prevention through Environmental Design (CPTED) Introducing and Learning from the "Dark Side"* (Cozens & Love, 2017); and *Geographical Juxtaposition: A New Direction in CPTED* (Cozens et al., 2019); Climate change and crime—exploring future risks, resilience and crime prevention through environmental design (CPTED) (Cozens & Stoks, 2024); Exploring Land Use, Crime and Perceptions of Crime in a Retail Setting: Implications for Crime Prevention Through Environmental Design (CPTED) (Cozens and Davern, 2025); Some Like it Hot: exploring crime cold spots, environmental criminology and CPTED (Cozens and Stoks, 2025a); About Time? Exploring the temporal dimensions and imperatives to CPTED using a construction industry-aligned methodology for more sustainability and resilient cities (Cozens and Stoks, 2025b); Crime prevention in the city: a current status and new directions for CPTED (Cozens and Stoks, 2025c).

Paul Ekblom: *The Conjunction of Criminal Opportunity—a Tool for Clear, 'Joined-Up' Thinking about Community Safetyand Crime Reduction'* (2000); and *Crime Prevention, Security and Community Safety Using the 5Is Framework* (2011). Professor Ekblom is a leading thinker on *Crime Science* issues and its application across a wealth of subjects. Concepts like the *Conjunction of Criminal Opportunity (CCO)* and the *5Is* (pronounced 'five eyes') are cutting edge and require some persistence to comprehend. However, this is vital as they illustrate how rather than just the three sides of the *Crime and PAT Triangles*, there are a multiplicity of *Crime Prevention* interventions—many available to the *DOCO*. He also authored the *Home Office Crime Reduction Toolkits* (2001); *Design Against Crime: Crime Proofing Everyday Products* (2012); and (with Rachel Armitage) *Rebuilding CPTED: Strengthening the Links with Crime Science* (Armitage & Ekblom, 2019).

Graham Farrell, Nick Tilley, Andromachi Tseloni and Jen Mailley: *The Crime Drop and the Security Hypothesis* (2011). The authors chart the reduction in *Vehicle Crime* in Australia, the USA, England and Wales from the early to mid-1990s and powerfully contend this resulted from an increasing range of security measures incorporated into motor vehicles. Tseloni et al. (2017) applied this same approach to *Residential Burglary* with similarly highly positive conclusions. Also recommended is Professors Pease and Farrell's Crime Prevention Unit Paper *Once Bitten, Twice Bitten: Repeat Victimisation and its Implications for Crime Prevention* (1993).

Marcus Felson and Ron Clarke: *Opportunity Makes the Thief: Practical Theory for Crime Prevention* (1998). These two prodigious authors and professors have written authoritatively about *Situational Crime Prevention* (as indicated repeatedly throughout this handbook). Treat this Home Office Police Research Paper as very much an introduction to their work and easy to comprehend.

Lorraine Gamman: Professor of Design at the Design Against Crime Research Centre, Central Saint Martins University of the Arts London, Lorraine has been pivotal in ensuring that 'design' is a key element of *Designing Out Crime*. This is most noticeably evident in the work of her department in respect of theft-resistant bicycles, chairs, handbags and ATM *Privacy Zone* art. See especially: *Built Environment: Sustainability* via*Security: A New Look* (2009—co-edited with Rachel Armitage); Johnson et al. (2010) *Theft of customers' personal property from cafes and bars*; and *Tricky Design: The Ethics of Things* (2018—co-edited with Tom Fisher).

Tony Holt and John Spencer: *A Little Yellow Box: The Targeting of Automatic Teller Machines as a Strategy in Reducing Street Robbery* (2005). A fascinating read and one which provides much-needed evidence for the effectiveness of *Symbolic (Psychological) Barriers* (see Sect. 2.12) in the form of *Privacy Zones/Personal Defensible Space* laid out in front of such ATMs.

Jane Jacobs: *The Death and Life of the Great American Cities* (1961). Perceived as the great guru of post-Second World War planning issues, Jacobs' beliefs are repeatedly misquoted by people who don't appear to have read this book—her magnum opus. Well worth reading and still in print.

C. Ray Jeffery: *Crime Prevention Through Environmental Design* (1971). The criminologist who invented CPTED and his major reference work.

Shane Johnson and Kate Bowers: *Permeabilityand Burglary Risk: Are Culs-de-Sac Safer?* (2010). Professors Johnson and Bowers convincingly challenge *Eyes Upon the Street*, demonstrating the link between *Permeability* and *Crime* and one that should spell the death knell for the *'Leaking' Cul-de-Sac*. Also and with Rob Guerette, *Crime displacement: what we know, what we don't know, and what it means for crime reduction* (2014). See also Sect. 2.1.

George L. Kelling and James Q Wilson: are best known for their *Broken Windows* theory (1982) and repeatedly referenced in this handbook. Their original article, first published in *The Atlantic Monthly*, is easy to locate on the internet. Moreover, it provides an authoritative read and strikes many chords with which the DOCO/CPTED Officer/Practitioner will agree.

Gloria Laycock: *The UK Car Theft Index: an example of government leverage* (2004). A rare example of a highly successful *Naming and Shaming* exercise—and a warning to those tempted to follow the same route. See Sect. 1.5.13. Also read Professor Laycock's *Property Marking: A Deterrent to Domestic Burglary?* (1985) and as relevant now as it was three decades ago.

Oscar Newman: *Defensible Space: Crime PreventionThrough Urban Design* (1972); *Defensible Space: People and Design in the Violent City* (1973); and *Creating Defensible Space* (1996). Newman and C.Ray Jefferey are perceived as the founders of *CPTED* and these are his seminal works.

Ken Pease OBE: Professor Pease is a massive authority on Situational Crime Prevention with an impressive list of easily comprehensible publications. Begin with his essay with Ivan Hill, *The 'wicked' issues: displacement and sustainability*, in Ballintyne et al. (2000). Also read the now legendary *The Kirkholt Project: Preventing burglaryon a British public housing estate* (1991); *Cracking Crime Through Design* (2001); and also with Professor Graham Farrell, *The Carbon Cost of Crimeand Its Implications* (2009—see Sect. 2.11). Another simultaneously enlightening and engaging read is Farrell et al. (2007) in *Imagination for Crime Prevention: Essays in Honour of Ken Pease* (2007)—a compendium of DOC topics and each written by an eminent Crime Scientist.

Barry Poyner: An early exponent of *Designing Out Crime,* recommended reading includes: *Design against Crime: beyond defensible space* (1983); *Crime Free Housing* (1991—with Barry Webb). And his final work, *Crime-Free Housing in the 21st Century* (2006).

Secured by Design: *Design and Access Statements: How to Use Them to Prevent Crime* (2010) and originally published by West Yorkshire Police (2009).

Amongst the raft of highly informative reference documents on the *SBD* web-site, the *Development Guides*, *SBD New Schools, SBD Residential 2025* and *SBD Non-Residential 2025* are the ones most regularly interrogated by *DOCOs*.

Henry Shaftoe and Sarah James: *Do symbolic barriers prevent crime and offer an increased sense of safety?* (2004) A thought-provoking read detailing the paucity of real evidence from *Crime Science* in support of *Symbolic (Psychological) Barriers* (see Sect. 2.12) and the necessity for more evidence-based research. See also Henry Shaftoe's *Convivial Urban Spaces: Creating Effective Public Spaces* (2008).

Nick Tilley: *Crime Prevention* (2009). Professor Tilley's book is easy to read, comprehend and provides a highly informative resumé of a broad spectrum of headings under the *Crime Prevention* banner. Section 1.5, 'Situational measures and mechanisms', is most relevant to the delivery of the *Designing Out Crime* function. Also acquire, *Not Rocket Science? Problem-solvingand crime reduction* (2000)by Tim Read and Nick Tilley (see Sect. 1.5 of this handbook). And most recently with Professor Graham Farrell the vitally significant, *Elegant security: Concept, evidence and implications* (2022). Professor Tilley (together with the Home Office) also founded the Tilley Awards, celebrating the best *POP* and *Problem-Solving* projects in the UK and which feed directly into the Goldstein Awards held in the USA.

Brandon Welsh and David Farrington OBE: these professors from opposite sides of the Atlantic have provided repeated meta-analysis regarding the effectiveness of CCTV (2002) and Street Lighting (2002, 2008) in reducing *Crime*. Digest and appreciate their analysis as providing the leading-edge understanding of these interventions.

Incidentally and where listed, you will find that these academics are highly approachable and it does not take an inordinate amount of effort to track them down on the internet and solicit a response to your inquiry. However, brevity is the watchword in the question you pose and it goes without saying, do not expect them to write your dissertation or thesis!

Compass Database: UK DOCOs are strongly advised to access this database of 214,000 (and growing) planning appeal decisions. Much of our DOC work overlaps with that of the Planning profession and these judgements by the Planning Inspectorate, provide both educational and fascinating reading. Moreover, crime, anti-social behaviour and fear of crime regularly appear as 'relevant planning considerations' within these Planning Appeal Judgements (see also Sect. 7. 5). Indeed, reading through these decisions, it is gratifying to realise how in the overwhelming majority of instances, Planning Inspectors behave as what we might describe as *Street Level Crime Scientists (see* Sect. 1.5.3*)* in their decision-making.

pop.center.org: All DOCOs and CPTED Officers/Advisers should be aware of and regularly consult the POP Centre database. It includes a vast array of research material and the results of what works (and doesn't work) in practice and in the real world of Crime Prevention initiatives. On occasion, cross-Atlantic descriptions might be lost in translation, e.g. Gasoline Drive-Offs equates with Garage Forecourt Bilking. But as always, persist and you should find.

10.3 Glossary

5Is—(pronounced 'Five Eyes') a process model of *Problem-Solving* based on Intelligence, Intervention, Implementation, Involvement and Assessment. See Sects. 2.7 and 2.14.

Accessibility—the ability of people to move around an area to reach places and facilities. Includes the elderly, less able-bodied, those in wheelchairs or with young children, luggage, etc. Since first the *Disability Discrimination Act 1995* and now the *Equality Act 2010* became law, building design must take account the needs of the less able-bodied.

Active Frontage—the front or edge of a building or space that has windows and doors, as opposed to blank walls, fences and garages.

Active Rooms—the word 'Active' in this context means rooms in building elevations from which there is direct and regular visual connection between the room and the street or parking court. Such visual connection can be expected from rooms such as kitchens and living rooms, but not from more private rooms e.g. bedrooms and bathrooms.

Anticipatory Benefits—refers to the key concept in a research paper published by Smith et al. (2002) and their findings that *Crime Prevention* often begins once an intervention has been announced, rather than having to wait for its implementation.

Anti-Social Behaviour (ASB)—(also referred to as 'Disorder') relates to conduct that is generally perceived as unwarranted and unwanted by the overwhelming majority of ordinary citizens. ASB often amounts to preparatory actions in the commission of crime.

Approved Document Q—issued under the Building Regulations 2010, Approved Document Q is designed to prevent unauthorised access to dwellings (including flats). This Building Regulation came into effect on 1 October 2015 and applies to new dwellings built in England. Windows and doorsets fitted to new dwellings must now possess proven security performance.

Architectural Liaison Officer (ALO)—see DOCO.

BOSS (British Oil Security Syndicate) Forecourt Watch—Crime Reduction partnership initiative designed to meet the requirements of both police and fuel retailers within a police area. BOSS established more than 130 Forecourt Watch schemes across the UK.

British Crime Survey (BCS)—the now-discontinued biannual assessment by way of public interview to assess the true extent of *crime* committed, as opposed to recorded *crime* (that which is reported to the police). The BCS always concluded that far higher levels of crime are actually committed, compared to those recorded. See also *Crime Survey for England and Wales*.

British Standards Institution (BSI)—originally incorporated by Royal Charter, the BSI is an independent testing house producing national standards e.g. for security products. It prepares and publishes British Standards and offers specialist services, e.g. the BSI library and Standardline database.

Broken Windows theory—invented by Kelling and Wilson (1982), describes how a failure to rapidly repair relatively minor issues (e.g. broken windows) can cause geographical areas of the built environment to enter a spiral of decay with associated crime, fear of crime (FOC) and anti-social behaviour (ASB).

Brownfield Land—land that has previously been developed.

BS 'Kitemark'—the British Standards Institution is the owner of a registered certification trademark. Usually associated with the words, 'approved to British Standard' followed by the relevant number (BS 3621 in the instance of five lever mortise deadlocks). This mark acts as an assurance that production has complied with a system of supervision, control and testing including periodic inspection of the manufacturing process.

Business Improvement District—defined geographical area in which businesses have agreed to pay an additional tax or fee so as to fund improvements. In return, these central government-inspired initiatives deliver tax reductions and other financial support.

Capable Guardian—crime preventers who are ready, willing and capable of *surveillance* and intervention. May include employers, park-keepers, bus conductors, passers-by, police, residents, security guards, etc.

Ceramic Marker—a specialist form of Property Marking that engraves a post-code (zip-code) and house number, or other identifying mark on items of pottery or porcelain.

Certification (of products)—independent, ongoing third-party monitoring of the manufacturing process of security tested products such as doors, windows and shutters, including periodic sample re-testing and factory audits to ensure consistency. Independent certification by an accredited body is a vital requirement of SBD.

Code for Sustainable Homes (2006)—produced by the UK government's Department for Communities and Local Government (DCLG), this now repealed check-list of requirements awarded a series of points to developers for incorporating sustainability elements within the construction process.

Code of Practice—a document that specifies the rules and procedures with which a particular initiative, activity and individuals will comply.

Commensurate Risk—the ability to deliver security recommendations that are believed to be appropriate to the risk perceived to exist in a specified location, at a given time and applied to a given task(s), function(s) and/or personnel/goods that need to be protected in that environment (see Sect. 1.5.15).

Community Safety—an aspect of the quality of life in which people, individually and collectively, are sufficiently free from or reassured about a range of real and perceived risks connected with crime and perceived risks connected with *crime* and related misbehaviour; are able to cope with the consequences of the incidents they experience; and if unable to cope alone are helped to do so. Taken together, this establishes the conditions for them to pursue their cultural, social and economic life.

Corporate Approach to Crime Prevention and Community Safety—in essence, the term is used to describe how it is the responsibility of everyone in society to prevent crime, rather than solely the police.

CPDA—see DOCO.

CPTED (Crime Prevention Through Environmental Design)—C. Ray Jeffery's description of Crime Prevention achieved through changing the design of the built environment. In much of the world, the term CPTED has evolved to include Environmental Elements, Technological Innovations, Physical Security 'Target Hardening' Measures, and both Second and Third Generation forms of CPTED, together with elements like Community Interaction.

CPTED Practitioner—see DOCO

Crime Impact Statement (CIS)—the report requested by planning authorities within the Greater Manchester Police area and other forces, as part of the consultation and evaluation process delivered by their Design for Security Consultants (DfSCs). The CIS process involves identifying, predicting, evaluating and mitigating the Crime and Disorder effects of a development proposal early in the design process, prior to planning decisions being taken and commitments made. The purpose of the CIS is to ensure that those involved in the design consider crime, disorder and fear of crime before determining whether to proceed with new projects. A CIS will include all the requirements for Secured by Design accreditation, should the client apply for the award.

Crime Pattern Analysis—the system by which primarily (although not exclusively) Recorded Crime is analysed.

Crime Pattern Theory (CPT)—Brantingham and Brantingham (1981, 2008) developed CPT as a means of explaining why crimes are committed in certain locations and their patterns. It is based on the presumption that crime is not random, but either opportunistic or planned. CPT contends that crime takes place when the Activity Space of a victim or target intersects with the Activity Space of an offender.

Crime Prevention/Reduction—essentially the same meaning can be applied whether the terms Prevention or Reduction are used. However, whilst both terms are in common usage and their meanings blurred, there is a more than subtle difference between the two (see below).

Crime Prevention—is the intervention in the causes of criminal activity to reduce their occurrence and potential seriousness.

Crime Prevention Design Adviser—see DOCO.

Crime Prevention Design Group (CPDG)—UK body of police DOCO representatives from England, Scotland, Wales, Northern Ireland, the Channel Islands, Isle of Man and Secured by Design. The CPDG meets on a quarterly basis to decide on the delivery of SBD and Designing Out Crime across the UK. In 2016, it approved the adoption of the description Designing Out Crime Officer (DOCO). CPDG superseded the former Crime Prevention Technical Committee in 1996.

Crime Reduction—this description came into popular use in the late 1990s and includes any action to reduce the frequency and seriousness of criminal events. It implies a holistic approach to reducing crime along the Crime Reduction Intervention Sustainability Spectrum (see Sect. 2.15) from DOC measures to the arrest and imprisonment of offenders.

Crime Prevention/Reduction Officer (CPO/CRO)—describes a police officer or police staff (non-sworn) employee whose role is to address a variety of issues regarding the management of Crime Prevention/Reduction and who has attended and passed a Crime Reduction Officer's Course. Their skills include the development of Partnership Working; reducing crime through analysis of patterns, trends and statistics; and the application of Problem-Solving approaches (including technological solutions) to all forms of criminal activity.

Crime Survey for England and Wales—biannual assessment by way of public interview to assess the true extent of crime committed, as opposed to recorded crime (that which is reported to the police). The CSEW repeatedly concludes that far higher levels of Crime are actually committed, compared to those recorded.

Defensible Space—an environment where the physical characteristics allow legitimate occupiers to assert influence and control to ensure their security (see Newman, 1992, 1993 and Sect. 4.4.1).

Design for Security Consultant (DfSC)—description used in Greater Manchester Police (GMP) area as an alternative to DOCO. GMP have always recruited from the professions of the built environment to fulfil this role. Developers are charged a locally negotiated levy for the production of a Crime Impact Statement (CIS).

Designing Out Crime/Architectural Liaison/Crime Prevention Design—the combination of traditional Physical Security 'Target Hardening', Measures, Technological Innovations and the Environmental Elements of Crime Prevention Through Environmental Design (CPTED—see Chapter 4), other Situational Crime Prevention interventions and most importantly to underpin this with developments within and evidence from the worlds of Crime Science and Environmental Criminology (see Chapter 2).

Designing Out Crime Officer (DOCO)—also known as Crime Prevention Design Adviser, Architectural Liaison Officer (ALO) or Design for Security Consultant (DfSC). Specially trained police officer or police staff (non-sworn) member, employed by the UK police services to deliver Designing Out Crime and administer, inspect and deliver the Secured by Design award. The term Designing Out Crime Officer was formally adopted by the UK Crime Prevention Design Group (CPDG) in 2016 as the preferred title for this role and designated as a specialist role by the National Police Chiefs' Council (NPCC) in 2021.

Diamond Tipped Marker or Scribe—marker used to etch the postcode (zip-code) onto metal surfaces. Usually supplied with a stencil for this purpose.

DICE (Design Improvement Controlled Experiment)—description used by Professor of Geography Alice Coleman in her book, Utopia on Trial (1985) and refers to her central government financed £50 million refurbishment project applied to six estates across England and Wales in the late 1980s/early 1990s.

Disorder—see Anti-Social Behaviour.

Distraction Burglary—an especially insidious form of Residential Burglary (house-breaking) in which the elderly and other vulnerable individuals are deliberately targeted by criminals posing as 'bogus officials' (often utility workers) or adopt a similar subterfuge to gain access and steal from the premises. *Distraction Burglars* often work in pairs, with one criminal talking to the victim whilst the other is engaged in ransacking the home. See also Operation Liberal.

DOCO—see Designing Out Crime Officer.

Door Chain or Restrictor—device that allows the dwelling main entrance door to be opened a few inches whilst the caller confirms their identity. However, it is not an extra lock and should only be used when there is a caller at the door. A growing body of opinion is hostile to such devices—on grounds of Fire safety and their capacity to allow forced entry by *Distraction Burglars* when not used correctly.

Door Viewer (Spyhole)—optical device installed at eye level, enabling the occupier to identify callers prior to opening the door. Permitting a wide angle of vision, it is most often simple to install and may also incorporate an electronic audible amplification element.

Easily Accessible—such windows or doorsets are those that can be accessed via a flat roof, balcony, or other similar structure e.g. external fire-escapes, supporting or decorative balcony detail.

Equality Act 2010—replaced much of the Disability Discrimination Act 1995 and in the context of the built environment, reasserts the legal requirement that all new development (including refurbishment of existing premises) must take into account the access requirements of the less able-bodied and incorporate measures to facilitate such Access and Movement.

Expert Witness—in the UK, an individual recognised by the judicial process as having a level of expertise in a given subject area and who can be relied upon to provide evidence in this regard. The role of DOCO might now be accepted within this definition.

Five Lever Mortise/Sash Deadlock—a lock in which the main body of the device is mortised, i.e. it is fitted into a hole cut into the edge of and contained within the door. Its mechanism is operated by five levers, and it will comply with the minimum standard BS 3621.

Formal Surveillance—meaning surveillance by persons employed or tasked with keeping watch, such as police, security guards and CCTV operators.

Hinge Bolts—fixed bolts fitted to the inside edge on the hinge side of the door. When the door closes these bolts protrude into holes morticed into the door frame. Ideal for outward opening doors where hinges are exposed.

Home Zone—in the United Kingdom (and further afield) translation of the Dutch Woonerf (living street/recreation area) and an attempt to reclaim for the local community a given area of public open space from the domination of motor vehicles, especially those travelling at speed.

Hot-Spot—locations in which incidents of crime and/or anti-social behaviour (disorder) are concentrated (described as 'clustered', 'dispersed', or 'hot-points').

Informal Surveillance—surveillance by those not tasked with such a role e.g. pedestrians and car drivers.

Internal Security—protection for the target/inside the premises or *Target Enclosure*. This includes security features such as *Intruder Alarms*, safes, concealment of articles and Property Marking.

Layered Security Approach—the manner in which a security review or analysis of the plans for a proposed development is undertaken, working from the outside inwards. Also known as the 'Onion Peeling' process.

'Leaking' Cul-de-Sac—is a road providing no through route for motor vehicles, yet at least one footpath (sometimes combined with a cycleway) links it to another cul-de-sac, road, public open space, bridleway, etc. See Sect. 6.2.

Local Development Plan—in the UK, these replaced Unitary Development Plans and Local Authorities are statutorily required to create, maintain and apply the local planning policies that identify how land is used and what will be built and where. Adopted Local Plans provide the framework for development.

Meta-analysis—describes a holistic and systematic review of all the available evidence and theories. Especially appropriate in the context of Crime Displacement and Diffusion of Benefits (see Sect. 2.1).

Mixed Use Development—multiple uses within a building, on a site, or within a specific area—residential, business, recreational, etc. More specifically, 'horizontal' mixed uses are side by side, whereas 'vertical' mixed uses are on different floors of the same building.

Movement—see *Permeability*.

National Planning Policy Framework—UK government planning policy introduced in March 2012, replacing 1800 pages of Planning Policy Guidance and Planning Policy Statements (see Sect. 7.1).

Natural Surveillance—design of a building(s) and/or streetscape, retail park, industrial estate, transport infrastructure, etc. that limits Crime Opportunities by enhancing the likelihood a potential offender will be seen. The effectiveness of such measures relies on witnesses reacting to and/or reporting what they have seen to law enforcers, combined with the potential offender's expectation of such a reaction (see Sect. 4.4.2).

Neighbourhood Watch—includes Home Watch, Community Watch, Church Watch (and a host of other schemes under the 'Watch' umbrella). These partnerships between the police and public aim to prevent crime in a geographical area, through Informal Surveillance, reporting suspicious incidents and developing social cohesiveness.

New Urbanism/Urban Villagers—an urban design movement and arguably the dominant planning ideology in the world. It advocates walkable neighbourhoods and increased *Permeability/Movement*, yet is prone to ignore how these have the capacity to produce increased Crime Opportunities, FOC and ASB, or the mass of evidence from Crime Science that contradicts such ideals. See especially: Sects. 6.2, *'Leaking'* culs-de-sac; 6.5, Home Zones and Shared Space; and 6.16, Pedestrianisation of Town and City Centres.

Onion Peeling Process—see Layered Security approach.

Operation Liberal—in the UK, this is a national police initiative that aims to combat *Distraction Burglary* (see above entry) by sharing intelligence, best practice and targeting known and especially serial offenders who commit this insidious form of crime, preying on the elderly and vulnerable members of society.

Operations Manual—the collection of rules and procedures under which a particular facility is governed. See also Sect. 8.3, Good Housekeeping Measures.

Opportunist Criminal—(see also Sect. 2.3) the term used to describe the type of criminal estimated to account for the majority of crime committed. Two types of opportunist criminal are often identified. The first is not particularly intent on committing crime, but when the circumstances present themselves will do so. The second actively seeks out opportunities. Both can be compared with 'professional, career or travelling criminals' who pursue a known target, resolving to overcome whatever obstacles are placed in their way. There is no clear distinction between these 'types', which simply serve as a convenient means of describing criminal activity. It is commonly understood that the opportunist can be more easily prevented from committing their crime.

'Park Mark'—the Safer Parking 'Park Mark' Award is conferred on parking facilities that have met the requirements of a risk assessment conducted by the police Accredited Assessor (AA) and the Development Manager (DM). The award confirms that the parking operator has put in place measures that can help deter crime and ASB. The scheme is managed by the British Parking Association on behalf of Secured by Design. Further information is available at https://www.britishparking.co.uk (see also Sect. 5.19).

PAS 24 Doors and Windows—PAS 24 (Publicly Available Specification) is a UK standard that assesses door and window security in a holistic fashion: door/window, frames and furniture (including security elements) are constructed away from site and installed into the building as a complete unit. PAS 24 doors/windows are inherently more secure as they do not suffer the problems associated with inappropriately installed door/window furniture (primarily locks). Other standards are detailed in Sects. 4.2.1, Doors and Door Security; 4.2.2, Windows and Glazing; and in Chapter 5, Secured by Design.

Passive Infra-Red Detectors (PIRs)—sensors designed to detect rapid thermal or heat changes, such as an intruder entering a room. PIRs receive but do not transmit energy and are commonly used to activate Intruder Alarm systems and Security Lighting.

Patio Door Locks—secure the sliding pane to the frame. The best patio doors will incorporate a multi-point locking system and anti-lift device, now standard features.

Perimeter Protection (Intruder Alarms)—includes any form of detection device that operates at the time of or prior to entry of the intruder into a protected area, e.g. a sensor located on the window frame of a dwelling.

Permeability—the extent to which an area can be accessed, most usually with reference to pedestrians via roads, footpaths, rights of way (official and unofficial), etc. (see Sect. 4.4.6). A 'movement network' can be shown on plans and in a variety of different printed and computer-generated images. Within the planning profession, *Permeability* is referred to as *Movement*.

Perimeter Protection—any physical barrier that protects the actual private space outside the shell of the building—namely the garden, rear yard, driveway and side access routes. Fences, walls or hedging most often perform this function. The perimeter is usually the first line of physical defence against the intruder.

Planning Permission—refers to the need for almost all development in the United Kingdom to pass scrutiny by the local Planning Authority before it is built.

Private Finance Initiative (PFI)—a means of funding public construction projects, whereby they are built by the private sector and then leased to the state (central or local government) usually for 30 years before transferring to full public ownership. Especially used for school building projects before being superseded by the Building Schools for the Future programme (2005–2010).

Problem-Oriented Policing and Partnerships—invented by Professor Herman Goldstein, Problem-Oriented Policing (POP) is a policing strategy that identifies and analyses specific Crime and Disorder problems with the aim of developing effective response strategies. POP requires the police to proactively identify the underlying problems which can be targeted to reduce Crime and Disorder at their roots. Goldstein's 1979 model was expanded in 1987 by Professors John E. Eck and William Spelman into the SARA (Scanning, Analysis, Response and Assessment) model of Problem-solving. More recently, the term Problem-Oriented Policing and Partnerships has been used in an attempt to encourage partner organisations to participate.

Property Crime—includes all offences where property is stolen or damaged, including all forms of theft, burglary, robbery, auto-crime and criminal damage.

Problem-solving—a key mechanism in delivering Goldstein's concept of Problem-Oriented Policing (POP), a number of different forms of Problem-solving have been developed including SARA and the 5Is. See Sects. 2.14 and 2.7.

Psychological Barriers—see Symbolic Barriers

Raid Control—the former anti-Commercial Robbery initiative that utilised CCTV, cash security, cash degradation and staff training, sponsored by a range of different companies including IBP. See Van Koppen and Jansen (1998); and Appendix D.

Rational Choice Theory (RCT)—devised by Professors Derek Cornish and Ron Clarke(1998). In the context of Situational Crime Prevention, RCT takes the utilitarian view that human beings consider means, ends, costs and benefits before making a rational choice. It is predicated on the basis that three elements must coincide: motivated offender(s), suitable target, and the absence of a capable guardian(s) See Sect. 2.4.

Reducing the Pay-Off—making the property stolen less valuable in the hands of thieves and thereby act as a deterrent. Dye-marking and Property Marking are examples.

Removal of the Means to Commit Crime—crime is often committed due to the availability (Crime Opportunity) of items that assist its commission. For example, ladders left insecure or tools in an unlocked garden shed used to force doors or windows open. By denying the criminal access, the Crime Opportunity can be prevented.

Rim Deadlock and Rim Automatic Deadlock—more secure than the rim night-latch, as the spring latch-bolt is converted into a dead-bolt. The dead-locking facility prevents opening from the outside. Ideal for doors too thin to enable a mortice lock to be installed.

Rim Night-Latch—a lock fitted to the inside face of the door with a spring-operated latch-bolt. The bolt engages automatically when the door is slammed and is usually operated by a key on the outside and a knob on the inside. Commonly found on many Front Doors to the dwelling, it has very little security value when fitted close to ordinary annealed/float or Toughened Safety Glass. Moreover, it can be easily released by:

(a) Breaking the glass panel, reaching through and turning the knob;

(b) Shoulder charging or kicking the door to force entry;
(c) Inserting a plastic card or thin foil strip between the frame and the 'bolt' so as to release the lock;
(d) Using a crow-bar or similar to force the door/lock.

Routine Activity Theory (RAT)—established by Professors Laurence Cohen and Marcus Felson (1979), it describes why crimes take place and is based on the contention that the following three things must coincide: Suitable Target; absence of a Capable Guardian; and presence of a Likely Offender (see Sect. 2.2).

SARA (Scanning, Analysis, Response, Assessment)—a process model of Problem-solving developed by Professors John E. Eck and William Spelman (1987). See Sects. 2.7 and 2.14.

Section 17 Crime and Disorder Act 1998—states that 'Without prejudice to any other obligation imposed on it, it shall be the duty of each authority to which this section applies to exercise its various functions with due regard to the likely effect of the exercise of those functions on, and the need to do all that it reasonably can to prevent Crime and Disorder in its area.'

Secured by Design (SBD) Award—launched in 1989, owned by the police services of the UK and managed by the Mayor's Office for London, SBD is the most successful, long-running and *Sustainable Crime Prevention* initiative in the history of policing. It is delivered by DOCOs across England, Scotland, Wales, Northern Ireland, Isle of Man and Channel Islands. SBD incentivises and rewards developers who incorporate police specified security measures and advice (see Chapter 5).

Secured by Design Licensing Scheme—managed by Secured by Design, this is a system of licensing the manufacturers of approved security products.

Secured by Design National Building Approval—SBD NBA provides developers (primarily housebuilders) with a more efficient and simplified route to compliance with Part Q of Schedule 1 (Security) to the Building Regulations 2010 and the Scottish Building Standard 4.13—Security. Launched in 2016, SBD NBA produces a bespoke Technical Schedule for each developer.

Secured Environments—a police certification scheme awarded to organisations that are able to demonstrate that they have adopted key principles for protecting themselves against Crime (see Sect. 5.12). The scheme is administered by Perpetuity Research and Consultancy International Ltd on behalf of Secured by Design. Further information available at https://www.securedenvironments.com

Shell Protection—the shell (Target Enclosure) is the main fabric or structure of the building and includes the walls, roof, external doors and windows. The weakest elements of the shell are usually the doors and windows. The target is most usually located within the shell and consequently, great importance is placed on preventing the shell from being breached—thus the value of Physical Security 'Target Hardening' Measures and Technological Innovations (see Sects. 4.1.1–4.2.4).

Site Specifics—describes how any development may possess additional, risk-associated characteristics. For example, its location within an existing High-Crime Area or Hot-Spot, or where specific types of Crime are known to be prevalent. SBD Development Guides list the minimum-security measures necessary to attain the award. However, site specifics may dictate that increased levels of security are necessary and the DOCO may insist on their inclusion before the SBD award can be conferred.

Situational Crime Prevention—the theory that seeks to: reduce the opportunities for criminals to commit crime; alters criminals' ideas about whether they can get away with a particular crime; makes it seem harder, riskier and less rewarding to commit crime. As such, the majority of recommendations made by the DOCO will fall under this heading.

Sold Secure—a test and certification body owned by the Master Locksmiths Association, that accredits security products such as motor vehicle and bicycle anti-theft devices, graded as bronze, silver, gold or diamond.

Stand-Off Area—most often a reference to the ground between the site perimeter and the fabric of the building. Primarily used in connection with mitigating the effects of vehicle-borne explosive devices (see Sect. 6.17, Counter Terrorism Design). It has distinct parallels with the Environmental Elements of CPTED and more specifically the creation of Defensible Space, with the overall intention of preventing those with criminal and/or ASB intent from approaching/being able to approach the building or other target.

Statutory Consultee—as part of the Planning Process in the UK, certain professions and bodies have a right of audience in law and to have their views considered.

Symbolic Barriers—describes changes in road surface colour or texture, ceremonial gates and gate-posts and any other feature that purports to be a physical barrier, yet is never or impossible to be used for that purpose. These are also sometimes referred to as Psychological Barriers (see Sect. 2.12).

Target Enclosure—e.g. safes, buildings, gated compounds situated in a wider environment like a shopping mall, park, housing estate, etc.). This is an important distinction, because in for example the crime of Residential Burglary, Ekblom (2000) is differentiating the dwelling that is being broken into (the Target Enclosure) from that which is stolen e.g. a cache of jewellery.

Target Hardening—placing physical barriers between the criminal and the target. The term used throughout this handbook is Physical Security 'Target Hardening' Measures. A golden rule is that the greater the attraction value of the target, the greater the required strength or resistance of the barriers/construction. Used as a description by Oscar Newman in 1972, 'Target Hardening' has equal application in respect of vehicles, boats and indeed, aircraft, ATMs, etc.

Target Removal—without a target the criminal cannot commit a Crime. For example: concealing valuable items so that they cannot be seen by passers-by/those breaking into the house; removing valuable items to a bank vault/safety deposit box; changing from pre-payment meters to credit meters.

Territoriality—the expression used to describe how residents and/or other occupants can exercise perceived or real control over a building (such as a dwelling, multi-storey block of flats), or in a neighbourhood, business or other community.

Terrorism—the use or threat of a specified action designed to influence the government, or to intimidate the public or a section of the public, where the use or threat is made for the purpose of advancing a political, religious or ideological cause. The action is a 'specified action' if it: involves serious violence against the person; serious damage to property; endangers a person's life, other than the person committing the action; creates a serious risk to the health or safety of the public; or is designed to seriously interfere with or disrupt an electronic system. Section 1, The Terrorism Act 2000.

Trap Protection (Intruder Alarms)—this includes any form of detection device that will not operate until an intruder actually enters the area protected by the device.

UKAS (United Kingdom Accreditation Service)—the sole national accreditation body recognised by government to assess against internationally agreed standards, organisations that provide certification, testing, inspection and calibration services.

UK DOCA (Designing Out Crime Association)—Former association of like-minded professionals (Academics, Architects, CPTED Advisers/Officers/

Practitioners, Crime Scientists, Criminologists, DOCOs. Planning Officers) who had a shared interest in how the built environment and particular products can be adapted to prevent or reduce crime.

Ultra-Violet Marker or Pen—uses a marking agent invisible to the naked eye but appears under ultra-violet light. Used to mark property with a postcode (zip-code) and house number or similar identifying feature. Needs to be reapplied every 6 months as it will fade.

Wider Police Family—in the UK, collective term used to describe police officers, unsworn police staff, police community support officers and members of the Special Constabulary (part time, voluntary police officers).

Window Locks—usually key-operated (or sometimes purely push-button) locks designed to provide secondary security to opening windows. Retro-fitted, surface-mounted window locks are of dubious reliability should the window be forced with a screw-driver, chisel, crow-bar or similar jemmying device. As an alternative, multi-point, non-key-operated locking systems contained within the body of the opening leaf, combine the best elements of Fire safety and security.

References

Armitage, R. (2007). Sustainability Versus Safety: Confusion, Conflict and Contradiction in Designing Out Crime. In G. Farrell, K. Bowers, S. Johnson, & M. Townsley (Eds.), *Imagination for Crime Prevention: Essays in Honour of Ken Pease. Crime Prevention Studies* (Vol. 21, pp. 81–110). Willan.

Armitage, R. (2013). *Crime Prevention Through Housing Design: Policy and Practice.* Palgrave Macmillan.

Armitage, R. (2016). Crime Prevention Through Environmental Design. In R. Wortley, & M. Townsley (Eds.), *Environmental Criminology and Crime Analysis.* Routledge.

Armitage, R. (2017). *Burglars' Take on Crime Prevention Through Environmental Design (CPTED): Reconsidering the Relevance from an Offender Perspective.* Macmillan.

Brantingham, P. J., & Brantingham, P. L. (1981). Notes on the Geometry of Crime. In P. J. Brantingham & P. L. Brantingham (Eds.), *Environmental Criminology.* Sage.

Brantingham, P., & Brantingham, P. (2008). Offender Mobility and Crime Pattern Formation from First Principles. In L. Liu & J. Eck (Eds.),*Artificial Crime Analysis Systems: Using Computer Simulations and Geographical Information Systems* (pp. 193–208). Idea Press.

Clarke, R. V. (1999). *Hot Products: Understanding, Anticipating and Reducing Demand for Stolen Goods* (Police Research Series 112). Home Office.

Clarke, R. V., & Eck, J. (2003). *Become a Problem-Solving Crime Analyst in 55 Small Steps.* Jill Dando Institute, University College London. Available online at: www.jdi.ucl.ac.uk/publications/manual/crime_manual_content.php

Clarke, R. V., & Eck, J. (2007). *Understanding Risky Facilities*. Center for Problem-Oriented Policing. Available online at: http://www.popcenter.org/tools/PDFs/risky_facilities.pdf

Clarke, R. V., & Lester, D. (2013). *Suicide: Closing the Exits*. Routledge: New York.

Clarke, R., & Mayhew, P. (1988). The British Gas Suicide Story and Its Implications for Prevention. In M. Tonry & N. Morris (Eds.), *Crime and Justice: A Review of Research* (Vol. 10). Chicago University Press.

Cohen, L. E., & Felson, M. (1979). Social Change and Crime Rate Trends: A Routine Activity Approach. *American Sociological Review, 44*, 488–608.

Cornish, D., & Clarke, R. (2003). Opportunities, Precipitators and Criminal Decisions: A Reply to Wortley's Critique of Situational Crime Prevention. In *Crime Prevention Studies* (Vol. 16). Criminal Justice Press.

Cozens, P., & Love, T. (2015). A Review and Current Status of Crime Prevention Through Environmental Design (CPTED). *Journal of Planning Literature*, 1–20.

Cozens, P., & Love, T. (2017). Crime Prevention Through Environmental Design (CPTED) Introducing and Learning from the "Dark Side". *Oxford Research Encyclopaedia of Criminology and Criminal Justice*. Available online at: 412 http://criminology.oxfordre.com/view/10.1093/acrefore/9780190264079.001.0001/acrefore9780190264079-e-2

Cozens, P., & Stoks, F. (2024). Climate Change and Crime - Exploring Future Risks, Resilience and Crime Prevention through Environmental Design (CPTED). *Crime Prevention and Community Safety, 26*(4):420–439. https://doi.org/10.1057/s41300-024-00221-8

Cozens, P., & Davern, B. (2025). Exploring the Relationship between Land Use, Perceptions of Crime and Crime Prevention through Environmental Design (CPTED) in a Retail Setting. *The CPTED Journal*. https://www.thecptedjournal.net/uploads/4/8/5/5/48559983/cozens_davern_2025-planning_land-use_and_environmental_criminology_final.pdf

Cozens, P., & Stoks, F. (2025a). Some Like it Hot: exploring crime cold spots, environmental criminology and crime prevention through environmental design (CPTED). *Crime Prev Community Saf, 27*, 48–69. https://doi.org/10.1057/s41300-025-00226-x

Cozens, P., & Stoks, F. (2025b). About Time? Exploring the Temporal Dimensions and Imperatives to Crime Prevention through Environmental Design (CPTED) Using a Construction Industry-Aligned Methodology for More Sustainable and Resilient Cities. Submitted to the Security Journal.

Cozens, P. & Stoks, F. (2025c). A Current Status and New Directions for Crime Prevention Through Environmental Design (CPTED). Chapter 23, Crime Prevention in the City, pp. 393–409, in *The Handbook on Cities and Crime*. Edited by Dietrich Oberwittler and Rebecca Wickes. Cheltenham, UK: Edward Elgar Publishing.

House of Commons Library, (2024). Research Briefing: Suicide Prevention Policy. https://researchbriefings.files.parliament.uk/documents/CBP-10090/CBP-10090.pdf. Accessed 5 August 2025

Cozens, P., Love, T., & Davern, B. (2019). Geographical Juxtaposition: A New Direction in CPTED. *Social Sciences, 8*, 252. Available online at: https://www.semanticscholar.org/paper

Davey, C. L., Wootton, A. B., & Marselle, M. (2012). Engaging Young People in Design Against Crime Design Research. *Swedish Design Research Journal, 1*(12), 29–38.

Davey, C.L., Hodge, M., & Wootton, A. B. (2005). Designing Against Terrorism. *Engineering Designer, 30*(1) (January/February), 10–13.

Department for Communities and Local Government. (2004). *Safer Places: The Planning System and Crime Prevention.* DCLG.
Department for Communities and Local Government. (2005). *Planning Policy Statement 1 (PPS1): Delivering Sustainable Development.* DCLG.
Eck, J. E., & Spelman, W. (1987). *Solving Problems: Problem-Oriented Policing in Newport News.* Police Executive Research Forum.
Ekblom, P. (2000). The Conjunction of Criminal Opportunity—A Tool for Clear, 'Joined-Up' Thinking and Community Safety and Crime Reduction. In S. Ballintyne, K. Pease, & V. McLaren (Eds.), *Secure Foundations: Key Issues in Crime Prevention, Crime Reduction and Community Safety.* Institute of Public Policy Research.
Ekblom, P. (2011). *Crime Prevention, Security and Community Safety Using the 5Is Framework.* Palgrave Macmillan.
Ekblom, P. (2012). *Design Against Crime: Crime Proofing Everyday Products.* Lynne-Reiner.
Farrell, G., & Tilley, N. (2022). Elegant Security: Concept, Evidence and Implications. *European Journal of Criminology, 19*(5), 932–953.
Farrell, G., Tilley, N., Tseloni, A., & Mailley, J. (2011). The Crime Drop and the Security Hypothesis. *Journal of Research in Crime and Delinquency, 48*, 147–175.
Felson, M., & Clarke, R. (1998). *Opportunity Makes the Thief: Practical Theory for Crime Prevention, Police Research Series Paper 98.* Home Office.
Gamman, L. (2013). *Shoplifting Scams and Designing Out Crime.* Loss Prevention Magazine.
Gamman, L., & Thorpe, A. (2012). *The Korean Experience—3rd Generation CPTED?* Lecture given to the Designing Out Crime Association at Central St Martins, London, on 21 March 2013.
Gamman, L., & Thorpe, A. (2014). Design for Democratic Crime Prevention. *Urban Pamphleteer: Design and Trust, 3*, 6–9.
Hawton, K., Townsend, E., Deeks, J., Appleby, L., Gunnell, D., Bennewith, O., & Cooper, J. (2001). Effects of Legislation Restricting Pack Sizes of Paracetamol and Salicylate on Self Poisoning in the United Kingdom: Before and After Study. *British Medical Journal, 322*, 1203.
Holt, T., & Spencer, J. (2005). *A Little Yellow Box: The Targeting of Automatic Teller Machines as a Strategy in Reducing Street Robbery.* Perpetuity Press.
Jeffery, C. R. (1971). *Crime Prevention Through Environmental Design.* Sage.
Johnson, S. D., & Bowers, K. J. (2010). Permeability and Burglary Risk: Are Culs-de-Sac Safer? *Journal of Quantitative Criminology, 26*(1), 89–111.
Johnson, S.D., Bowers, K., Gamman, L., Mamerow, L., & Warne, A. (2010). *Theft of Customers' Personal Property in Cafes and Bars* (Problem-Oriented Guides for Police, Problem-Specific Guides Series No. 60). Washington DC: U.S Department of Justice, Office of Community Oriented Policing Services.
Johnson, S. D., Guerette, R. T., & Bowers, K. (2014). Crime Displacement: What We Know, What We Don't Know, and What It Means for Crime Reduction. *Journal of Experimental Criminology, 10*(4), 549–571.
Laycock, G. (1985). *Property Marking: A Deterrent to Domestic Burglary?* Crime Prevention Unit Paper 3. Home Office.
Laycock, G. (2004). The UK Car Theft Index: An Example of Government Leverage. In *Crime Prevention Studies* (Vol. 17, pp.25–44). Verlag.
Newman, O. (1972). *Defensible Space: Crime Prevention Through Urban Design.* Macmillan.

Newman, O. (1973). *Defensible Space: People and Design in the Violent City*. Architectural Press.

Newman, O. (1996). *Creating Defensible Space*. U.S. Department of Housing and Urban Development.

Pease, K., & Farrell, G. (2009). *The Carbon Cost of Crime and Its Implications*. Secured by Design.

Pease, K., & Farrell, G. (2014). What Have Criminologists Done for Us Lately? In *The Handbook of Security* (pp. 65–88). Palgrave.

Pease, K., & Farrell, G. (1993). *Once Bitten, Twice Bitten: Repeat Victimisation and its Implications for Crime Prevention*. Home Office Crime Prevention Unit.

Pease, K. (2001). *Cracking Crime Through Design*. Design Council.

Poyner, B. (2006). *Crime Free Housing in the 21st Century*. Jill Dando Institute of Crime Science.

Poyner, B. (1983). *Design Against Crime: Beyond Defensible Space*. Butterworth.

Read, T., & Tilley, N. (2000). *Not Rocket Science? Problem-solving and Crime Reduction* (Crime Reduction Research Series 6). Home Office.

Secured by Design. (2010). *Design and Access Statements: How to Use Them to Prevent Crime*. Secured by Design.

Shaftoe, S. (2008). *Convivial Urban Spaces: Creating Effective Public Spaces*. Earthscan.

Shaftoe, H., & James, S. (2004). *Do Symbolic Barriers Prevent Crime and Offer an Increased Sense of Safety?* Routledge.

Smith, M. J., & Cornish, D. B. (2012). *Secure and Tranquil Travel: Preventing Crime and Disorder on Public Transport*. UCL Jill Dando Institute of Crime Science.

Smith, M., Clarke, R. V., & Pease, K. (2002). Anticipatory Benefits in Crime Prevention. *Crime Prevention Studies, 13*, 71–88. Criminal Justice Press.

Stokes, M. (2022). *An Investigation of the Sustainability of Crime Prevention in the Built Environment: Impact and Implementation Factors* (Published Doctoral Thesis). University of Huddersfield.

Stokes, M. (2025). *A Tale of 3 Bridges: Barriers to Suicide Prevention*. Presentation given at the conference of the International CPTED Association in Palm Springs CA on 11 May 2025.

The Washington Post. (2025). *New Golden Gate Safety Nets are Reducing Suicide Deaths, Study Finds*. Article published on 22 March 2025.

Tilley, N. (2009). *Crime Prevention*. Willan.

Tseloni, A., Farrell, G., Thompson, R., Evans, E., & Tilley, N. (2017). *Domestic Burglary Drop and the Security Hypothesis*. Springer Open.

Van Koppen, P., & Jansen, R. (1998). The Road to Robbery: Travel Patterns in Commercial Robberies. *British Journal of Criminology, 38*(2), 230–246.

West Yorkshire Police. (2009). *Design and Access Statements: How to Use Them to Prevent Crime*. West Yorkshire Police.

Wootton, A. B., Marselle, M., & Davey, C. (2009, April 1–3). City Centre Crime: Design Thinking for Safer City Centres. In *8th European Academy of Design Conference*. Robert Gordon University.

Appendices

Appendix A: Garage Forecourt Bilking, Crime Prevention/Reduction Matrix

M. Stokes, *Designing Out Crime*, Crime Prevention and Security Management,
https://doi.org/10.1007/978-3-031-99840-9

GARAGE FORECOURT BILKING CRIME PREVENTION/REDUCTION MATRIX

Reducing Garage Forecourt Bilking (making off without payment) by using the 25 techniques of Situational Prevention

Increase Efforts

1. TARGET HARDEN:

Promote use on garage forecourt premises of fully tested and approved SBD licensed security products detailed at **www.securedbydesign.com**

Encourage installation of full colour, high specification, CCTV systems with 31 day digital recording on garage forecourts subject to repeat Bilking modus operandi

2. CONTROL ACCESS:

Advocate installation of enhanced physical security measures including PAS 24 door-sets and windows containing Laminated Safety Glass at garage forecourt premises

3. SCREEN EXITS:

Advise garage forecourt operators to minimise and narrow the number of entrances and exits to and from the forecourt and to ensure the perimeter is secure, in order to prevent vehicles 'escaping' by driving across grass verges, pavements, etc

4. DEFLECT OFFENDERS:

Use 'Tooled Up for School', 'If Only I'd Known' and 'Trouble' education programs to deter criminal inclination at most vulnerable ages. See **www.west-midlands.police.uk/oneknifeonelife/html/index.html**

5. CONTROL TOOLS & WEAPONS

Promote use of 'Thief-Proof' number-plates by car owners as at **www.theaa.com/motoring_advice/security/number_plates.html**

Utilise ANPR (Automatic Number-Plate Recognition) technology on those garage forecourts being repeatedly targeted for Bilking

Increase Risks

6. EXTEND GUARDIANSHIP:

Encourage garage forecourt operators to join together and establish BOSS Forecourt Watch or other Crime Prevention schemes see **www.bossuk.org/Forecourt-Watch**

Promote staff training in dealing with Bilkings and via the highly successful 'Raid Control' initiative to reduce incidence of Robbery. See **www.raid-control.org**

7. ASSIST NATURAL SURVEILLANCE:

Liaise with garage forecourt operators to ensure clear lines of sight from all public areas and especially around the pumps, in order to maximise the potential for Natural Surveillance from cash-till area

8. REDUCE ANONYMITY

Utilise information sharing protocols to circulate offender/vehicle details amongst members of BOSS Forecourt Watch and Facewatch for investigative purposes: **www.ico.gov.uk/upload/dpo_conference_2010/workshop_d_presentation.ppt**

Advise garage forecourt operators to install bespoke CCTV equipment in order to provide high quality recorded images for investigative and evidential purposes

9. UTILISE PLACE MANAGERS:

Encourage garage forecourt operators to invest in staff training and proven crime reduction initiatives like 'Raid Control'

10. STRENGTHEN FORMAL SURVEILLANCE:

Use Facewatch, 'Domehawk', 'Sherpa' or similar redeployable CCTV cameras for preventive, investigative and evidential purposes at garage forecourt premises regularly targeted for Bilkings

Reduce Rewards

11. CONCEAL TARGETS:

Persuade garage forecourt operators to 'cone-off' pumps furthest away from the payment kiosk at times of minimum usage and especially during hours of darkness. Where multiple pumps are available, regularly change those that are 'coned off' to disrupt potential offenders

12. REMOVE TARGETS:

Encourage garage forecourt operators to insist on payment in advance for fuel delivery at the pump, ideally as a blanket requirement or between specific times of the day/night, or to 'customers' exhibiting suspicious behaviour

13. IDENTIFY PROPERTY:

Advise garage forecourt operators to invest in Retail Radio Link networks to circulate details of vehicles known to be involved in Bilkings and/or suspected of using/stealing false number plates to execute same

14. DISRUPT MARKETS:

Investigate the opportunity for blanket insistence on payment in advance during known periods of vulnerability at local networks of garage forecourt sites, in order to minimise displacement activity

15. DENY BENEFITS:

To prevent garage forecourt criminality migrating into the payment kiosk, encourage installation of 'Fog Bandit', 'Smokecloak' or similar attack-activated fogging agents with a proven track record of deterrence/prevention. Locate such devices close to the counter, in order to 'shepherd' offenders back towards the entrance door. See **www.bandituk.co.uk & www.assured-ltd.co.uk/smoakcloak.php**

Reduce Provocations	Remove Excuses

16. REDUCE FRUSTRATIONS & STRESS:

Ensure positive police response to information and intelligence regarding garage forecourt Bilkings and where necessary, share with other operators by way of BOSS Forecourt Watch **www.bossuk.org/Forecourt-Watch** and any other local networks

17. AVOID DISPUTES:

Solicit garage forecourt sites to conduct regular risk assessments and invest in staff training that incorporates conflict resolution, to ensure processes and physical security measures in place are appropriate to the identified risk to staff members and customers

18. REDUCE EMOTIONAL AROUSAL:

Encourage installation of Symbolic Barriers in garage forecourt design, so as to delay/hamper escape by vehicles involved in Bilkings and thereby deny emotional 'buzz' when committing the offence

19. NEUTRALISE PEER PRESSURE:

Invest in 'Deflect Offender' campaigns (as at '4') in order to prevent entry into criminality via the offence of Bilking

20. DISCOURAGE IMMITATION:

Utilise partnership working with BOSS, the supermarkets and other garage forecourt operators to Design Out Crime from the built environment (see **www.popcenter.org**) utilising CPTED (Crime Prevention Through Environmental Design) methodology to minimise entrance and exit dimensions, incorporate rumble strips, increase lighting levels and its quality and use Symbolic Barriers like painted hash markings

Censor details of modus operandi in subsequent press statements/dealing with the media, etc.

Conduct regular trawls at **www.popcenter.org** and **www.whatworks.college.police.uk** for latest solutions to preventing Bilking

21. SET RULES:

Initiate a local review of practices and procedures re crime investigation, recording and policing Bilking offences

22. POST INSTRUCTIONS:

Encourage garage forecourt site managers to display signage/notices in respect of (for example) time-delay locks, removal of motorcycle helmets/balaclavas and making payment prior to the delivery of fuel - either as a blanket requirement or during specific times of the day/night

Promote the display of signage in respect of BOSS Forecourt Watch, Facewatch and other schemes, CCTV surveillance and the resort to civil recovery of debt by those who drive off without making payment

23. ALERT CONSCIENCE:

Utilise 'Crimestoppers' and media campaigns (**www.crimestoppers-uk.org**) to highlight seriousness of Bilking as an offence, and the exemplary nature of sentencing when offenders are convicted

24. ASSIST COMPLIANCE:

Promote media and publicity campaigns that reinforce the duty of all citizens to report suspicious individuals/conduct and be prepared to act as bona fide witnesses in court

Ensure Problem Solving approach is used as a constituent of Problem Oriented Policing and mandatory tool to deliver reductions in Garage Forecourt Bilking

25. CONTROL DRUGS & ALCOHOL:

Where garage forecourt kiosks sell alcohol, cigarettes and other higher value products (like meat and coffee), encourage site managers to locate such items behind/close to the counter and away from the entrance/ exit door

Appendix B: Cannabis Farms, Crime Prevention/ Reduction Matrix

CANNABIS FARMS CRIME PREVENTION/REDUCTION MATRIX

Reducing Commercial Cannabis Cultivation by using the 25 techniques of Situational Prevention

Increase Efforts

1. TARGET HARDEN:

Promote use by landlords and letting agencies of Secured by Design licensed security products as detailed at **www.securedbydesign.co.uk**.

Encourage landlords and letting agents to prioritise the necessity for higher levels of security at long term void and unoccupied properties

2. CONTROL ACCESS:

Advocate installation of enhanced physical security measures including PAS 24 (and other standards) doorsets and windows in void and derelict premises known to be targeted for Commercial Cannabis Cultivation. See SBD web-site **www.securebydesign.co.uk**

3. SCREEN EXITS:

Mindful of '7', encourage landlords and letting agencies to conceal (by way of shrubs, trees, fencing, hedges, walls, etc.) and minimise number of entrances and exits (ideally to one) so as to deny key decision-making multiple attack and escape routes

4. DEFLECT OFFENDERS:

Invest in 'Tooled Up for School', 'If Only I'd Known' and 'Trouble' education programmes to deter criminal inclination at most vulnerable of ages. See **www.west-midlands.police.uk/oneknifeonelife/html/index.html**

5. CONTROL TOOLS & WEAPONS

Enforce legislation at 'Head' Shops' and internet suppliers of equipment and seeds blatantly being sold for domestic and Commercial Cannabis Cultivation

Increase Risks

6. EXTEND GUARDIANSHIP:

Encourage landlords and letting agents to regularly visit and inspect their premises to ensure they are not being used for Commercial Cannabis Cultivation

Promote training workshops for new and existing private residential landlords re vetting of potential tenants and reporting suspicious activity to the police

7. ASSIST NATURAL SURVEILLANCE:

Liaise with landlords and letting agents to ensure clear lines of sight from public areas at front of premises, in order to maximise the potential for Natural Surveillance into the property

8. REDUCE ANONYMITY

Establish information sharing protocols to enable circulation amongst landlords' associations and Local Authority partners to share details of persons renting properties suspected of Commercial Cannabis Cultivation. See **www.ico.gov.uk/upload/dpa_cconference_2010/workshop_d_presentation.ppt**

9. UTILISE PLACE MANAGERS:

Encourage landlords associations and Local Authority partners to regularly visit and inspect their premises and report suspicious activity to the police

10. STRENGTHEN FORMAL SURVEILLANCE:

Utilise Facewatch and 'Domehawk', 'Sherpa' and other redeployable CCTV cameras at premises known or suspected to be used for Commercial Cannabis Cultivation

Deploy and publicise use of infra-red detection equipment by police helicopters to identify Commercial Cannabis Cultivation in roof spaces, cellars, etc.

Reduce Rewards

11. CONCEAL TARGETS:

Persuade landlords and letting agents to deploy measures that suggest void or vacant premises are occupied, e.g. curtains, furniture and both in-house and exterior security lighting

12. REMOVE TARGETS:

Advise landlords, letting agents and Local Authority partners re inadvisability of leaving void or vacant premises for long periods of time and necessity to find alternative uses for such properties when difficult/unable to let

13. IDENTIFY PROPERTY:

Utilise proprietary marking systems such as 'Datatag', 'RedWeb', 'Smartwater' **www.datatag.co.uk**, **www.redwebsecurity.com**, **www.smartwater.com** to identify in advance equipment and products with the capacity to be used in Commercial Cannabis Cultivation - and publicise the message that this is being done

14. DISRUPT MARKETS:

Investigate the opportunity for 'sting' operations allowing discovered sites of Commercial Cannabis Cultivation, to continue in business until such a time that more than just the 'gardeners' can be arrested and prosecuted

15. DENY BENEFITS:

Use Proceeds of Crime Act 2002 asset seizure to prosecute criminals believed to be living off the earnings from Commercial Cannabis Cultivation

Following discovery of Commercial Cannabis Cultivation, ensure prompt removal and destruction/preservation of plants and equipment in accordance with Force and evidential procedures

Reduce Provocations	Remove Excuses
16. REDUCE FRUSTRATIONS & STRESS: Ensure police response to information and intelligence regarding the existence of both domestic and Commercial Cannabis Cultivation, including updating nearby residents, landlords and partners	**21. SET RULES:** Develop and evolve consistent practices and procedures in respect of investigation and policing of Commercial Cannabis Cultivation, including the seizure and disposal of plants and equipment
17. AVOID DISPUTES: Advise landlords, letting agents (e.g. via the West Midlands Arson Task Force booklet, 'Management of Void Premises') not to negotiate with those involved in Commercial Cannabis Cultivation (whatever the temptation of a £30,000 pay-off!) Reinforce this message with the fact that such negotiation would amount to conspiracy to supply drugs and thereby attract a substantial custodial sentence	**22. POST INSTRUCTIONS:** Advise police officers to desist from using misnomer 'Hydroponics', as in 99% of instances cultivation is by conventional means of compost Raise awareness with our partners (LAs, Fire Service, utility companies) to notify police about premises with blacked out windows, curtains permanently drawn, dumped compost, sweet smell of cannabis/ air freshener, excessive power usage, or other suspicious conduct
18. REDUCE EMOTIONAL AROUSAL: Investigate the identity of different strains of Cannabis, their potency and the potential to target the seeds of same through the overt and covert use of herbicides and genetic manipulation, in order to reduce the profitability of Commercial Cannabis Cultivation	**23. ALERT CONSCIENCE:** Utilise 'Crimestoppers' (**www.crimestoppers.co.uk**) and media campaigns to highlight seriousness of Commercial Cannabis Cultivation as an offence, and the exemplary nature of sentencing when offenders are convicted
19. NEUTRALISE PEER PRESSURE: Identify groups believed to be involved in Commercial Cannabis Cultivation and target those on the periphery for MALVA (multi-agency interventions) via the likes of Birmingham Reducing Gang Related Violence (BGRV), CRASBOs, ASBOs and new powers relating to gangs	**24. ASSIST COMPLIANCE:** Promote media and publicity campaigns that reinforce the duty of all citizens to report suspicious individuals/conduct suspected of being involved in Commercial Cannabis Cultivation and be prepared to act as bona fide witnesses in court Ensure Problem Solving approach is used as a constituent of Problem Oriented Policing and mandatory tool to deliver reductions in Commercial Cannabis Cultivation
20. DISCOURAGE IMMITATION: Censor details of modus operandi regarding Commercial Cannabis Cultivation in subsequent press statements, interviews with the media, etc. Conduct regular trawls at **www.popcenter.org** and **www.whatworks.college.police.uk** for latest solutions to prevent Commercial Cannabis Cultivation	**25. CONTROL DRUGS & ALCOHOL:** Obtain intelligence regarding and target high profile offenders suspected of involvement in Commercial Cannabis Cultivation for resale, dealing and/or money laundering

Appendix C: Car-Key Burglary, Crime Prevention/ Reduction Matrix

CAR KEY BURGLARY CRIME PREVENTION/REDUCTION MATRIX

Reducing Car Key Burglary by using the 25 techniques of Situational Prevention

Increase Efforts

1. TARGET HARDEN:

Promote use by all householders of Secured by Design fully tested and approved PAS 24 front and all exterior doorsets and windows as detailed at **www.securedbydesign.com**

Encourage adoption of Secured by Design Homes initiative by private house-building industry so as to emulate success of social house-builders in reducing all forms of Residential Burglary

2. CONTROL ACCESS:

Where appropriate, persuade house-holders of efficacy in gating front driveway in order to reinforce layered security approach

Where unable to garage, promote use of intruder alarm zoning when the house is occupied e.g. ground floor areas when the occupiers are asleep upstairs

3. SCREEN EXITS:

Conceal (by way of shrubs, trees, fencing, hedges, walls, etc) and minimise number of entrances and exits (ideally to one) so as to deny key decision-making, multiple-attack and escape routes

4. DEFLECT OFFENDERS:

Promote installation of Alarm Receiving Centre (as opposed to 'bells only') Intruder Alarms by householders who own makes and models of high profile targeted vehicles

5. CONTROL TOOLS/WEAPONS:

Advocate necessity to stop and search known Car Key Burglary (CKB) suspects en route to addresses of potential victims

Increase Risks

6. EXTEND GUARDIANSHIP:

Revisit, revise and advocate deterrent capacity of Cocoon, Wraparound and Neighbourhood Watch schemes

Task Neighbourhood Teams to provide high visibility policing at homes owning vehicles targeted/vulnerable to Car Key Burglars

7. ASSIST NATURAL SURVEILLANCE:

Encourage householders to install security lighting to front and rear of their properties. Security lighting at the front to be controlled by movement-activated PIR detectors – where believed appropriate

Utilise Designing Out Crime and Secured by Design measures to ensure LAs and developers include proven deterrent characteristics of clear and overlooking windows/lines of sight with no blind corners

8. REDUCE ANONYMITY:

Liaise with partner agencies to move public seating, unrestricted parking, bus-stops and other opportunities for anonymity from the vicinity of homes known to possess vehicles attractive to Car Key Burglars (CKBs)

9. UTILISE PLACE MANAGERS:

Encourage concierge staff and vehicle owners to adopt preventive measures: e.g. garaging, installing security devices, reporting suspicious activity to the police

10. STRENGTHEN FORMAL SURVEILLANCE

Publicise value of high quality CCTV recorded images in domestic locations (e.g. front driveways/hallways) for investigative and evidential purposes

Install 'Memocams' in homes of repeat victims that intelligence suggests may be targeted

Reduce Rewards

11. CONCEAL TARGETS:

Advise via Crime Prevention/Reduction campaigns all car owners (especially of those makes and models attractive to thieves) to garage cars overnight and when not in use

12. REMOVE TARGETS:

Advise all car owners to lock and remove keys from vehicles and the immediate vicinity of the dwelling's main entrance door, storing same in Faraday bags

However, car owners must NOT be advised to take car keys to the bedroom, or make them difficult to locate once intruders have accessed the dwelling

13. IDENTIFY PROPERTY:

Encourage owners of cars known to be targeted by Car Key Burglars, to install 'Tracker' or similar identification systems

Encourage property marking of Sat Navs, CD cassettes and all removable and high value electrical items

14. DISRUPT MARKETS:

Investigate potential for intelligence-led 'sting' operations using vehicles known to be targeted by Car Key Burglars (CKBs)

Raise public and media awareness of 'ringed' cars being sold by criminals and necessity to authenticate provenance of same by potential buyers

15. DENY BENEFITS:

Use Proceeds of Crimes Act 2002 against criminals believed to be committing CKBs

Encourage installation of 'Fog Bandit', 'Smokecloak' or similar attack activated fogging agents with a proven record of deterrence/disruption, in specific locations such as garages and bedrooms

Reduce Provocations	Remove Excuses
16. REDUCE FRUSTRATIONS AND STRESS: Produce guidance for all car owners and especially those known to be targeted by Car Key Burglars/Car Jacking, on driving safely with the door locks engaged and not stopping in isolated or vulnerable locations	**21. SET RULES:** Liaise with and update Crown Prosecution Service, Magistrates, Judiciary and Legislators as to seriousness of Car Key Burglary as a specific crime type and to treat as such/pass exemplary sentencing
17. AVOID DISPUTES: Advise householders to avoid confrontation with Car Key Burglars, inside or outside the dwelling, on the basis that avoiding injury that might otherwise result is more important than losing property – whatever the monetary value	**22. POST INSTRUCTIONS:** Promote the display of notices/signage indicating the presence of intruder alarms, CCTV, overt security devices, 'Tracker', car alarms, guard dogs, etc Distribute targeted Designing Out Crime advice regarding home security, emphasising the necessity to utilise security devices and garage vehicles at all times when not in use
18. REDUCE EMOTIONAL AROUSAL: By way of Motor Vehicle Theft Index, continue to lobby motor vehicle and insurance industries to reduce (through design and enhanced security measures) attractiveness of vehicles targeted by Car Key Burglars (CKBs)	**23. ALERT CONSCIENCE:** Utilise media campaigns, 'Crimestoppers' (**www.crimestoppers-uk.org**) and other appropriate initiatives to highlight seriousness of Car Key Burglary as an offence and the exemplary nature of sentencing when offenders are convicted - especially when they confront householders within their own dwellings
19. NEUTRALISE PEER PRESSURE Invest in school education/teaching programmes that steer potential offenders away from crime, especially types perceived as victimless or as 'rite of passage' to more serious offences like Car Key Burglary (CKB)	**24. ASSIST COMPLIANCE** Promote media and publicity campaigns that reinforce the duty of all citizens to report suspicious individuals/conduct and be prepared to act as bona fide witnesses in court, if required Ensure Problem Solving approach is used as a constituent of Problem Oriented Policing and mandatory tool to deliver reductions in Car Key Burglary
20. DISCOURAGE IMMITATION: Utilise partnership working with LAs to Design Out Crime from built environment by mandating off-road, gated and garage parking within curtilage of dwellings, where appropriate Censor details of modus operandi in subsequent press statements, interviews with the media, public presentations, etc. Conduct regular trawls at **www.popcenter.org** and **www.whatworks.college.police.uk** for latest solutions to prevent Car Key Burglary	**25. CONTROL DRUGS & ALCOHOL:** Target high profile offenders known to be committing Car Key Burglary to finance lifestyles they could not otherwise afford (e.g. high value sports cars, conspicuous consumption of recreational drugs, property)

Appendix D: Cash/Valuables in Transit (C/ViT), Crime Prevention/Reduction Matrix

C/ViT ROBBERY CRIME PREVENTION/REDUCTION MATRIX

Reducing Cash/Valuables in Transit Robbery by using the 25 techniques of Situational Prevention

Increase Efforts

1. TARGET HARDEN:

Promote use of Secured by Design licensed security products and design guides detailed at: **www.securedbydesign.com**

Advocate installation of cash-dock/airlock cash delivery systems to eliminate Across Pavement Transfers (APTs) by C/ViT delivery and collection security guards

Encourage adoption of 'Raid Control' anti-robbery initiative, to prevent displacement of crime into retail and other commercial premises

2. CONTROL ACCESS:

Solicit installation of electronic door access and time control systems at all vulnerable delivery points used by C/ViT security guards

3. SCREEN EXITS:

Minimise number of entrances and exits (ideally to one) in order to deny criminals key decision-making feature of multiple-attack and escape routes

4. DEFLECT OFFENDERS:

Install 'Inferno' or similar high pitched screeching alarm sirens

Invest in school-based 'Tooled Up for School', 'If Only I'd Known' and 'Trouble' education programmes to deter criminal inclination at most vulnerable of ages

5. CONTROL TOOLS/ WEAPONS

Investigate capacity to prohibit sale (as per proscribed list of chemicals) of bleaching agents used to launder dye-stained notes

Increase Risks

6. EXTEND GUARDIANSHIP:

Revisit and formalise procedures in respect of Operation Vanguard 'Follow That Van' and initiate innovative campaigns designed to enhance deterrent capabilities of C/ViT transfers

Promote staff training to deal with C/ViT Robbery as per 'Raid Control' initiative and Crime Prevention/Reduction advice

7. ASSIST NATURAL SURVEILLANCE:

Liaise with C/ViT industry to ensure clear windows/lines of sight and no blind corners in immediate vicinity where cash/valuables are delivered or collected from premises

8. REDUCE ANONYMITY

Liaise with partner agencies (primarily Local Authority departments) to move potential crime generators like public seating, unrestricted parking, bus-stops, etc., from the vicinity of C/ViT delivery and collection locations

9. UTILISE PLACE MANAGERS:

Encourage town/city centre managers and Neighbourhood Police Teams to provide greater support and surveillance of C/ViT delivery and collection locations

10. STRENGTHEN FORMAL SURVEILLANCE:

Launch (and where necessary re-launch) police policy to improve CCTV recorded image quality for investigative and evidential purposes, concentrating on repeat victims and those locations previously targeted

Promote membership of Banknotewatch, Pubwatch, Safer Gems and other Crime Prevention/Reduction initiatives

Reduce Rewards

11. CONCEAL TARGETS:

Persuade C/ViT industry and other partners to invest in/permit covert delivery systems, practices and methodology

12. REMOVE TARGETS:

Actively discourage routine delivery times, days and routes for C/ViT delivery vehicles

13. IDENTIFY PROPERTY:

Encourage use by C/ViT industry of glue, dye and proprietary marking and tagging systems for all C/ViT luggage, storage and delivery mechanisms and vehicles

14. DISRUPT MARKETS:

Investigate potential for and employ 'sting' operations using C/ViT deliveries to locations repeatedly targeted for Robbery

Raise public and media awareness that dye-stained notes in circulation are proceeds of crime and of the action to be taken when in receipt of same

15. DENY BENEFITS:

Use Proceeds of Crime Act 2002 legislation to prosecute criminals suspected of any involvement in C/ViT Robberies and living off such 'earnings'

Encourage installation of 'Fog Bandit', 'Smokecloak' and similar attack activated fogging agents with proven track record of deterrence/prevention in the commission of Robbery

Reduce Provocations	Remove Excuses
16. REDUCE FRUSTRATIONS & STRESS: Encourage banks and all premises where cash or valuables are routinely delivered and collected, to eliminate queuing by C/ViT security staff once inside the building and engaged in these processes	**21. SET RULES:** Advocate C/ViT delivery crew and receiving staff comply with operating principles at all times, including pre-notification of arrival/ departure at delivery premises
17. AVOID DISPUTES: Solicit retail and commercial sites to conduct regular Threat and Risk Assessments (TARAs) to ensure processes and physical security measures in situ are appropriate and commensurate to the identified risk associated with cash or valuables being delivered or collected	**22. POST INSTRUCTIONS:** Promote the display of signage in respect of (for example) time-delay locks and prohibiting the wearing of motorcycle helmets and balaclavas in vicinity of C/ViT premises and vehicles Encourage local authorities to prohibit parking in C/ViT delivery bays and to prosecute offenders
18. REDUCE EMOTIONAL AROUSAL: Investigate instances of offenders being high on drugs (including alcohol) when committing C/ViT Robbery offences and wherever possible, prosecute suppliers of said controlled drugs	**23. ALERT CONSCIENCE:** Utilise 'Crimestoppers' and media campaigns to highlight seriousness of C/ViT Robbery as an offence and the exemplary nature of sentencing when offenders are convicted - unlike in some other countries
19. NEUTRALISE PEER PRESSURE: Invest in 'Deflect Offender' campaigns (as at '4') to prevent 'rite of passage' Street Robbery occurring and evolving into C/ViT Robbery offences	**24. ASSIST COMPLIANCE:** Promote media and publicity campaigns that reinforce the duty of all citizens to report suspicious individuals and/or conduct and be prepared to act as bona fide witnesses and attend court if necessary in respect of the crime of C/ViT Robbery Ensure Problem Solving approach is used as a constituent of Problem Oriented Policing and mandatory tool to deliver reductions in C/ViT Robbery
20. DISCOURAGE IMMITATION: Utilise partnership working with local authorities to Design Out Crime from the built environment, using CPTED methodology to minimise the distance between delivery premises and where C/ViT vehicles can park and/or wait Censor details of modus operandi in subsequent press statements, interviews and any other interaction with the media Conduct regular trawls at **www.popcenter.org** and **www.whatworks.college.police.uk** for latest solutions to prevent C/ViT Robberies	**25. CONTROL DRUGS & ALCOHOL:** Target high profile offenders known to be committing C/ViT Robberies to purchase large quantities of drugs for resale/dealing/ money laundering

Appendix E: Metal Theft, Crime Prevention/Reduction Matrix

METAL THEFT CRIME PREVENTION/REDUCTION MATRIX

Reducing Metal Theft by using the 25 techniques of Situational Prevention

Increase Efforts

1. TARGET HARDEN:

Promote use by utility companies and all owners of metal targeted by thieves of Secured by Design licensed security products detailed at **www.securedbydesign.com**

Encourage landlords and letting agents to prioritise higher levels of security at long term void properties where metal items (like copper piping, boilers and lead flashing) are present

2. CONTROL ACCESS:

Advocate installation of enhanced physical security measures including PAS 24 (and other standards) doorsets and windows with laminated glass, in void premises targeted by metal thieves

Strictly monitor and manage distribution and return of utility company/other commercial Sat Navs pre-loaded with locations of sites where metal is readily available

3. SCREEN EXITS:

Using shrubs, hedges, trees, fencing, walls, etc. conceal and restrict number of entrances and exits (ideally to one) in order to deny key decision-making multiple attack and escape routes

4. DEFLECT OFFENDERS:

Invest in 'Tooled Up for School', 'If Only I'd Known' & 'Trouble' education programmes to deter criminal inclination at most vulnerable of ages. See **www.west-midlands.police/oneknifeonelife/html/index.html**

5. CONTROL TOOLS & WEAPONS

Adopt policies to monitor distribution, management and retrieval of keys to vulnerable locations/assets where metal is stored or otherwise readily available to be stolen

Increase Risks

6. EXTEND GUARDIANSHIP:

Raise awareness of serious nature of Metal Theft to wider police family including Neighbourhood/Business Watch schemes and the necessity to report suspicious behaviour

Encourage landlords and letting agents to regularly visit and inspect their premises to ensure no instances of Metal Theft and report any thefts to the police

7. ASSIST NATURAL SURVEILLANCE:

Liaise with utility providers and managers of other locations containing easily accessible metal, advising clear lines of sight within and from public areas at front of premises in order to maximise the potential for Natural Surveillance into the property

8. REDUCE ANONYMITY

Enforce wearing of photo ID badges at all sites, together with a culture amongst staff of challenging those without same

Establish information sharing protocols to circulate amongst utility providers, etc. details of persons and vehicles believed to be involved in Metal Theft. See **www.ico.guv.uk/upload/dpo_conference_2010/workshop_d_presentation_ppt**

9. UTILISE PLACE MANAGERS:

Foster increased staff vigilance, regular visits to vulnerable/isolated premises and reporting of suspicious activity to the police

10. STRENGTHEN FORMAL SURVEILLANCE:

Encourage managers of vulnerable premises to install a fully effective CCTV system with 31 day recording capability

Utilise redeployable CCTV cameras like 'Domehawk' and 'Sherpa' in preventive and investigative roles near premises suspected of dealing in stolen metal

Reduce Rewards

11. CONCEAL TARGETS:

Persuade utility providers, landlords and letting agents to disguise or deploy measures that suggest void or vacant premises are occupied: e.g. drawn curtains, furniture, intruder alarms and both internal and external time-controlled security lighting

12. REMOVE TARGETS:

Encourage householders, landlords, letting agents, businesses, Local Authorities, etc. to securely store metal items and maintain ongoing campaigns to ensure continued safe storage

13. IDENTIFY PROPERTY:

Utilise targeted property marking initiatives (e.g. 'Enigmatag', 'Redweb', 'Selectamark', 'Smartwater') on specific metal products.

Maintain publicity campaigns advertising property marking remains active, including news of success stories in preventing/detecting Metal Theft and prosecuting offenders

14. DISRUPT MARKETS:

Conduct joint operations with Local Authority departments, Trading Standards, the Environment Agency, Kegwatch (**www.kegwatch.co.uk**) etc., to target those scrap metal dealers and other businesses that intelligence suggests are dealing in stolen metal

15. DENY BENEFITS:

Utilise Proceeds of Crime Act 2002 legislation to strip convicted criminals of their illegally-acquired assets obtained through Metal Theft

Encourage the use of alternatives to metals such as lead, copper and bronze through the use of polymer based roofing materials, glass fibre and acrylic compounds

Reduce Provocations	Remove Excuses
16. REDUCE FRUSTRATIONS & STRESS: Foster media and public awareness of how Metal Theft is both financially expensive and a serious threat to everyday life, causing disproportionate criminal damage and delays to the transport infrastructure, telecommunications and utility providers in the form of power disruptions	**21. SET RULES:** Encourage robust staff screening, vetting and training of new employees Enforce strict procedures in respect of visitor arrival, escort and removal of materials and equipment from sites
17. AVOID DISPUTES: Encourage all scrap metal dealers to adopt strict codes of practice when receiving suspicious salvage and display notices to this effect within premises Utilise the Problem Analysis Triangle (PAT) or Conjunction of Criminal Opportunity (COCO) as a means to reduce Metal Theft and in disputes concerning ownership	**22. POST INSTRUCTIONS:** Promote display of notices/signage re CCTV, property marking and staff procedures to prevent Metal Theft Encourage scrap metal dealers to display notices/signage describing the strict procedures that will always be enforced concerning the receipt of scrap metal, including 'No ID - No Sale' and 'No cash transactions'
18. REDUCE EMOTIONAL AROUSAL: N/A to Metal Theft (which includes lead, copper, steel, aluminium, brass, bronze, nickel and platinum - but not gold or silver)	**23. ALERT CONSCIENCE:** To highlight seriousness of Metal Theft, use 'Crimestoppers' (**www.crimestoppers-uk.org**) and media campaigns Use message boards, advertising campaigns, etc. with key message Metal Theft is a serious crime and not a nebulous, victim-free issue
19. NEUTRALISE PEER PRESSURE: Target known offenders and ensure such practices as surveillance, prosecution and conviction are known to their associates. Work with partner agencies to close down scrap-yards that are repeat offenders in dealing in stolen metal	**24. ASSIST COMPLIANCE:** Promote media and publicity campaigns that reinforce the duty of all citizens to report suspicious individuals/conduct and be prepared to act as bona fide witnesses in cases of Metal Theft Ensure Problem Solving approach is used as a constituent of Problem Oriented Policing and mandatory tool to deliver reductions in Metal Theft
20. DISCOURAGE IMMITATION: Utilise partnership working and liaison with utility companies, etc. to Design Out Crime from the built environment and ensure damage caused by Metal Theft is repaired promptly. See especially **www.popcentre.org/guides** Censor details of modus operandi in subsequent press statements/dealing with the media, etc. in respect of Metal Theft Conduct regular trawls at **www.popcenter.org** and **www.whatworks.college.police.uk** for latest solutions to prevent Metal Theft	**25. CONTROL DRUGS & ALCOHOL:** Through Arrest Referral and similar post arrest interview procedures, establish link between Metal Theft and the desire to fund drugs or alcohol dependency

Appendix F: Night-Time Economy, Crime Prevention/ Reduction Matrix

NIGHT TIME ECONOMY CRIME PREVENTION/REDUCTION MATRIX

Reducing Night Time Economy Crime by using the 25 techniques of Situational Prevention

Increase Efforts

1. TARGET HARDEN:

Promote use at all licensed premises serving alcoholic drinks of SBD licensed security products as detailed at **www.securedbydesign.com**

Persuade owners of void and derelict licensed premises of need for higher levels of security to prevent Arson, Metal Theft and other criminal activity

2. CONTROL ACCESS:

Advocate installation of enhanced physical security measures in licensed premises e.g. PAS 24 (and other standards) doorsets and windows (especially toilet windows – a repeat means of access)

Solicit use of licensed door staff at pubs, clubs, etc. with existing issues of disorder or in the vicinity of same

3. SCREEN EXITS:

Counsel use of licensed door staff at main entrance and ideally the sole entrance exit to premises (excluding emergency exits)

Encourage use of 'Body Cams' (body worn CCTV cameras) by door staff at licensed premises with history of disorder or in close proximity of such venues

4. DEFLECT OFFENDERS:

Persuade Local Authority partners to install public drinking water fountains, proven to reduce acts of disorder in environments with high density of licensed premises

Advise against provision of public seating in immediate vicinity of licensed premises unless under their immediate control and supervision

5. CONTROL TOOLS/ WEAPONS

Solicit use of plastic or 'toughened' glasses at all licensed premises with history of Disorder and Violence, or in immediate vicinity of such environments, events, etc.

Increase Risks

6. EXTEND GUARDIANSHIP:

Ensure Police Neighbourhood Teams liaise with licensed premises, Chambers of Trade, Business Improvement Districts, Federation of Small Businesses, etc.

Encourage membership of Facewatch, Pub Watch, Club Watch, etc. schemes creating information sharing networks for licensed premises

Promote training workshops for new and existing bar staff and door security employees

7. ASSIST NATURAL SURVEILLANCE:

Liaise with managers and designers of licensed premises to ensure clear lines of sight from the bar across all public areas, in order to maximise the opportunities for Natural Surveillance

8. REDUCE ANONYMITY:

Establish information sharing protocols amongst licensed premises to circulate details and photos of individuals barred or convicted of disorder. See Facewatch. **www.ico.gov.uk/upload/dpo_conference-2010/workshop_d_presentation.ppt**

9. UTILISE PLACE MANAGERS:

Solicit assistance from town, city and other place managers by updating them about new trends in disorder and other crime taking place at licensed premises

10. STRENGTHEN FORMAL SURVEILLANCE:

Utilise Facewatch and deploy 'Domehawk', 'Sherpa' and other redeployable CCTV cameras to prevent/investigate crime and disorder in vicinity of licensed premises

Establish Retail Radio Link schemes to enable communication between licensed premises and other retailers, in order to share information regarding potential instances of anti-social behaviour and worse

Reduce Rewards

11. CONCEAL TARGETS:

Advise owners of licensed premises regarding inadvisability of leaving such void or vacant property for long periods of time and of the need to find alternative uses, or at least give the impression that they are occupied or otherwise in use

12. REMOVE TARGETS:

Recommend that no cash should be stored on licensed premises. Display notices to this effect and empty till drawers should be left open when the premises are closed

Counsel how fruit machines/one armed bandits, ATMs (cashpoint machines) etc. should be locked within secure cabinets when licensed premises are closed and display notices to this effect

13. IDENTIFY PROPERTY:

Utilise covert and overt proprietary property marking and tagging systems, to identify in advance that kegs, bottles, cans, etc. of alcoholic drinks are so marked, will set off alarms and can be traced. Publicise this is being done
See **www.kegwatch.co.uk**

14. DISRUPT MARKETS:

Initiate 'Test Purchasing' of alcoholic drinks from premises believed to selling to persons under age or otherwise in breech of licensing or taxation laws

15. DENY BENEFITS:

Instigate robust response by police and partners to offending behaviour that relates to licensed premises and/or the consumption of alcoholic drinks in the arena of the Night Time Economy

Use Proceeds of Crime Act 2002 asset seizure legislation to prosecute criminals believed to be benefitting from the sale of tax avoided alcoholic drinks

Reduce Provocations	Remove Excuses
16. REDUCE FRUSTRATIONS & STRESS: Liaise with Local Authority partners regarding permanent and temporary provision of portable urinals and more permanent hydraulic 'Uriflo' or 'Urinoir' systems, on grounds of both public health and preventing disorder, as in cities like Amsterdam, Bruges and now London	**21. SET RULES:** Develop and evolve consistent practices and procedures in respect of policing and investigation of the Night Time Economy and related alcohol/drug-fuelled offences
17. AVOID DISPUTES: Establish Taxi Marshalling Schemes in town/city centres with high concentration of licensed premises and where sudden increase in customers for genuine taxis occurs as pubs and clubs close - most especially on Friday and Saturday nights Encourage and assist in enforcing banning orders across pubs, clubs and all other licensed premises in a given area	**22. POST INSTRUCTIONS:** Persuade licensed premises to display signage/notices demanding proof of age, unacceptable behaviour and warning about the CCTV recording of images Liaise with Local Authority partners and where appropriate, utilise legal powers to ban drinking in public places
18. REDUCE EMOTIONAL AROUSAL: Promote alcohol free parties for under 18s or where under 18s are present Support 'Drink Sensibly' campaigns that include advice to consume lower strength alcoholic drinks and how individuals are responsible for their drinking and behaviour	**23. ALERT CONSCIENCE:** Maintain year round 'Don't Drink and Drive' campaigns, with special emphasis at holiday times and during national celebrations Utilise 'Crimestoppers' and media campaigns to highlight seriousness of alcohol related Violence and Disorder (**www.crimestoppers-uk.org**)
19. NEUTRALISE PEER PRESSURE: Investigate potential and where appropriate, utilise capacity to name and shame individuals convicted of disorder or other alcohol fuelled criminal offences Warn men and women never to hail private hire cars and wherever possible to travel in pairs from pubs and clubs, most especially during the early hours of the morning	**24. ASSIST COMPLIANCE:** Support media and publicity campaigns that reinforce the message not to drink and drive (or having consumed recreational drugs) together with the duty of all citizens to report suspicious individuals/conduct and be prepared to act as bona fide witnesses of those involved in Disorder or worse at licensed premises, or sale of illicit drink Ensure Problem Solving approach is used as a constituent of Problem Oriented Policing and mandatory tool to deliver reductions in Night Time Economy Crime
20. DISCOURAGE IMMITATION: Censor details of modus operandi re offences fuelled by alcohol in pubs, clubs and other licensed premises, appearing in subsequent press statements and interviews with the media Conduct regular trawl at **www.popcenter.org** and **www.whatworks.college.police.uk** for latest solutions to preventing crime and disorder in the Night Time Economy	**25. CONTROL DRUGS & ALCOHOL:** Instigate procedures to enforce the law in respect of under age drinking in pubs and clubs and in the sale of alcoholic drinks at off licenses, supermarkets, etc.

Appendix G: Serious Acquisitive Crime, Crime Prevention/Reduction Matrix

SERIOUS ACQUISITIVE CRIME PREVENTION/REDUCTION MATRIX

Reducing Serious Acquisitive Crime by using the 25 techniques of Situational Prevention

Increase Efforts

1. TARGET HARDEN:

Promote use of Secured by Design licensed security products and design guides, as detailed at **www.securedbydesign.com**

Encourage householders, businesses and other commercial interests to conduct TARAs (Threat and Risk Assessments) of premises and tasks targeted for Serious Acquisitive Crime and apply resulting recommendations

2. CONTROL ACCESS:

Advocate enhanced physical security measures including PAS 24 (and other standards) doorsets and windows with laminated glass, in premises suspected of being targeted for Serious Acquisitive Crime

Recommend electronic door entry control systems at vulnerable locations and the necessity to prevent 'tail-gating'

3. SCREEN EXITS:

Reduce and minimise number of entrances and exits (ideally to one) so as to deny key decision-making, multiple attack and escape routes.

Avoid use of signage such as 'staff entrance' and 'cashiers office' that inadvertently assists those involved in Serious Acquisitive Crime

4. DEFLECT OFFENDERS:

Invest in 'Tooled Up for School', 'If Only I'd Known' & 'Trouble' education programmes to deter criminal activity. See **www.west-midlands.police.uk/oneknifeonelife/html/index/htm**

5. CONTROL TOOLS & WEAPONS

Lobby for proscribed listing of bleaching agents used to recirculate dye-stained banknotes that are the proceeds of crime

Increase Risks

6. EXTEND GUARDIANSHIP:

Encourage membership of Business Watch, Retail Radio Link, Facewatch and other crime reduction schemes

Promote 'Raid Control' anti-commercial robbery initiative. See **www.raidcontol.org.uk**

Advocate 'Secured Environments' safety and security management initiative at universities, schools, hospitals, etc. to reduce vulnerability to criminal behaviour. See **www.securedenvironments.com**

7. ASSIST NATURAL SURVEILLANCE:

Advise retailers and other businesses regarding store layout design and position of cash-till, to deter Robbery and Theft by maximising Natural Surveillance

8. REDUCE ANONYMITY

Establish Information Sharing Protocols amongst businesses targeted for Serious Acquisitive Crime (e.g. licensed premises and jewellers) to circulate details including photos of convicted suspects. See **www.ico.guv.uk/upload/dpo_conference_2010/workshop_d_presentation_ppt**

9. UTILISE PLACE MANAGERS:

Liaise with and update shopping centre, retail park, town and city managers regarding new crime trends, especially those relating to Serious Acquisitive Crime

10. STRENGTHEN FORMAL SURVEILLANCE:

Solicit managers of vulnerable premises and public area sites to install CCTV systems with routine recording of images as per the ICO CCTV Code of Practice

Utilise Facewatch and 'Domehawk', 'Sherpa' and similar redeployable CCTV cameras in preventive, investigative and evidential roles near premises which intelligence suggests may be subject to Serious Acquisitive Crime

Reduce Rewards

11. CONCEAL TARGETS:

Encourage installation of cash-dock/air-lock cash delivery systems to eliminate Across Pavement Transfers (APTs)

Design 'Safer Routes' for students who walk/cycle from university campuses to residential, town and city centres, etc

12. REMOVE TARGETS:

Promote systems that replace cash handling with debit and credit cards

Solicit 'Designing Out Crime' and 'Design Against Crime' initiatives that create products such as the anti-bag snatch chair to frustrate Serious Acquisitive Crime

13. IDENTIFY PROPERTY:

Utilise 'Datatag', 'Enigmatag', 'Redweb', 'Smartwater', etc. proprietary marking systems to identify banknotes/valuables subject to Serious Acquisitive Crime

Encourage installation of proven vehicle identification systems, such as 'Tracker'

14. DISRUPT MARKETS:

Raise public awareness of 'ringed' cars being sold by criminals and the necessity for buyers to authenticate the vehicle's provenance

Maintain surveillance at pawnbrokers, scrap metal yards, car boot sales, etc. suspected of dealing in stolen goods

15. DENY BENEFITS:

Use Proceeds of Crime Act 2002 legislation to seize assets of criminals believed to be living off the earnings of Serious Acquisitive Crime

Install 'Fog Bandit', 'Smoakcloak' or similar attack activated fogging agents with proven record of deterrence. See **www.bandituk.co.uk** & **www.assured-ltd.co.uk/smokecloak.php**

Reduce Provocations	Remove Excuses
16. REDUCE FRUSTRATIONS & STRESS: Solicit taxis and private hire vehicles to install on board CCTV with mandatory recording facilities and display notices/signage to this effect Encourage reduced crowding by control of numbers in pubs and clubs to avoid disputes escalating into Robbery, Violence and worse	**21. SET RULES:** Enforce strict procedures in respect of visitor escort and removal of materials and equipment from sites
17. AVOID DISPUTES: Advocate robust staff vetting/employment screening and training Advise householders and public not to confront Armed Robbers, Car Key Burglars, etc. as Personal Injury and worse are more important than Property Theft - whatever the monetary value	**22. POST INSTRUCTIONS:** Counsel the elderly and vulnerable to always engage the door chain or restrictor before answering the door - and fix a notice to this effect in the immediate vicinity outside Display notices/signage demanding balaclavas and motorcycle crash helmets must be removed in vicinity of banks and cash handling premises
18. REDUCE EMOTIONAL AROUSAL: Use initiatives like the Motor Vehicle Theft Index to lobby vehicle/insurance industries to reduce through design, speed limiters and security measures, the attractiveness of high value cars and plant equipment to Car Key Burglars and specialist criminals	**23. ALERT CONSCIENCE:** Utilise 'Crimestoppers' and encourage the media to pro-actively report Serious Acquisitive Crime. See **www.crimestoppers-uk.org** Work with partner agencies like Trading Standards and FACT to raise public awareness of counterfeit goods/video piracy
19. NEUTRALISE PEER PRESSURE: Support education programmes as at '4' aimed at school pupils, with the message that 'Taxing' fellow students is Robbery and will attract the full sanction of the law Invest in 'Deflect Offender' campaigns to prevent 'rite of passage' crime of Street Robbery evolving into C/ViT Robbery	**24. ASSIST COMPLIANCE:** Promote media and publicity campaigns that reinforce duty of all citizens to report suspicious conduct/individuals regarding Serious Acquisitive Crime and be prepared to act as bona fide witnesses in court Ensure Problem Solving approach is used as a constituent of Problem Oriented Policing and as a mandatory tool to deliver reductions in Serious Acquisitive Crime
20. DISCOURAGE IMMITATION: Censor details of modus operandi about Serious Acquisitive Crime in subsequent press statements and interviews with the media Encourage rapid repair of premises subjected to 'smash and grab' or 'ram-raid' style Commercial Burglary or Robbery Regularly interrogate internet sites like **www.popcenter.org** and **www.whatworks.college.police.uk** for new techniques to prevent Serious Acquisitive Crime	**25. CONTROL DRUGS & ALCOHOL:** Obtain intelligence regarding and target high profile offenders involved in dealing/ laundering money that is the proceeds of Serious Acquisitive Crime Maintain inspection of pharmacies and similar premises to ensure legal distribution of controlled drugs such as Methadone to registered users

Appendix H: Violent Crime, Crime Prevention/Reduction Matrix

VIOLENT CRIME PREVENTION/REDUCTION MATRIX

Reducing Violent Crime by using the 25 techniques of Situational Prevention

Increase the effort

1. TARGET HARDEN:

Promote use by landlords and letting agencies of Secured by Design licensed security products as detailed at **www.securedbydesign.co.uk**

Encourage landlords and letting agents to prioritise the necessity for higher levels of security at long term void and derelict properties

2. CONTROL ACCESS:

Advocate enhanced physical security measures including PAS 24 (and other standards) doorsets and windows within premises known to be targeted for Commercial Cannabis Cultivation and any potential venue for Violent Crime (e.g. booby-trapped locations)

3. SCREEN EXITS:

Counsel use of licensed door staff at main entrance door to pubs and clubs and ideally the sole entrance/exit used (excluding emergency exits)

Encourage use of 'Body-Cams' (body worn CCTV cameras) by door staff at licensed premises with history of Violent Crime or in close proximity to such premises

4. DEFLECT OFFENDERS:

Invest in 'Tooled Up for School', 'If Only I'd Known' and 'Trouble' education programmes to deter criminal inclination at most vulnerable of ages

5. CONTROL TOOLS & WEAPONS

Enforce legislation re sale of knives and firearms - including safe keeping and licensing of same, as appropriate

Solicit use of plastic or 'toughened' glasses at all licensed premises with history of disorder and wounding, or in immediate vicinity of such premises

Increase the risks

6. EXTEND GUARDIANSHIP:

Encourage landlords and letting agents to regularly visit/inspect premises to ensure no Commercial Cannabis Cultivation or other serious crime is taking place

Utilise Facewatch and establish Pub Watch, Club Watch and other information sharing networks aimed at reducing Violent Crime at licensed premises

7. ASSIST NATURAL SURVEILLANCE:

Provide Designing Out Crime advice to ensure clear windows/lines of sight and no blind corners in immediate vicinity of where cash/valuables are delivered and collected by C/ViT security staff

8. REDUCE ANONYMITY:

Establish information sharing protocols for pubs and clubs to circulate details of those barred for disorder or Violent Crime. See **www.ico.gov.uk/upload/dpo_conference-2010/workshop_d_presentation**

9. UTILISE PLACE MANAGERS:

Solicit assistance from managers of town and city centres, licensed premises, banks and other targeted locations, by updating them re Violent Crime trends taking place at such locations and within the vicinity

10. STRENGTHEN FORMAL SURVEILLANCE:

Use Facewatch and deploy 'Domehawk', 'Sherpa' and other redeployable CCTV cameras to prevent, investigate and evidence premises known or suspected to be locations for Violent Crime

Instigate intelligence-led high visibility policing and encourage same by security guards, at locations known to be targeted by criminals engaged in Violent Crime - with full strategic risk assessment

Reduce the rewards

11. CONCEAL TARGETS:

Persuade landlords and letting agents to deploy measures that suggest void or vacant premises are occupied and therefore not available for criminal activity e.g. Human Trafficking or Child Sexual Exploitation

12. REMOVE TARGETS:

Provide personal safety advice to potential or known victims of Violent Crime and if required, arrange for a Domestic Violence Sanctuary Room to be created within their home

13. IDENTIFY PROPERTY:

Continue to map and regularly update intelligence regarding repeat use of firearms in Violent Crime and whenever possible, remove them from circulation

Encourage use by C/ViT industry of glue, dye and proprietary marking and tagging systems for all C/ViT storage and delivery mechanisms/vehicles

14. DISRUPT MARKETS:

Liaise with Inland Revenue, Customs and Excise, Trading Standards, Environment Agency, Local Authorities and target those criminals involved in Violent Crime emanating from Human Trafficking, Slave Labour, Child Sexual Exploitation, unlawful sale of alcohol, drugs and counterfeit goods

15. DENY BENEFITS:

Use Proceeds of Crime Act 2002 to prosecute criminals believed to be living off the earnings from Violent Crime

Following discovery of illicit alcohol production or Commercial Cannabis Cultivation, ensure prompt removal and destruction/preservation of equipment and plants in accordance with police and evidential procedures

Reduce provocations	Remove excuses
16. REDUCE FRUSTRATIONS & STRESS: At locations/events with history of alcohol-fuelled Violent Crime, liaise with partners to provide water fountains and portable urinals, proven to reduce instances of disorder	**21. SET RULES:** Develop and evolve consistent practices and procedures re policing and investigating Violent Crime - most especially Homicide, Robbery, Serious Assaults/Woundings, Serious Fraud and Commercial Cannabis Cultivation
17. AVOID DISPUTES: Establish Taxi Marshalling Schemes in town/city centres with high concentration of licensed premises and sudden increase in demand for genuine taxis, as pubs and clubs close during early hours of Friday, Saturday and Sunday mornings Encourage and assist in enforcing banning orders (especially in respect of those individuals with convictions for Violent Crime) across all pubs, clubs and other licensed premises in a given area	**22. POST INSTRUCTIONS:** Invest in continuing professional development of all police staff and those in associated agencies regarding new forensic procedures and the necessity to avoid contamination of Violent Crime scenes and equipment used in crime Encourage display of Data Protection Act signage in pubs, clubs and public areas regarding existence of CCTV cameras and recording equipment used for their protection, safety and any restriction on consumption of alcohol in public
18. REDUCE EMOTIONAL AROUSAL: Maintain vigilance and enforce the law in respect of new designer drugs Investigate identity and potency of different strains of Cannabis and potential to reduce the profitability of Commercial Cannabis Cultivation Target illegal alcohol production, the sale of same and that of pornography (e.g. at car boot sales)	**23. ALERT CONSCIENCE:** Utilise 'Crimestoppers' and media campaigns to publicise instances of Violent Crime, the necessity for information from the public and exemplary nature of sentencing when offenders are convicted
19. NEUTRALISE PEER PRESSURE: Identify groups (e.g. gangs) perceived to be involved in Violent Crime and target those on the periphery or believed to be about to engage in such offences with MALVA style interventions such as Gang targeting powers	**24. ASSIST COMPLIANCE:** Promote media and publicity campaigns that reinforce the duty of all citizens to report suspicious individuals or conduct suspected of being involved in Violent Crime and act as bona fide witnesses at court Ensure Problem Solving approach is used as a constituent of Problem Oriented Policing and mandatory tool to deliver reductions in Violent Crime
20. DISCOURAGE IMMITATION: Censor details of modus operandi regarding commission of Violent Crime in information released to the media, press statements and interviews with reporters and on TV/radio Conduct regular trawls at **www.popcenter.org** and **www.whatworks.college.police.uk** for latest solutions to prevent Violent Crime	**25. CONTROL DRUGS & ALCOHOL:** Obtain intelligence and target high profile offenders known/suspected of Violent Crime, dealing in drugs or alcohol and using same to finance their lifestyles and/or launder money

Appendix I: West Midlands Arson Task Force Poster

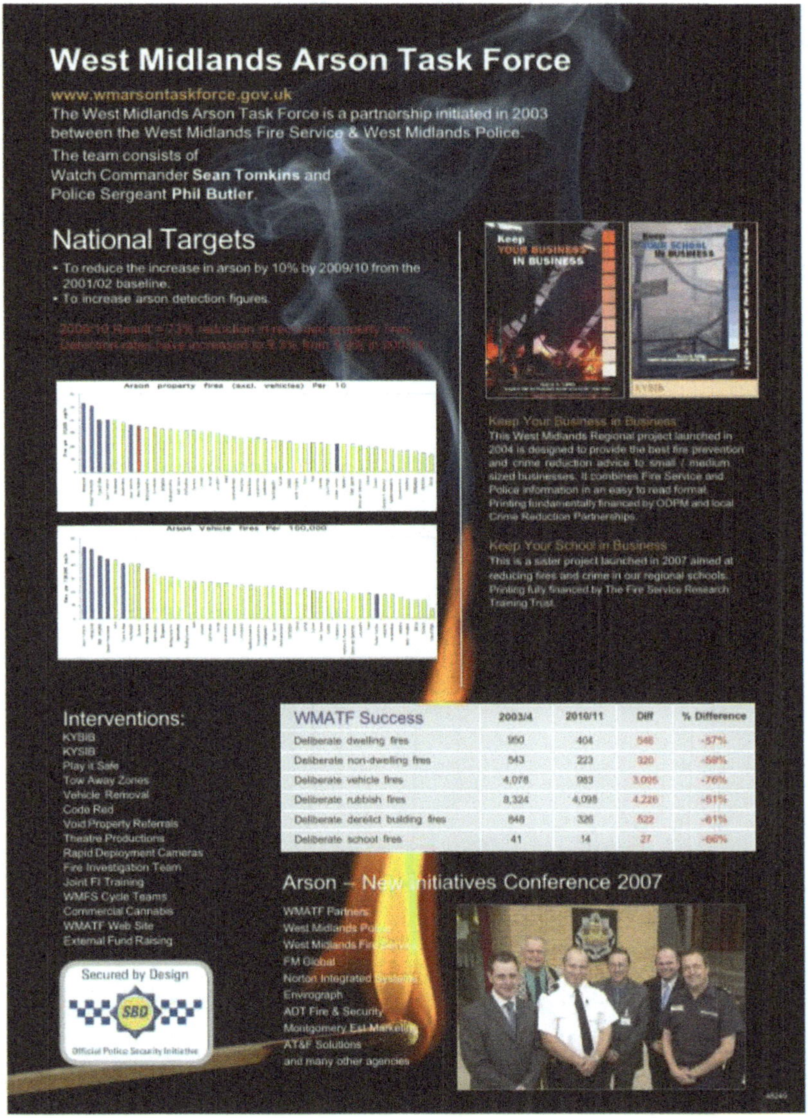
West Midlands Arson Task Force
www.wmarsontaskforce.gov.uk
The West Midlands Arson Task Force is a partnership initiated in 2003 between the West Midlands Fire Service & West Midlands Police.
The team consists of
Watch Commander Sean Tomkins and
Police Sergeant Phil Butler.
National Targets
• To reduce the increase in arson by 10% by 2009/10 from the 2001/02 baseline.
• To increase arson detection figures.
Arson property fires (excl. vehicles) Per 10
Arson Vehicle fires Per 100,000
Keep YOUR BUSINESS IN BUSINESS
Keep YOUR SCHOOL IN BUSINESS
Keep Your Business in Business
This West Midlands Regional project launched in 2004 is designed to provide the best fire prevention and crime reduction advice to small / medium sized businesses. It combines Fire Service and Police information in an easy to read format. Printing fundamentally financed by ODPM and local Crime Reduction Partnerships.
Keep Your School in Business
This is a sister project launched in 2007 aimed at reducing fires and crime in our regional schools. Printing fully financed by The Fire Service Research Training Trust.
Interventions:
KYBIB
KYSIB
Play it Safe
Tow Away Zones
Vehicle Removal
Code Red
Void Property Referrals
Theatre Productions
Rapid Deployment Cameras
Fire Investigation Team
Joint FI Training
WMFS Cycle Teams
Commercial Cannabis
WMATF Web Site
External Fund Raising
Secured by Design
SBD
Official Police Security Initiative
WMATF Success | 2003/4 | 2010/11 | Diff | % Difference
Deliberate dwelling fires | 950 | 404 | 546 | -57%
Deliberate non-dwelling fires | 543 | 223 | 320 | -59%
Deliberate vehicle fires | 4,078 | 983 | 3,095 | -76%
Deliberate rubbish fires | 8,324 | 4,098 | 4,226 | -51%
Deliberate derelict building fires | 848 | 326 | 522 | -61%
Deliberate school fires | 41 | 14 | 27 | -66%
Arson – New Initiatives Conference 2007
WMATF Partners:
West Midlands Police
West Midlands Fire Service
FM Global
Norton Integrated Systems
Envirograph
ADT Fire & Security
Montgomery Exl Marketing
AT&F Solutions
and many other agencies

Appendix J: Birmingham Domestic Violence Sanctuary Scheme Poster

History of the scheme
The current Sanctuary scheme was established as a city wide scheme in September 2008.
The Birmingham Sanctuary Scheme receives around 30 referrals for the service each quarter. These referrals come from a range of sources and include the Police, third sector agencies, neighbourhood offices, housing teams and BASBU. All those referred to the scheme are assessed by West Midlands Police Crime Prevention Design Advisers who survey properties and make recommendations on the security measures needed.
Drivers for change – Government Targets
- reduce homelessness
- reduce the amount of repeat domestic violence incidents
Secured by Design
SBD
Official Police Security Initiative
Referral Process
Award winning scheme
The Birmingham Sanctuary scheme has won two awards since its inception:- 'West Midlands Police Partnership Award 2009', and 'ACPO Secured by Design - Innovation Award 2010'.
The picture shows the Birmingham - based Crime Prevention Design Advisors receiving their award at the 2010 ACPO-SBD conference in Nottingham.
The scheme was 'Highly Commended' in the ACPO-SBD Innovation Awards.
Conference 2011
Scheme Objectives
The long-term objectives include:
- Reduce homelessness,
- Sustain communities by reducing the need for people experiencing domestic violence to move away from home,
- Increase safety and security for victims of domestic violence,
- Reduce repeat victimisation and risk of violence.
What is different about this scheme?
Emphasis is placed upon a complete package of support:
- Secured by Design standards,
- Access to civil orders and interventions.
Scheme partners:
Birmingham City Council
West Midlands Police
Birmingham & Solihull Women's Aid (BSWAID)
West Midlands Arson Task Force
WM Fire Service
Theam Security
and other referral agencies
There is an agreed list of approved products and interventions that are fitted. Birmingham City Council, West Midlands Police and Theam security regularly discuss and review other suitable security measures.
Data/Results
2008-09 (Starting September 2008) 25 properties had Sanctuary work completed.
2009-10 - 74 properties had work completed; 17 included a full 'Sanctuary Room'.
2010-11 - 73 properties had work completed; 9 included a full 'Sanctuary Room'.
2011 – (up to end of October) 45 properties; including one full 'Sanctuary room'.
Since the start of the city-wide project in September 2008, the scheme has improved the lives of 217 victims/potential victims by providing enhanced security measures.
Press and publicity - Radio and TV
In March 2011 the partnership organised a conference to promote the scheme. The event was part-funded by the 'SBD Innovation Award' and the International Police Association. Police officers, PCSOs and other interested partner agencies attended from around the country to learn about the scheme and share good practice.
The conference received a lot of media attention from local radio stations and newspapers, plus a five-minute slot on BBC TV.
Birmingham City Council
THEAM SECURITY

Appendix K: Redevelopment of Warwickshire Country Cricket Club Poster

Redevelopment of Warwickshire County Cricket Club
History of the site - Warwickshire started playing at Edgbaston, under lease from Lord Calthorpe, in 1886. The first match was against a wandering MCC XI. The ground was given Test status in 1902 and hosted England v Australia in May of that year. It remains one of the premier test grounds.
Secured by Design
SBD
Official Police Security Initiative
History of the Planning Application
The force domestic extremism assets (including Counter Terrorist Security Advisors) engage, contribute to and benefit from involvement with regionally based and national reduction and prevention partners, including protective security arrangements.
The force can provide evidence of joint working to promote preventative measures with the other agencies and external partners.
The new stand shall not be brought into use until an amended hard landscaping scheme has been submitted to and approved in writing by the Local Planning Authority. The scheme shall address hostile vehicle mitigation measures, and the scheme shall be provided and maintained as approved during the life of the development, unless otherwise approved in writing by the Local Planning Authority.
There are styles of vehicle borne threat that the CPNI advise sites to consider protecting themselves against.
(with or without warning)
Phase 2 Residential Development (225,000 sq Feet Of Space). Application submitted (Summer 2011). CPDA and CTSA currently advising the site and planners.
Future development planned for the site (during next decade). To include hotel (11 storeys), commercial office space (150,000 sq feet), retail (45,000 sq feet), and night time economy /leisure units (11,000 sq feet).
CTU
Birmingham City Council
Warwickshire County Cricket Club

References

Ackerman, J., & Rossmo, K. (2015). How Far to travel? A Multilevel Analysis of the Residence-to-Crime Distance. *Journal of Quantitative Criminology, 31*, 237–262.

Adamson, S. (2005). Burglary Reduction in Action: The Hartlepool Experience. *Crime Prevention and Community Safety: An International Journal, 7*(2).

Aldridge, J. (1994). *Operational Requirement Manual for CCTV.* Police Scientific Development Branch.

Aldridge, J. (2000). *Digital Imaging in the Criminal Justice System—Policing Issues.* Police Scientific Development Branch.

Al-Kahtani, S. (1996). *Travel to Crime Distances of Burglars in Saudi Arabia.* University of Manchester.

Alston, J. D. (1994). *The Serial Rapist's Spatial Pattern of Target Selection* (Unpublished thesis). Burnaby, B.C: Simon Fraser University.

Anderson, J. M., Bluthenthal, R., & Scott Ashwood, J. (2013). Reducing Crime by Shaping the Built Environment with Zoning: An Empirical Study of Los Angeles. *University of Pennsylvania Law Review, 161*, 699–756.

Angel, S. (1968). *Discouraging Crime Through City Planning* (Working paper no. 75). University of California, Berkeley: Institute of Urban & Regional Development.

Applied Criminology Centre, University of Huddersfield. (2011). *Residential Design and Crime Project.* Commissioned by ACPO SBD, Home Office and Campaign for Architecture in the Built Environment (CABE). Huddersfield University: Applied Criminology Centre.

Armitage, R. (1999). *An Evaluation of Secured by Design Housing Schemes Throughout the West Yorkshire Area.* The Applied Criminology Group, The University of Huddersfield.

Armitage, R. (2002). *To CCTV or Not to CCTV? Community Safety Practice Briefing.* NACRO.

Armitage, R. (2005). *Secured by Design—An Investigation of Its History, Development and Future Role in Crime Prevention* (Unpublished PhD Thesis). University of Huddersfield, Huddersfield.

Armitage, R. (2006). Predicting and Preventing: Developing a Risk Assessment Mechanism for Residential Housing. *Crime Prevention and Community Safety: An International Journal, 8*(3), 137–149.

M. Stokes, *Designing Out Crime*, Crime Prevention and Security Management,
https://doi.org/10.1007/978-3-031-99840-9

Armitage, R. (2007). Sustainability Versus Safety: Confusion, Conflict and Contradiction in Designing Out Crime. In G. Farrell, K. Bowers, S. Johnson, & M. Townsley (Eds.), *Imagination for Crime Prevention: Essays in Honour of Ken Pease. Crime Prevention Studies* (Vol. 21, pp. 81–110). Willan.

Armitage, R. (2013). *Crime Prevention Through Housing Design: Policy and Practice.* Palgrave Macmillan.

Armitage, R. (2016). Crime Prevention Through Environmental Design. In R. Wortley, & M. Townsley (Eds.), *Environmental Criminology and Crime Analysis.* Routledge.

Armitage, R. (2017). *Burglars' Take on Crime Prevention Through Environmental Design (CPTED): Reconsidering the Relevance from an Offender Perspective.* Macmillan.

Armitage, R. (2018). *Urban Design and Crime—Pro-Social Design Would Reduce Crime.* Available online: https://www.hud.ac.uk/news/2018/july/pro-social-design-reduces-crime

Armitage, R., & Gamman, L. (2009). *Sustainability via Security: A New Look. Built Environment, 35*(3), 297–301. Available at http://eprints.hud.ac.uk/5690

Armitage, R., & Monchuk, L. (2009). *Re-evaluating Secured by Design (SBD) Housing in West Yorkshire.* University of Huddersfield.

Armitage, R., Smyth, G., & Pease, K. (1999). Burnley CCTV Evaluation. In K. Painter & N. Tilley, (Eds.), *Crime Prevention Studies* (Vol. 10). Criminal Justice Press.

Armitage, R., Monchuk, L., & Rogerson, M. (2010). It Looks Good, But What Is It Like to Live There? Assessing the Impact of Award Winning Design on Crime. *Special Volume of European Journal on Criminal Policy and Research, 17*(1), 29–54.

Armitage, R., & Ekblom, P. (2020). *Rebuilding CPTED: Strengthening the Links with Crime Science.* Routledge.

Association of British Insurers. (2006). *Securing the Nation.* Association of British Insurers.

Atkins, S., Hussain, S., & Storey, A. (1991). *The Influence of Street Lighting on Crime and Fear of Crime* (Crime Prevention Unit Paper No. 28). Home Office.

Atlas, R. (1999, September 20–22). *The Alchemy of CPTED: Less Magic, More Science!* Paper presented at the 1999 International CPTED Conference, Mississauga, Canada.

Atlas, R. (2008). *21st Century Security and CPTED.* Taylor & Francis Group.

Avery, J. and Avery, P. (1982). Scandinavian and Dutch Lessons in Childhood Road Traffic Accident Prevention. *British Medical Journal (Clinical Research ed.), 285*(6342).

Balantyne, S., Pease, K., & McLaren, V. (Eds.). (2000). *Secure Foundations: Key Issues in Crime Reduction and Community Safety.* Institute for Public Policy Research.

Barclay, G. C., & Tavares, C. (1999). *Digest 4: Information on the Criminal Justice System in England and Wales.* Home Office.

Barr, R., & Pease, K. (1990). Crime Placement, Displacement and Deflection. In M. Tonry & N. Morris (Eds.), *Crime and Justice* (Vol. 12, pp. 28–37). Chicago University Press.

BBC Radio 4. (2008). *Gardeners' Question Time (GQT)* Broadcast on 21 September 2008. BBC.

Beavon, D. J. (1984). *Crime and the Environmental Opportunity Structure: The Influence of Street Networks on the Patterning of Property Offences.* Masters thesis, Simon Fraser University. Burnaby, British Columbia.

Beavon, D. J. K., et al. (1994). The Influence of Street Networks on the Patterning of Property Offences. In R. V. Clarke (Ed.), *Crime Prevention Studies* (Vol. 2, pp. 115–148). Criminal Justice Press.

Beckford, C. (1994). *The Alleygater's Guide to Gating Alleys*. Metropolitan Police Service 2 Area CPDA Group.

Beckford, C., Hankins, M., & Stagg, C. (1996). *Under Lock and Quay. Reducing Criminal Opportunities by Design*. Metropolitan Police Service 2 Area CPDA Group.

Bennewith, O., Nowers, M., & Gunnell, D. (2007). Effect of Barriers on the Clifton Suspension Bridge, England, on Local Patterns of Suicide: Implications for Prevention. *British Journal of Psychiatry, 2007*(190), 266–267.

Ben-Joseph, E. (1995). Changing the Residential Street Scene: Adapting the Shared Street (Woonerf) Concept to the Suburban Environment. *Journal of the American Planning Association, 61*(4), 504–515.

Bennett, T., & Durie. (1999). *Preventing Residential Burglary in Cambridge: From Crime Audits to Targeted Strategies* (Police Research Series, Paper 108). Home Office Policing and Reducing Crime Unit.

Bennett, T., & Wright, R. (1984). *Burglars on Burglary*. Gower.

Bernasco, W., & Block, R. (2011). Robberies in Chicago: A Block-Level Analysis of the Influence of Crime Generators, Crime Attractors, and Offender Anchor Points. *Journal of Research in Crime Delinquency, 48*, 33–57.

Bernasco, W., & Luykx, F. (2003). Effects of Attractiveness, Opportunity and Accessibility to Burglars on Residential Burglary Rates of Urban Neighborhoods. *Criminology, 41*, 981–1001.

Berry, G., Briggs, P., Erol, R., & Van Staden, L. (2011). *The Effectiveness of Partnership Working in a Crime and Disorder Context: A Rapid Evidence Assessment* (Research Report 52). Home Office.

Bevis, C., & Nutter, J. B. (1977). *Changing Street Layouts to Reduce Residential Burglary*. Paper Presented at the Annual Meeting of the American Society of Criminology, Atlanta, Georgia.

Biddulph, M. (2001). *Home Zones: A Planning and Design Handbook*. Policy Press.

Birmingham City Council. (1993). *The Gun Quarter: Planning and Urban Design Framework*. Llewellyn-Davies.

Birmingham City Council, Department of Planning and Architecture. (1996). *Shopfronts Design Guide*. Birmingham City Council.

Bowers, K. (2014). Risky Facilities: Crime Radiators or Crime Absorbers? A Comparison of Internal and External Levels of Theft. *Journal of Quantitative Criminology, 30*, 389–414.

Bowers, K. J., & Johnson, S. D. (2003). Measuring the Geographical Displacement and Diffusion of Benefits Effects of Crime Prevention Activity. *Journal of Quantitative Criminology, 19*, 275.

Bowers, K., Johnson, S., & Hirschfield, A. (2004). Closing Off Opportunities for Crime: An Evaluation of Alley-Gating. *European Journal on Criminal Policy and Research, 10*, 285–308.

Braga, A. A., Hureau, D. M., & Papachristos, A. V. (2012). An Ex Post Facto Evaluation Framework for Place-Based Police Interventions. *Evaluation Review, 35*(6), 592–626.

Brand, S., & Price, R. (2000). *The Economic and Social Costs of Crime* (Home Office Research Study No. 217). Home Office.

Brantingham, P. J., & Brantingham, P. L. (1975). The Spatial Patterning of Burglary. *The Harvard Journal, 14*(2), 11–23.

Brantingham, P. J., & Brantingham, P. L. (1981). Notes on the Geometry of Crime. In P. J. Brantingham & P. L. Brantingham (Eds.), *Environmental Criminology*. Sage.

Brantingham, P., & Brantingham, P. (1993). Nodes, Paths and Edges: Considerations on the Complexity of Crime and the Physical Environment. *Journal of Environmental Psychology, 13*, 3–28.

Brantingham, P., & Brantingham, P. (1995). Criminality of Place: Crime Generators and Crime Attractors. *European Journal on Criminal Policy and Research 3*,3, special issue on Crime Environments and Situational Prevention, 5–21.

Brantingham, P., & Brantingham, P. (2008). Offender Mobility and Crime Pattern Formation from First Principles. In L. Liu & J. Eck (Eds.),*Artificial Crime Analysis Systems: Using Computer Simulations and Geographical Information Systems* (pp. 193–208). Idea Press.

Bridgestock, M. (2000). *Homezones—Good Practice in Britain and Henbury, Bristol Feasibility Study*. University of the West of England and Sustrans.

Brisbane City Council. (2004). *Graffiti Prevention Guidelines*. Brisbane City Council.

British Standards Institution (BSI). (1986). *BS 8220: Part 1: 1986 British Standard Guide for Security of Buildings Against Crime. Part 1. Dwellings*. BSI.

British Standards Institute. (2020). *BS 5489–1:2020—Lighting*. BSI.

Bromley, R. D. F., & Nelson, A. L. (2002). Alcohol-Related Crime and Disorder Across Urban Space and Time: Evidence from a British City. *Geoforum, 33*, 239–254.

Brown, B. (1995). *CCTV in Town Centres: Three Case Studies, Police Research Group, Crime Detection and Prevention Series, Paper 68*. Home Office.

Brown, J. (1999). *An Evaluation of the Secured by Design Initiative in Gwent, South Wales* (Unpublished MSc dissertation). Scarman Centre for the Study of Public Order. Leicester: University of Leicester.

Brown, L. A., Malecki, E. J., & Philliber, S. J. (1977). Awareness Space Characteristics in a Migration Context. *Environment and Behavior, 9*(3), 335–348.

Brown, B. B., & Altman, I. (1983). Territoriality, Defensible Space and Residential Burglary: An Environmental Analysis. *Journal of Environmental Psychology, 3*, 203–220.

Brown, B., & Bentley, D. (1993). Residential Burglars Judge Risk: The Role of Territoriality. *Journal of Environmental Psychology, 13*, 51–61.

Brown, R., & Scott, M. (2007). *Implementing Responses to Problems*. Problem-Oriented Guides for Police (Problem-Solving Tools Series No. 7). Centre for Problem-Oriented Policing.

Building Research Establishment. (2002). *Client Report: Trigger Factors and Tipping Points of Crime*. Building Research Establishment.

Building Research Establishment. (2008). *BREEAM BRE Environmental and Sustainability Standard, BREEAM, Education, BES 5051: Issue 1.0, 3.0, Assessor Manual*. http://www.designlighting.co.uk/pages/technical/breeam/docs/BREEAM

Building Research Establishment. (2014). *BREEAM UK New Construction: Non-domestic Buildings Technical Manual: BES 5052: Issue 4.1*. SD5076: 2.1 2014.

Bullock, K., Erol, R., & Tilley, N. (2006). *Problem-Oriented Policing and Partnerships: Implementing an evidence-based Approach to Crime Reduction*. Willan.

Burgess, E. W. (1925). The Growth of the City. In R. E. Park, E. W. Burgess, & R. D MacKenzie (Eds.), *The City*. University of Chicago Press.

Campaign for Architecture in the Built Environment (CABE). (2007). *Design and Access Statements: How to Write, Read and Use Them*. CABE.

Carstensen, N., & Frederiksen, K. (1997). *Situational Crime Prevention*. In L. Ravn (Ed.), *Kriminalistisk Arbog*. Criminal Procedure Institute.

Caulkins, J. (1992). *Thinking About Displacement in Drug Markets: Why Observing Change of Venue Isn't Enough*. Journal of Drug Issues.

Chenery, S., Holt, J., & Pease, K. (1997). *Biting Back II: Preventing Repeat Victimisation in Huddersfield* (Crime Detection and Prevention Paper 82). Home Office.

Children's Play Council. (1997). *Home Zones: Reclaiming Residential Streets*. National Children's Council.

Church of England General Synod. (1978). *Parochial Registers and Records Measure*. Church of England General Synod.

Clare, J., & Morgan, F. (2009). In J. D. Freilich, & G. R Newman, *Reducing Terrorism Through Situational Crime Prevention in Crime Prevention Studies* (Vol. 25). Lynne Rienner.

Clarke, R. V. (1983). Situation Crime Prevention: Its Theoretical Basis and Practical Scope. *Crime and Justice, 4*, 225–256.

Clarke, R. V. (1990). *Armed Robbery Interim Report 2: Target Hardening of Banks and Displacement of Robberies*. National Police Research Unit.

Clarke, R. V. (1995). Situational Crime Prevention. In Crime and Justice Vol. 19, *Building a Safer Society: Strategic Approaches to Crime Prevention* (pp. 91–150). University of Chicago Press.

Clarke, R. V. (1999). *Hot Products: Understanding, Anticipating and Reducing Demand for Stolen Goods* (Police Research Series 112). Home Office.

Clarke, R. V. (2004). *Closing Streets to Reduce Crime: Should You Go Down This Road? Problem-Oriented Guides for Police Response Guides Series No. 2*. US Department of Justice, Office of Community Oriented Policing Services.

Clarke, R. V. (2010). *Situational Crime Prevention Successful Case Studies*. Lynne Rienner.

Clarke, R. V., & Eck, J. (2003). *Become a Problem-Solving Crime Analyst in 55 Small Steps*. Jill Dando Institute, University College London. Available online at: www.jdi.ucl.ac.uk/publications/manual/crime_manual_content.php

Clarke, R. V., & Eck, J. (2007). *Understanding Risky Facilities*. Center for Problem-Oriented Policing. Available online at: http://www.popcenter.org/tools/PDFs/risky_facilities.pdf

Clarke, R. V., & Eck, J. E. (2009). *Crime Analysis for Problem Solvers in 60 Small Steps*. Center for Problem-Oriented Policing and U.S. Department of Justice, Office of Community Oriented Policing Services.

Clarke, R., & Mayhew, P. (Eds.). (1980). *Designing Out Crime*. Home Office.

Clarke, R., & Mayhew, P. (1988). The British Gas Suicide Story and Its Implications for Prevention. In M. Tonry & N. Morris (Eds.), *Crime and Justice: A Review of Research* (Vol. 10). Chicago University Press.

Clarke, R. V., & Weisburd, D. (1994). Diffusion of Crime Control Benefits: Observations on the Reverse of Displacement. In R. Clarke (Ed.), *Crime Prevention Studies* (Vol. 2). Criminal Justice Press.

Cockfield, C. (1998, revised 2003). *Public Conveniences Problem Reduction Guide*. Staffordshire Police.

Cohen, L. E., & Felson, M. (1979). Social Change and Crime Rate Trends: A Routine Activity Approach. *American Sociological Review, 44*, 488–608.

College of Policing. (2015a). *National Policing Crime Prebention Strategy*. College of Policing.

Coleman, A. (1985). *Utopia on Trial: Vision and Reality in Planned Housing*. Hilary Shipman.

Cornish, D. (1994). The Procedural Analysis of Offending and its Relevance for Situational Prevention. In R. V. Clarke (Eds.), *Crime Prevention Studies* (Vol. 3). Lynne Rienner.

Cornish, D., & Clarke, R. (1986). *The Reasoning Criminal*. Springer-Verlag.

Cornish, D., & Clarke, R. (2003). Opportunities, Precipitators and Criminal Decisions: A Reply to Wortley's Critique of Situational Crime Prevention. In *Crime Prevention Studies* (Vol. 16). Criminal Justice Press.

Cornish, D., & Clarke, R. (2008). The Rational Choice Perspective. In R. Wortley & L. Mazarolle (Eds.), *Environmental Criminology and Crime Analysis*. Willan Publishing.

Cornish, D. B., & Clarke, R. V. (2011). *Understand Crime Displacement: An Application of Rational Choice Theory*. Routledge.

College of Policing. (2015b). *National Policing Crime Prevention Strategy*. College of Policing.

Cox, K. (2008). The Application of Crime Science to the Prevention of Elder Abuse. *British Journal of Nursing, 17*(13).

Cozens, P. (2014). *Think Crime! Using Evidence, Theory and Crime Prevention Through Environmental Design (CPTED) for Planning Safer Cities* (1st ed.). Praxis Education.

Cozens, P. M. (2016). *Think Crime! Using Evidence, Theory and Crime Prevention Through Environmental Design (CPTED) for Planning Safer Cities*. Praxis Education.

Cozens, P. (2016b). *Think Crime: Using Evidence, Theory and Crime Prevention Through Environmental Design (CPTED) for Planning Safer Cities* (2nd ed.). Praxis Education.

Cozens, P., & Babb, C. (2018). *What's Up on Beaufort Street? Exploring Crime, Traffic Risks and Crime Prevention Through Environmental Design (CPTED)*. Available online at: https://inglewoodonbeaufort.com/2018/12/03/inglewood-cpted-report

Cozens, P., Hillier, D., & Prescott, G. (1999). The Sustainable and the Criminogenic. *The Case for New-Build Housing Projects in Britain. Property Management, 17*, 252–261.

Cozens, P., Hillier, D., & Prescott, G. (2001). Crime and the Design of Residential Property. Exploring the Perceptions of Planning Professionals, Burglars and other Users. *Property Management, 19*, 222–248.

Cozens, P., & Hillier, D. (2012). Revisiting Jane Jacobs 'Eyes on the Street' for the 21st Century: Evidence from Environmental Criminology. In S. Hirt & D. Zahm (Eds.), *The Urban Wisdom of Jane Jacobs* (pp. 196–214). Routledge.

Cozens, P., & Love, T. (2015). A Review and Current Status of Crime Prevention Through Environmental Design (CPTED). *Journal of Planning Literature*, 1–20.

Cozens, P., & Love, T. (2017). Crime Prevention Through Environmental Design (CPTED) Introducing and Learning from the "Dark Side". *Oxford Research Encyclopaedia of Criminology and Criminal Justice*. Available online at: 412 http://criminology.oxfordre.com/view/10.1093/acrefore/9780190264079.001.0001/acrefore9780190264079-e-2

Cozens, P., Love, T., & Davern, B. (2019). Geographical Juxtaposition: A New Direction in CPTED. *Social Sciences, 8*, 252. Available online at: https://www.semanticscholar.org/paper

Cozens, P., & Van der Linde, T. (2015). Perceptions of Crime Prevention Through Environmental Design (CPTED) at Australian Railway Stations. *Journal of Public Transportation, 18*, 73–92.

Cromwell, P., Olson, J., & Avary, D. (1991). *Breaking and Entering: An Ethnographic Analysis of Burglary*. Sage.

Crouch, S., Shaftoe, H., & Fleming, R. (1999). *Design for Secure Residential Environments*. Longman.

Crowe, T. (1991). *Crime Prevention Through Environmental Design: Applications of Architectural Design and Space Management Concepts*. Butterworth-Heinemann.

Crowe, T. (2000). *Crime Prevention Through Environmental Design*. Butterworth-Heinemann.

Daly, H. (2002, April 30). *Sustainable Development: Definitions, Principles, Policies*. Invited Address World Bank.

Data Protection Registrar. (1998). *Data Protection Guidance on CCTV*. Office of the Data Protection Registrar.

Data Protection Registrar. (2000). *CCTV Data Protection—Codes of Practice*. Office of the Data Protection Registrar.

Davey, C.L., Hodge, M., & Wootton, A. B. (2005). Designing Against Terrorism. *Engineering Designer, 30*(1) (January/February), 10–13.

Davey, C. L., Wootton, A. B., & Marselle, M. (2012). Engaging Young People in Design Against Crime Design Research. *Swedish Design Research Journal, 1*(12), 29–38.

Deakin, J., Medina-Ariza, J. J., Spencer, J., & Ainsworth, P. B. (2003). *Both Sides of the Coin*. Greater Manchester Police.

de Andrade, D., Homel, R., & Mazerolle, L. (2016). Boozy Nights and Violent Fights: Perceptions of Environmental Cues to Violence and Crime in Licensed Venues. *Journal of Interpersonal Violence*, 1–22. https://doi.org/10.1177/0886260516657910.

Department for Communities and Local Government. (2004). *Safer Places: The Planning System and Crime Prevention*. DCLG.

Department for Communities and Local Government. (2005). *Planning Policy Statement 1 (PPS1): Delivering Sustainable Development*. DCLG.

Department for Communities and Local Government. (2006). *Code for Sustainable Homes*. DCLG.

Department for Communities and Local Government (DCLG). (2007). *Manual for Streets*. DCLG.

Department for Communities and Local Government (DCLG). (2012a). *National Planning Policy Framework (NPPF)*. DCLG.

Department for Communities and Local Government (DCLG). (2012b). *National Planning Practice Guidance (NPPG)*. DCLG.

Department for Communities and Local Government. (2014). *Housing Standards Review: Summary of Responses*. DCLG.

Department of the Environment. (1994). *Circular 5/94 Planning Out Crime*. DOE.

Department for Transport, Local Government and the Regions (DTLR). (2001). *Home Zones—Planning and Design, Traffic Advisory Leaflet 10/01*. DTLR.

Department for Transport, Local Government and the Regions (DTLR) and Campaign for Architecture in the Built Environment (CABE). (2001). *Better Places to Live: By Design: A Companion Guide to PPG3*. DTLR.

Department for Transport. (2012). *Secure Stations Scheme*. DfT.

Design Council. (2003). *Think Thief: A Designer's Guide to Designing Out Crime*. Design Council.

Ditton, J., & Short, E. (1998). Evaluating Scotland's First Town Centre CCTV Scheme: Airdrie's Crime Data and Airdrie's Offenders. In C. Norris & G. Armstrong (Eds.), *Surveillance, CCTV and Social Control—The Rise of Closed Circuit Televisual Surveillance in Britain*. Gower.

Dubourg, R., & Hamed, J. (2005). *The Economic and Social Costs of Crime against Individuals and Households 2003/4* (Home Office RDS Online Report 30/05).

Dudley Metropolitan Borough Council. (2002). *Design for Community Safety: Supplementary Planning Guidance*. Design and Development Division.

Eck, J. E. (1993). The Threat of Crime Displacement. *Criminal Justice Abstracts, 25*, 527–546.

Eck, J. E., & Spelman, W. (1987). *Solving Problems: Problem-Oriented Policing in Newport News*. Police Executive Research Forum.

Eck, J. W. (1997). Preventing Crime in Places. In L. W. Sherman, et al. (Eds.), *Preventing Crime: What works, What Doesn't, What's Promising?* National Institute of Justice.

Eck, J., & Weisburd, D. L. (2015). Crime Places in Crime Theory. *Crime and Place: Crime Prevention Studies*(Vol. 4, pp. 1–33). Criminal Justice Press.

Eck, J., Chainey, S., Cameron, J. G., Leitner, M., & Wilson, R. E. (2005). *Mapping Crime: Understanding Hot-Spots*. National Institute of Justice.

Ekblom, P. (1994). Proximal Circumstances: A Mechanism-Based Classification of Crime Prevention. *Crime Prevention Studies*(Vol. 2). Criminal Justice Press.

Ekblom, P. (1996). Towards a Discipline of Crime Prevention: A Systematic Approach to its Nature, Range and Concepts. In T. Bennett (Ed.), *Preventing Crime and Disorder: Targeting Strategies and Responsibilities*. Cropwood Series. Institute of Criminology.

Ekblom, P. (1997). Gearing up Against Crime: A Dynamic Framework to Help Designers Keep up with the Adaptive Criminal in a Changing World. *International Journal of Risk, Security and Crime Prevention, 214*, 249–265.

Ekblom, P. (2000). The Conjunction of Criminal Opportunity—A Tool for Clear, 'Joined-Up' Thinking and Community Safety and Crime Reduction. In S. Ballintyne, K. Pease, & V. McLaren (Eds.), *Secure Foundations: Key Issues in Crime Prevention, Crime Reduction and Community Safety*. Institute of Public Policy Research.

Ekblom, P. (2001). How to Think about Crime and its Reduction—Rigorously, Systematically and Inclusively. In P. Ekblom, C. Lehman, & K. Pease (Eds.), *Reducing Offending II: An Updated Assessment of Research Evidence on Ways of Dealing with Offending Behaviour*. Home Office Research Study. Home Office.

Ekblom, P. (2002). Towards a European Knowledge Base' and 'The Five Is: Experimental Framework for a Knowledge Base for Crime Prevention Projects. In *European Crime Prevention Conference 2002* (Vol.1, pp. 62–97). Copenhagen: Danish Crime Prevention Council. https://www.designagainstcrime.com/web/towards_european_k_pease.pdf

Ekblom, P. (2003a, April 11). *Delivering Joined-Up Policy Solutions at Local Level: Developing the Good Practice Guidance for Planning Out Crime*. Paper presented at a conference on 'Designing Out Crime'. London, 11 April 2003.

Ekblom, P. (2003b). Organised Crime and the Conjunction of Criminal Opportunity Framework. In A. Edwards & P. Gill (Eds.), *Transnational Organised Crime: Perspectives on Global Security*. Routledge.

Ekblom, P. (2011a). *Crime Prevention, Security and Community Safety Using the 5Is Framework*. Palgrave Macmillan.

Ekblom, P. (2011b). Deconstructing CPTED… and Reconstructing It for Practice. Knowledge management and research. *European Journal on Criminal Policy and Research, 17*(1), 7–28.

Ekblom, P. (2012). *Design Against Crime: Crime Proofing Everyday Products*. Lynne-Reiner.

Ekblom, P., & Tilley, N. (2000). Going Equipped: Criminology, Situational Crime Prevention and the Resourceful Offender. *British Journal of Criminology, 40*(3), 375–398.

Eras, V. J. M. (1957). *Locks & Keys Throughout the Ages*. Bailey Bros and Swinfen Ltd.

Essex Police. (1995). *Bugularis Disembowlis: Your Garden Helping You to Bite Back Against Burglars*. Essex Police.

Farrell, G., Tilley, N., Tseloni, A., & Mailley, J. (2011). The Crime Drop and the Security Hypothesis. *Journal of Research in Crime and Delinquency, 48*, 147–175.

Farrell, G., & Tilley, N. (2022). Elegant Security: Concept, Evidence and Implications. *European Journal of Criminology, 19*(5), 932–953.

Farrington, D., & Painter, K. (2002, November). *How to Evaluate the Inpact of CCTV on Crime*. Paper given at Conference on Evaluating the Real Evidence on CCTV in Leicester.

Farrington, D., & Welsh, B. (2002). *Effects of Improved Street Lighting on Crime: A Systematic Review* (Home Office Research Study No. 251). Home Office.

Farrington, D. P., & Welsh, B. C. (2010). *The Future of Crime Prevention: Developmental and Situational Strategies*. National Institute of Justice.

Felson, M. (1986). Routine Activities, Social Controls, Rational Decisions, and Criminal Outcomes. In D. Cornish & R. V. G. Clarke (Eds.), *The Reasoning Criminal*. Springer-Verlag.

Felson, M. (2002). *Crime and Everyday Life* (3rd ed.). Sage.

Felson, M., & Clarke, R. (1998). *Opportunity Makes the Thief: Practical Theory for Crime Prevention, Police Research Series Paper 98*. Home Office.

Fisher, T., & Gamman, L. (2018). *Tricky Design: The Ethics of Things*. Bloomsbury.

Fleming, R., & Burrows, J. (1986) *The Case for Lighting as a Means of Preventing Crime* (Home Office Research Bulletin, 22) (pp. 14–17). Home Office.

Flint, A. (2009). *Wrestling with Moses: How Jane Jacobs Took on New York's Master Builder and Transformed the American City*. Random House.

Forrester, D., Chatterton, M., & Pease, K. (1988). *The Kirkholt Burglary Prevention Project*, Rochdale (Crime Prevention Unit: Paper 13). Home Office.

Forrester, D., Frenz, S., & Pease, K. (1990). *The Kirkholt Burglary Prevention Project: Phase II* (Crime Prevention Unit Paper 23). Home Office.

Frace, C. (2020). *Importance of Safety in Public Parks for Crime Prevention*. California Polytechnic State University.

Frank, R., Dabbaghian, V., Reid, A. A., Singh, S. K., Cinnamon, J., & Brantingham, P. (2011). Power of Criminal Attractors: Modelling the Pull of Activity Nodes. *Journal of Artificial Societies and Social Simulation, 14*(1), 6.

Fry, P. (2008, April). Paper presented at *CCTV User Group Conference*, Manchester.

Gabor, T., & Gottheil, E. (1984). Offender Characteristics and Spatial Mobility: An Empirical Study and Some Policy Implications. *Canadian Journal of Criminology, 26*, 267–281.

Gamman, L. (2013). *Shoplifting Scams and Designing Out Crime*. Loss Prevention Magazine.

Gamman, L., & Thorpe, A. (2012). *The Korean Experience—3rd Generation CPTED?* Lecture given to the Designing Out Crime Association at Central St Martins, London, on 21 March 2013.

Gamman, L., & Thorpe, A. (2014). Design for Democratic Crime Prevention. *Urban Pamphleteer: Design and Trust, 3*, 6–9.

Garside, R. (2004). *Crime, Persistent Offenders and the Justice Gap*. Crime and Society Foundation.

Gawande, A. (2009). *The Checklist Manifesto: How to Get Things Right*. Picador.

Gerrard, G. (2007). *United Kingdom National CCTV Strategy*. Accessible at https://www.viseum.co.uk/downloads/uk-cctv-government-strategy-recommendations.pdf

Gill, M. (2000). *Commercial Robbery*. Blackstone.

Gill, M. (2001). The Craft of Robbers of Cash-in-Transit Vans: Crime Facilitators and the Entrepreneurial Approach. *Internation/al Journal of the Sociology of Law, 29*(3), 277–291.

Gill, M., Spriggs, A., Allen, J., Argomaniz, J., Bryan, J., Hemming, M., Jessiman, P., Kara, D., Kilworth, J., Little, R., Swain, D., & Waples, S. (2005). *The Impact of CCTV: Fourteen Case Studies*. Home Office Online Report. Home Office www.homeoffice.gov.uk/rds/pdfs05/rdsolrl505.pdf

Gladwell, M. (2000). *The Tipping Point, How Little Things Can Make a Big Difference*. Little Brown.

Glensor, R. (1996). Community Problem Solving in the USA: A New Synergy. In *Focus on Police Research and Development, No. 7*(pp. 14–15). Home Office Police Research Group.

Glowacki, R., Lojek, K., Ostrowska, E., Tyburska, A., & Urban, A. (2010). *CPTED as an Element of Local Safety*. Polish Department of Internal Safety and Administration: Szczytno.

Goldstein. (1979, April). Improving Policing: A Problem-Oriented Approach. *Crime and Delinquency* (pp. 235–58).

Goldstein, H. (1990). *Problem-Oriented Policing*. McGraw-Hill.

Goldstein, H. (1997). The Pattern of Emerging Tactics for Shifting Ownership of Prevention Strategies in the Wave of Change in Policing: Their Implications for both Environmental Criminology and the Police. Paper presented at the sixth *Annual Seminar on Environmental Criminology and Crime Analysis*. National Police Academy, Oslo.

Green, M. (1999). *The Appropriate and Effective Use of Security Technologies in U.S. Schools*. National Institute of Justice.

Greenburg, S., & Rohe, W. M. (1983). External Solutions to Neighborhood-Based Problems: A Critical Review of the Relationships Between Fear of Crime, Bystander Surveillance, and Changes in the Crime Rate. *Victimology, 5*, 133–151.

Groff, E. R. (2005). *The Geography of Juvenile Crime Place Trajectories*(Masters Thesis). University of Maryland.

Groff, E. R., & La Vigne, N. G. (2001a). Forecasting the Future of Predictive Crime Mapping. *Crime Prevention Studies, 13*, 29–57.

Groff, E. R., & La Vigne, N. G. (2001b). Mapping an Opportunity Surface of Residential Burglary. *Journal of Research in Crime and Delinquency, 38*(3), 257–278.

Groff, E., & McCord, E. S. (2011). The Role of Neighborhood Parks as Crime Generators. *Security Journal*1–24.

Guerette, R. T., & Bowers, K. (2009). Assessing the Extent of Crime Displacement and Diffusion of Benefits: A Review of Situational Crime Prevention Evaluations. *Criminology, 47*(4).

Hamilton-Baillie, B. (2008). Shared Space: Reconciling People, Places and Traffic. *Built Environment, 34*(2), 161–181.

Harding, D. (2004, July 29). The Graffiti Gaffe—Council Given £2m Urban Art Grant Rips Down Work. In *Metro National Newspaper*. Metro.

Hawton, K., Townsend, E., Deeks, J., Appleby, L., Gunnell, D., Bennewith, O., & Cooper, J. (2001). Effects of Legislation Restricting Pack Sizes of Paracetamol and Salicylate on Self Poisoning in the United Kingdom: Before and After Study. *British Medical Journal, 322*, 1203.

Hearnden, I., & McGill, C. (2004). *Decision-Making by House Burglars: Offenders' Perspectives* (Home Office Research Findings 249). Home Office.

Hertfordshire Constabulary Crime Prevention Design Service and the British Toilet Association. (2010). *Publicly Available Toilets: Problem Reduction Guide.* Welwyn Garden City: Hertfordshire Constabulary.

Hesseling, R. (1994) Displacement: A Review of the Empirical Literature. In *Crime Prevention Studies* (Vol. 3). Criminal Justice Press.

Hesseling, R., & Aron, U. (1995). *Car Burglaries: Reduction or Displacement? The Results of a Rotterdam Campaign to Curb Thefts from Cars.* Gouda Quint Bv.

Hill, I., & Pease, K. (2000). The 'Wicked' Issues: Displacement and Sustainability. In S. Ballantyne, et al. (Eds.), *Secure Foundations: Key Issues in Crime Prevention, Crime Reduction and Community Safety.* Institute for Public Policy Research.

Hilborn, J. (2009). *Dealing With Crime and Disorder in Urban Parks. Problem-Oriented Guides for Police Response Guides Series No. 9.* US Department of Justice, Office of Community Oriented Policing Services.

Hillier, B. (1999). *Space is the Machine: A Configurational Theory of Architecture.* Cambridge University Press.

Hillier, B. (2004). Can Streets be Made Safer? *Urban Design International, 9*, 31–45.

Hillier, B., & Hanson, J. (1984). *The Social Logic of Space.* Cambridge University Press.

Hillier, B., & Shu, S. (2000). Crime and Urban Layout: The Need for Evidence. In S. Ballintyne, K. Pease, & V. McLaren (Eds.), *Secure Foundations: Key Issues in Crime Prevention, Crime Reduction and Community Safety.* Institute of Public Policy Research.

Hillier, B., & Sahbaz, O. (2005). High Resolution Analysis of Crime Patterns in Urban Street Networks: An Initial Statistical Sketch from an Ongoing Study of a London Borough. In *Proceedings of the Fifth International Space Syntax Symposium*, held in Delft, The Netherlands.

Holt, T. (2000). *Baits His Hook and Takes Your Cash.* Greater Manchester Police: Unpublished Report.

Holt, T., & Spencer, J. (2005). *A Little Yellow Box: The Targeting of Automatic Teller Machines as a Strategy in Reducing Street Robbery.* Perpetuity Press.

Home Office. (2001). *Crime Prevention Toolkits.* Home Office.

Home Office. (2016). *Modern Crime Prevention Strategy.* Home Office.

Home Office Crime Prevention Centre. (1987). *Police Architectural Liaison Manual of Guidance.* HOCPC.

Home Office Crime Prevention Centre. (1994). *Police Architectural Liaison Manual of Guidance.* HOCPC.

Home Office Crime Reduction College. (2001). *Home Security: An Introduction to Domestic Surveying.* HOCRC

Home Office Scientific Development Branch. (2009). *CCTV Operational Requirements Manual.* Home Office Scientific Development Branch.

Home Office Standing Conference on Crime Prevention. (1991). *Safer Communities: The Local Delivery of Crime Prevention Through the Partnership Approach.* Home Office.

Home Office and Campaign for Architecture in the Built Environment (CABE). (2011). *Residential Design and Crime Project*. Home Office.

Houghton, G. (1992). *Car Crime in England and Wales: The Home Office Car Theft Index; Crime Prevention Unit Paper 33*. Home Office.

House of Commons Library. (2023). *Statistics on access to cash, bank branches and ATMs Lorna Booth for House of Commons Library*. Accessible at: https://researchbriefings.files.parliament.uk/documents/CBP-8570/CBP-8570.pdf

IBISWorld. (2023). *Security Systems Services Industry Research Report* Accessed on 28 September 2023 at my.ibisworld.com/uk/en/industry/n80.200/products-and-markets

Information Commissioner's Office. (ICO–2008). *CCTV Code of Practice*. Office of the Information Commissioner.

Institution of Civil Engineers. (1998). *Design Bulletin 32, Layout of Residential Roads and Footpaths*. HMSO.

Intergovernmental Panel on Climate Change. (2008). *Climate Change: Synthesis Report. Summary for Policymakers*. IPCC.

International Energy Agency. (2007). *CO2 Emissions from Fuel Combustion 1971–2005*. IEA.

Jacobs, J. (1961). *The Death and Life of Great American Cities*. Random House.

Jeffery, C. R. (1971). *Crime Prevention Through Environmental Design*. Sage.

Jeffery, C., & Zahm, D. (1993). Crime Prevention Through Environmental Design, Opportunity Theory and Rational Choice Models. In R.V. Clarke & M. Felson (Eds.), *Routine Activity Theory and Rational Choice. Advances in Criminological Theory*(Vol. 5). Transaction Publishers.

Jencks, C. (1984). *The Language of Post-Modern Architecture*. Rizzoli.

Jenkins, S. (1999). *England's Thousand Best Churches*. Penguin Group.

Johnson, S. D., & Bowers, K. J. (2010). Permeability and Burglary Risk: Are Culs-de-Sac Safer? *Journal of Quantitative Criminology, 26*(1), 89–111.

Johnson, S.D., Bowers, K., Gamman, L., Mamerow, L., & Warne, A. (2010). *Theft of Customers' Personal Property in Cafes and Bars* (Problem-Oriented Guides for Police, Problem-Specific Guides Series No. 60). Washington DC: U.S Department of Justice, Office of Community Oriented Policing Services.

Johnson, S. D., Guerette, R. T., & Bowers, K. (2012). Crime Displacement and Diffusion of Benefits. In B. C. Welsh & D. P. Farrington (Eds.), *The Oxford Handbook of Crime Prevention* (pp. 337–353). Oxford University Press.

Johnson, S. D., Guerette, R. T., & Bowers, K. (2014). Crime Displacement: What We Know, What We Don't Know, and What It Means for Crime Reduction. *Journal of Experimental Criminology, 10*(4), 549–571.

Jones, A., Valero-Silva, N., & Lucas, D. (2016). *The Effects of 'Secure Warm Modern' Homes in Nottingham: Decent Homes Impact Study*. Nottingham City Homes.

Kaplankiran, T., Mailley, J., Whitehead, S., & Farrell, G. (2008). Mobile Phone Reprogramming: Its Extent and Prevention. *Crime Prevention and Community Safety, 10*, 271–279.

Kelling, G., & Wilson, J. (1982). Broken Windows: The Police and Neighbourhood Safety. *The Atlantic Monthly* (pp. 29–38). Washington DC.

Kershaw, C., Nicholas, S., & Walker, A. (2008). Crime in England and Wales 2008/8. *Home Office Statistical Bulletin 07/08.*

Key Note Market Intelligence. (2014). *Market Report 2014 Closed-Circuit Television.* Key Note Ltd.

Kirby, S. (2018, July 24). Chair's Opening Address Given to UK Wide Vehicle Crime Problem Solving Event hosted by West Midlands Police in Birmingham on 24 July 2018.

Kirby, K. & Reed, D. (2004) *Developing a Problem-oriented Organisation.* London: Association of Chief Police Officers.

Kitchen, E. (2007). Speech given at the *National Training Event for Architectural Liaison Officers and Crime Prevention Design Advisers* held at Hinckley, Leicestershire May 2007.

Knights, B., Pascoe, T., & Henchley, A. (2002). *Sustainability and Crime: Managing and Recognising the Drivers of Crime and Security.* Building Research Establishment.

Kuo, F., & Sullivan, W. (2001). Environment and Crime in the City. Does Vegetation Reduce Crime? *Environment and Behaviour, 33*(3), 343–367.

Lancaster, D. (2010). *Carbon Reduction and Cost Efficiency Saving Model.* Secured by Design.

Lancaster, D. (2011). *Carbon Cost of Crime Calculator*. Secured by Design.

Laycock, G. (1985). *Property Marking: A Deterrent to Domestic Burglary?* Crime Prevention Unit Paper 3. Home Office.

Laycock, G. (1989). *An Evaluation of Domestic Security Surveys.* Crime Prevention Unit Paper 18. Home Office.

Laycock, G. (1997). Operation Identification, or the Power of Publicity? In R. Clarke (Ed.), *Situational Crime Prevention: Successful Case Studies* (2nd ed.). Harrow and Heston.

Laycock, G. (2004). The UK Car Theft Index: An Example of Government Leverage. In *Crime Prevention Studies* (Vol. 17, pp.25–44). Verlag.

Le Corbusier. (1950). *Le Modulor*. Edition de l'architecture d'auhourd'hui.

Leicestershire Constabulary and Leicester City Council. (1995). *Crime Prevention by Planning and Design.* Leicester City Council Information Services Division.

Lindsey, B., & McGillis, D. (1988). Citywide Community Crime Prevention: An Assessment of the Seattle Program. In D. P. Rosenbaum (Ed.), *Community Crime Prevention: Does It Work?* Sage.

Llewelyn-Davies. (1993). *The Gun Quarter: Planning 7 Urban Design Framework.* Birmingham City Council.

Lonely Planet. (2016). *New York City.* Lonely Planet.

Lottman, H. (1976). *How Cities are Saved*. Universe.

Lottman, H. (1991). *Colette: A Life*. Secker.

Lynch, K. (1960). *The Image of the City*. MIT Press.

Mahalingham, V. (1996). Sutton Town Centre Public Perception Survey. In M. Bulos & D. Grant, (Eds.), *Towards a Safer Sutton? CCTV One Year On.* London Borough of Sutton.

Macdonald, J. E., & Gifford, R. (1989). Territorial Cues and Defensible Space Theory: The Burglars Point of View. *Journal of Environmental Psychology, 9*, 193–205.

Martin, G. H., & Williams, A. (Eds.). (2003). *Domesday Book: A Complete Translation.* Penguin Classics.

McKay, D., & Cox, A. (1979). *The Politics of Urban Change*. Croom Helm.

Mativat, F., & Tremblay, P. (1997). Counterfeiting Credit Cards: Displacement Effects, Suitable Offenders and Crime Wave Patterns. *British Journal of Criminology, 37*, 165–183.

Mawby, R. (2016). Policing Policies in Cities Dependent on Tourism: Developing an Evidence-Based Approach. *International Journal of Police Science and Management, 18*(3), 195–205.

Mayhew, P. (1979). Defensible Space: The Current Status of a Crime Prevention Theory, article published in *The Howard Journal of Criminal Justice*(Vol. XVIII, pp. 150–159). Howard League for Penal Reform.

Mayhew, P., Clarke, R., Sturman, A., & Hough, J. (1976). *Crime as Opportunity*(Home Office Research Study 34). HMSO.

Mayhew, P., & Clarke, R. (1976). *Crime as Opportunity* (Home Office Research Study No. 34). Home Office.

Mayne, R., & Rowan, C. (1956). Nine Principles of Policing–1829. In *A New Study of Police History by Charles Reith*. Oliver and Boyd.

Meades, J. (1999). We're on the Road to Nowhere. Article in *The Times, Weekend*(pp.10–11), dated Saturday September 11, 1999. News International.

Metropolitan Police. (1996). *The Garden Strikes Back—A Guide to Plants that Burglars Hate*. Metropolitan Police Service.

Metropolitan Police. (1999). *Operation Bumblebee—A Secure Garden*. Metropolitan Police Service.

Ministry for Housing, Communities and Local Government (2024). *National Planning Policy Framework*. MHCLG.

MIT SENSeable Lab. (2013). *3rd Generation CPTED and the Eco-Friendly City*. Presentation given by Gregory Saville, Mateja Mihinjac and Tarah Hodgkinson at the 2013 ICA (International CPTED Association) Conference held in Mexico City on September 11/12 2013.

Morgan, J. (1991). *Safer Communities: The Local Delivery of Crime Prevention Through the Partnership Approach*. Home Office.

Morris, S. (1996). *Policing Problem Housing Estates* (Police Research Group Crime Detection and Prevention Series Paper 74). Home Office.

Morrison, S. A., & O'Donnell, I. (1996). *Armed Robbery: A Study in London* (Occasional paper no. 15). Oxford: Centre for Criminological Research.

Moseley Society and Birmingham City Council. (1992). *Moseley Shopfront Design Guide*. The Moseley Society.

Moss, K., & Pease, K. (1999). Crime and Disorder Act 1998, Section 17: A Wolf in Sheep's Clothing. *Crime Prevention and Community Safety, 1*(4), 15–19.

Mott, G., & Townsend, A. (2011). *About the ATM Industry*. ATM Security Working Group (ATMSWG).

NaCTSO. (2009). *Counter Terrorism Protective Security Advice for Places of Worship*. NaCTSO.

NaCTSO. (2020). *Protecting Crowded Places from Terrorism*. NaCTSO.

Nee, C., & Meenaghan, A. (2006). Expert Decision Making in Burglars. *British Journal of Criminology, 46*, 935–949.

National Police Chiefs' Council and Police Scotland. (2015). *Police Response to Security Systems* National Police Chiefs' Council; Kincardine: Police Scotland.

Network Rail. (2009). *Graffiti using Hydrofluoric Acid, Infrastructure Group Safety Bulletin*. Network Rail.

Newlands, M. (1983). *Residential Burglary Patterns in a Vancouver Neighbourhood* (Unpublished Honor's thesis, Department of Criminology). Burnaby, British Columbia: Simon Fraser University.

Newman, O. (1972). *Defensible Space: Crime Prevention Through Urban Design*. Macmillan.

Newman, O. (1973). *Defensible Space: People and Design in the Violent City*. Architectural Press.

Newman, O. (1976). *Design Guidelines for Creating Defensible Space*. Washington: National Institute of Law Enforcement and Criminal Justice, Law Enforcement Assistance Administration. Washington D.C: U.S. Dept. of Justice.

Newman, O. (1996). *Creating Defensible Space*. U.S. Department of Housing and Urban Development.

North Yorkshire Police (undated leaflet) *Crime Prevention Measures and Advice relating to the Misuse of Drugs in Public Toilet Areas, particularly within Retail and Fast Food business areas*. Northallerton: North Yorkshire Police.

Nubani, L., & Wineman, J. (2005). The Role of Space Syntax in Identifying the Relationship Between Space and Crime. In A. van Nes (Ed.), *Proceedings of the Fifth International Space Syntax Symposium*(pp. 413–422). The Netherlands.

Nuttgens, P. (1989). *The Home Front*. BBC Books.

Office of the Deputy Prime Minister. (1998). *Places, Streets and Movement: A Companion Guide to Design Bulletin 32 Residential Roads and Footpaths*. ODPM.

Office of the Deputy Prime Minister. (ODPM) and the Home Office. (2004). *Safer Places: The Planning System and Crime Prevention*. Thomas Telford.

Office of the Surveillance Camera Commissioner. (2021). *Surveillance Camera Code of Practice*. https://www.gov.uk/government/publications/update-to-surveillance-camera-code

Orwell, G. (1949). *Nineteen Eighty-Four*. Secker and Warburg.

Painter, K. (1994). The Impact of Street Lighting on Crime, Fear and Pedestrian Street Use. *Security Journal, 5*, 116–124.

Painter, K. A., & Farrington, D. P. (1997). The Crime Reducing Effect of Improved Street Lighting. The Dudley Project. In R. V. Clarke (Ed.), *Situational Crime Prevention: Successful Case Studies* (2nd ed., pp. 209–226). Harrow and Heston.

Painter, K. A., & Farrington, D. P. (1999a). Improved Street Lighting: Crime Reducing Effects and Cost-Benefit Analyses. *Security Journal, 12*(1), 17–32.

Painter, K. A., & Farrington, D. P. (1999b). Street Lighting and Crime: Diffusion of Benefits in the Stoke-on-Trent Project. In K. A. Painter & N. Tilley (Eds.), *Surveillance of Public Space: CCTV, Street Lighting and Crime Prevention* (pp. 77–122). Criminal Justice Press.

Pascoe, T. (1993a). Domestic Burglaries: The Burglar's View. *Building Research Establishment Information* (Paper IP 19/93). Garston, Watford: Building Research Establishment.

Pascoe, T. (1993b). Domestic Burglaries: The Police View. *Building Research Establishment Information* (Paper IP 20/93). Garston, Watford: Building Research Establishment.

Pascoe, T. (1996). *Predicting the targets of household burglars—developing risk-based models from a case study of recent experience in Redditch, United Kingdom* (PhD Thesis). Cranfield University School of Management. Accessible at: www.vellig-ontwerp-beheer.nl/publicaties/predicting-the-targets-of-household-burglars

Pascoe, T. (1999). *Evaluation of Secured by Design in Public Sector Housing—Final Report.* Building Research Establishment.

Pascoe, T., & Topping, P. (1997). Secured by Design: Assessing the Basis of the Scheme. *International Journal of Risk, Security and Crime Prevention, 2*(3), 161–173.

Payne, B. K., & Gainey, R. R. (2006). The Criminal Justice Response to Elder Abuse in Nursing Homes: A Routine Activities Perspective. *Western Criminology Review, 7*(3), 67–81.

Pease, K. (1991). The Kirkholt Project: Preventing Burglary on a British Housing Estate. *The Security Journal, 2*, 73–77.

Pease, K. (1992). Preventing Burglary on a British Housing Estate. In R. Clarke (Ed.), *Situational Crime Prevention: Successful Case Studies.* Harrow and Heston.

Pease, K. (1999). A Review of Street Lighting Evaluations: Crime Reduction Effects. In K. Painter & N. Tilley (Eds.), *Surveillance of Public Space: CCTV, Street Lighting and Crime Prevention. Crime Prevention Studies*(Vol. 10). Criminal Justice Press.

Pease, K. (1997). Predicting the Future: The Roles of Routine Activity Theory and Rational Choice Theory. In G. Newman, R. Clarke, & S. Giora Shoham (Eds.), *Rational Choice and Situational Crime Prevention.* Ashgate.

Pease, K. (2001). *Cracking Crime Through Design.* Design Council.

Pease, K., & Farrell, G. (1993). *Once Bitten, Twice Bitten: Repeat Victimisation and its Implications for Crime Prevention.* Home Office Crime Prevention Unit.

Pease, K., & Farrell, G. (2009). *The Carbon Cost of Crime and Its Implications.* Secured by Design.

Pease, K., & Farrell, G. (2014). What Have Criminologists Done for Us Lately? In *The Handbook of Security* (pp. 65–88). Palgrave.

Philips, P. D. (1980). Characteristics and Typology of the Journey to Crime. In D. E. George-Abeyie & K. D. Harris (Eds.), *Crime: A Spatial Perspective.* Columbia University Press.

Police Crime Prevention Academy. (2021). Level 5 Diploma Crime Prevention—Designing Out Crime Recognised Prior Experience & Learning (RPEL) Pathway. Police Crime Prevention Academy.

Power, A. (1997). *Estates on the Edge: The social Consequences of Mass Housing in Northern Europe.* Macmillan Press Ltd.

Poyner, B. (1983). *Design Against Crime: Beyond Defensible Space.* Butterworth.

Poyner, B. (1988). Video Cameras and Bus Vandalism. *Journal of Security and Administration, 11*, 44–51.

Poyner, B., & Webb, B. (1991). What Works in Crime Prevention: An Overview of Evaluations, presented at *the British Criminology Conference.* York.

Poyner, B. (1992). Situational Crime Prevention in Two Parking Facilities. In R. Clarke (Ed.), *Situational Crime Prevention: Successful Case Studies* (pp. 42–57). Harrow and Heston.

Poyner, B. (2006). *Crime Free Housing in the 21st Century.* Jill Dando Institute of Crime Science.

President's Commission on Law Enforcement and Administration of Justice. (1967). *The Challenge of Crime in a Free Society.* Dutton.

Project for Public Spaces. (2005). *How to Turn a Place Around: A Handbook for Creating Successful Places.* Project for Public Spaces.

Protect UK. (2023). *Protect UK Advice and Guidance.* https://www.protectuk.police/guidance

Pyle, G. (1974). *'The Spatial Dynamics of Crime', Department of Geography Research Paper 159*. University of Chicago Press.
Ramsay, M. (1982). *City Centre Crime: A Situational Approach to Prevention* (Research and Planning Unit Paper 10). Home Office.
Ramsay, M., & Newton, R. (1991). *The Effect of Better Street Lighting on Crime and Fear: A Review* (Crime Prevention Unit, Paper 29). Home Office.
Rapoport, A. (1982). *The Meaning of the Built Environment. A Nonverbal Communication Approach*. Sage.
Read, T., & Tilley, N. (2000). *Not Rocket Science? Problem-solving and Crime Reduction* (Crime Reduction Research Series 6). Home Office.
Rengert, G. (1990). *Drug Purchasing as a Routine Activity of Drug Dependent Property Criminals and the Spatial Concentration of Crime*, paper presented at the American Society of Criminology 1990 Annual Conference.
Rengert, G. F. (2004). The Journey to Crime. In G. Bruinsma, H. Elffers, & J. D. Keijser (Eds.), *Punishment, Places and Perpetrators: Developments in Criminology and Criminal Justice Research*. Willan.
Rengert, G. F., & Wasilchick, J. (1985). *Suburban Burglary: A Time and a Place for Everything*. Charles C. Thomas.
Rengert, G., & Wasilchick. (2000). *Suburban Burglary: A Tale of Two Suburbs* (2nd ed.). Charles C. Thomas.
Repetto, T. A. (1974). *Residential Crime*. Ballinger.
Reynald, D. M. (2009). Guardianship in Action: Developing a New Tool for Measurement. *Crime Prevention and Community Safety: An International Journal, 11*(1), 1–20.
Reynald, D., & Elffers, H. (2009). The Future of Newman's Defensible Space Theory: Linking Defensible Space and the Routine Activities of Place. *European Journal of Criminology, 6*(1), 25–46.
Rhodes, W. M., & Conly, C. C. (1981). Crime and Mobility. In P. J. Brantingham & P. L. Brantingham (Eds.), *Environmental Criminology*. Sage.
RIBA. (2010). *RIBA Guidance for Counter-Terrorism*. Royal Institute of British Architects (RIBA).
Rio Earth Summit. (1992, June 3–14). *The United Nations Conference on Environment and Development (UNCED)*. Held in Rio de Jeneiro.
Roberts, J. (1981). *Pedestrian Precincts in Britain*. Transport and Environmental Studies.
Roberts, M. (2004). *Good Practice in Managing the Evening and Late Night Economy: A Literature Review from an Environmental Perspective*. Office of the Deputy Prime Minister.
Robinson, M. B. (1996). The Theoretical Development of CPTED: 25 Years of Responses to C. Ray Jeffery. *Advances in Criminological Theory, 8*.
Rosenbaum, D. (1988). Community Crime Prevention: A Review and Synthesis of the Literature. *Justice Quarterly, 5*(3), 325–395.
Rubenstein, H., Murray, C., Motoyama, T., & Rowse, W. (1980). *The Link Between Crime and the Built Environment. The Current State of Knowledge*. National Institute of Justice.
Rudlin, D., & Falk, N. (1995). *Sustainable Urban Neighbourhoods: Building the 21st Century Home*. Architectural Press.

Sampson, R., & Eck, J. E. (2008, September). Super Controllers: Can I be your Superman? Presentation given at the 19th POP (Problem-Oriented Policing) Conference held in Bellevue. Washington.

Sarno, C. (1996). The Impact of Closed Circuit Television on Crime in Sutton Town Centre. In M. Bulos & D. Grant (Eds.), *Towards a Safer Sutton? CCTV One Year On*. London Borough of Sutton.

Saville, G., & Cleveland, G. (1997, December 3–5). 2nd Generation CPTED: An Antidote to the Social Y2 K Virus of Urban Design. Paper presented at the *2nd Annual International CPTED Conference,* Orlando, FL. Accessed 24 December 2024, http://www.veilig-ontwerp-beheer.nl/publicaties/2nd-generation-cpted-antidote-to-the-social-y2k-virus-of-urban-design

Scates, L. (2007). Incorporating CPTED into the Palmdale Planning Process, presentation given to the *Conference of the National Institute of Crime Prevention*, Las Vegas, on 1 August 2007.

Schneider, A. (1988). Neighborhood-Based Antiburglary Strategies: An Analysis of Public and Private Benefits from the Portland Program. In D. P. Rosenbaum (Ed.), *Community Crime Prevention: Does It Work?* Sage.

Schneider, R., & Kitchen, T. (2002). *Planning for Crime Prevention: A Transatlantic Perspective*. Routledge.

Scott, M. S. (2018). *How Can We Collaborate More Effectively?* Presentation given at Problem Solving Conference hosted by Durham Constabulary in the City of Durham on 26 February 2018.

Scott, M. S., & Kirby, S. (2012). *Implementing POP: Leading, Structuring and Managing a Problem-Oriented Police Agency*. Community Oriented Policing Services, U.S. Department of Justice.

Scottish Executive Development Department. (2006). *Designing Safer Places—Planning Advice Note (PAN 77)*. Scottish Executive Development Department.

Secured by Design. (1989). *Secured by Design—The Concept,* South East Region Senior Crime Prevention Officers Conference. Rea Publicity.

Secured by Design. (1989). *Secured by Design—Estate*, South East Region Senior Crime Prevention Officers Conference. Rea Publicity.

Secured by Design. (2002). *Thames Valley Police: Youth Centres and Sports Systems* youth shelter doc for pdf 2002 (https://www.securedbydesign.com)

Secured by Design. (2005). *SBD Hospitals*. London: Secured by Design. SBD_hospitals_2005.pdf (https://www.securedbydesign.com)

Secured by Design. (2009). *The Carbon Cost of Crime and your Home*. Secured by Design.

Secured by Design. (2010a). *The Capital Costs of Secured by Design accreditation, by David Langdon Associates*. Secured by Design.

Secured by Design. (2010b). *Design and Access Statements: How to Use Them to Prevent Crime*. Secured by Design.

Secured by Design. (2011a). *Carbon Reduction and Cost Efficiency Saving Model*. Secured by Design.

Secured by Design. (2011b). *Lighting Against Crime: A Guide for Crime Reduction Professionals*. Secured by Design.

Secured by Design. (2014). *SBD The Safer Tram Stop Award*. Secured by Design. The Safer Tram Award (https://www.securedbydesign.comsecuredbydesign.com)

Secured by Design. (2014). *SBD New Schools 2014*. Secured by Design. New_Schools_ 2014.pdf (https://www.securedbydesign.comsecuredbydesign.com)
Secured by Design. (2015). *SBD Commercial Developments 2015*. Secured by Design. Accessible at www.securedbydesign.com/industry-advice-and-guides/
Secured by Design. (2016). *SBD Homes 2016*. Secured by Design. Accessible at www.sec uredbydesign.com/industry-advice-and-guides/
Secured by Design. (2018). *The Sanctuary Scheme Installation Manual: A Place of Safety for Those At Risk*. Secured by Design.
Secured by Design. (2019). *SBD Self Build*. Secured by Design. SELF_BUILD_2019_ BROCHURE_LR.pdf (https://www.securedbydesign.com)
Secured by Design. (2021a). *Construction Site Security Guide*. Secured by Design. CONSTRUCTION_SITE_SECURITY_GUIDE_ A4_8pp.pdf (https://www.securedbydesign.com)
Secured by Design. (2021b). *The Police Crime Prevention Initiatives Guide to Lighting*. PCPI.
Secured by Design. (2022a). *Security Overlay to the RIBA Plan of Work*. Security_Overlay_ to_RIBA_Plan_of_Workpdf.pdf (https://www.securedbydesign.com)
Secured by Design. (2022b). *SBD Specialist and Supported Housing Schemes*. Secured by Design. Specialist_and_Supported_ Housing_Schemes.pdf (https://www.securedby design.com)
Secured by Design. (2023). *Security Systems Policy*. Secured by Design. Accessible at www. securedbydesign.com/security-systems-policy/
Secured by Design, Bicycle Association Sustrans and (2023). *SBD Cycle Parking and Security Standards*. Secured by Design. 05132-Cycle-Parking-and-Security-Standards-June-2021-REV-6.pdf (https://www.securedbydesign.com)
Secured by Design. (2023a). *SBD Homes 2023*. Secured by Design. HOMES_GUIDE_2023_ web.pdf (https://ww.securedbydesign.com)
Secured by Design. (2023b). *Secured Environments*. Secured byDesign. https://www.secure denvironments.com
Secured by Design. (2023c). *SBD Commercial Guide 2023*. London: Secured by Design. COMMERCIAL_GUIDE_2023_web.pdf (https://www.securedbydesign.com)
Secured by Design and the British Parking Association. (2004). *Park Mark Safer Parking Award*. Secured by Design; Haywards Heath, Sussex: BPA.
Secured by Design and Perpetuity Research. (2015). *Secured Environments*. Secured by Design; and Leicester: Perpetuity Research and Consultancy.
Shaftoe, H., & James, S. (2004). *Do Symbolic Barriers Prevent Crime and Offer an Increased Sense of Safety?* Routledge.
Shaftoe, S. (2008). *Convivial Urban Spaces: Creating Effective Public Spaces*. Earthscan.
Sherman, L., Shaw, J., & Rogan, D. (1995). *The Kansas City Gun Experiment, Research in Brief Series*. National Institute of Justice.
Sherman, L., Farrington, D., Welsh, B., & MacKenzie, D. (Eds.). (2002). *Evidence-Based Crime Prevention*. Routledge.
Shu, C. F. (2000). Housing Layout and Crime Vulnerability. *Urban Design International, 4*, 177–188.
Shu, S., & Huang, J. (2003). Spatial Configuration and Vulnerability of Residential Burglary: A Case Study of a City in Taiwan. In *Proceedings of the 4th International Space Syntax Symposium*. London.

Sidebottom, A., Tilley, N., & Eck, J. (2012). Towards Checklists to Reduce Common Sources of Problem-Solving Failure. In *Policing: A Journal of Policy and Practice, 6*(2), 194–209.

Skinner, B. F. (1953). *Science and Human Behavior*. Macmillan.

Skinns, D. (1998). Crime Reduction, Diffusion and Displacement: Evaluating the Effectiveness of CCTV. In C. Norris, J. Moran, & G. Armstrong (Eds.), *Surveillance, Closed Circuit Television and Social Control* (pp. 116–127). Ashgate.

Skogan, W. (1990). *Disorder and Decline: Crime and the Spiral of Decay in American Neighborhoods*. Free Press.

Skudder, H., Druckman, A., Cole, J., McInnes, A., Brunton- Smith, I., & Ansaloni, G. P. (2016). Addressing the Carbon-Crime Blind Spot: A Carbon Footprint Approach. *Journal of Industrial Ecology*. https://doi.org/10.1111/jiec.12457

Smith, M., Clarke, R. V., & Pease, K. (2002). Anticipatory Benefits in Crime Prevention. *Crime Prevention Studies, 13*, 71–88. Criminal Justice Press.

Smith, M. J., & Cornish, D. B. (2012). *Secure and Tranquil Travel: Preventing Crime and Disorder on Public Transport*. UCL Jill Dando Institute of Crime Science.

Spickenheuer, J. (1983). untranslated, Voetsurveillance en Preventie- voorlichting in Amsterdam Osdorp: een Onderzoek naar de Effecten op de Criminaliteitsontwikkeling, Gravenhage, Netherlands: Staatsuitgeverij.

Squires, P. (1998). *An Evaluation of the Ilford Town Centre CCTV System*. Health and Social Policy Research Centre, University of Brighton.

Squires, P., & Measor, L. (1996). *Closed Circuit TV Surveillance and Crime Prevention in Brighton: Half Yearly Report*. Health and Social Policy Research Centre, University of Brighton.

Stapley F. (2016). *The Good Pub Guide 2016*. Ebury Press.

Stenhouse, D. (1977). *Understanding Towns*. Canlat.

Stern, N. (2007). *The Economics of Climate Change*. Cambridge University Press.

Stokes, M. (1993). *Guide to Place of Worship Security*. West Midlands Police.

Stokes, M. (2016). *Guide to Place of Worship Safety and Security*. Unpublished update of 1993 design guide, originally written by the author for West Midlands Police.

Stokes, M. (2022). *An Investigation of the Sustainability of Crime Prevention in the Built Environment: Impact and Implementation Factors* (Published Doctoral Thesis). University of Huddersfield.

Stokes, M. (2025). *A Tale of 3 Bridges: Barriers to Suicide Prevention*. Presentation given at the conference of the International CPTED Association in Palm Springs CA on 11 May 2025.

Strong, R. (1990). *The Lost Treasures of Britain*. Guild Publishing.

Strong, R. (2007). *A Little History of the English Country Church*. Vintage Books.

SURF Centre. (2002). *Hulme, Ten Years on*. University of Salford.

Tanghe, J., Vlaeminck, S., & Berghoef, J. (1984). *Living Cities: A Case for Urbanism and Guidelines for Re-urbanism*. Pergamon.

Taub, R., Taylor, D., & Dunham, J. (1984). *Paths of Neighborhood Change: Race and Crime in Urban America*. University of Chicago Press.

Taylor, L. (1973). The Meaning of the Environment. In C. Ward (Ed.), *Vandalism*. Architectural Press.

Taylor, R., & Gottfredson, S. D. (1987). Environmental Design, Crime and Prevention: An Examination of Community Dynamics. In A. Reiss & M. Tonry (Eds.), *Communities and Crime* (pp. 387–416). University of Chicago Press.

Taylor, R. B. (2002). Crime Prevention Through Environmental Design (CPTED): Yes, No, Maybe, Unknowable and all of the Above. In R. B. Bechtel (Ed.), *Handbook of Environmental Psychology* (pp. 413–426). Willey.

Teedon, P., Reid, T., Griffiths, P., Lindsay, K., Glen, S., McFayden, A., & Cruz, P. (2009). *Secured by Design Impact Evaluation Final Report.* Glasgow Caledonian University.

Teeman, T. (1999). Traffic-Free Zone. Article in *The Times, Weekend* (p. 11), dated Saturday September 11, 1999. News International.

Telep, C. W., Mitchell, R. J., & Weisburd, D. (2014). How Much Time Should the Police Spend at Crime Hot Spots? Answers from a Police Agency Directed Randomized Field Trial in Sacramento, California. *Justice Quarterly, 31*, 905–933.

Thames Valley Police. (2010). *The Planning Compendium of Crime Prevention and Reduction in the Planning System.* Secured by Design.

Thornton, A., Hatton, C., Malone, C., Fryer, T., Walker, D., Cunningham, J., & Durrani, N. (2003). *Distraction Burglary Amongst Older Adults and Ethnic Minority Communities* (Home Office Research Study 269). Home Office.

Tien, J., O'Donnell, V., Barnett, A., & Mirchandani, P. (1979). *Street Lighting Projects: National Evaluation Program, Phase 1 Report.* National Institute of Law Enforcement and Criminal Justice, US Department of Justice.

Tilley, N. (1993). Understanding Car Parks, Crime and CCTV: *Evaluation Lessons from Safer Cities* (Police Research Group Crime Prevention Series Paper 42). Home Office.

Tilley, N. (2009). *Crime Prevention.* Willan.

Tilley, N., Pease, K., Hough, M., & Brown, R. (1999). *Burglary Prevention: Early Lessons from the Crime Reduction Programme, Policing and Reducing Crime Unit, Crime Reduction Research Series Paper 1.* Home Office.

Tilley, N., Thompson, P., Farrell, G., Grove, L., & Tseloni, A. (2015). Do Burglar Alarms Increase Burglary Risk? A Counter-Intuitive Finding and Possible Explanations. In *Crime Prevention and Community Safety*(Vol. 17, 1, pp.1–19). Macmillan.

Toy, S. (1955). *A History of Fortification: 3000 B.C. to A.D. 1700.* Heinemann.

Transport 2000/Streets for People/Children's Play Council. (1999). *At Home in My Street: Exploring Home Zones in the Netherlands and Germany.* Transport 2000 and the Children's Play Council.

Tseloni, A., Farrell, G., Thompson, R., Evans, E., & Tilley, N. (2017). *Domestic Burglary Drop and the Security Hypothesis.* Springer Open.

Tseloni, A., et al. (2017). *Crime Science, 6*, 3. https://doi.org/10.1186/s40163-017-0064-2

UK Government. (1824). *Vagrancy Act.* The National Archives. Accessible at https://www.legislation.gov.uk/ukpga/Geo4/5/83/section/4

UK Government. (1957). *Occupiers' Liability Act.* The National Archives. Accessible at www.legislation.gov.uk/ukpga/Eliz2/5-6/31/contents

UK Government. (1974). *Health and Safety at Work Act.* The National Archives. Accessible at www.legislation.gov.uk/ukpga/1974/37

UK Government. (1977). *Unfair Contract Terms Act* (1977). The National Archives. Accessible at www.legislation.gov.uk/ukpga/1977/50

UK Government. (1980). *Highways Act*. The National Archives. Accessible at www.legislation.gov.uk/ukpga/1980/66

UK Government. (1984). *Data Protection Act*. The National Archives. Accessible at http://www.legislation.gov.uk/ukpga/1984/35/contents

UK Government. (1984). *Occupiers' Liability Act*. The National Archives. Accessible at www.legislation.gov.uk/ukpga/1984/3/contents

UK Government. (1986). *Public Order Act*. The National Archives. Accessible at www.legislation.gov.uk/ukpga/1986/64/contents

UK Government. (1990). *Town and Country Planning Act* (1990) Section 106 Agreements. The National Archives. Accessible at www.legislation.gov.uk/ukpga/1990/8/section/106

UK Government. (1992). *The Management of Health and Safety at Work Regulations*. The National Archives. Accessible at www.legislation.gov.uk/uksi/1992/2051/contents/made

UK Government. (1995). *Disability Discrimination Act*. London. Accessible at www.legislation.gov.uk/ukpga/1995/50/contents

UK Government. (1998). *Data Protection Act*. The National Archives. Accessible at www.legislation.gov.uk/ukpga/1998/29/contents

UK Government. (1998). *Human Rights Act*. The National Archives. Accessible at www.legislation.gov.uk/ukpga/1998/48/contents

UK Government. (1998). *Crime and Disorder Act*. The National Archives. Accessible at www.legislation.gov.uk/ukpga/1998/1998/37/contents

UK Government. (2000). *Countryside and Rights of Way Act*. The National Archives. Accessible at www.legislation.gov.uk/ukpga/2000/37/contents

UK Government. (2000). *Freedom of Information Act*. The National Archives. Accessible at www.legislation.gov.uk/ukpga/2000/36/contents

UK Government. (2000). *Regulation of Investigatory Powers Act*. The National Archives. Accessible at www.legislation.gov.uk/ukpga/2000/23/contents

UK Government. (2000). *Terrorism Act*. The National Archives. Accessible at www.legislation.gov.uk/ukpga/2000/11/contents

UK Government. (2000). *Transport Act*. The National Archives. Accessible at legislation.gov.uk/ukpga/2000/38/contents

UK Government. (2003). *Licensing Act*. The National Archives. Accessible at www.legislation.gov.uk/ukpga/2003/17/contents

UK Government. (2005). *Clean Neighbourhoods and Environment Act*. The National Archives. Accessible at www.legislation.gov.uk/ukpga/2005/16/contents

UK Government. (2006). *The Railways and Other Guided Transport Systems (Safety) Regulations*. The National Archives.

UK Government. (2008). *Climate Change Act*. The National Archives. Accessible at www.legislation.gov.uk/ukpga/2008/27/contents

UK Government. (2010). *Equality Act*. London. Accessible at www.legislation.gov.uk/ukpga/2010/15/contents

UK Government. (2012). *Protection of Freedoms Act*. The National Archives. Accessible at www.legislation.gov.uk/ukpga/2012/9/contents

UK Government. (2014). *Anti-Social Behaviour, Crime and Policing Act*. The National Archives. Accessible at www.legislation.gov.uk/ukpga/2014/12/enacted

United Nations Interregional Crime and Justice Research Institute (UNICRI) and MIT, Senseable, Lab. (2012). *Improving Urban Security Through Environmental Design.* UNICRI.

University of Salford and University of Huddersfield. (2009). *NPCPS Implementation Planning Research Project: Final Report.* Design Against Crime Solution Centre.

Urban Task Force. (1999). *Towards an Urban Renaissance.* Urban Task Force.

Van der Voordt, T. J. M., & Van Wegen, H. B. R. (1990). Testing Building Plans for Public Safety: Usefulness of the Delft Checklist. *Netherlands Journal of Housing and Environmental Research, 5*(2), 129–154.

Van Koppen, P., & Jansen, R. (1998). The Road to Robbery: Travel Patterns in Commercial Robberies. *British Journal of Criminology, 38*(2), 230–246.

Warren, F., & Stollard, P. (1988). *Safe as Houses: A Working Paper Reviewing the Importance of Housing Design and Layout in Achieving Security.* Institute of Advanced Architectural Studies.

The Washington Post. (2025). *New Golden Gate Safety Nets are Reducing Suicide Deaths, Study Finds.* Article published on 22 March 2025.

Weideman, B. P., Thrane, M., Christensen, P., Schmidt, J., & Lokke, S. (2008). Carbon Footprint: A Catalyst for Life Cycle Assessment. *Journal of Industrial Ecology, 12*(1), 3–6.

Wiedmann, T., & Barrett, J. (2011). A Greenhouse Gas Footprint Analysis of UK Central Government, 1990–2008. *Environmental Science and Policy, 14*(8), 1041–1051.

Weight, A. (2009). *Best Practice for Physical ATM Security.* ATM Security Working Group (ATMSWG).

Weisburd, D., Wyckoff, L., Ready, J., Eck, J. E., Hinkle, J. C., & Gajewski, F. (2006). Does Crime Just Move Around the Corner? A Controlled Study of Spatial Displacement and Diffusion of Crime Control Benefits. *Criminology, 44*, 549–591.

Weisburd, D., Telep, C., Hinkle, J. C., & Eck, J. E. (2010). Is Problem-Oriented Policing Effective in Reducing Crime and Disorder? Findings from a Campbell. *Criminology and Public Policy, 9*(1), 139–172.

Weisburd, D., Britt, C. Wilson, D.B. and Wooditch, A. (2020) *Basic Statistics in Criminology and Criminological Justice.* Springer Nature.'

Welsh Government. (2016). *TAN (Technical Advice Note) 1: Design.* Welsh Government.

Welsh, B., & Farrington, D. (2002). *Crime Prevention Effects of Closed Circuit Television: A Systematic Review*(Home Office Research Study No. 252). Home Office.

Welsh, B. P., & Farrington, D. C. (2008). Effects of Improved Street Lighting on Crime. *Campbell Systematic Reviews, 2008*, 13. https://doi.org/10.4073/csr.2008.13

West Midlands Arson Task Force. (2005). *Keeping Your Business in Business.* West Midlands Fire Service and West Midlands Police.

West Midlands Arson Task Force. (2006). *Keeping Your School in Business.* West Midlands Fire Service and West Midlands Police.

West Midlands Arson Task Force. (2009). *Keeping Your Business in Business.* West Midlands Fire Service and West Midlands Police.

West Midlands Police. (1990). *Video Working Party Report.* West Midlands Police.

West Midlands Police. (1993). *Guide to Place of Worship Security.* West Midlands Police.

West Midlands Police. (1994). *CCTV—Its Interface with the Police.* West Midlands Police.

West Midlands Police. (1995). *Knowing Your Home: Straightforward Advice and a Simple DIY Survey to Help you Manage the Risk to Your Home and Keep Your Home Contents Where They Belong.* West Midlands Police.

West Midlands Police. (1996). *Knowing Your School: Straightforward Advice and a Simple DIY Survey to Help You Manage the Risk to Your School.* West Midlands Police Reprographics Department.

West Midlands Police. (2007). *CCTV Policy.* West Midlands Police.

West Midlands Police and West Midlands Fire Service. (1993). *On the Third Day.* West Midlands Police.

West Yorkshire Police. (2009). *Design and Access Statements: How to Use Them to Prevent Crime.* West Yorkshire Police.

White, R. C. (1932). The Relations of Felonies to Environmental Factors in Indianapolis. *Social Forces, 10,* 498–509.

Wignal, T. (2008, August). Murals list communities—The Opposite of Graffiti. *CPTED Perspective, 7*(3).

Wikstrom, P.-O. (2006). Individuals, Settings, and Acts of Crime: Situational Mechanisms and the Explanation of Crime. In P.-O. Wikstrom & R. Sampson (Eds.), *The Explanation of Crime: Context, Mechanisms and Development.* University Press.

Wilcox, P., & Eck, J. (2011). Criminology of the Unpopular Implications for Policy Aimed at Payday Lending Facilities. *Criminal Public Policy, 10,* 473–482.

Wiles, P., & Costello, A. (2000). *The 'Road to Nowhere': The Evidence for Travelling Criminals* (Home Office Research Study 207). Home Office Research, Development and Statistics Directorate.

Willis, K. (2006). Armed Robbery: Who Commits It and Why? *Trends and Issues in Crime and Criminal Justice* No. 328. Australian Institute of Criminology.

Winchester, S., & Jackson, H. (1982). *Residential Burglary: The Limits of Prevention*(Home Office Research Study No. 74). Home Office Research and Planning Unit.

Wood, E. (1967). *Social Aspects of Housing in Urban Development.* Department of Economic and Social Affairs. United Nations.

Wootton, A. B., Marselle, M., & Davey, C. (2009, April 1–3). City Centre Crime: Design Thinking for Safer City Centres. In *8th European Academy of Design Conference.* Robert Gordon University.

Wright, R., Heilweil, M., Pelletier, P., & Dickinson, K. (1974). *The Impact of Street Lighting on Crime.* University of Michigan.

Zimbardo, P. (1973). A Field Experiment in Auto Shaping. In C. Ward, (Ed.) *Vandalism.* Architectural Press.

Zimring, F. (1972). The Medium is the Message: Firearm Caliber as a Determinant of Death from Assault. *Journal of Legal Studies.*

Index

M. Stokes, *Designing Out Crime*, Crime Prevention and Security Management,
https://doi.org/10.1007/978-3-031-99840-9

D

R

Y

Z

The manufacturer's authorised representative in the EU is Springer Nature Customer Service Centre GmbH, Europaplatz 3, 69115 Heidelberg, Germany. If you have any concerns regarding our products, please contact ProductSafety@springernature.com

Printed and bound by CPI Group (UK) Ltd, Croydon, CR0 4YY
16/09/2026
02217783-0006